Russell S Black

RARE BIRDS WHERE AND WHEN

AN ANALYSIS OF STATUS & DISTRIBUTION
IN BRITAIN AND IRELAND

Dedicated to P and our beautiful girls India and Ruby

RARE BIRDS
Where and When

*An Analysis of Status & Distribution
in Britain and Ireland*

Ray Scally

Volume 1
sandgrouse to New World orioles

Russell Slack
with historical perspectives by Ian Wallace

Published 2009 by Russell Slack, Rare Birds Books, York
Copyright © 2009 text, figures and tables Russell Slack

Design, layout and typsetting by Julian R. Hough
Illustrations by the following artists:
James Gilroy, Julian R. Hough, Ray Scally and Ian Wallace

ISBN 978-0-9562823-0-9

Printed and bound in Great Britain by the MPG Books Group, Bodmin and King's Lynn.

The author would like to thank *British Birds* and *Birding World*
for their generous assistance with this book. The author would
also like to thank the many finders of rare birds who kindly let me
reproduce the stories behind their discoveries that enhance the species accounts within this book.

www.britishbirds.co.uk   www.birdingworld.co.uk

CONTENTS

PREFACE

The seeds for this book were unwittingly sown many years ago. As a youngster desperate to gather birdwatching acumen it was my mother who fuelled my learning through a regular supply of books from our local library. One evening a copy of *Discover Birds* by Ian Wallace appeared. Like many of my generation the fusion of images and words were motivational. Ian's book was an inspiration, and I feel most privileged that that the author who so enthused me all those years ago has been kind enough to contribute a rich historical backdrop to the species accounts contained herein.

Rarities and vagrancy patterns have a special interest for me. I would label myself as a self-finder of birds who derives the greatest pleasure from vagrants by stumbling across my own rarities. My local patches in Yorkshire are the inland floods of the Lower Derwent Valley and the coastal gullies that punctuate the coastline around the Gothic backdrop of Whitby. Through geography and good fortune I'm lucky that both areas have greatly facilitated my quest to find 'my own'. These underwatched areas are overshadowed by much better scrutinised and better placed sites elsewhere in the county, but I thrive on the challenge of proving that most places can and do turn up rare birds given enough coverage. Countless hours spent wandering around prime rarity habitat left me with much time to ponder why rare birds seem to appear in some places but not others. This is especially so during days when sites elsewhere seem to be flooded with a deluge of scarcities and yet the bushes in front of me are depressingly empty.

Interestingly, many species that I once yearned to find, such as Pallas's Warbler and Radde's Warbler, have since been demoted to scarcities. Our perception of what constitutes a rarity changes constantly. After all a rare bird is simply a common bird out of place. Although I've managed to fulfill the quest for a self-found Pallas's many times over, Raddes's has so far eluded all my best efforts at enticing a stray from the undergrowth. The apparent upturn in fortunes of many species once considered rare has also occupied a disproportionate amount of silent reflection while sitting out quiet periods looking over a coastal ravine. Many changes can be attributed to observer effort and improved field identifications, but countless other factors have also played a part.

It took until autumn 2007 for the seeds that would eventually grow into this book to begin to germinate. A career change fired up my enthusiasm to pick up my pen and the publication you are now reading slowly began to appear on paper. Although several authors have covered rarities in the past none had coalesced analysis and detailed interpretation. Added to that I felt that finders accounts always make a good read. I soon decided that putting the British and Irish records against a European or Western Palearctic backdrop was another essential. The final component was the inclusion of taxonomic material and with it the teasing prospect of informing the reader of the status of potential splits.

Perhaps naïvely, I envisaged a single book that could encapsulate all of this material. I opted from the outset for a self-publishing route, so I was constrained by neither word nor page counts. It soon became apparent that covering all species and subspecies treated as rarities was not going to fit tidily into a single tome. I made the decision that rather than compromise my ideas and reduce the content I would splice the book in two: volume 1 would cover sandgrouse to New World orioles and volume 2 swans to auks. I hope that

Sardinian Warbler by Ian Wallace

the fact that nearly 500 pages are dedicated solely to passerines and near passerines is sufficient justification.

At the outset I envisaged that I could comfortably take care of generating all the content. How wrong could I have been? I take full responsibility for the material within the book, but the quality of the finished product could not have been achieved without the help of a large number of contributors. There are far too many to mention here, but to appreciate their input please refer to the acknowledgements section that follows.

Regarding content there are two people to whom I am most indebted. Firstly, I would like to thank Ian Wallace for recognising the potential of the project at the outset. I am delighted that this book has returned some of the inspiration he gave me and caused him to contribute over 50 years worth of experience to the texts. Secondly, Tim Melling – former secretary of the *BOURC* and a birder with an encyclopaedic knowledge of everything – has assisted hugely through numerous comments and suggestions to the draft texts. I'd also like to thank all of the individuals involved in commenting upon the individual records from each of the European countries and for their patience with me in answering question after question on this and that.

To John Jackson who waded with enthusiasm through the final manuscript editing as he went I owe a massive debt, as I do to Julian Hough who painstakingly designed and typeset the book. I also offer my gratitude to the authors who offered additional sections: Alex Lees and James Gilroy for their superb chapter on vagrancy; and Adam Rowlands and Bob McGowan for their insights into the workings of the *BBRC* and *BOURC*. Finally to Ray Scally who not only produced a superb piece of artwork for the cover but also found time in his busy schedule to do numerous internal illustrations as well.

Last but not least, I turn to my family. To my parents thank you for acting as a taxi service during my early birding apprenticeship and bankrolling my early book collection and choice of optics. For Linda, my wonderfully supportive wife, I can not thank you enough for backing me as both a birder and an author. To my beautiful girls, India and Ruby, hopefully in times to come you will understand why your father spent so much time staring at a computer night after night. I hope you too will discover an all-consuming and absorbing pastime upon which to focus your attention. I love you all. This book is dedicated to you for your love and support.

ACKNOWLEDGEMENTS

I would like to extend my thanks to the numerous representatives of rarities commit-tees throughout Europe and beyond who frequently responded to the many requests for clarification over various records. These include Yann Kolbeinsson (Iceland), Søren Sørensen (Faeroe Islands), Vegard Bunes (Norway), Anders Blomdahl (Sweden), Aleksi Lehikoinen (Finland), Sebastian Klein & Morten Bentzon Hansen (Denmark), Jan Lontkowski & Tadeusz Stawarczyk (Poland), Peter H. Barthel (Germany), Arnoud B van den Berg (the Netherlands), Marnix Vandegehuchte (Belgium), Gilles Biver (Luxembourg), Marc Duquet (France), Mark Lawlor (Channel Islands), Paul Milne and Kieran Fahy (Ireland), Jon Green (Wales), Peter Knaus (Switzerland), Johannes Laber (Austria), Jiri Horacek (Czech Republic), Alfréd Trnka (Slovakia), Zalai Tamás (Hungary), Vadim Ryabitsev (Russia), Nikos Probonas & Filios Akriotis (Greece), Guy Kirwan (Turkey), Gordan Lukac (Croatia), Andrea Corso (Italy), Colin Richardson (Cyprus), Joe Sultana & John Attard Montalto (Malta), Yoav Perlman (Israel), George Gregory (Kuwait), Patrick Bergier (Morocco), Ricard Gutiérrez (Spain) and João Jara (Portugal). I would also like to thank Mike Wilson for going through the Russian literature on my behalf.

Special thanks go out to the various people who had to painstakingly go through the drafts. To Ian Wallace and Tim Melling especially for numerous comments and suggestions plus the following also ploughed through material: Tim Barker, Jon Green, Kieran Fahy, Jono Leadley and Andy Wilson. Finally to John Jackson for such a great editing job.

I thank my co-authors for their contributions, to Alex Lees and James Gilroy for a superb and thought-provoking vagrancy section, plus Bob McGowan (*BOURC*) and Adam Rowlands (*BBRC*). Thanks also to Keith Naylor for allowing me access to his database of rarities and Nigel Hudson (secretary of the *BBRC*) for numerous exchanges over the course of my research. Without the skill and expertise of Julian Hough to design and typeset the book there would have been no book and the artwork of Ray Scally, Julian Hough, Ian Wallace and James Gilroy provide superb accompaniment to the species accounts.

Many others have assisted greatly along the way including: Ian Andrews, Dawn Balmer, Graham Catley, Martin Collinson, Dave Dunford, Mark Grantham, Peter Kennerley, Andrew Lassey, Anthony McGeehan, Tom McIlroy (picture of Ian Wallace), Mike Pennington, Adrian Pitches, Roger Riddington, Stuart Winter and Peter Whelerton.

Finally, but not least many thanks to *British Birds*, *Birding World* and Spurn Bird Observatory for granting permission for the use of material in the book and to the indi-vidual finders who willingly granted me use of their material.

FOREWORD

Bimaculated Lark by Ian Wallace

In Britain and Ireland, rare birds have been entering recorded natural history since the late 17th century. The very first of these recorded identifications were of striking non-passerines. Witness our first 12 Black-winged Stilts, dated from 1684, seven of which were mown down. Then it was the fate of most animals with curiosity value to be collected for exhibition. A harvest of specimens, which often changed hands at growing values, was stuffed and set up for viewing in the decorative glass cabinets of stately homes and nat-ural history museums. From the mid-18th century passerines began to appear. Unsur-prisingly, none were little and brown puzzles. The first ever in 1742 was a still-stunning Rosy Pastor (Rose-coloured Starling). The second in 1753 was a Nutcracker and the third in 1792 a Wallcreeper of which two primaries were sketched by an artistic lady and sent to Gilbert White for verification.

Move the clock hands of written ornithology and its changing morals on 325 years and rare birds now meet human observers of very different behaviour. Gone in Europe are the shotguns and chloroform jars; everywhere are birdwatchers capable of field identifications, armed only with the marvels of the new observation and information technologies. The certain recording and mass collection of rarities, nearly all offered as immediate targets by the birdlines and pagers, is an everyday event.

The recent history of rare birds has been subject to two forms of national discipline and a fast-growing literature. Its cardinal pieces began with *The Handbook of British Birds 1938-41*, whose senior editor H.F. Witherby was the final judge of rarity claims from 1907 to 1943. Writing to H.G. Alexander on 11th February 1939, he noted that *"doubting people's opinions is not a job I like at all but it has to be done!"* After his death in 1943, the supervision of rare bird claims passed to other specialist editors of *British Birds*, first B.W. Tucker and second P.A.D. Hollom who published the then electrify-ingly *Popular Handbook of Rarer British Birds* in 1960. During its preparation, P.A.D. H and *BB*'s new Executive Editor, I. J. Ferguson-Lees, realised that if the growing finds of the more numerous post-war observers were to be reviewed properly and published centrally, a new and more democratic national discipline was needed.

Hence in 1959, with the co-operation of county societies, came the *British Birds* Rarities Committee. Its 10 members – the original *"Ten Rare Men"* – were chosen for their mix of identification lore and knowledge of the observer universe. With its temporal remit eventually extended back to 1950, the *BBRC* has been the judge and archivist of nearly all rare bird records for more than 50 years and counting. For a fuller history, see particularly the original constitution that accompanied its 1958 report and *BB*'s 10-year review (Wallace 1970) and 50-year appraisal (Dean 2008).

The committee's annual report for 2007 featured over 700 judgements on at least 130 species and subspecies. The 584 acceptances form a testament to the expertise of the 1,000 or so rarity hunters who supply most of the committee's workload. In their wake and dependant on the degree of rarity, up to 4,000-5,000 rarity collectors (alias listers and in extreme form twitchers) travel fast and queue patiently for the joys of new species (alias more ticks). How many more rarities would be found if all 5,000 were prime hunters is an unanswered question!

BBRC and the rather more august Records Committee of the British Ornithologists' Union (BOURC), which has still the last words on first admissions to the *British List* and taxonomic issues, work hard. Commendably also, BBRC has published its annual report almost always on time. It has not, however, hit one of its original targets: the regular unfolding of specific occurrence patterns. This task has, however, been attempted by a series of industrious authors or small teams. The most important of their books have been *Rare Birds in Britain and Ireland* (Sharrock and Sharrock 1976), an updated version of the same (Dymond, Fraser and Gantlett 1989), the *Photographic Handbook to the Rare Birds of Britain and Europe* (Mitchell and Young 1997, updated 1999), *A Reference Manual of Rare Birds in Great Britain and Ireland* (Naylor, 1996, Vol I, 1998 Vol II) and *Rare Birds in Britain and Ireland: A Photographic Record* (Cottridge and Vinicombe 1996). The second last has been adopted by the BBRC as the soundest historical archive of both accepted and rejected records. Beware, however, its occasional out-dating by later reviews of individual records and the shifting of taxonomic sands.

Alternative catalogues of rarities and free speech on their politics have come from the *UK400 Club*, *Birding World* and *Birdwatch*, most notably in *The Status of Rare Birds in Britain and Ireland 1800-1990* (Evans 1994), but none has been accepted as "holy writ" by BBRC or BOURC. Thus two decades has passed without an updated summary and intelligent review of the official British and Irish rarities.

Enter happily Russell Slack who, changing up several gears from his efforts in *Rare and Scare Birds in Yorkshire* (Wilson and Slack 1996), decided to fill the gap with a definitive guide. The result is this mighty book. Turn a few pages and its promise will be self-evident. It is a veritable Baedeker to *Rare Birds, Where and When*. It covers all the near passerines and passerines officially accepted for Britain and Ireland. Innovatively its sets them in their modern taxonomy and against their home range and European status.

I cannot think of a recent solo effort greater than Russell's and I have much enjoyed adding some historical colour to his research and letting off some steam over the few remaining messes. So, to learn about rarities as avian beings and not just tick them off, make sure that his book is on your nearest bookshelf. Above all, it is the boon that would-be self-finders have been awaiting.

Ian Wallace.

VAGRANCY MECHANISMS IN PASSERINES AND NEAR-PASSERINES
Alexander C. Lees and James J. Gilroy

Britain is exceptionally well placed to receive vagrant birds. Its position on the edge of Europe combined with constantly changing weather patterns and army of amateur and professional ornithologists conspire to produce perhaps the richest vagrant avifauna anywhere on Earth.

Blue-winged Warbler by James Gilroy

Britain and Ireland receive vagrants from every possible direction: species from the high Arctic, Nearctic, North Africa, Central Asia and Siberia, and wanderers from the world's oceans. Some species that breed no nearer than the Urals, such as Yellow-browed Warbler, are regular passage migrants; others, such as Crested Lark, breed as close as Northern France yet remain extreme vagrants here.

Of the 580 bird species officially recorded in Britain up to the end of July 2008, 50 per cent are rare vagrants. That 289 full species are considered here is testament to the reliance on vagrants of our avifaunal lists. There are a further 32 subspecies and undefined visitors, for example albatrosses. Sixteen per cent of recorded birds have occurred on fewer than five occasions: an amazing 92 full species in Categories A to D have occurred on one to four occasions.

Since 1980 more than 80 new species have been added to the British and Irish lists, including new vagrants and those resulting from taxonomic splits. New species are recorded on an annual basis. There are many mechanisms by which vagrant birds can come to find themselves apparently hopelessly lost in this part of Europe. Some arrivals are related to population growth or range expansion. Others are driven by food shortages or unusual weather events. Many birds undoubtedly arrive here as a result of internal errors in their migration apparatus, causing them to depart on deviant headings. These broad mechanisms have been discussed in detail in several previous works on vagrancy (for example, Cottridge and Vinicombe 2001). In this account, we endeavour to take a more direct look at the patterns of vagrancy occurring in these islands, addressing questions relating to causes of rare bird occurrence on a regional basis. The mechanisms discussed here relate to passerines and near passerines; vagrancy in non passerines will be covered in Volume 2.

Lees, A. C. and Gilroy, J J. 2009. Vagrancy Mechanisms in Passerines and Near-Passerines. In: Slack, R. *Rare Birds, Where and When: An analysis of status and distribution in Britain and Ireland. Volume 1: sandgrouse to New World orioles.* Rare Bird Books, York.

NEARCTIC VAGRANCY

Autumn Vagrancy Patterns

The appearance of Nearctic landbirds in Western Europe represents one of the most amazing feats of vagrancy performed by any species. The regular arrival of small-bodied passerines and near-passerines from North America on our shores is truly remarkable. The best explanation for their occurrence lies close to the old and oft-quoted adage that vagrant birds are simply blown off course. Arrivals of North American passerines show a strong peak in autumn, and are almost always associated with a specific set of weather conditions described comprehensively by the work of Norman Elkins (Elkins 1979, 1988 and 1999).

Elkins explains how many species of migratory bird breeding in the northern part of North America follow a relatively similar migratory pattern in autumn: moving across eastern North America and funneling east towards the Atlantic seaboard. The birds then turn southwards to move en masse along the coast towards their winter quarters. During this coast-hugging – for some species trans-oceanic – phase of the migration, birds become vulnerable to the changeable weather patterns that occur in the Atlantic region. Storms are relatively frequent throughout the autumn on the east coast of North America. Many weather systems are fierce and fast moving enough to entrain and displace birds from their normal flight path. When a specific set of weather conditions are met, large numbers of migrating birds may be displaced in our direction.

Migrants typically depart to the southeast on coast-hugging flights in relatively clear weather associated with high-pressure systems. Such weather patterns offer ideal conditions for normal migration: light tail winds and clear skies allow easy navigation. Given the changeable nature of the Atlantic weather, these conditions are often short-lived. Southbound birds are frequently caught up in cold frontal zones associated with offshore low-pressure systems. Deep cloud often builds up around these cold fronts, reducing the birds' capacity to navigate using the stars.

In such conditions, birds are likely to be drifted farther and farther out to sea, particularly as strong westerly winds often occur along such cold fronts. The weather systems typically move east from coastal areas towards the edge of the warm waters of the Gulf Stream, where they strengthen and deepen, and generate frontal waves. These waves are formed around pockets of warm air on the southern side of low-pressure systems. The associated cloud, precipitation and winds are intensified. Disorientated migrants within these eastbound weather systems can be drifted rapidly eastwards across the open ocean. With favourable tailwinds the birds may eventually complete a transatlantic crossing (Elkins 1979).

This weather-based theory is strongly supported by the pattern of occurrence of recorded autumn North American passerine vagrants in the British Isles. Virtually every multiple arrival of passerine and near-passerine vagrants from the west occur following spells of prolonged westerly gales at mid latitudes, associated with Atlantic depressions that cross from the North American eastern seaboard at high speed. Most individuals brought to the British Isles by such weather systems are discovered on western headlands and islands. It has also been suggested that some individuals may reach us by traveling in the high-level jet stream, which generally flows west to east across the Atlantic at similar latitudes (Newton 2007).

Wood Thrush by James Gilroy

Westerly jet stream winds reaching 250 km an hour could potentially bring migrants over 5,000 km from Atlantic Canada to the UK in less than 24 hours. The problem with this theory is that conditions within the jet stream, including extreme wind speeds and low air temperatures, are likely to threaten the survival of small birds. Very few passerines would be able to survive such conditions long enough to make a crossing. Moreover, in some years North American landbirds arrive concurrently with monarch butterflies, which could not possibly survive jet-stream conditions, again hinting that most bird species probably travel below jet-stream altitudes.

Further evidence that virtually all of our autumn North American vagrants are brought by Atlantic depressions comes from analysis of the range of species that occur here and their relative frequency in comparison to the overall community of birds passing down the North American eastern seaboard. McLaren *et al* (2006) carried out the most detailed analysis of Nearctic passerine vagrancy in Europe yet published. They found that the range of species occurring on this side of the Atlantic includes a significant over-representation of over-ocean migrants: species such as Blackpoll Warbler and Grey-cheeked Thrush that are known to routinely take the transoceanic short-cut directly between the northeastern seaboard and their winter quarters or staging grounds in the Caribbean. This finding supported previous suggestions (for example Robbins 1980 and Elkins 1979) that most Nearctic vagrants reaching the UK originate in flights off southeastern USA and are displaced downwind (northeastward) across the North Atlantic. Given favourable tailwinds, it is estimated that this journey could be completed in around two days.

Many of the long-distance migrants involved in vagrancy events to Europe carry enough stored fat to enable them to fly non-stop for 30 to 50 hours or longer (Nisbet 1963) and consequently are capable of making such a crossing unaided. Many of these vagrants are found in a similar state of exhaustion, with no remaining fat reserves, to individuals arriving in their normal winter quarters. That some do not recover suggests that they have not only burnt up all fat reserves but also metabolised vital tissues. McLaren *et al* (2006) found that the most commonly occurring autumn transatlantic vagrants were larger-bodied long-distance migrants (mostly boreal breeders) that were common in late autumn in the North American migration watch points.

Typically these included longer-distance migratory species that winter in the Amazon basin (Yellow-billed Cuckoo, Grey-cheeked Thrush, Red-eyed Vireo and Blackpoll Warbler), the Andes (Swainson's Thrush and Rose-breasted Grosbeak) or even farther south in South America (Bobolink).

More rarely, representatives of a group of species that winter farther north – predominantly in the Caribbean and Central America – visit the British Isles. These include Northern Parula, Northern Waterthrush, Common Yellowthroat and a number of species that have been recorded on just a handful of occasions. Given the relatively short distance of their normal migration in comparison to those species wintering in South America, these species are evolutionarily less well-equipped for making long-distance sea crossings. It is unsurprising that they survive trans-Atlantic crossings only very rarely.

The occurrence of the more southerly breeding species is thought to be associated with a specific set of meteorological conditions causing reversed northeastward passage – known as downwind flight or retrograde migration – along the North American Atlantic seaboard. This phenomenon, revealed by radar studies, occurs when flights of southbound migrants are blown backwards by strong opposing winds (Gehring 1963; Larkin and Thompson 1980). Such wind vectors off southern parts of the US east coast can carry disorientated birds north into the region off the northeast seaboard where most of our more regular northern vagrants are thought to originate. From here, flights of retrograde migrants can get caught up in eastbound Atlantic depressions and are consequently brought to Europe. This phenomenon is likely to have been the principal cause of major influxes in 1976, 1985, 1986 and 1995, all of which involved a wide range of species (and were dominated by more southern species such as Red-eyed Vireo and Baltimore Oriole). In more typical years this strong southerly vector does not develop and arrivals on this side of the Atlantic tend to be of lower magnitude and dominated by trans-oceanic migrants such as Grey-cheeked Thrush and Blackpoll Warbler intercepted by eastbound frontal waves during their normal southward passage over the ocean.

Previous authors (Cottridge and Vinicombe 1996; Newton 2007) have speculated about apparent northern or southern biases in occurrence of some species noting, for instance, that all our Tennessee Warblers have occurred in Scotland and most of the Red-eyed Vireos have occurred in the south. A plausible explanation for this pattern is that many individuals travel along innate deviant headings caused by some form of genetic mutation or other in-built anomaly.

It has been argued that if a proportion of individuals set off on a deviant heading – for example, a 180° reversal of the normal route – vagrants might move along species-specific trajectories, giving rise to patchy distributions of records (Cottridge and Vinicombe 1996). Although theoretically possible, there are several problems with this hypothesis. Firstly, for a small bird to maintain such a specific direct heading across the Atlantic in stormy conditions is likely to be near impossible, particularly given the unpredictable movement of the storm systems that evidently carry these birds here.

Secondly, the hypothesis is based on a small sample size and data from elsewhere in Western Europe do not necessarily support the assertion. Although the Red-eyed Vireo is decidedly rare in Northern Britain, it is the commonest transatlantic vagrant to Iceland (where there have been 19 birds) and records extend south to the Azores (10 records, all in recent autumns). Likewise, although the first five records of Tennessee Warbler were all from northwestern points (Fair Isle, Orkney, Iceland and the Faeroes) the most recent was on the Azores. Until a fuller pattern emerges, it remains tenuous to speculate on patterns with so few occurrences in a system with so much potential noise created by changing distribution patterns of observer effort, weather patterns and population fluctuation in the breeding range.

The spectre of ship-assistance often hangs over the occurrence of Nearctic vagrant landbirds, but for long-distance over-water migrants at least it seems probable that most occur here entirely unassisted. Falls of passerine migrants occur quite regularly on ships, particularly during unfavourable weather conditions. But most accounts of such falls state that the majority of individuals do not remain on board for long periods (Durand 1972). Transit times for fast vessels are in the order of five to seven days. Both insectivorous and granivorous species have been recorded surviving such crossings, but granivorous species are much more likely to remain on deck for long periods, particularly in spring (Durand 1963, 1972).

Durand (1972) recorded 58 species of landbirds on transatlantic crossings, including 21 species not yet recorded in the British Isles. Some regular transatlantic vagrants were never recorded landing on vessels (for example Grey-cheeked Thrush), while the most abundant species recorded was White-throated Sparrow. This latter species routinely incriminates itself by appearing near ports. Many other Nearctic species, particularly sparrows, have turned up at major ports such as Seaforth, Felixtowe and Southampton, suggesting that a proportion of individuals arrive here via ship-assistance. Many more are likely to leave their carrier boats within sight of land, and might therefore arrive well away from major ports.

Durand (1963) lists an interesting example of a Baltimore Oriole in October 1962 which he noted *"did not join the Mauretania until 40°W (3,000 km from New York), stayed for several days, pecking at limes and toast on the open decks, and left in very good shape within an hour or two of the Irish coast to make a very probable, though unrecorded, landfall"*.

Despite such evidence, the fact that the vast majority of autumn transatlantic arrivals are associated with specific weather events suggests that the most species – long-distance migrants at least – arrive here unaided.

Several of the Nearctic species that have occurred here in autumn, particularly those arriving later in the season, are not classed as long-distance Neotropical migrants. These include Mourning Dove, Belted Kingfisher, Yellow-bellied Sapsucker, Buff-bellied Pipit, Northern Mockingbird, Brown Thrasher, Varied Thrush, American Robin, Red-breasted Nuthatch, Savannah Sparrow, Song Sparrow, White-crowned Sparrow and White-throated Sparrow.

The occurrence pattern of some species – in particular Mourning Dove, Yellow-bellied Sapsucker, Buff-bellied Pipit, American Robin and Savannah Sparrow – mirrors that of the more expected long-distance migratory vagrants from North America. Records in western Britain and Ireland match the typical time window and follow the passage of deep Atlantic depressions. Although we do not know the extent of their physiological capabilities, the pattern of occurrence suggests that many may have crossed the Atlantic unassisted.

Speculation on the origins of other species is constrained by small sample sizes. Brown Thrasher in Dorset and Red-breasted Nuthatch in Norfolk arrived around the time one would expect unassisted vagrants to occur, but they were recorded at atypical locations reasonably close to major ports. Neither species has much of a history of long-distance vagrancy in North America and both are short-distance migrants at best, although Red-breasted Nuthatch is irruptive and has also occurred in Iceland (and Bermuda, around 1,000 km from the mainland, at least three times 1975, 1977 and 1978). Such irruptive boreal species can on occasion make huge trans-continental movements, but are not evolutionarily adapted to long water crossings. Unassisted vagrancy would therefore seem unlikely on current knowledge, and in both cases birds were likely to have crossed the Atlantic at least partly aboard ship; Durand (1972) recorded both on ships.

The case for unassisted vagrancy for Northern Mockingbird is even more tenuous. This species is largely sedentary and rarely occurs as a vagrant even into the Canadian interior. The existing accepted autumn record, although from a likely location (Cornwall), was very early in the autumn. However, it occurred at the same time as Black-billed Cuckoo and Black-and-white Warbler in the region and the three non-accepted records of Northern Mockingbird (Kent 1851, Norfolk 1971 and Wales 1978) were also in August, highlighting how difficult it is to ascertain origins in such species.

WINTER VAGRANCY

Every so often an American landbird is found wintering in the UK. Such events invariably send birders into a furore of excitement during the normally dull winter period. Such birds are typically granivorous or frugivorous species – such as White-crowned and White-throated Sparrows, Dark-eyed Junco and Baltimore Oriole – found attending garden bird feeders. More typical autumn vagrants such as Catharus thrushes and Dendroica warblers are found much more rarely in the winter, suggesting that most either continue moving south after making landfall or perish while attempting to winter when they fail to find sufficient food. Many individuals could also remain undetected away from garden feeding stations, where they are less likely to cross paths with observers. It may be that most Nearctic vagrants perish after using up their pre-migratory fat deposits during the ocean crossing. Certainly this seems to be the case for both Yellow-billed and Black-billed Cuckoos, which are typically found exhausted.

Temperate winter conditions are probably not suitable for wintering by insectivorous species adapted to spend the winter in the Neotropics, although singles of four species of wood warbler have been discovered surviving well past mid winter. Survival rates may be higher in the mildest winters. In the winter of 1988-1989, one of the mildest on record, three species of Nearctic vagrant attempted to winter (Golden-winged Warbler, Common Yellowthroat and Baltimore Oriole). The preceding autumn was a particularly good one for vagrants (Elkins 1999).

While it can be safely assumed that most of the wintering Nearctic passerines discovered here arrived in the preceding autumn, some species could conceivably cross the Atlantic during winter. American Robin is abundant in the southeastern USA in winter and regularly undertakes cold weather movements en masse. A coasting bird could easily be swept eastwards into developing wave depressions along the Gulf Stream boundary, from which a transatlantic crossing in warm sectors may be initiated (Elkins 1979). Dark-eyed Junco is another regular late-season vagrant that might arrive by a similar crossing, but as with the other sparrows (and American Robin) ship-assistance may be the norm rather than the exception.

SPRING VAGRANCY

Spring vagrancy of North American passerines is more rare and less predictable than in autumn. The species composition is also markedly different, with Nearctic sparrows predominating. Indeed, the fact that American sparrows are generally rare or absent within major autumn arrivals of other Nearctic vagrants suggests that something different occurs with this group. The paucity of records of these species from classic sites for transatlantic vagrants, such as the Isles of Scilly, and the bias towards areas in close proximity to major ports suggests that many sparrows arrive on boats. Considering the lack of suitable wave depressions across the Atlantic during spring, generally clear sunny weather along the US east coast, the predominantly overland spring passage of most species in North America and a bias in records towards the east coast and Shetland, it seems likely that birds do not routinely make unassisted transatlantic crossings in spring.

Elkins (1979) found little relationship between the occurrences of suitable wave depressions and the arrival of Nearctic migrants in spring. Many individuals were recorded during or after blocking situations, such as prolonged easterlies, where unaided passage would be very unlikely. The only exception is an apparent spring fall of Nearctic vagrants in 1977, when conditions were conducive to transatlantic vagrancy. Four Nearctic passerines were recorded: Yellow-rumped Warbler on Fair Isle; Dark-eyed Junco in Highland; and two White-crowned Sparrows, in Yorkshire and on Fair Isle. The possibility of partial ship assistance seems likely for some or all of these birds (Elkins 1979).

The lack of spring records of the typical autumn vagrants such as cuckoos, Grey-cheeked Thrush, Red-eyed Vireo and Bobolink is interesting. Perhaps they suffer a high degree of mortality after arrival in autumn or attempt to re-orientate and maybe even attempt a sea crossing at lower latitudes. Most of the Nearctic species that occur in spring are those that typically winter in central or North America. These may be more adapted to the conditions occurring in a typical western European winter (Elkins 1999).

Some spring vagrants might overshoot up the eastern seaboard and continue across the Atlantic, resulting in landfall in northern Scotland along a rhumb line (a straight line rather than a great circle) route. This hypothesis is unlikely, as the subset of species involved is not of typical long-distance migrants. Perhaps more importantly, the birds are not likely to be assisted by favourable winds.

Spring records of some strongly migratory species – for example Eastern Phoebe, Tree Swallow, Yellow-rumped Warbler and Cape May Warbler – are perhaps the best candidates for overshooting. However, these records could also have involved birds that arrived in the previous autumn and moved north in the following spring. Ship-assistance could also conceivably have occurred in these cases, although all are medium-range migrants that did not occur in close proximity to major ports (unlike spring records of Blackpoll Warbler and Lark Sparrow).

VAGRANCY FROM THE EASTERN PALEARCTIC

Autumn Vagrancy

One of the most prominent features of the British Isles avifauna is the extensive roll call of passerine vagrants from the Eastern Palearctic. We are visited regularly by species from across Asia from the northern extremity of the Siberian tundra to the arid steppes and deserts of southern central Asia. Among them are species with ranges so far removed from western Europe that one would perhaps never expect them to occur here of their own volition. The distances travelled by some of these errant individuals are staggeringly huge, and their occurrence represents one of the most interesting quirks of our natural history.

The causes behind autumn long-range vagrancy from the east remain poorly understood and a matter of much debate. Of all forms of vagrancy discussed here, the arrival of eastern species is perhaps the least likely to be caused directly by weather patterns. Unlike passerine vagrants from the Americas that are brought to the British Isles primarily by weather events, species from Asia are very unlikely to be influenced by the weather in such a significant way. Our weather systems are seldom linked to those of the regions where these birds originate. It is extremely rare that a single wind connects our islands to the parts of Asia typically occupied by our eastern vagrants (Elkins 1988).

Many of our most regular Asian visitors breed no nearer than central Russia, and in these cases it is simply impossible for wind drift to be responsible for bringing birds across the entirety of their journey. Winds inevitably cause migrants to deviate from their desired course, but such effects are only likely to last for long periods when individuals have simply no choice but to fly with the wind, as with sea crossings. Over the land most migrants can avoid prolonged drifting by simply grounding and waiting for more suitable weather conditions before continuing their journey.

Weather systems are unlikely to be the root cause of vagrancy from Asia, but they certainly play a significant role in influencing the number of vagrants that arrive here. Any birder who has spent time on the east coast of Britain will know that the weather has an enormous impact on the arrivals of migratory species, both common and rare.

The effects of weather can be rapid and dramatic. In the right conditions, our coastlines can be transformed from birdless wastelands to bountiful havens teeming with continental migrants. There are almost always a few vagrants caught up in such movements, and there can be no doubt that these weather systems play an important role in carrying vagrants to us from the continent. It is likely that many eastern vagrants passing through Europe get entrained in movements of more local migrants and become affected by the same weather patterns that give rise to common migrant falls on our coastlines. Several vagrant species that breed on the near continent, such as Thrush Nightingale, probably arrive primarily as a result of wind drift in such conditions in autumn (Williamson 1959).

Interestingly, the biggest arrivals of eastern vagrants tend to occur when anticyclonic conditions prevail over the continent and eastwards towards Siberia (P. Harvey in litt). These anticyclones are usually associated with clear weather and light winds over Scandinavia and central Europe. Such conditions are not particularly conducive to drifting birds off their normal headings. These weather patterns are, however, suitable for facilitating the passage of birds that are already moving westward across Europe on a deliberate heading. This again suggests that although weather conditions play a role in assisting the arrival of eastern vagrants, wind drift is not the ultimate causal factor. Indeed, a good proportion of vagrants arrive in weather conditions that are far from ideal for falls of continental migrants (Nisbet 1962, Elkins 2005).

It is not infrequent that eastern vagrants are discovered on days when few other migrants are found, particularly at remote migration watchpoints or small islands, where coverage is regular and birds are easier to detect due to a lack of cover. In some cases, vagrant birds seem to have actively battled against the weather in order to arrive on our shores. Given that observer coverage decreases dramatically on our coasts when conditions are not ideal for drifting continental migrants, it must be assumed that many of these eastern vagrants go undiscovered during periods of apparently unsuitable weather.

If the weather is not directly responsible for bringing many eastern vagrants to the British Isles, what is? This question has been a cause of great debate and interest throughout the history of ornithology. A fully satisfactory answer remains elusive. There are numerous theories relating to types of navigational error that could give rise to the patterns of vagrancy that we encounter, and there remains considerable argument over their relative merit. Before considering these issues in detail, it is worth considering a few broader points that have significance to the overall frequency of vagrancy from the east.

Pallas's Reed Bunting by James Gilroy

A starting point in understanding why so many eastern vagrants reach our shores is to consider a basic feature of geography: landmass. Asia is a very big place. Not only is Asia large, but much of the continent remains relatively untouched by man, a vast and biodiverse wilderness. Northern Asia remains dominated by wild primary habitats: forests, swamps and tundra that stretch unbroken across many millions of square kilometres. As a consequence, the region supports truly enormous populations of birds, and passerines in particular. Although figures are difficult to estimate, common species such as the Yellow-browed Warbler and Pallas's Warbler must have world populations numbering in the tens of millions. Another important element is the climate. The region endures extremely cold winters and consequently most of the passerines breeding in the mid and high latitudes are strongly migratory. In combination, these factors contribute to the propensity for vagrancy among passerines from the east. With such large populations performing long distance movements, it is inevitable that a significant number of individuals will make mistakes during the course of their migration. It is unsurprising that the species occurring here most frequently tend to be those that are most abundant within their home range.

Another simple geographical feature that predisposes the British Isles to receive far-eastern vagrants is our position at the western extremity of the Eurasian landmass. Unlike most Nearctic vagrants, eastern species face few genuinely significant barriers on their way here. There are no great oceans or deserts to cross. Birds can exploit constant re-fuelling opportunities in order to continue their great journey. The only really significant obstacle before arriving on the British east coast – the North Sea – could even be helpful in terms of increasing the proportion of eastern vagrants that are discovered by birders after arrival. Although only a short crossing, it presents enough of a barrier that passerines making the crossing are often tired enough to seek refuge in the first patches of coastal scrub available. Thus they make landfall in fringe coastal areas where they can be found more readily, rather than melting away into the interior landscape.

These basic principles contribute to the high propensity for vagrancy from the east, but what are the underlying factors causing vagrancy to occur in the first place? Studies of the navigational capacities of migratory birds have revealed that they are increibly complex. Experiments have revealed that birds can draw on a wide range of cues, stimuli and internal mechanisms to determine their position, route and timing (see Newton 2007). They may use multiple sensory organs in order to navigate, including vision, olfactory receptors and even specialised eye cells that can detect magnetic fields. With such a complicated array of inter-related navigational machinery, it is perhaps unsurprising that many things can go wrong.

It is likely that within each migratory population variation will exist in the accuracy of navigational tools as a result of genetic mutation or other factors influencing a bird's development. This variation could take many forms, given the wide range of variables associated with migration behaviour.

One particular navigational error that has attracted particular attention as a likely cause of vagrancy is the phenomenon of reverse migration. This occurs when an individual makes a 180° miscalculation in orientation, causing it to perform the exact opposite of its intended migration route (Rabol 1980). It is a relatively simple error for an individual to mistake north for south. The potential significance of this navigational error becomes clear when we examine the normal migratory trajectories of most eastern vagrants that regularly reach western Europe. On the whole, these species breed in Siberia and migrate to winter in southern Asia. In many cases, a reversal of this normal journey brings the birds directly to Western Europe (Figure i).

The most frequently cited example in support of reverse migration as a primary cause of vagrancy concerns the occurrence patterns of Red-breasted and Collared Flycatchers in the British Isles (Cottridge and Vinicombe 1996). Red-breasted Flycatchers breed abundantly from central Europe eastwards. They migrate southeast from these breeding grounds to spend the winter in southern Asia. A reversal of this normal migration route brings birds straight to western Europe, where they are relatively common as autumn vagrants.

The breeding range of Collared Flycatcher is similar, but this species winters exclusively in East Africa. Consequently, the entire population migrates almost due south in the autumn. As such, the reverse migration shadow for Collared Flycatcher barely brushes the British Isles, being orientated more towards Scandinavia. As expected under reverse migration theory, Collared Flycatcher is practically unknown in the British Isles in autumn. With this compelling evidence, the reverse migration shadow has been used widely as an explanation for patterns of autumn vagrancy. It has even been suggested as a tool to predict which vagrants could occur here naturally and which are likely to occur only as escapes from captivity (Cottridge and Vinicombe 1996).

Being a relatively simple error to make, one would expect reverse migration to be a relatively common cause of vagrancy, as suggested by many authors. However, the extent to which it can be used as a blanket explanation for patterns of vagrancy from the east is debatable. Many patterns of vagrant occurrence do not follow the predictions of the theory. Even the classic Red-breasted and Collared Flycatchers example does not bear up to close scrutiny. It is indisputable that Red-breasted Flycatcher is the commoner of the pair in the British Isles in autumn. However, the same is also emphatically true across the whole of Scandinavia, even within Collared Flycatcher's reverse migration shadow: there are no autumn records of Collared Flycatcher in Norway and only one from Finland, but many hundreds of records of Red-breasted Flycatcher. The rarity of Collared Flycatcher could be explained by the fact that they are extremely difficult to distinguish from the much commoner Pied Flycatcher in autumn. Never the less, the fact that there is no evidence for Collared Flycatcher being any more common within the reverse migration shadow than outside greatly weakens the strength of the example in support of the reverse migration theory (Gilroy and Lees 2003).

Many other occurrence patterns concerning eastern species also indicate that reverse migration is not the sole cause of autumn vagrancy. Several eastern species that occur regularly in the British Isles follow a north-south migration in autumn, and therefore cannot occur here as a result of simple reverse migration (Figure ii). These include Richard's, Blyth's and Olive-backed Pipits and Black-throated Thrush. Several of our other regular autumn vagrants, including Pied Wheatear and Isabelline Shrike, migrate from central Asia to eastern Africa in a south-westerly direction, such that a 180° reversal would take them far to the east of Britain (Figure iii).

Peaks in occurrence of vagrants may occur in certain directions as a result of regularly occurring errors, including reverse migration, but it is clear that vagrancy in most migratory species can occur in almost any direction outside the normal migration route (Gilroy and Lees 2003). The causes of these aberrations remain poorly understood, but presumably often relate to genetic mutations or abnormalities that influence any of the many components of birds' navigational instincts.

Another intriguing possibility is that local irregularities in Earth's magnetic field can influence the development of navigational apparatus. Birds reared in areas with anomalous magnetic field patterns could theoretically develop navigational mechanisms that differ from the population average, potentially leading to localised vagrancy. This hypothesis is given tantalising support by the fact that parts of Siberia, where so many of our regular vagrants originate, are known to have high levels of magnetic irregularity (Alerstam 1990). However, the true influence of these features on bird migration remains unknown, and any speculation is purely conjectural.

Whatever the underlying cause, it is clear that autumn vagrancy is a relatively common phenomenon amongst migratory species from the Eastern Palearctic, and that there is a significant element of randomness in the course that such birds can take. Indeed, records of species from the very far east, for example Chestnut-eared Bunting, underline that distance is no barrier to such movements.

All evidence points to the conclusion that there are few limits to the realms of possibility concerning long-distance vagrancy. There remains considerable potential for further species to reach us from the east, given that such a diverse community of long-distance migrants exists at the opposite end of the Palearctic landmass from us. As birder coverage and knowledge increases, we can look forward to more additions to our avifauna. This extensive shopping list of tantalising and exotic species such as

Forest Wagtail, Siberian Bush Warbler, Dark-sided and Siberian Flycatchers and White-throated Rock Thrush cannot be too far away from delivery to our door.

Winter and Spring vagrancy
Despite their relative abundance during the autumn migration period, passerine and near passerine vagrants from the East Palearctic are surprisingly infrequent in the British Isles during winter and spring. A handful of species, particularly those most abundant in autumn – such as Richard's Pipit and Yellow-browed Warbler – are now found wintering almost annually in very small numbers, a phenomenon that may be increasing in response to a recent series of mild winters. Most of our other regular Siberian vagrants have been found over-wintering occasionally, particularly granivorous or frugivorous species (especially buntings and thrushes) that are more likely to find sufficient food resources during the British winter than insectivorous species.

Overall, though, the frequency of occurrence of Siberian species in winter is surprisingly low in light of the number of birds that must occur each autumn. This is likely to be related to the relative harshness of our winter climate being unfavourable for species that generally winter in the tropics. In addition, the likelihood of detection for wintering passerines must decrease once they have moved away from our well-watched coastlines.

The same arguments also apply to many of the North American passerines over-wintering in the British Isles, many of which are extremely rare as autumn vagrants. It is remarkable that tropical-winterers such as Black-and-White Warbler, Baltimore Oriole and even Golden-winged Warbler have been discovered over-wintering here, and yet much commoner Siberian species such as Citrine Wagtail and Radde's Warbler have not. This discrepancy is difficult to account for, but could perhaps be related to the more cryptic appearance of these Siberian species in relation to their gaudy North American counterparts.

In most years a small trickle of passerine vagrants from the East Palearctic is recorded in the early spring. Most are short-stay appearances at coastal watch points in southern areas of the British Isles. These records almost certainly relate to vagrants that arrived in Western Europe in the preceding autumn and which are performing a return passage after wintering somewhere farther south.

Later in spring, records of passerine vagrants from Siberia generally tail off. It is surprising just how rare East Palearctic species are during the late spring period. A handful of eastern species that breed relatively close to the British Isles in Scandinavia – for example Thrush Nightingale, Greenish Warbler and Rustic Bunting – are recorded quite frequently. These species are found most often when anticyclonic weather conditions give rise to easterly winds over the near continent, conditions that are conducive to drifting northbound migrants across the North Sea. However, the suite of East Palearctic species that occurs so frequently during similar weather conditions in autumn remains extremely rare here in spring.

VAGRANCY FROM THE NEAR CONTINENT

Migratory Species from the Mediterranean

Vagrants from the Mediterranean occur regularly in the British Isles. Most records concern long-distance migrants that breed across the region and winter in tropical Africa. Interestingly, patterns of vagrancy in species from this region are distinctly different from those of the other geographical areas we have dealt with. One principal difference is timing. Vagrants from points east and west tend to show a strong peak in autumn, but many Mediterranean species occur just as frequently or even more regularly in spring. The regularity of spring vagrancy from areas south of the British Isles can be explained by the phenomenon of overshooting.

The arrival of spring vagrants from the Mediterranean is often linked closely to weather patterns. Most arrivals occur when high-pressure systems over the continent bring warm southerly or southeasterly winds up from the Mediterranean. Such conditions are ideal for encouraging northward migratory flights to take place. It seems likely that with a strong backing wind, favourable weather patterns can carry a proportion of individuals much farther north than they would otherwise intend, bringing them beyond their normal breeding grounds and into northwest Europe (Elkins 2005). It is quite common for such weather systems to bring multiple arrivals of Mediterranean species to southern parts of the British Isles, often with many individuals arriving synchronously.

Although it is tempting to assume that spring overshooting from the Mediterranean is principally a weather-driven phenomenon, this is unlikely to be fully the case. The distances travelled by some individuals are very large, particularly in the case of species from the eastern Mediterranean, and are unlikely to be the result of wind carriage alone. Another possible cause of overshooting is that a proportion of individuals simply fail to turn off their migratory instinct on arrival in their normal breeding grounds, and continue moving along their original migratory trajectory (Cottridge and Vinicombe 1996). This scenario implies that an innate error can occur within an individual's migratory apparatus in a similar way that internal navigational errors are thought to contribute to long-distance vagrancy by birds in autumn.

A further possibility is that overshooting may occur when individuals fail to find a mate or suitable territory within their breeding grounds. Such individuals might continue moving ever farther along their original migratory trajectory in search of breeding opportunities. This scenario is perhaps most likely to occur in conjunction with short distance weather-driven overshooting; individuals that pass over the bulk of their breeding range may consequently find themselves having difficulty locating a mate or suitable breeding habitat. If these individuals fail to compensate for their original overshoot by backtracking, they may continue moving along their original route and venture well beyond their normal range.

There are several lines of evidence supporting this hypothesis. One is that for most species the majority of spring overshoots are males (Newton 2007). Populations of most bird species are male-biased (Donald 2007) and, therefore, it is normal for a surplus of unmated males to be present on the breeding grounds. These unmated males are perhaps the most likely individuals to be driven beyond their normal range in the continuing search for mates.

Secondly, arrivals of overshoots in the British Isles often peak somewhat later than arrival times in the Mediterranean. Although many southern vagrant species can occur any time from late February onwards, there is a general peak in late May and early June. This is at least a month later than the principal arrival period for most migrants in the Mediterranean. The suggestion is that overshooting peaks at a time when most birds on the breeding grounds will be already paired up, driving remaining individuals to move beyond their normal range in a vain search for breeding opportunities.

Some species whose centre of abundance lies in the Mediterranean region number among our more regular vagrants: species such as Alpine Swift, Hoopoe, Short-toed Lark, Red-rumped Swallow, Subalpine Warbler and Woodchat Shrike. Many other abundant and strongly migratory species from the region are surprisingly rare in the British Isles. Geographically the Mediterranean basin and the British Isles lie very close together, at least in the context of avian vagrancy.

As the crow flies, the English coastline lies just 750 km from the nearest stretch of the Mediterranean Sea in France. Countries as close as Spain, Italy and Greece host a whole suite of common migratory passerines that remain extremely rare stragglers to Britain. Given the huge distances travelled by many of the most regular Siberian vagrants that reach here, one would expect that the British Isles would be well within range for any Mediterranean species adapted for long-distance migration. Pallas's Warblers must travel at least 5,000 km from their nearest breeding grounds to reach us; the length of journey travelled by the much rarer Western Bonelli's Warbler from France may be only a few hundred kilometres. Other examples include both eastern and western forms of Black-eared Wheatear, Olivaceous and Orphean Warblers, as well as Rufous-tailed Rock Thrush, Rufous-tailed Scrub Robin and many more. Most of these species migrate broadly due south to winter in Africa, some moving extremely long distances into the far south of the continent.

Invoking the standard explanation for patterns in autumn vagrancy – reverse migration likelihood – does not help to explain why some Mediterranean species are so rare in northern Europe. The British Isles is well placed to receive reverse migrants from species migrating south from the western Mediterranean to Africa (Figure iv). One theory to explain this pattern was put forward by Thorup (2004), who argued that navigation along direct north-south migration routes is easier than it is for species with more complex routes involving an east-west component, as occurs in most Asian migratory species.

The relative rarity of migratory species from the Mediterranean may also be explained by global population sizes. In contrast to most of our eastern vagrants, Mediterranean specialists have small ranges and inevitably smaller global population sizes. In particular, species limited to the eastern Mediterranean – such as Rüppell's Warbler,

Olive-tree Warbler, Cretzschmar's Bunting – are all limited to a relatively small breeding area and have comparatively small global populations. Consequently it is unsurprising that such species are some of the most rare of our vagrants.

Population size can explain several other apparently confusing patterns of vagrancy from the Mediterranean. That British records of Western Bonelli's Warbler greatly outnumber those of Eastern Bonelli's ties in neatly with the fact that the global population of Western is estimated to be 10 times higher than that of Eastern (Snow and Perrins 1997). The same is true for Western and Eastern Olivaceous Warblers, but this time in reverse: the eastern form has a much larger world population and is thought to account for all records to date.

In other cases, population size seems to shed little light on vagrancy patterns. Black-eared Wheatear is abundant in areas as close as Iberia, and yet remains very rare in the British Isles. Black-eared Wheatear is rarer than its closest relative, the Pied Wheatear, which has a similar world population size but breeds much farther away. Perhaps the most extreme example is the Orphean Warbler, with a population of up to 10,000 pairs breeding as close as France and yet only four accepted British records.

Plausible explanations for the surprising rarity of these species are lacking. These patterns serve to underline how little is understood about the underlying causes of avian vagrancy.

It is evident from ringing recoveries and observations at migration watchpoints that some bird species migrate across a much wider spread of directions than others, depending at least partly on the location and distribution of their wintering areas (Busse 2001). The underlying causes of this variation both within and between species remain unclear.

VAGRANCY IN SEDENTARY SPECIES OR SHORT-DISTANT MIGRANTS

There are a few species that are extremely rare vagrants to the British Isles and yet are common residents on the near continent as little as 41 km from the Kent coast. Crested Lark and Short-toed Treecreeper are prime examples, breeding commonly almost within sight of our southern coastline. These birds' rarity here can be explained fairly simply by their highly sedentary habits. Physiologically, there is no reason why they should not cross the Channel, but rates of long-distance dispersal are so low in these species that they are apparently strongly disinclined to do so. Short-toed Treecreeper is probably under-recorded here on account of its cryptic similarity to Common Treecreeper, which by and large precludes the identification of non-vocal individuals.

Zitting Cisticola is another largely sedentary species, but one that occasionally makes irruptive movements. It has spread north in Europe over many years, and is regularly touted as a candidate species to colonise the UK in the wake of continued climatic amelioration. However, this species is extremely prone to local extinction following harsh winters. It spread as far north and east as Belgium in the 1970s, subsequent cold periods extirpated these peripheral populations. Despite expanding its range up as far as the Channel Coast, it remains extremely rare in the UK.

The case of the Penduline Tit is similar, although this species is also a true migrant, with central European and Scandinavian populations wintering in Southern Europe. At least some of the occurrences in the UK originate from these migratory populations, as is evident from ringing recoveries. Formerly an extreme vagrant to the UK, this species also underwent a large range expansion north and west and is now a regular breeder as close as the Netherlands. British records are unsurprisingly concentrated along the south and southeast coasts adjacent to the breeding areas.

The occurrence of each of these largely sedentary species in the British Isles is likely to be associated with exploratory dispersal. In almost all species, full-time residents included, a proportion of individuals will undertake some amount of dispersive movement when trying to locate a suitable territory or home range. Dispersal usually takes the form of post-fledging movement of juveniles in the late summer and autumn, and to a lesser extent post-breeding dispersal of adults. The distances involved are typically small, and the direction taken apparently random.

Dispersal in migratory species often occurs towards the wintering area and is typically over longer distances (Paradis *et al* 1998). It is uncertain what the adaptive value of such flights might be, but potential reasons include prospecting for a suitable territory site, fluctuations in food supply and avoidance of competition and inbreeding (Williamson 1959; Baker 1978; Greenwood 1980). Failure to switch off this dispersive urge might cause nominally 'resident' birds to move hundreds or even thousands of kilometres from their place of birth.

The lack of records of several resident species from the near continent in the British Isles – most notably Black, Middle-spotted and Grey-headed Woodpeckers – can be attributed to their extremely low rates of dispersal and avoidance of sea crossings apparent in most typical woodpeckers. Dispersal distances have been found to be greater in species living in wet habitats than those living in dry habitats, perhaps because of the greater patchiness of wet habitats both temporally and spatially (Paradis *et al* 1998). Low rates of dispersal in insectivorous woodland species such as the woodpeckers, tits and Short-toed Treecreeper may reflect this trend, particularly given that historically most of Western Europe was covered in forest, rendering pointless long-distance dispersal to find new forest patches.

SEDENTARY SPECIES FROM THE MEDITERRANEAN & NORTH AFRICA

Among the species recorded occasionally in the British Isles are a number of largely sedentary species that breed in the Mediterranean basin and farther south in arid North Africa. This group includes Lesser Short-toed Lark, Blue Rock Thrush, White-crowned Black Wheatear, Marmora's Warbler, Trumpeter Finch, Spanish Sparrow and Rock Sparrow. None breed any closer than southern France and all are exceptional vagrants to the British Isles and elsewhere in Northern Europe. Differences in the relative frequency of vagrancy amongst these species appear to be correlated with the magnitude of migratory tendencies they show.

Some species – for example Sardinian, Spectacled and Marmora's Warblers – are at least partially migratory and are unsurprisingly more common as vagrants than the highly sedentary species such as Rock Sparrow. Records of these partially migratory species tend to be concentrated in spring and often occur during periods of weather suitable for encouraging overshooting. This suggests that most vagrant individuals come from migratory populations that are more likely to be prone to overshoot their normal breeding grounds.

Among typically sedentary species, vagrancy is most likely to arise through extreme dispersal events, particularly involving young and inexperienced individuals that may fail to suppress the urge to move randomly away from their natal area in search of suitable territories or mates. Such movements may be assisted by weather. The first and only British record of White-crowned Black Wheatear and first British record of Marmora's Warbler occurred within a few weeks of each other in spring 1982, associated with a period of abnormally dry and sunny weather from a very warm air stream originating in North Africa; the first week of June was the hottest for 35 years (Brown 1986).

Some species from this region, most notably Trumpeter Finch, are known to be partially nomadic within their normal range. Nomadism is most common in ecosystems with high variation in resources such as food and water over time and space, forcing birds to undertake unpredictable movements that track these fluctuating resources. Nomadic birds tend to occur in ecosystems associated with low, variable, and unpredictable rainfall patterns such as deserts and semi-deserts. Periodic irruptions of species such as Trumpeter Finch often coincide with periods of extreme climatic conditions within their normal range. Various other desert or semi-desert specialists are known to make similar irruptive movements periodically, including Desert Lark and Temminck's Horned Lark; these should be considered potential vagrants to the British Isles in future years.

Alpine Vagrants

Five species of largely resident alpine birds occur as vagrants to the UK: Alpine Accentor, Wallcreeper, Citril Finch and Rock Bunting. Of these, Alpine Accentor occurs the most frequently and also undertakes the most pronounced altitudinal migrations, with birds wintering widely but uncommonly in the Iberian and Balkan lowlands. Wallcreeper shares a similar range and movements within Europe, but has a smaller population size and thus a limited pool of individuals available for vagrancy (assuming that 'our' vagrants come from Europe and not farther east). Citril Finch is more abundant than either of these two species but occurs at lower altitudes and does not undertake pronounced movements, although it does apparently winter regularly south to southern Spain and even North Africa (Benoit and Märki 2004, Navarrete et al 1991).

Rock Bunting breeds closer to the UK than others of this group, but is still exceedingly rare. As with Citril Finch, it occurs at lower elevations in winter and northern populations are most prone to dispersal, so its rarity in the UK and elsewhere in northwest Europe is quite puzzling. These vagrancy events as far as the UK may not be the results of extreme weather conditions but extreme exploratory movements of the nature of those described under Vagrancy from the near continent. The establishment of a population of White-winged Snowfinch in the Corsican mountains (Thibault and Bonaccorsi 1999) also demonstrates that long-distance movement by sedentary montane birds may occur from time to time. This last species, which has occurred as a vagrant north to Helgoland (Coues 1895), should also be expected as an extreme vagrant to the UK.

Boreal and Arctic Irruptive Species.

A major driver of population trends and hence vagrancy potential of birds breeding in the Boreal and Arctic zones is high annual variability in food resources: voles and lemmings for raptors and owls; berry and pinecone crops for a variety of passerines. Both of these groups specialise on food resources that may fluctuate regionally more than one hundred-fold from one year to the next. These fluctuations are often regionally independent, such that poor food supplies in one region may coincide with good supplies in another. In such cases birds can move hundreds or thousands of kilometres from one breeding area to another in search of suitable breeding conditions (Newton 2007).

An irruptive migration occurs in years of widespread food shortage or when population levels outstrip food supply. Irruptions may extend across millions of square kilometres, resulting in population-level displacement to lower latitudes. The UK is within the regular irruptive range of many northern species – Rough-legged Buzzard, Common Crossbill and northern Bullfinch to name three – and lies on the periphery of the range of others.

Some irruptive species that breed very close to us on the near continent remain extremely rare here because they have low population sizes, do not tend to disperse far or show a reticence to cross large water bodies. Most boreal owls appear to fall within this category.

Small mammals, voles in particular, are known for their regular multi-annual population fluctuations, with a periodicity of between three and seven years documented in several countries. In Northern Scandinavia, abundance cycles of microtine voles generally increase in length and magnitude as a function of latitude. Most predator species occurring in the northern part of Fennoscandia are specialised in utilising this resource; more generalist predators dominate farther south. The greater proportion of generalist predators is assumed to stabilize rodent populations in the south, as predators can switch to several alternative prey species when rodent numbers fall (Hanski *et al* 1991, Turchin and Hanski 1997).

In the far north, where fewer alternative prey types are available, predators will continue to target rodents even when the populations are crashing. Once rodent numbers are sufficiently low, the predators have no alternative but to move off in search of new feeding grounds. Winter irruptions of species such as Tengmalm's Owl will occur during such periods of low abundance of small mammals within the northern part of their range (Mikkola 1983). In contrast, Tengmalm's Owl populations in southern areas (as close to the UK as Belgium) have a wider range of bird and mammal species on which to prey and are, therefore, much less prone to long-distance irruptive movements. As such we can assume that the Tengmalm's Owls occurring extralimitally in the UK and elsewhere in Western Europe come not from the proximate resident populations but from the distant irruptive ones. The extralimital occurrence of Hawk Owl is tied to similar fluctuations in prey abundance. At least one individual of this species has been assigned to the Nearctic subspecies, demonstrating that food shortages may even prompt transatlantic vagrancy.

Snowy Owl is a strongly nomadic inhabitant of the tundra, moving frequently in search of areas of high lemming population density and breeding only in peak years of lemming abundance (Watson 1957, Portenko 1972). Snowy Owl irruptions in North America have been documented since about 1880 and occur every three to five years (Newton 2002). Irruptions into the British Isles seem to occur over a longer timescale, probably as sea crossings are only likely to be attempted during the largest irruptions. Arrivals here appear to fluctuate according to the relative severity of the winter: records peaked during the relatively cold period in the 1960s and 1970s when an invasion lead to extralimital breeding in Shetland. Historically most records come from the Northern Isles, but Snowy Owls also reach our western coasts, suggesting that many may be of Nearctic origin. Transatlantic ship assistance is known to have occurred on at least one occasion, in Suffolk in 2001.

The second important group of boreal irruptive vagrants is passerines that are dependent on tree-fruit or seed crops. Trees of many species require more than one annual cycle to accumulate sufficient nutrient reserves to produce a fruit crop; in any given area most of the trees of a given species (or even different species) fruit simultaneously, being exposed to the same weather conditions. This results in a profusion of tree fruits in some years and a near-complete absence in others. As a result, these boreal irruptive migrants typically show a wide spread of random dispersive headings, as is apparent both from observations and from ring recoveries.

Some of the long-distance movements involving boreal irruptive passerines are truly spectacular: a ringed Bohemian Waxwing was recorded travelling 6,000 km from Ukraine to Siberia; and a Eurasian Siskin moved 3,000 km from Sweden to Iran. Most striking were a ringed Common Redpoll that moved 8,350 km from Belgium to China and another that moved 10,200 km from Siberia to Canada (Newton 2007). Such long-distance movements in Siberia may be more likely to occur on an east-west axis as movement to the south takes dispersers into the arid lands of Central Asia where suitable foraging opportunities for these species would be severely limited.

Some species have very narrow diet specialisations within their normal ranges. Irruptions may consequently depend on seed or fruit crops of a single species, for example, irruptions of Spotted Nutcracker tend to be triggered by failures in the seed crop of the Siberian Stone Pine. Epic irruptive movements also occur in boreal-breeding species in the Nearctic: a ringed Pine Siskin travelled 3,950 km between Quebec and California; and an Evening Grosbeak travelled 3,400 km between Maryland and Alberta.

Such movements hint at the possibility of transatlantic vagrancy in species such as Pine Siskin and support the already documented occurrence of Cedar Waxwing, Red-breasted Nuthatch, Evening Grosbeak and Varied Thrush here in Britain. However, as these species are not adapted to long over-water sea-crossings, such records could be related at least partially to ship-assisted passage.

Steppe Irruptive Species

Three species restricted to the grassy steppes of central Russia occur as vagrants to the UK and elsewhere in Western Europe. These species – Pallas' Sandgrouse and White-winged and Black Larks – are among the most charismatic of all British vagrants. All are extremely scarce, poorly known and rarely seen by western ornithologists even on the breeding grounds. White-winged Lark is the most migratory of the three. It winters in Ukraine, Crimea, Caucasus, Transcaspia and Iran (although some remain within the breeding range all year). Black Lark is more sedentary, but a proportion of the population moves a short distance to the west or southwest in September and October, with some wintering in Ukraine and southeastern regions of European Russia (Lindroos and Tenovuo 2002).

Both these larks show a bias towards spring occurrences in Europe, but considering the location of their breeding and wintering grounds and migratory heading are very unlikely to occur as overshoots. Koistinen (2002) compared the European records of the two species and could only find one instance of simultaneous vagrancy, although several recent records of both species occurred during periods of prolonged south-easterly winds. Koistinen postulated that extreme winter weather conditions within the normal wintering area (heavy snow or deep frost) could inhibit foraging and spur long-range dispersal to alternative feeding areas.

The vagrancy pattern of Pallas's Sandgrouse is arguably even more enigmatic. Exceptional numbers arrived in Western Europe in eight years between 1859 and 1908. It even bred in the UK. Subsequently it has returned to the status of an extreme vagrant. These mass irruptions have been attributed to a lack of food within the normal range, particularly the abundance of the seed-bearing Orache plant *Agriophyllum squarrosum*. Collapses in the abundance of this food source are thought to result from either prolonged drought (Newton 2007) or heavy snowfalls (Dementiev and Gladkov 1951). A trend towards the gradual desiccation of the Aralo-Caspian region at the western part of the sandgrouse's range may have reduced the chances of regular vagrancy to the west (it also occurs as an irruptive vagrant to the east, to Japan for instance). However, heavy snowfalls in the source region for vagrancy may still periodically produce small westerly incursions of this species: such conditions were associated with the occurrence of a Pallas's Sandgrouse in Kent in 1964.

Figure i: Red-flanked Bluetail reverse migration projection. On normal migration, westernmost populations head on an initial heading of almost due east in order to avoid crossing inhospitable desert and mountain regions in Central Asia. A 180° reversal of this route would bring birds directly to Western Europe.

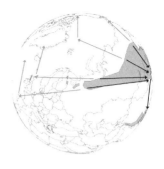

Figure ii: Black-throated Thrush reverse migration projection. This strong-flying species moves directly south from the breeding grounds in autumn. Reverse migrants would move into northern Siberia and beyond. In fact, vagrants of this species have been recorded in almost all compass directions outside the normal route. (Gilroy and Lees 2003)

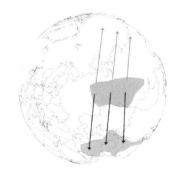

Figure iii: Rufous-tailed Scrub Robin reverse migration projection. Despite being a relatively common long-distance migrant with an apparently ideal reverse migration shadow, this species is phenomenally rare in Northern Europe.

Figure iv: Pied Wheatear reverse migration projection. This regular vagrant is one of several central Asian species that winter predominantly in East Africa. Individuals in Western Europe in autumn appear to have made 90° misorientation away from their normal route.

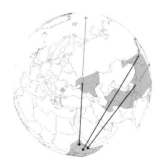

SUMMARY

There can be little doubt that long-distance vagrancy, particularly among small-bodied passerine birds, represents one of the most fascinating and intriguing features of the natural world. Birders with an interest in vagrancy in Britain & Ireland are extremely lucky on several counts. Not only are we ideally placed to receive vagrants from across the globe but also we have the benefit of a vast databank of previous records, thanks to the efforts of hundreds of amateur and professional enthusiasts, dating back hundreds of years. Despite this wealth of data, there is much to be learned about the underlying causes of these remarkable movements. The issues raised in this account represent the mere tip of an iceberg, and a great number of questions remain to be answered. It is hoped that with the ever-increasing popularity of birding, particularly in continental Europe, our understanding of bird movements and occurrence patterns will improve greatly in future years, bringing new insights into these amazing phenomena.

THE BRITISH ORNITHOLOGISTS' UNION RECORDS COMMITTEE (*BOURC*)
Bob McGowan, Chairman BOURC

Rock Sparrow by Ian Wallace

Professor Alfred Newton founded the British Ornithologists' Union (*BOU*) in 1858. It is one of the oldest and most respected ornithological societies. Its aim is to promote ornithology to the scientific and birdwatching communities, which it achieves primarily through publication of its quarterly journal *Ibis* (available online at *http://www.ibis. ac.uk*). *BOU* also organises regular meetings and conferences, and funds ornithological research projects.

Function, History and Operation
The Records Committee (*BOURC*) is a standing committee of the *BOU*. Its function is to maintain the official list of birds occurring in Britain. The first official *British List* of birds was produced in 1883; the 7th edition (*Ibis* 148: 526-563) was published in 2006. A *BOU* List Committee existed prior to the Records Committee, publishing its first report in *Ibis* in 1956. Prior to 1956 there were 22 reports from the Committee on the nomenclature and records of occurrences of rare birds in the British Isles, the first

in 1918 and the last in 1950. The current official *British List* can be viewed on line at *http://www.BOU.org.uk/recbrlst1.html*

The Committee consists of a Chairman, Secretary and eight members. The Chairman and Secretary serve four-year terms in that role, but membership of the Committee is usually for a ten-year period. The skills base of the Committee includes expertise in the fields of identification, taxonomy, historical research, museum work, and knowledge of the bird trade. The Committee meets twice each year.

Reviewing the validity of subspecies on the *British List* – whether vagrants, migrants or endemics – may form part of the ongoing work of *BOURC*. Information on naturalised populations is monitored for consideration for inclusion in Category C (see *Ibis* 147: 803–820). Reviews are periodically undertaken of older records, particularly Category B species that have not occurred in the wild since 1949.

Anyone can ask for old or rejected records to be reviewed by *BOURC* if they provide fresh evidence to justify re-examination. However, a simple difference of opinion with the *BOURC* decision on a record is not new evidence. New evidence usually involves further information on vagrancy or status in captivity.

TSC

BOURC has a Taxonomic Sub-committee (TSC), which advises *BOURC* on taxonomic issues affecting the *British List* via its own reports published in *Ibis*. In October 2002, the TSC published a paper outlining the basis on which it will make its species-level taxonomic decisions (Guidelines for assigning species rank. Helbig *et al Ibis* 144: 518-525). This paper outlines evidence required by the TSC to decide whether a taxon merits species rank. The TSC soon followed this publication with its first recommendations for taxonomic changes (*Ibis* 144: 707–710).

Category F Sub-committee

A Category F Sub-committee was set up in 2007 to develop a list of species known to occur in Britain prior to 1800.

First Records

Ask most birders about the work of the *BOURC* and it will be its role of admitting new species and subspecies to the *British List* that they know best. Records of birds potentially new to Britain are passed to the *BOURC* by the *British Birds* Rarities Committee (*BBRC*) after that committee has examined the evidence. For records relating to new species and subspecies for Britain – the fabled 'firsts' – *BOURC* considers the identification, taxonomy and origin of the bird. Detailed investigations into racial and species identification, escape likelihood and vagrancy potential are undertaken to determine the validity of the record before admission to the *British List*.

The *BOURC* Secretary prepares a file summarising each record. The file contains original descriptions and supporting documentation, including *BBRC* comments, correspondence from independent specialists, an analysis of the captive status of the species and its escape likelihood, and extracts from books and journals referring to migration and vagrancy patterns. Files are usually circulated by post, occasionally by email, and usually take several months for all 10 members to form a considered opinion on a record. *BOURC* also uses its own internal website for storing photographs and other material that supports each review or circulation. The committee's first task is to confirm the identification, which requires unanimous agreement. A two-thirds majority is required on categorisation; files are re-circulated if this majority is not achieved. All files are archived for future reference.

BOURC alone decides which species are to be admitted to the *British List* and how they are to be categorised. For first records, *BBRC* is concerned solely with identification. However, *BBRC* also assesses large numbers of subsequent records of major rarities after 1949. The workload of both committees is substantial and complementary. *BOURC* works closely with *BBRC*, whose function is to collect, investigate and apply uniform standards of identification to claimed records of rare birds in England, Scotland and Wales, and at sea within the British Economic Zone, which now extends to 200 nautical miles (370 km) or the half-way line between neighbouring countries where this is less than 200 nautical miles.

Species Categories

BOURC maintains the *British List* on behalf of the *BOU*, legislators and the international birdwatching and ornithological communities. Part of the function of the List is to indicate the status of each species recorded in Britain. This is achieved by categorisation (which was first introduced by the *BOU* in the 1970s and widely followed by other national list committees).

The *British List* uses the following species categories:

- **A** Species that have been recorded in an apparently natural state at least once since 1 January 1950.
- **B** Species that were recorded in an apparently natural state at least once between 1 January 1800 and 31 December 1949, but have not been recorded subsequently.
- **C** Species that, although introduced, now derive from the resulting self-sustaining populations.
- **D** Species that would otherwise appear in Category A except that there is reasonable doubt that they have ever occurred in a natural state. Species placed only in Category D form no part of the *British List*, and are not included in the species totals.
- **E** Species that have been recorded as introductions, human-assisted transportees or escapees from captivity, and whose breeding populations (if any) are thought not to be self-sustaining. Species in Category E that have bred in the wild in Britain are designated as E*. Category E species form no part of the *British List* (unless already included within Categories A, B or C).
- **F** Records of bird species recorded before 1800.

Some species are placed in multiple categories. For example, those species occurring in Category A which now have naturalised populations recognised in Category C (e.g. Red Kite). Any species known to have escaped from captivity is added to Category E, but may still be in Category A (e.g. European Goldfinch).

Only species in Categories A, B and C are counted for British List totals.

For full details see the *British List* pages of the *BOU* website or *The British List 7th Edition* (*Ibis* 148: 526-563).

Reporting

BOURC decisions are published in the Committee's regular reports in *Ibis*. At one time few birders had access to *Ibis*, but with the arrival of online publishing that has changed: all *BOURC* reports are now available free to view online via the *BOU* website (www.*bou*.org.uk). *BOURC* members may also write longer papers on species reviews and individual decisions for publication in the popular birding press.

The Official *British List*

With its long-established and trusted maintenance of the list of birds recorded in Britain, the following organisations have indicated their support of *BOU* and that the decisions on both status and taxonomy reached by *BOURC* are accepted by them as comprising the official *British List*: British Trust for Ornithology, Countryside Council for Wales, English Nature, Joint Nature Conservation Committee, Royal Society for the Protection of Birds, Scottish Natural Heritage, Scottish Ornithologists' Club, Wildfowl and Wetlands Trust, and The Wildlife Trusts. *BOURC* also liaises with the Association of European Rarities Committees (AERC), both to share information and to contribute to the compilation of an official European List of birds.

Responsibility for the Irish and Northern Irish Lists lies with the Irish Rare Birds Committee (IRBC) and the Northern Ireland Birdwatchers' Association (NIBA) respectively. The Isle of Man (which is not a legislative part of the UK) also keeps its own list, which is maintained by the Manx Ornithological Society (MOS). Decisions relating to the *British List* will continue to be published by *BOURC* in its annual reports in *Ibis*, and decisions relating to the Isle of Man are summarised in these reports.

Northern Waterthrush by Ian Wallace

All the published items referred to above can be viewed online via the *British List* pages of the *BOU* website. This also has details of changes to the List and gives access to recent Committee reports.

THE BRITISH BIRDS RARITIES COMMITTEE (*BBRC*): RECORDING RARE BIRDS IN BRITAIN FOR 50 YEARS
Adam Rowlands, Chairman BBRC

Rustic Bunting by Ray Scally

A comprehensive analysis of the origins and history of the *British Birds Rarities Committee* (*BBRC*) by Alan Dean was published in *British Birds* in 2007 (*BB* 100: 149-176) and provides a definitive source of reference on the subject. I have used that reference extensively in preparing this write up, but any errors or inconsistencies are my own. The establishment of the Rarity Records Committee was announced in an editorial of the August 1959 issue of *British Birds*. The justification for setting up a national committee to review rare bird claims was *"a growing realisation that a large number of birds formerly thought to be rarities are reaching the British Isles regularly, and even in some numbers."* This was considered to be due in part to changes in habitat and expansion of breeding distribution, but it was also suspected that many had been overlooked in the past and the discovery of increasing numbers was thought to be a consequence of larger numbers of bird-watchers, considerable advances in field identification and the increase in bird observatories and trapping.

It was demonstrated that some species had increased dramatically in recorded occurrence during the 1950s. Two examples – Woodchat Shrike (1990) and Melodious Warbler (1963) dropped from the list of species considered by *BBRC* illustrate this phenomenon. Prior to the founding of the Committee records of rare birds in Britain up until 1940 had been documented in *The Handbook of British Birds* (Witherby *et al* 1938-1941). The records collected in these five volumes had been sourced from a variety of journals including *British Birds*, *The Zoologist*, *Ibis* and the *Bulletin of the British Ornithologists' Club*, together with national and regional avifaunas. Post-Witherby *et al* the records from 1940-1957 were published in a number of different places, including a growing number of newly established county bird reports. Throughout this period, the assessment of the validity of these claims rested with the editors of these publications.

The records of the rarest species were published in *British Birds*, so not only did the editorial team of that journal assess the records but also the published descriptions were then available for public scrutiny.

One of the main aims of the newly formed Rarity Records Committee was to overcome the inconsistency presented by different individual bodies determining the validity of the records and to provide a more uniform standard of assessment to rare bird records nationally. The Committee also aimed to bring together all the records in one place by publishing an annual rarity report.

The first annual report's introduction emphasised that each individual record of a rare bird has comparatively little value on its own, but the collective value of all records taken together may offer explanations to the origins of migratory movements, while illustrating trends in range expansion and the differences in occurrence patterns between adult and first-year birds.

It became clear within the first five years of the Committee's operation that the collection of records of lesser rarities to aid assessment of range expansion and migration phenomena was not achievable if it involved the assessment of each individual record by the national committee (*BB* 57:305). Another function of the Committee, the intention to pass on the knowledge gained from its work to observers by reappraising the identification criteria of particular groups, remained important.

In terms of the assessment process, the first annual report (for 1958) identified that a field description compiled before reference to books was of paramount importance, accompanied by details of the circumstances of the observation. It was also acknowledged that some knowledge of the observers' reliability was almost as important as the account of the observation itself. Consequently, at that time if the observer was not known to at least one member of the Committee, steps were taken to find out what experience of identification they had. In addition, prevailing weather conditions and simultaneous appearances of the species concerned (or ones of similar distribution) elsewhere in Britain and Ireland and the near continent were also taken into account when assessing the validity of a record.

Records were accepted only if the equivalent regional organisation was in agreement or had been consulted, although it was recognized that the criteria applied by some regional bodies did not correspond to the national committee's approach and there would inevitably be a difference of opinion in some cases. The value of photographic or sound-recorded evidence to objectively document the record of a rarity was expressed in a letter responding to a 1960 *British Birds* editorial on bird recording which referenced the Rarity Committee. This went so far to suggest that a sight record should only be regarded as ideal only if supported by a photograph and/or sound recording.

More recently there is a growing suspicion that records will not be considered acceptable if they are unsupported by such evidence, but that is certainly not the case at the present time. Field descriptions are still considered on their merits, employing the general principles that have remained in place for the 50 years of the *BBRC*'s operation. The increase in documenting rare bird records via photographic means and publishing these images on the Internet is radically affecting the manner in which rare birds are identified and has resulted in some significant changes to the level of detail and nature of the information available to rarity assessors.

Since 1959, *BBRC* has assessed records relating to more than 340 species. The committee has also adjudicated records of at least 26 distinctive races and claims of a number of species yet to be accepted onto the *British List*. It has also considered records that cannot be accepted to a specific species, but unequivocally relate to rare taxa; Fea's/Zino's Petrel and southern skuas are examples.

The original list of species considered by the Committee was published in the 1959 announcement (*BB* 52: 242-243), where the criteria for selection were described as "necessarily somewhat arbitrary". At that stage, some species were determined as national rarities outside their regular native or established ranges in Britain at the time. Red-crested Pochard outside London, Golden Eagle outside Scotland, Red Kite outside Wales, Bearded Tit outside East Anglia and Firecrest outside England and Wales are examples, but these were removed from the list of species considered following the 1962 annual report.

As early as 1964 it was recognised publicly that there was a difficulty in encouraging observers to document records of birds that they were seeing relatively frequently, in this case referring to the decision to drop species that were being recorded more frequently than 10 times per annum (*BB* 57:305). A perceived difficulty in encouraging observers to document their sightings has continued to impact on the work of many national and regional committees.

Attempts to define the abundance threshold of a rarity took a long time to establish, with Robert Mengel (then editor of *The Auk*) arguing that a frequency value should be derived by measuring the number of accepted occurrences over time divided by the number of observers searching for rarities (*BB* 56:423-425). Obviously that last variable has been the one that is most difficult to establish, but the remainder of the equation has formed the basis for establishing whether a species meets the threshold to be considered by the *BBRC*.

With advances in identification and an increasing number of observers dedicated to finding rare birds, various species have continued to show significant increases in their pattern of occurrence. These factors were responsible for the annual total of records exceeding 500 in the 1970s and 1,000 by 1990. To ensure that the workload remains realistic for a national body reliant on volunteer effort, periodic reviews of the species considered led to 52 taxa being removed between 1959 and 2006. In 1982 a threshold of 150 records in the last 10 years with at least 10 records in eight of those years was introduced to trigger removal from the list of species considered (*BB* 75: 375).

In 2006 a one-off exercise was made to remove species that had occurred more than 200 times with over 100 occurrences in the last 10 years. This review reduced the list of rarities and enabled the committee to spend more time on rare subspecies that had historically been omitted. Despite these changes, the committee currently assesses between 700 and 800 records per annum.

Table i: Taxa removed from the list considered by the *BBRC* since 1976.

1976   Cetti's Warbler

1979   Long-tailed Skua

1982   Cory's Shearwater, Purple Heron, White Stork, Buff-breasted Sandpiper, Richard's Pipit, Tawny Pipit, Savi's Warbler, Aquatic Warbler, Serin and Common Rosefinch

1987   Common Crane and Ring-billed Gull

1990   Surf Scoter, Little Egret, European Bee-eater, Pallas's Warbler and Woodchat Shrike

1992   Green-winged Teal

1993   Ring-necked Duck, Short-toed Lark and Little Bunting

1998   White-tailed Eagle and Kumlien's Gull

2001   American Wigeon, Black-crowned Night Heron and Rose-coloured Starling

2005   Black Brant

2006   Ferruginous Duck, Wilson's Storm-petrel, Great White Egret, Black Kite, Red-footed Falcon, American Golden Plover, White-rumped Sandpiper, White-winged Black Tern, Alpine Swift, Red-rumped Swallow, Red-throated Pipit, Subalpine Warbler, Greenish Warbler, Dusky Warbler, Radde's Warbler, Coues's Arctic Redpoll and Rustic Bunting.

In order to assist with documenting rare bird occurrences, the concept of presenting records that could be assigned to one of two species that were considered difficult to separate in the field was enshrined from the first report, where records of birds considered to have been either Icterine or Melodious Warblers and Long-billed or Short-billed Dowitchers were published alongside records identified to species.

Although Icterine and Melodious Warblers were not considered by the Rarities Committee after 1962 and Long-billed and Short-billed Dowitchers are now considered to be separable in the field, either given reasonably good views or by way of vocalisations, the principle of considering records as either/or remains to this day. Species pairs such as Booted/Syke's and Western/Eastern Bonelli's Warbler are published under this category when appropriate. In our times of changing taxonomies and pioneering efforts to record more species (particularly at sea), it would appear that we may be entering a period where records of aggregated species may have to be onsidered to maintain the national record.

The Committee in its first report recognized that no equivalent body is infallible and that some correct records would be rejected because the evidence was considered to be insufficient, not because the identification was necessarily considered to be wrong. In 1969 (*BB* 62: 44) it was emphasised that while acceptance of a record denotes that the Committee is fully satisfied with the proof offered, non-acceptance does not necessarily imply any more than the evidence so far produced falls short of that necessary for acceptance.

It was identified at that stage that subsequent good evidence can carry a record across the threshold of qualifying for acceptance and the Committee have always regarded it as quite natural to reassess rejected records if significant new evidence comes to light. This requirement for fresh evidence remains the cornerstone of the process that enables records to be reconsidered.

Observers may feel that the Committee has come to a wrong conclusion in its deliberations, but without new evidence there can be little justification to reassess the record. There is inevitably a degree of subjectivity in reaching a decision on some records where the information provided is considered to be on the borderline of the thresholds for acceptance.

BBRC members have in recent times tried to quantify levels of confidence when reaching a decision, with some applying an instinctive 95 per cent confidence level with the decision they reach. However, it is clear that quantitative analogies are not feasible in such circumstances and reflect the inherent difficulties in coming to consistent conclusions. Fortunately the size of the Committee, with at least five members voting on every record and all 10 voting on many, should help to remove some of the potential imbalance caused by this relative subjectivity.

For the 50 years of its operation the Committee has comprised 10 voting members. The reason for establishing such a large committee was to ensure that it brought wide experience in as many aspects of field identification problems and pitfalls as possible, to compensate for the reduced publication of descriptions of rarities that enabled public scrutiny of the records. Originally this component included the Secretary and Chairman, but more recently these have become non-voting members (the former since the late 1970s and the latter since the 1990s).

The qualities required for membership of the Committee were not detailed clearly before 1998. In 1975 it was documented that no current members of the committee had seen less than half the species subject to adjudication and that most had good working libraries to hand. The increased opportunities for world travel from the early 1980s meant that the degree of field experience available to prospective members grew significantly. Current members have field experience of most of the species they are assessing.

In 1991 an international meeting of rarities committees in the Netherlands initiated the Association of European Rarities Committees (AERC) and established guidelines for national committees. The guidelines largely followed the model established by BBRC. Within these guidelines it identified that every committee member should have qualifications that contribute to the work of the committee, examples being extensive field experience, knowledge of the current literature, skills in ringing or in examining museum skins and knowledge of the current birding scene.

In 1998 it was described that the new appointments to BBRC for that year had extensive experience, sound judgement and a proven ability in the field, in addition to the capacity to analyse the records and deal with the paperwork involved in rarity assessment.

These characters continue to be important for committee membership and have been further clarified to the following criteria:

- A widely acknowledged expertise in identification
- Proven reliability in the field
- A track record of high-quality submissions of descriptions of scarce and rare birds to county records committees and *BBRC*
- Considerable experience of record assessment
- The capacity to work quickly and efficiently, and handle the considerable volume of work involved
- Easy access to and knowledge of information technology
- Regional credibility

The criteria have remained relatively stable since the formation of the committee (albeit not explicitly documented), although an increasing emphasis on electronic information systems knowledge reflects the committee's intentions to move with technological advances. These are increasingly influencing the documentation of rare bird occurrences. One hundred years ago, a specimen was required to document the occurrence of a rarity, leading to the oft-quoted maxim *"what's hit is history, what's missed is mystery"*. The *BBRC* was formed amid the post-war movement towards field identification, with the first field guides and improvements in optical equipment allowing a pioneering band of ornithologists to record observations of rare birds in the field.

During the 1960s, under the Chairmanship of Phil Hollom, the working procedures of the committee and the annual Report on rare birds in Great Britain became firmly established and the role of the committee in assessing and documenting records of national rarities was seldom questioned. The main concerns voiced in the letters pages of *British Birds* were in relation to species that were dropped by the Committee in the early 1960s as their occurrence was too frequent and the Committee did not have the capacity to deal with their assessment (*BB* 57: 303 –307). Concerns were expressed that the absence of assessment by the national body would reduce the opportunity to evaluate occurrence patterns that could be more easily interpreted when the records were collected in the annual report.

During the 1970s, initially with Ian Wallace in the Chairman's role and then with Peter Grant taking this position, the Committee focused more attention on its stated aim of sharing its discoveries pertaining to revised identification criteria for the species considered. The result was a number of groundbreaking identification articles published in *British Birds*. By this stage *BBRC* and *British Birds* were at the cutting edge of bird identification in Britain. This phase continued at least until the end of Peter's time as Chairman in 1986.

The 1970s and 1980s saw significant improvements in the quality of optical equipment available to field birders, particularly prismatic telescopes. This in turn led to the opportunity to study fine details of plumage that had not been possible previously in the field. This approach was reflected in the quality of the identification literature being produced, with some seminal papers, most particularly Peter Grant and Lars Jonsson's *Identification of stints and peeps*, which graced the pages of *British Birds* in 1984. This approach engaged an up and coming generation of birders and the focus turned to being able to observe and describe minutiae of feather detail in the field to enable objective and reliable identification.

A perception grew that some of the previous generation's skills employed to confirm the identification of rarities – principally jizz-related features such as shape, proportions and prominent plumage characters - were less reliable than the new approach. The New Approach to Identification gained support through a series of papers by Peter Grant and Killian Mullarney, published in *Birding World* and a subsequent pamphlet.

In truth, birders frequently employ both new and old methods to identify birds in the field. The jizz-based approach often allows an initial identification to be made with a high degree of confidence. But to prove identification beyond reasonable doubt for subtle species the more detailed analysis provided by The New Approach is required to convince *BBRC* that the record is safe to accept to the national record. This has led to a higher degree of stringency associated with record assessment during the last 20 years, enabled by further improvements in optical equipment and photographic equipment and techniques combined with an increased knowledge of identification features. The growth of world travel and a substantial increase in the literature available has played a considerable part in increasing knowledge available to birders.

The New Approach to identification evolved from a new generation of birders who had a significant interest in rare bird identification and vagrants. As well as developing this advanced approach to identification, new methods of disseminating rare bird news were enabled by advances in technology and the enterprising skills of a number of leading birders. The profile of vagrants was raised further. Topicality was also improved with articles documenting the occurrence of the rarest species in newly launched magazines such as *Birding World* and *Birdwatch*. Identification texts that appealed to the growing number of birders whose main interest lay in rarities became far more common.

These advances led to a degree of despondency about the *BBRC*, particularly if records published in the other journals did not appear in the annual report of the rarities committee. At the same time there was a perception that record assessment was leaning heavily towards the identification principles proposed by The New Approach. This led a number of observers, some with considerable experience, to believe that the thresholds for acceptance for rare bird claims were becoming too stringent. This applied particularly to seawatching, where the holistic approach to identification is applied to identifying birds at long range from coastal headlands.

The committee needed to strive to address the concerns of the disaffected and the *BBRC* Chairmen of the time, Peter Landsdown, and Rob Hume attempted to forge better links and understanding. They talked at several conferences and held meetings to both explain the committee's functions and to encourage observer's continued cooperation. Such cooperation remains vital to maintaining an accurate database of rare bird occurrences. A new series of articles in *British Birds*, entitled From the Rarities Committee's files helped disseminate this information to a wider audience.

In the early 1990s *BBRC*'s procedures were reviewed to identify ways of dealing with an ever-increasing workload posed by the growing number of rarity records. It needed to ensure that records could be processed and decisions published more quickly and efficiently, which lead to further reviews of the species considered by the committee. When species are dropped by *BBRC*, responsibility for assessing claims passes to county records committees. In 1993 a fast-track system of circulation was introduced for more straightforward records, where a sub-committee of five members assessed submissions in the first instance and a full circulation was only required if the record was not unanimously accepted.

This period also saw the rise and rapid development of digital photography and video recording to document rarity occurrences. During the 1990s the availability of photographic evidence to support rarity claims increased steadily. The value of this supporting documentation was recognised through the annual Carl Zeiss award, introduced in 1991 in recognition of the most valuable photographic evidence supplied to the committee.

By the end of the 1990s, digital formats were becoming increasingly commonplace. Websites were established on the Internet where photographers could upload their images of rare and scarce migrants and share them with a worldwide web audience. The ability to connect digital cameras and video recorders to telescopes and produce high magnification images of birds increasingly provided more observers with an opportunity to capture still or moving footage of rare birds. This enabled significant details to be recorded even on relatively distant birds, allowing the principles of The New Approach to be applied in circumstances where this would previously have not been possible due to the distances of observation and the magnifications available.

The *BBRC* responded to these changes by moving to accommodate the technological advances. Under the Chairmanship of Colin Bradshaw a move from the traditional paper-based assessment system to an electronic one began. This was assisted ably, initially by Pete Fraser and finally by Nigel Hudson who completed the conversion to a fully electronic system when he was appointed Secretary in 2007.

Converting from a written paper archive based on field descriptions and photographs to a digital era of electronic submissions and images has been challenging, but has been achieved almost completely. It is now established that images uploaded to web sites such as *Birdguides* and *Rare Bird Alert* will be available to *BBRC* for assessment.

An online submission form has also been created to allow submission of details in support of photographs to be simply undertaken over the web. A further online form to enable submissions that are not supported by photographs is currently under development. All records are now assessed by *BBRC* electronically, with all members' comments and votes being recorded without paper documentation.

Another big change in recent times has been the re-evaluation of species limits, particularly as a consequence of advances in molecular studies and DNA sequencing. This has led to a period where a number of subspecies have been elevated to specific status. Further splits are anticipated. Taxonomic decisions rest with the *British Ornithologists' Union Records Committee* (*BOURC*) on the recommendation of its Taxonomic Subcommittee (TSC), the changes – both recognized and anticipated – have led birders and ornithologists to focus their attention on diagnosable taxa, regardless of whether they are currently considered species or subspecies.

BBRC has long considered a selection of rare subspecies, but a more comprehensive review of the forms assessed by the committee was required and a working group was established in 1999 to investigate Racial Identification Among Changing Taxonomy (RIACT) and to develop criteria for assessing claims of subspecies rare enough to be considered by *BBRC*. RIACT work involves current and previous members of the committee together with outside specialists.

Chris Kehoe presented a document summarizing the findings of RIACT and outlining the criteria established by its researches thus far in 2006 (*BB* 91: 619-645). This established the option of informal submissions, where documentation was invited even if it was felt that the level of detail would not be sufficient to enable acceptance of the record with the aim of assisting the committee with its research in this area. This research not only remains a key area of *BBRC* activity but also presents a significant workload over and above the core remit of processing and publishing rarity records.

In 2004, 45 years after it had set out its initial intentions, *BBRC* re-examined its constitution and objectives. These were summarized in *British Birds* (97: 260–263) and are repeated below:

BBRC aims to maintain an accurate database of records of the occurrence of rare taxa in Britain in order to enable individuals or organisations to assess the current status of and any changes in the patterns of occurrence and distribution of these taxa in Britain.

To support this aim, *BBRC* will strive to:

- work closely with County Recorders, Bird Observatories and observers to ensure that all records of rare taxa are submitted to this database;
- provide interested parties with an accurate and complete annual report detailing records of rare taxa in Britain;
- continue to vet all records of rare taxa in an independent, open, rigorous and consistent manner, and to provide observers with feedback on the assessment process as appropriate;
- continue to develop and publish criteria for the identification of rare taxa and to provide relevant information to other observers who wish to do this in partnership with the Committee.

BBRC's remit has only ever covered records from Britain. Records from Ireland have appeared in the report in various formats through much of the committee's history, but problems have beset their inclusion from as early as 1960 *(Brit. Birds* 54: 173–200). After a chequered history, Irish records were formally excluded from the species statistics from 2001. This followed representations from the *Irish Rare Birds Committee* and a decision by the *BOURC* to exclude Irish records from its reports *(Brit. Birds* 95: 477). Records from the Isle of Man are still considered by *BBRC* and included in the statistics, despite the fact that these records are not considered by *BOURC*. In 1981 it was announced that records from the Channel Islands would be considered by *BBRC* (*BB* 74:314), with accepted records published in the species comment and not included in the statistics. This arrangement was not maintained because the islands are zoo-geographically part of France.

In 2006 the committee published the results of a review of records preceding its formation (*BB* 99: 460-464). The review of records from 1950-57 had been pre-empted by a decision of the AERC that 1 January 1950 should act as the standard date to distinguish between Category A and B records for national lists. Although it proved impossible to have a comprehensive review of all rarity records, the rarest species were considered, extending *BBRC*'s historical tenure for adjudicating on records of rare birds in Britain.

BBRC has always maintained close liaison with the *BOURC*, whose functions include maintaining the *British List* and adjudicating on first records for Britain. No record rejected by the *BOURC* has been published by the *BBRC*, even if *BBRC* considered it acceptable. Since 1983 financial support for the Committee has been provided by sponsorship from the optical company *Carl Zeiss Ltd*. This provides a very welcome arrangement that supports the work of the Committee to the present day.

The proportion of national records considered by the Committee has come in for some criticism over the years, so it is valuable to attempt to identify the proportion of records that may have been overlooked. In the first annual report in 1958 it was stated that details for all but 16 of the 360 or so records known for that year had been obtained, a total of just over 95 per cent. Fifty years later, the introduction to the 2007 annual report established that all but 30 of the 800 or so anticipated records had been received, again indicating that over 95 per cent were considered. This level of consistency feels accurate when reviewing the national archive and indicates that the *BBRC* database represents an accurate assessment of the status of rare birds in the UK, thus fulfilling the main objective of the Committee.

As we enter the year of the 50th anniversary of the formation of the committee that became the *BBRC* some birders continue to question the relevance of such a national body in light of the tremendous transformations that have taken place in rare bird observations, documentation, identification and sharing of information.

Russell Slack's book reinforces the value of a national body that collates records of rare birds and produces annual reports. Without that effort, the present work would have been significantly more challenging or even impossible to produce. It would have required far more decisions from the editor in terms of what to include and exclude.

Decisions of the *BBRC* may not be welcomed all of the time, but at least they present a relatively consistent decision process from a reviewing committee rather than the more ambiguous decisions of a single individual or disparate local groups. The need to move and adapt with the ever-evolving world of rare bird recording will continue to challenge the committee. It is essential that these changes be embraced with new procedures and operations. I am certain *BBRC* can adapt over the next 50 years to achieve its centenary and provide a database for a future thorough review of rarities in the UK to match the present fine example.

SPECIES ACCOUNTS

1. Species (and scientific) names, and order, are in keeping with *The British List* (Dudley *et al*. 2006). English names for subspecies are those used most popularly within recent publications or the birdwatching community.

2. Author citations are presented throughout the species accounts. Some contain parentheses around the citation and others do not. Parentheses surrounding the author citation indicate that this is not the original taxonomic placement.

Taking Pallas's Sandgrouse as an example, Pallas originally published the name of the type specimen from the southern Tartarian Desert as *Tetrao paradoxa* Pallas, 1773. It was later moved to the genus *Syrrhaptes* and is now referred to as *Syrrhaptes paradoxus* (Pallas, 1773). The brackets reflect the subsequent change of genus.

Alternatively, Red-necked Nightjar was first described by Temminck as *Caprimulgus ruficollis* in 1820 from a specimen taken at Algeciras, Spain. There have been no subsequent changes to its placement and the author citation has no parentheses.

3. British and Irish records are accepted at a national level to the end of 2007 – the author has made no attempt to reassess any of these records, even though some appear anomalous, or possibly fraudulent, in light of current knowledge.

4. European and Western Palearctic records are those accepted at a national level, or likely to be so. The views of members of national rarities committee's and national lists have been consulted in order to produce as definitive a list of records as possible. For some areas, a number of records that have not yet been assessed are included (such as Azores records yet to be assessed by the *Comité Português de Raridades*).

5. Limits of the Urals/Western Siberia region: Annual reports (*Materialy* 1995–2008) edited by V.K. Ryabitsev and his *Identification Handbook* (2008) are concerned with the birds of the Urals/Western Siberia region. To the west of the Urals (Western Palearctic), this comprises a narrow band from the Kara and Barents Seas and taking in the Pechora and Kama rivers, Syktyvkar and Izhevsk, south to the westward-flowing Ural at Orenburg. Extending east from the Urals to the River Yenisey and south to the extreme north of Kazakhstan, the vast area of Western Siberia includes the Yamal and Gydanskiy Peninsulas in the north and the River Ob' and its tributwaries.

Pallas's Sandgrouse *Syrrhaptes paradoxus*

(Pallas, 1773). Breeds from Kazakhstan and Uzbekistan to Mongolia and north Central China. Partially migratory and dispersive, with northern populations moving south in winter; populations prone to periodic irruptions.

Monotypic.

An exceptionally rare vagrant from central Asia. Formerly much more regular with irruptions noted in 12 years between 1859 and 1909, the largest in 1863 and 1888. Has occasionally bred in the Western Palearctic, including Britain, following westerly irruptions.

Status: Formerly regular, with the last remnants of periodic irruptions in 1909. The next sighting was not until 1964, since when there have been six records comprising seven birds, the last of which was in 1990. The cause of these movements is unknown. The largest, in 1888, was thought to be triggered by heavy snowfalls and a hard snow crust causing drinking and feeding difficulties across the breeding areas.

Historical review: The first British records comprised a male taken at Walpole St. Peter, Norfolk, in early July 1859 and three, one of which was shot, on 9th July 1859 at Tremadoc, Caernarfonshire.

In the 19th century periodic large emigrations occurred both east and west of the species' core range. Notable arrivals in Western Europe occurred during May 1863 and May 1888. A smaller incursion in 1908 involved just under 100 birds. Sightings were documented in 10 other years between 1859 and 1909.

The first recorded major influx started in mid May 1863. The forerunners were located on 13th May at Port Clarence, Cleveland, and near Stirling, Forth, followed by many and widespread arrivals through May and June. By early November large flocks were reported in Northumberland, about 120 birds were in Yorkshire, 100 in Lincolnshire and at least 60 were killed in Norfolk. Smaller numbers were recorded in many other counties (Brown and Grice 2005). Around 125 were recorded in Scotland (Forrester *et al* 2007). At least 18 reached Ireland.

The immigration of 1888 is the largest and most widespread on record. The first birds arrived in early May. Eventually the influx involved a minimum of 5,100 birds (Ian Wallace *pers comm*). Many remained through to the following year. The magnitude of this particular event is incomprehensible for birdwatchers of the modern era, who grant this nomadic inhabitant of the central Asian steppe a near-mythical status.

The scale of the movement makes it one of the most amazing ornithological events ever seen in Western Europe.

The British Isles total for the 1888 invasion includes 1,500-2,000 birds in Scotland, 1,100-1,200 in Norfolk, 950-970 in Yorkshire, several hundred in Northumberland and 500 in Suffolk. The invasion penetrated west as far as the Isles of Scilly in the south and the Outer Hebrides in the north. More than 110 birds reached Ireland and around 20 made it to Wales. In Europe birds pushed as far southwest as Spain (where one was shot from a party of six at the Albufera coastal dunes on 9th June 1888).

The scale of the immigration led to birds breeding on British soil for the first time. Two nests each with two eggs were both found near Beverley, East Yorkshire, in June and July 1888 (Mather 1986) and a pair with seven young was reported from near Newmarket, Suffolk, in August 1888 with breeding "possibly" at Eriswell, Suffolk, in 1889 (Piotrowski 2003). Breeding also took place in Scotland. Chicks were found at Binsness, Moray and Nairn in late June 1888 and again close to the same place in early August 1889 (Forrester *et al* 2007). Given the scale of the invasion other pairs most likely bred unrecorded. Elsewhere in Western Europe breeding took place in Sweden, Denmark, Germany, Belgium and the Netherlands.

As befitted the culture of the time this bountiful arrival was greeted with extensive killing. A flock of 40 on Fair Isle in early June 1888 was reduced through shooting to just five a fortnight later (Forrester *et al* 2007). As a result of the prevalent slaughter an act of parliament was passed to protect the species. This came into force on 1st February 1889, by which time few birds remained (Brown and Grice 2005).

Compared with the 1888 influx, the 1908 arrival was a modest affair involving fewer than 100 birds. The first arrived in Hampshire in mid April. Records continued to early July and birds persisted to December. The last of the influx was a group of nine in Cleveland on 17th May 1909. These birds were the last representatives of a 50-year era of obvious irruptions. The next bird was not recorded for a further 55 years, with a brief fly-through in 1964. Britain was also on the periphery of a small European arrival in 1969.

There have been just six recent records involving seven birds, all since 1964.

1964 Kent: Stodmarsh 28th December
1969 Shetland: Foula 26th-30th May
1969 Northumberland: Seahouses male shot 5th September
1969 Northumberland: Elwick 6th September
1975 Fife: Isle of May two on 11th May
1990 Shetland: Loch of Hillwell and Quendale male, 19th May-4th June, 1990; also Loch of
     Spiggie on 22nd May

Additionally two sandgrouse, considered to be either Pallas's or Black-bellied Sandgrouse *Pterocles orientalis*, were seen at Wexford Harbour, Co. Wexford, in May 1954.

Participants in a sponsored bird race chanced upon the most recent record, a male on Shetland in 1990. Its protracted stay and accessible location mobilised many observers to make the trek north to pay homage during its residence.

At 6:00am we were walking along the Loch of Hillwell burn, hoping for a Wood Sandpiper, when we flushed a bird from an adjacent bare field. It flew across in front of us and landed in another field about 50 feet away. The comment "that pigeon looks like a Sandgrouse" gave way to incredulity as we realised it was indeed a Sandgrouse! After a brief interlude we concluded that it was a very pertinent male Pallas's Sandgrouse Syrrhaptes

paradoxus! DS left to phone the news out (even informing our opposing bird race team who had just arrived at the same place!), while the rest of the group kept it in view and took a detailed description. The first Shetland birders arrived in less than 15 minutes and for the next 3 hours we had excellent views down to 40 feet. Thereafter we continued our bird race and, of course, we won (95 species)!
Kevin Osborn and Dave Suddaby. Birding World 3:161-163.

Discussion: Elsewhere in Europe, there have been relatively few records since 1950, when a female was in Denmark on 11th November. One was present in Belgium on 2nd February 1956. In 1960 a small influx reached France, with one found dead and another seen at Picardy (northern France) on 25th January, a pair in Champagne (north-eastern France) on 15th March and 1st May and one shot at Pas-de-Calais (northern France) on 22nd March. One was in Belgium from 29th May-9th June 1961. Two appeared in the Netherlands in 1964, with one at Zandvoort from 21st November-25th December and a 'fly-past' at Amsterdam on 24th December; could this have been the bird seen in Kent four days later?

In 1969 a small influx into Western Europe included three in Britain. Continental records comprised one in the Netherlands from 17th-26th May, three in Finland on 22nd May (to 4th June), a male trapped in Denmark on 25th May and another in Finland on 31st May. Two males were in the Netherlands from mid-June to 6th July, one in Sweden from 23rd-25th June, another found dead in the Netherlands on 3rd September and, the last of the influx, an adult female was shot in Denmark on 5th October.

Since this influx there have been just a handful of European records. An adult female was in Denmark on 1st January 1972. A bird found dead in the Netherlands on 11th December 1972 was thought to be an illegal import. One was on Helgoland, Germany, on 31st July 1983. In 1990, in addition to the bird on Shetland, there were seven in Poland on 29th April and one in Norway from 20th July-11th November. A small arrival in 1992, the last of recent times, included four in Poland on 25th April and one in Finland on 10th June.

Arrivals are almost exclusively in late spring from May onwards with records continuing through the summer months. Outside of this period records during recent times are rare, but not unprecedented.

The reason for the cessation of such westward movements since the early part of the 20th century is uncertain, but is presumably associated with the contraction of the central Asian breeding range coupled with the development of agriculture and the expansion of cereal cultivation in Kazakhstan and Central Asia, though these are nothing more than speculation (Ryabitsev 2008)

For modern birdwatchers the magnitude of these historic incursions is difficult to conceive, as are the numbers that modern communications and increased mobility would have amassed during these colossal immigrations.

Our birding forefathers at the end of the 19th century would no doubt be surprised at the extreme rarity of the species nowadays. Modern birders can do little other than dream for the occasional displaced waif to reach our shores.

Oriental Turtle Dove *Streptopelia orientalis*

(Latham, 1790). Breeds from southern Urals, east to Japan and south to the Himalayas, Central China and Taiwan. Northern populations migratory, wintering in Southeast Iran, the Indian subcontinent, and from Southern China to Northern Thailand and Indochina.

Polytypic with six subspecies recognised. The two highly migratory races – *orientalis* (Latham, 1790) of the central Siberian taiga and *meena* (Sykes, 1832) of open woodland in central Asia – have been recorded as vagrants. Three records have been attributed to each race with the racial identity undetermined in two. Nominate *orientalis* differs from *meena* in a number of features: it is slightly larger and more bulky, with a darker plumage and grey-blue tips to the tail feathers; in *meena*, the tail feathers are tipped white (*BWP*).

An exceptionally rare vagrant from Asia, occurring mainly in late autumn and winter. Increasing numbers of records in northern Europe correlate with the recent upsurge of sightings in Britain.

Status: Eight records.

1889 North Yorkshire: Scarborough immature *orientalis* obtained 23rd October
1946 Norfolk: Castle Rising female *orientalis* shot 29th January
1960 Scilly: St. Agnes 2nd-3rd and 6th May probably *orientalis*
1974 Shetland: Fair Isle 1st-year thought to be *orientalis* 31st October-1st November
1975 East Yorkshire: Spurn juvenile/1st-winter *meena* 8th November
2002 Highland: Portmahomack Bay juvenile/1st-winter 9th November
2002 Orkney: Stromness, Mainland juvenile/1st-winter *meena* 20th November-20th December
2003 Caithness: Hill of Rattar 1st-winter *meena* 5th December 2003 to at least 24th March 2004; same St John's Brough 23rd February, 6th and 24th March 2004

Discussion: Six of the British records have been assigned to race. Both of the recent well-watched birds were attributed to the western race *meena*. This form breeds in the southern part of west Siberia south to Turkestan, Iran, Afghanistan, Kashmir and the Himalayas east to western Nepal. It winters mostly in India. Three older records have been assigned to the nominate race *orientalis* of central Siberia and southeast Asia from the Yenisey basin south to the eastern Himalayas and northern Vietnam. This race winters in southeast Asia. Two other records are unassigned to race. Lars Svensson (in Wilson and Korovin 2003) suggests that *orientalis* (Oriental Turtle Dove) and *meena* (Rufous Turtle Dove) may merit species status based on morphological and vocal differences. There is a narrow band of integration between the upper Ob' and upper Yenisey river valleys and in the east Russian Altai mountains (Gibbs *et al* 2001).
   Formerly regarded solely as a rare vagrant to the Western Palearctic, this subtle dove is now known to breed just within the Western Palearctic's eastern boundary in the Ural Mountains of Russia (Wilson and Korovin 2003). The species is surprisingly rare in Britain considering how many have been recorded as vagrants elsewhere in Europe. The total of more than 50 birds

recorded in northern Europe includes vagrants of both *orientalis* and *meena*, though most are unassigned to subspecies.

Nearly all *meena* have been in October and November; *orientalis* is spread evenly between October and April. The upsurge of records in the last 20 years probably reflects a better understanding of identification criteria. North European records have included wintering (and sometimes returning) birds in Finland, Sweden, Norway, Denmark and the Faeroe Islands.

Totals from northern European countries include 15 in Sweden (four *orientalis*; others undetermined), 13 in Finland (seven *meena*, three *orientalis* and three undetermined; a photographed bird from January 2009 still to be assessed), 10 in Denmark and five in Norway (three *orientalis*, two *meena*; one from January-April 2007 is pending; four other records not yet submitted from 2007-2008). The firsts for Poland (*orientalis* April 2002), Faeroe Islands (26th February-7th April 2006, probably since December 2005) and Estonia (*meena* November 2007) are all recent records. A further five have been recorded from France (including two *meena*) and one from Austria (undetermined, September 1995). There are three pending German reports, but the species is common in captivity (Peter Barthel *pers comm*). One in the Netherlands in late December 2008 bore a metal ring with 'Berlin' inscribed upon it. The bird had been bought from Berlin Zoo by a bird importer at a nearby village and then escaped; the species breeds in Berlin Zoo (Arnoud van den Berg *pers comm*).

Farther south in Europe there have been three from Spain (3rd December 1994, 30th September 2001 and another on 30th October and 6th November 2005, all presumed *meena*). In southeast Europe there are nine records from Greece (18th September 1948, 22nd April 1963, 21st August 1965, September 1966, 27th April 1986, September 1996, 2nd May 1998, *meena* on 4th May 2005 and two *meena* on 26th December 2008), one from Turkey (*orientalis* on 21st July 2004; possibly an escape), singles from Italy (*orientalis* collected on 25th September 1901 and one pending) and Hungary (*meena* on 18th December 1985). In the Middle East four have reached Israel (three *meena*, one *orientalis*) along the Rift Valley during October-November. It is considered a rare passage migrant in Kuwait with a high daily count of six on 1st October 2001.

The timing and location of the British records mostly fall in areas and at times where any European Turtle Dove *S. turtur* would receive a second look. As expected for an Asian vagrant the majority of British records have been in autumn and early winter, the one exception a spring bird on Scilly. After an absence of 27 years, the Orkney individual in 2002 enticed several hundred birdwatchers to the islands to enjoy views of this rare dove.

The trio of recent Scottish sightings mirrors the increase in sightings across northern Europe. In light of such statistics it is very surprising that there have been just eight in Britain, but clearly any report of an early winter 'turtle dove', perhaps consorting with Collared Doves *S. decaocto*, is more than worth following up.

Mourning Dove *Zenaida macroura*

(Linnaeus, 1758). Breeds across southern Canada from southern British Columbia east to Nova Scotia and south to Central America; locally in the Caribbean east to Puerto Rico; Revillagigedo (Clarion) and Tres Marias I. off Pacific coast of Mexico.

Polytypic with five subspecies recognised. Wing length increases from south to north, plumage tone darkest in east and palest in west, and toe length decreases from east to west. The racial identity of vagrants is undetermined, but probably involves *carolinensis* (Linnaeus, 1766), the dark, long-winged, long-toed and short-billed population of the eastern United States and Canada, and Bermuda and Bahamas (Aldrich 1993).

Very rare late autumn vagrant from the Nearctic.

Status: Four records, including two almost simultaneous arrivals in 2007.

1989 Isle of Man: Calf of Man 1st-winter trapped 31st October, found dead on 1st November
1999 Outer Hebrides: Carinish, N. Uist 1st-winter, 13th-15th November
2007 Outer Hebrides: Carnach, N. Uist 1st-winter, 29th October-7th November
2007 Co. Galway: Inishbofin 1st-winter, 2nd-15th November

The 2007 birds were popular attractions; the finder describes the excitement surrounding the first for Ireland in vivid detail.

A story always goes with the finding of a rare bird. Sometimes published accounts skip the turn of events surrounding the bird's discovery and concentrate on tedious descriptive details. For something as distinctive as a Mourning Dove there was never going to be a prize for identification perception, although it was important to try and establish the bird's age.

I had been on Inishbofin for part of October and got into the habit of keeping a diary, mainly to link weather conditions to migration. Little did I know what lay ahead in the final chapter. Only through a convulsion of luck did I happen to be on hand when the bird appeared. The following excerpt was written as events unfolded, its conclusion worryingly unclear until the very end.

'Friday 2nd November. Low cloud and grey, dreary light. SE wind. Lying under a duvet and waiting for signs of daylight, it took just three calls to get me outdoors. A Robin ticked, Siskins twittered and then a Yellow-browed Warbler hit high notes just outside the bedroom window. As ever with this magical island, there was hardly time to wonder if there might be something fresh in, when a male Blackcap flitted past. A Willow Warbler (rare in November) was nearby and by dusk the warbler totals stood at seven Chiffchaffs and five

Blackcaps. That made 15 individuals of four species. The day went well, despite not rising above an enjoyable mix of common migrants. When, about 1700hrs, I flushed a dapper male Brambling from a crop patch which perched and gave perfect views, I named it Bird of the Day. I hurried on, walking briskly along the road above Lough Teampaill, keen to squeeze birding time from the embers of daylight. I was trying to reach Regina's garden before dark, thinking that there might be just enough time to glimpse warblers making final insect sorties before going to roost.

Coming fast towards me but dropping low and out of sight, I saw what I speculated might be a Rock Dove. Its fast erratic flight said pigeon but I sensed that, if I saw it again, it could prove to be something else; maybe nothing more than a kamikaze Blackbird zooming off to bed. I was, nonetheless, curious. I got to the top of a rise and there it was – standing in the middle of the road about 30 metres ahead. Before raising binoculars I hoped that its small size was going to materialise into a Turtle Dove – new for Inishbofin – but as soon as I looked I knew what I beheld. The pointed tail, short red legs and especially the big black 'fingerprints' on the rear of its folded wings screamed MODO!

That second was a long one. In the space of it, I switched from Turtle Dove to Mourning Dove and experienced a pulse-rate rise from average to imminent coronary. I had a new bird for Ireland parked on the road in front of me. I suppose a kind of panic set in. I desperately wanted to get a picture – but in semi-darkness? I needed to reconfigure the ISO setting on the camera in order to boost shutter speed. I heard voices. People were on the road ahead of me and any moment now would walk around a bend and almost step on the dove. I shouted but it was too late; it sprang into flight with the speed of a clay pigeon. I fired the camera in what I knew was a vain, futile attempt to immortalise an epic rarity. As the group filed past me – virtually the only human beings I had seen all afternoon – I was in meltdown. I should have been in seventh heaven but it felt more like purgatory. The prize had been snatched away. It had flown off in the direction I was walking, so I continued. Incredibly, it had settled beyond the bend. Proving that cruelty can be limitless, this time a cat was stalking it. In a flash, the cat saw me and bolted. That action panicked the dove and it was off again, hanging right and dropping out of sight towards the hostel. I didn't pursue it, preferring instead to leave it in peace and pray that I might relocate it tomorrow.

Writing these words hours later, I don't know why I am not in a celebratory mood. I saw it briefly but well enough to identify it. Surprisingly, the hopelessly blurred photographs – handheld at one eighth of a second – show enough information to confirm the identification. I guess that I am less than ecstatic because, when you wait 40 years for such moments to arrive, you want them to consist of something better than 40 seconds of blind panic.

Saturday 3rd November. Totally overcast and perfectly calm. It was a night of one long post-mortem. Sometime in the wee small hours I became completely satisfied with the ability of my three photographs to resolve the indisputable image of a Mourning Dove. I started remembering Photoshop settings to boost shadows and bring up contrast. In my mind's eye I could visualise the bird's outline popping into a recognisable figure with short sandgrouse legs and a spiked appendage for a tail. There would be sufficient detail to upgrade it from smoke to substance. Job done, I began to pore over tactics for daybreak. Should I position myself on a height and hope to see it fly from roost and return roughly to where I last saw it? What, exactly, was it up to on the road? If it were digesting grit, then maybe it would not return to a particular spot. Was the best policy to walk and walk and hope that our paths crossed once more?

At 0730hrs the calls of two Yellow-browed Warblers were almost ignored. Regina's trees would have to wait. It felt like heresy, but nothing mattered until there was a conclusion to the hunt for Al-Zenaida. I walked slower than the pace of American tourists and made long sweeping scans over roads, telegraph wires, stone walls and fence posts. It was a rare morning of perfect calm. Calls could be heard at long range.

One call that will forever transport me back to childhood drifted from afar. An unseen Great Northern Diver was wailing from somewhere on the sea between Inishbofin and Inishlyon. I have heard the sound just twice before in Ireland, each time on flat calm November mornings. The eerie yet beautiful call formed a backdrop in Tales from the Riverbank, the TV of my toddlerhood. What a time to hear it again. Redwings, Fieldfares, a pair of Bramblings, and two apiece of Chiffchaff and Blackcap were the pick of migrants making their way through East Quarter.

Now events overtook solo birdwatching. The Island Discovery had been chartered and was due at 0945hrs. At 1000hrs a 50-strong column of birders filled the road at Regina's garden gate. Thus far, my searching had been in vain. Could reinforcements find it? Like a cell-dividing bacterium the posse split and headed in more and more directions at successive road junctions. With troops deployed and six hours at our disposal, surely somebody would bump into it? That depending on it still being alive. Well, it was. At the Clossy road junction I was with a splinter group that had turned right. Scouts among those that had turned left contained Michael O'Keefe who struck off on an unknown bearing. From an invisible location in West Quarter he phoned Seamus Enright who passed on the happy news that lit up my world.

What followed was a gravitational surge in the direction of others who had line of sight of Michael. It took ten minutes of speed walking to reach him along the track leading to the construction site of the new airstrip. And there it was, a phantom no more, the first transatlantic arrival at Inishbofin's Mourning Dove International Airport. What a moment.'
Anthony McGeehan. Birds Ireland website (www.birdsireland.com)

Discussion: A nest and two eggs, one containing a half developed embryo, was found on a steel structure imported from Texas, USA, via Rotterdam on board the Euroclipper that docked at Montrose, Angus and Dundee, on 21st-22nd September 1983 (Forrester *et al* 2007). Another bizarre record involved an adult found at London Heathrow airport on 9th February 1998. It had arrived in the hold of a plane from Chicago, USA, that day. Neither, of course, have a place on the *British List*.

One from Sweden (3rd-9th June 2001) was accepted on to category D. The only other Western Palearctic category A records are from Iceland (19th October 1995) and the Azores (2nd

November 2005; another on 23rd October 2008 is not yet assessed by the Portuguese Rarity Committee). Plumage details suggest that singles in spring 2008 in Germany (4th May; not yet reported to the RC) and Denmark (19th-21st May) were possibly the same individual as in Co. Galway in November 2007.

This attractive dove is one of the most abundant and widespread terrestrial birds of North America. Its population is estimated to be approximately 350 million. Twenty million are shot annually (US Fish and Wildlife Service 2007). Two principal subspecies occupy most of the range. Large grey-brown *carolinensis* from the east of its range is presumably responsible for British and Irish records. Eastern birds generally migrate south and southwest to wintering areas at lower latitudes within the breeding range.

Migration is mostly diurnal and primarily over land. Birds have been observed in trans-Gulf migration and sometimes land on boats. It is perhaps curious that it took until 1989 for one to reach the Western Palearctic. Mourning Dove was one of the most frequent North American landbirds recorded on transatlantic voyages made by Durand (1972) from 1961 to 1965. He recorded a total of 14 individuals on four occasions, including a party of 10 six hours from New York on 1st October 1963. The tired condition of the first two British records point at a direct transatlantic crossing; the two 2007 birds arrived following a strong westerly air stream.

Great Spotted Cuckoo *Clamator glandarius*

(Linnaeus, 1758). Common summer migrant to Spain; rare and local breeder in Portugal, southern France and east to Greece. Breeds discontinuously and uncommonly from central Turkey, Cyprus, Israel and Jordan to Northern Iraq and southwest Iran. Palearctic breeders winter in sub-Saharan Africa; range uncertain owing to resident African populations.

Monotypic.

Rare vagrant. Most are sub-adults overshooting north from the Mediterranean region in spring and early summer; juveniles wander from Europe in late summer and autumn.

Status: There is a total of 49 birds: 43 in Britain and six in Ireland (five records); 44 of these were between 1941 and 2007.

Historical review: The first was a 1st-year caught alive on Omey Island, Co. Galway, in March 1842. The next, the first British record, was a 1st-year that was shot at Clintburn in the Wark Forest near Bellingham, Northumberland, on 5th August 1870. The site is well inland and this is the only individual to have been recorded from such an unusual location. Further records followed in 1896 (Norfolk) and 1918 (one picked up dead near Caherciveen, Co. Kerry, in early spring and an additional bird present there for about a week). There followed a long gap until one in 1941 (Norfolk).

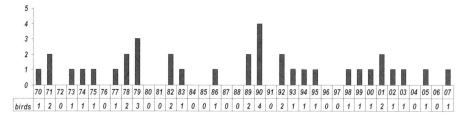

70	71	72	73	74	75	76	77	78	79	80	81	82	83	84	85	86	87	88	89	90	91	92	93	94	95	96	97	98	99	00	01	02	03	04	05	06	07
birds	1	2	0	1	1	1	0	1	2	3	0	0	2	1	0	0	1	0	0	2	4	0	2	1	1	1	0	0	1	1	1	2	1	1	0	1	0	1

Figure 1: Annual numbers of Great Spotted Cuckoos 1970-2007.

The 1950s mustered a further three, including the first for Wales found dead near Aberdovey, Meirionnydd, on 1st April 1956 and the first and only record for Scotland on Orkney (Rendall, juvenile, 14th-30th August 1959). The 1960s produced four more, three of them between Cornwall and the Isle of Man and one in Sussex.

A sharp rise in records – 12 more – during the 1970s possibly reflected increased numbers of observers. These included a then record of three in the southwest in 1979, with simultaneous spring birds in Cornwall and Scilly and a rare west coast autumn individual in Cornwall. Numbers fell in the 1980s, with six documented from just four years. The 1990s provided an upturn in fortunes with 11 birds, including a record four between late February and early April 1990, two of which were in Devon. A further seven birds arrived between 2000 and 2007.

Where: There is a clear seasonal disparity for occurrences. Most birds make landfall in the southwest. There have been seven from Cornwall (all but one in spring), four spring records from Scilly, three in Devon and two in Dorset, plus singles in spring from Hampshire and the Isle of Wight. Farther east, West Sussex has received three (two in spring) and Kent five (three in spring). The spread of spring records suggests that this striking cuckoo makes landfall on a broad front, though the species usually appears reluctant to extend its journey much farther north than the counties adjacent to the English Channel.

East coast records exhibit a more autumnal arrival pattern. Four of five in Norfolk during the review period were in autumn, as were the sole Suffolk and Cleveland records. The two Lincolnshire records comprise one each in spring and summer. Two from the Spurn area were autumn finds, as was the sole Scottish record.

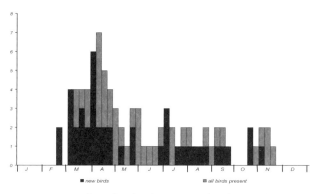

Figure 2: Timing of British and Irish Great Spotted Cuckoo records 1941-2007.

48

Table 1: Distribution of Great Spotted Cuckoos by region and season 1941-2007.

REGION	SPRING	AUTUMN*
South coast (Scilly to Kent)	22 (88%)	3 (12%)
East coast (Orkney to Suffolk)	2 (17%)	10 (83%)
West coast (Ireland,Wales,Northwest)	5 (71%)	2 (29%)
Total number of records	29	29

* Autumn defined as 1st July onwards.

Both Welsh records were in spring (Anglesey and Meirionnydd). The only Isle of Man record was a spring bird. One from Cheshire was in autumn. In Ireland, the three recent records comprise two in spring (Co. Wexford and Co. Dublin) and one in autumn (Co. Down).

When: The species remains an unpredictable overshoot. Periods with mild southerly airflows in early spring are conducive for spring overshoots bound for Iberia and responsible for most records since 1941 (66%). Autumn records (1st July onwards) make up 34 per cent of records: 9 per cent of them in the first week of July and the rest from late July onwards.

The south coast counties of England have hosted the majority of spring arrivals. Around 88 per cent of spring records on the English south coast have occurred in spring (see Table 1).

The east coast accounts for the overwhelming majority of autumn records, with 83 per cent of records for this region occurring at this season. The differing distributions suggest that birds during each period arrive from differing areas, with the late summer and autumn birds perhaps originating from the southeast rather than Iberia. Records from Scandinavia in autumn lend weight to this theory.

For such a rare bird and a summer migrant it is surprising that birds have arrived in every month from February and November. This distribution presumably reflects the species' protracted migration periods. The return to southern Europe from African wintering areas begins in early February. The main spring passage in North Africa and southern Spain occurs in March and the first half of April. Migration continues well into May. Return migration begins with the departure of adults in mid June, about two months earlier than most juveniles (Soler *et al* 1994).

Juvenile passage occurs mostly during August; some remain north of the Mediterranean into November.

The two earliest records were both from Devon in late February: an adult at Dawlish Warren on 22nd February 1998 and a 1st-year caught inside a building and released on Lundy on 23rd February 1990. However, both were superceded by a sighting that occurred after the review period of a 1st-summer near Ringaskiddy, Co. Cork, on 15th February 2009*. The peak period for records since 1941 lies between early March and early April: 17 finds falling between 5th March and 8th April; two-thirds of them in March. Three have arrived in May.

Juveniles begin to appear over a month later: four recorded between 1st and 8th July; six more between late July and late August; three more between 9th and 23rd September; and another three between 16th and 29th October. The latest remained at Aldeburgh, Suffolk, to 12th November 1992.

Overshoots to Britain and Ireland in spring are dominated by 1st-years. Just three birds were aged as adults by Lansdown (1995) and in subsequent rarities reports: Kent in March 1990, Devon in February 1998 and Norfolk in March 1999. In autumn, the majority are either juveniles or first-years; ageing is difficult as juveniles moult over an extended period of time (Lansdown, 1995). In both spring and autumn the age group with the least experience of migration is most susceptible to displacement.

Discussion: Elsewhere in northern Europe this species remains rare. Fourteen were recorded in the Netherlands in 1970-2006 (eight in March-May and six July-October). Five of a total of seven records in Belgium occurred from 1979-2007 (one in March, two in May and singles in July and October). There are three Channel Island records (two in February and one in April), about 20 German records and one from Poland (August). There are nine records from Switzerland, including an extraordinary record of a juvenile in the Bernese Alps from 17th-25th July 2007 that was fed by a male Black Redstart *Phoenicurus ochruros*. In Scandinavia there are just five from Sweden (all calendar-year birds in July), four Norway (one in April, two in August and one in September), two Finland (August and October) and six from Denmark (one May, three August and two September).

The stronghold population in Spain makes up over 98 per cent of the European breeding community. Great Spotted Cuckoo is increasing and expanding its range in Spain, France, Portugal and Italy (Hagemeijer and Blair 1997). Despite these increases there does not appear to have been a corresponding surge in records here, and sightings remain less than annual.

Black-billed Cuckoo *Coccyzus erythrophthalmus*

(A. Wilson, 1811). Breeds from Alberta eastwards across southern Canada and through northern and central United States south to Oklahoma and eastwards to North Carolina. Winters in northern South America, but distribution poorly known. Appears to winter primarily from Colombia east to western Venezuela and south to central Peru, possibly also in eastern Peru and north and east Bolivia.

Monotypic.

Very rare, markedly irregular and decreasing vagrant across the North Atlantic, mainly in October.

Status: A total of 14 records; 13 in Britain and one in Northern Ireland.

1871 **Co. Antrim:** Killead killed 25th September
1932 **Scilly:** Tresco 1st-winter picked up dead 27th October
1950 **Argyll:** Near Southend, Kintyre 1st-winter 6th November, found dead 8th
1953 **Shetland:** Foula picked up exhausted 11th October, died 12th
1965 **Cornwall:** Gweek found moribund 30th October
1967 **Devon:** Lundy 1st-winter male 19th October, found dead 20th
1975 **Cleveland:** Redcar trapped 23rd September, kept overnight and released 24th
1982 **Scilly:** St. Agnes 1st-winter 29th August, found dead 30th
1982 **Scilly:** St. Mary's 1st-winter 21st-23rd October, found dead 24th
1982 **Devon:** Barnstaple 1st-winter caught 21st October, released 22nd
1982 **Cheshire & Wirral:** Red Rocks, Hoylake 1st-winter 30th October
1985 **Scilly:** St. Mary's 1st-winter 12th October
1989 **At sea:** Sea area Forties oil platform Maureen, found exhausted 30th September, kept over
      night on the oil platform and released 1st October
1990 **Scilly:** St. Mary's 1st-winter 10th October, found dead on 11th

Discussion: Elsewhere in the Western Palearctic it is equally rare. There are two records from Iceland (found dead at the end of 1935, 21st-25th October 1982 when found dead). Others have reached Germany (far inland in the west on 8th October 1952) and Denmark (shot on 16th October 1970). There is an old record from Italy (1858) and a category E record from France which is considered to be of doubtful origin (female collected at Nissan, Hérault, on 20th July 1886). There have been three on the Azores (undated bird in 1902, 15th October 1969, 28th-29th October 2006).

All sightings from Britain and Ireland have been in the autumn. The earliest on Scilly (29th August 1982) preceded the next by over three weeks. Most have been in the latter half of October. The latest is dated 6th November. Five have graced the Isles of Scilly, including two in 1982. Devon is the only other county to have accommodated the species twice. Records from Cleveland, Cheshire and Sea area Forties point towards ship assistance. Durand (1972) recorded this species at sea some 1,650 km east of New York on 30th August 1965.

Although both Black-billed and Yellow-billed Cuckoos occur sympatrically through much of their ranges, the Black-billed Cuckoo has a slightly more northerly distribution and is the rarer of the two. Formerly much more common in North America, population densities have declined across its range throughout the 20th century, with particularly severe decreases in the 1980s and 1990s. This species is much the rarer of the two Nearctic cuckoos in the Western Palearctic (there are four Yellow-billed for every Black-billed in Britain and Ireland) an inequality explained by both relative scarcity and their differing autumn migration routes. Black-billed heads in a south or southwesterly direction, shunning Atlantic regions and the lengthy ocean crossing undertaken by Yellow-billed, rendering it less vulnerable to transatlantic vagrancy.

In many cases both Nearctic cuckoos are moribund when found. Most succumb shortly after arrival, presumably following the stress of the crossing and a subsequent inability to find sufficient food on arrival.

The glut of sightings in the 1980s briefly transformed the status of this species for a generation of rarity seekers, especially so for annual trippers to Scilly. This comparative excess is now a distant memory, with the last British record making landfall for two days almost two decades ago in 1990. Against a backdrop of declines in North America and changing weather patterns at key times of year for transatlantic vagrancy this species seems destined to remain a rarity of considerable magnitude for the foreseeable future. One reached the Azores in 2006. In doing so it offered a sliver of hope to a modern generation of birdwatchers that a future weather system might deliver another to Britain.

Yellow-billed Cuckoo *Coccyzus americanus*

(Linnaeus, 1758). Breeds across southern Canada from British Columbia to New Brunswick, and throughout the USA to central Mexico. Winters in South America south to Argentina.

Monotypic.

Rare, markedly irregular and decreasing vagrant across the North Atlantic, mostly in late autumn.

Status: The total of 68 birds recorded to the end of 2007 comprises 59 in Britain and nine in Ireland.

Historical records: A total of 17 occurred prior to 1950, nearly all of which were shot or found dead. The first was killed near Youghal, Co. Cork, in autumn 1825. This was followed by nine more between 1832 and 1899, including the first British record shot in autumn 1832 at Stackpole Court, Pembrokeshire. Seven further records fell between 1901 (two) and 1940.

The first of the modern recording era was a 1st-winter female found moribund at Exnaboe, Shetland, on 1st November 1952. The 1950s produced seven birds, no fewer than five of them from Scotland plus one from Yorkshire and one in East Sussex, a distribution of records at variance with expected arrival locations during recent times. These included a peculiar arrival of four in 1953, three of which occurred between 3rd and 11th October, including one on the Inner Hebrides on the west coast of Scotland and singles in Angus and Dundee, and Moray and Nairn which coincided with a Black-billed Cuckoo on Shetland. A bird in mid-November of that year frequented a garden close to the Yorkshire coast and was the first individual to delight an audience.

A Yellow-billed Cuckoo (Coccyzus americanus) occurred at Cloughton, near Scarborough, Yorkshire, from November 14th to 17th, 1953, and was seen by E. H. Ramskir, R. S. Pollard, E. A. Wallis, R. M. Garnett, myself and others. It was photographed in colour, and filmed, by W. R. Grist. There had been a local plague of white butterflies, the larvae of which had crawled up the sides of Mr. Ramskir's house to pupate. The cuckoo frequently

fluttered to the house walls and sides, and was watched as it searched for pupae, it even took insects from the frames of windows through which it was watched a few feet away.
Ralph Chislett. *British Birds 47: 173*

The 1960s added five more birds to the total, all bar one in West Sussex in western locations with singles in Scilly, Co. Cork, Co. Mayo and Argyll.

Ten more birds arrived in the 1970s (at that time easily the best tally of any decade). Records again came from far-reaching points of the compass. One was on the east coast of Scotland. Three were in England between East Yorkshire and Suffolk, including an inland bird found dead at Welton le Marsh, Lincolnshire, on 30th October 1978. Five were found between Hampshire and Scilly and one in Ireland.

The 1980s – a classic decade for Nearctic vagrants – broke all records by delivering no fewer than 18 birds, including a record five in 1985 and four in 1989. These birds conformed to modern expectations: five in Scilly, three from Devon and two for Cornwall. Singles were logged from Hampshire and the Isle of Man. In Ireland, Co. Waterford, Co. Cork and Co. Donegal shared the spoils. The solitary east coast record was from coastal Suffolk. Most noteworthy were inland surprises in South Yorkshire (a moribund bird at Armthorpe on 14th November 1981) and in Lincolnshire (Rauceby Warren from 18th-19th October 1987).

The 1980s boom failed to extend into subsequent decades. Nine arrived in the 1990s and just two since 2000. The records from the 1990s were another disparate bunch with three in the southwest (two from Scilly; one in Cornwall) and one in Pembrokeshire, plus singles in Orkney and Co. Antrim. Another east coast bird dropped in to Morpeth, Northumberland, on 22nd October 1995. Other inland birds included one in Bedfordshire (recently dead, Sandy, 6th December 1990) and in Surrey (caught and ringed, Oxted, 17th October 1991).

Both individuals of the new millennium conformed to modern geographical expectations, appearing in Scilly and Cornwall.

Where: As would be expected for a Nearctic vagrant the bulk of sightings have been from the southwest, though there are a good number of records from counties adjoining the English Channel and a decent tally along the east coast. The Isles of Scilly has amassed 12 records and Cornwall eight (five since 1971). Four have graced Lundy and four more were recorded in

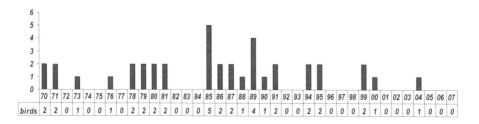

	70	71	72	73	74	75	76	77	78	79	80	81	82	83	84	85	86	87	88	89	90	91	92	93	94	95	96	97	98	99	00	01	02	03	04	05	06	07
birds	2	2	0	1	0	0	1	0	2	2	2	2	0	0	0	5	2	2	1	4	1	2	0	0	2	2	0	0	0	2	1	0	0	0	1	0	0	0

Figure 3: Annual numbers of Yellow-billed Cuckoos, 1970-2007.

Hampshire. Co. Cork and Orkney had three each. Dorset and Sussex have two apiece, as do Pembrokeshire, Argyll and Highland.

Surprisingly, the east coast counties of Lincolnshire and Suffolk each have two records; Yorkshire has no fewer than three. Other east coast records have come from Northumberland, Angus and Dundee, Moray and Nairn, and Highland. A further 11 counties have recorded one bird. Individuals in Bedfordshire, Surrey, Lincolnshire and South Yorkshire had wandered far inland.

Records in Ireland are distributed unusually widely for a Nearctic vagrant, with nine records shared between seven counties, including three in the northwest. The Scottish total of 10 records includes just one since 1970, a freshly dead bird picked up on North Ronaldsay in 1991.

When: Sightings between 1901 and 2004 condense into a period from late September onwards; the earliest on St. Mary's from 23rd-24th September 1981. The peak time for finds is late September through October, the latter month accounting for 67 per cent of all records. Arrivals are near constant throughout each week of the month, with a slight peak in the last

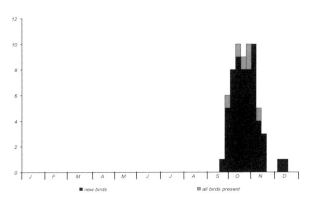

Figure 4: Timing of British and Irish Yellow-billed Cuckoo records, 1901-2007.

week. November records are not unusual. Seventeen per cent were found in the early part of that month, the latest both in Yorkshire and both on the 14th: a moribund bird at Armthorpe, South Yorkshire, in 1981 and one at Cloughton, North Yorkshire, to 17th November 1953. The two December records involve individuals found dead: on 6th at Sandy, Bedfordshire, in 1990 and freshly dead on 14th at Middleton-on-Sea, West Sussex, in 1960.

Since 1985 just three of the live records lasted for longer than a day: two survived for two days; and one persisted on Tresco from 12th–20th October 1999.

Discussion: The Azores accounts for the bulk of records elsewhere in the West Palearctic with 27 birds (although just three have so far been approved by the CPR). These include seven in 1965 and 12 between 2005-2007. Neither one of the spring records listed – 13th May 1966 and 11th April 1977 – is approved by the CPR.

In northern Europe, the three records from Iceland were all found dead (3rd January 1954, 5th October 1954 and 13th October 1987). There are two records from France comprising three birds (two shot on 31st October 1957 and one from 5th-6th October 2005). Norway also has two (7th October 1978, at sea in late February 1981 or 1982) as does Belgium (22nd October 1874, found moribund on about 31st October 2008) and there is one from Denmark (23rd October 1936).

Seven have been reported from Italy (28th October 1883, 25th November 1927, 2nd November 1932, 20th November 1932, 22nd October 1953, 7th June 1954 and 25th November 1968), including the first spring record for the Western Palearctic. The only other bird in southern Europe was in Spain (28th October 1994). There is one record from Morocco (25th October 1977).

Under normal migration birds may lose more than half their body weight during ocean crossings (Hughes 1999). Transatlantic crossings clearly represent a great expenditure of body reserves for cuckoos and most Yellow-billed are moribund when found. Some are fit enough for short-term stays, but many and perhaps all expire shortly afterwards. As with Black-billed Cuckoo the diet primarily comprises large insects, though with less reliance on caterpillars. The long-staying individual on Tresco in 1999 fed readily on Great Green Bush-crickets (*Tettigonia viridissima*) and introduced stick insects (*Phasmidae*).

The difference in numbers between Yellow-billed and Black-billed Cuckoos is explained by their differing migration strategies. Northerly populations of Yellow-billed migrate to South America via Central America and West Indies, which for some involves a substantial sea crossing, making them more susceptible to displacement from fast-moving Atlantic depressions than their congener.

It is tempting to suggest that ship assistance has played its part in a large proportion of east and south coast records in Britain. Durand (1972) twice recorded the species during eastbound transatlantic crossings. The high incidence of inland finds defies explanation, but the finders of such individuals no doubt welcomed this apparent propensity for inland diversions.

Populations declined throughout their range during the 1970s and 1980s, but have stabilised since then, particularly in the east (Sauer *et al* 2008) from where European vagrants are most likely to hail. Modern generations of birders will most probably struggle to find either of the two *Coccyzus* cuckoos on the British and Irish lists with anything like the regularity that their peers enjoyed in the early 1980s, a time when both birds featured prominently in the dazzling roll call of Nearctic rarity highlights from the Isles of Scilly.

Barn Owl Tyto alba

(Scopoli, 1769). These are distributed widely across Europe, Africa, southern Asia, Australasia, USA, Central and South America. Nominate alba (Scopoli, 1769) is found in Western and Southern Europe including the Balearics and Sicily to Northern Turkey; also Western Canaries and from Morocco to Egypt.

Polytypic, with 28 subspecies recognised by Bruce (1999) and 32 by Dickinson (2003).

On the basis of DNA evidence König & Weick (2008) gave specific rank to three groups of barn owls, each comprising several races: Common Barn Owl *Tyto alba* (Europe, Africa, Madagascar, Asia south to India and Malaysia); American Barn Owl *Tyto furcata* (North, Central and South America); and Australian Barn Owl *Tyto deliculata* (Australia, New Zealand and Polynesia).

Further taxa are also given specific rank but most authorities adopt a rather more conservative approach (for example, *BWP*; Bruce 1999). Clearly this diverse species will perplex taxonomists for some time.

Dark-breasted Barn Owl *Tyto alba guttata*

Central Europe east to Latvia, Lithuania and Ukraine, southeast to Albania, Macedonia, Romania and northeast Greece. T.a.alba and T.a.guttata intergrade from Netherlands, Belgium, north and eastern France to the Rhine valley and central Switzerland.

Probably a scarce migrant, but status uncertain as many less distinctive individuals are probably overlooked.

Status: The occurrence of this taxon in Britain has been confirmed by ringing data involving 16 nestlings from within the range of *guttata*. Many of these were found dead in the following year, reflecting the more dispersive nature of *guttata*.

British *alba* are less dispersive, though BTO-ringed birds have been found in Germany, Belgium and France. A BTO-ringed bird found at a RAF base in Afghanistan in 2006 had been ringed near a RAF base in Oxfordshire and had presumably hitched a lift on a transport plane (Grantham 2008).

Six birds ringed in Germany have been recovered here: ringed June 1988 seen on 17th November 1989 on the Buchan A oil platform in the North Sea and caught on 19th; ringed 7th June 1990 found on 3rd January 1991 in Cornwall; ringed 3rd June 1993 found 18th July 1994 in Argyll; ringed 11th July 1995 found 6th March 1996 in East Yorkshire; ringed 14th July 1997 found 22nd November 1997 on Orkney; and ringed 10th July 2000 found 22nd April 2001 in Caithness.

Eight birds ringed in the Netherlands have subsequently been recovered in Britain: ringed 15th June 1983 found 26th September 1984 in Kent; ringed 20th July 1990 found 11th March 1991 in East Yorkshire; ringed 27th June 1997 found 10th November 1997 in Hereford and Worcestershire; ringed 5th July 1997 found 11th March 1998 in Cornwall; ringed on 28th August 1999 found 31st October 1999 in Norfolk; ringed 4th September 1999 found 24th October 1999 in Suffolk; ringed 9th June 2005 found 17th February 2006 in Grampian; and ringed 4th June 2007 found dead 12th July 2008 in Norfolk - see below).

There have been single birds from Belgium (ringed 11th August 2000 found 25th October 2000 in Wiltshire) and Denmark (ringed 30th June 2003 found 1st December 2003 in North Yorkshire).

Dark-breasted Barn Owl is a rare visitor to Scotland with 24 records involving 27 birds: 21 were on the Northern Isles; two dropped in to North Sea oil platforms; two reached the Isle of May and there are two records from the mainland. Most have been in late autumn and early winter (Forrester *et al* 2007). There have been a number of further claims of dark-breasted birds from English east-coast counties, though the origin of these birds is not supported through ringing data. From 1970-2000, 17 dark-breasted birds were recorded in Norfolk (Taylor *et al* 1999; Brown and Grice 2005). Suffolk has eight records (Piotrowski 2003) but records are not restricted to the east coast and three (two dead) have reached the Isles of Scilly in October (Flood *et al* 2007). Ruttledge (1975) listed three records from Ireland, all of them shot in 1932: one in Co. Tipperary in January and two in Co. Kerry at the beginning of April.

Mixed breeding between the two forms has been proven in Britain. The most recent ringing return in 2008 (from Norfolk) was one of six nestlings ringed at Bingerden, the Netherlands, in June 2007. This dark female was discovered during routine ringing of Barn Owl boxes in East Anglia and was found to be incubating three eggs. It was subsequently picked up freshly dead on the A10 at Southery on 12th July 2008 (Grantham 2008).

Reports of typically pale British birds in Devon rearing dark *guttata*-like chicks have complicated the identification process (French 2006). Likewise a small number of birds in Dumfries & Galloway in the late 1990s contained birds showing plumage characteristics resembling *guttata* (Forrester *et al* 2007). This raises a strong possibility that some *guttata*-like birds are simply dark *alba* birds; looking like a *guttata* is no longer appears sufficient to support an unequivocal continental origin.

Discussion: Probability implies that many dark Barn Owls in Britain are likely to be *guttata*, particularly the darkest birds in areas suitable for receiving immigrants, such as the Northern Isles and on the east coast. But, overlap between the palest *guttata* and darkest *alba* makes not only the origin of many dark birds unproven but also many pale *guttata* overlooked. Establishing the provenance of an unringed *guttata* in Britain, especially in atypical locations, is difficult to ascertain with certainty (French 2006). However, the ringing data suggests that immigration in atypical locations is not unusual, even on less-expected mid-summer dates.

BBRC assesses claims and will publish records of birds resembling classic *guttata* with a caveat that some might be unusually dark *alba* (Kehoe 2006). No such records have yet been published in *BBRC* reports.

Eurasian Scops Owl *Otus scops*

(Linnaeus, 1758). Common summer migrant to North Africa and southern Europe from Iberia north to central France and east to Greece; also breeds across Ukraine, southern Russia and southern Siberia to western Mongolia, Kazakhstan and Iran. Most winter in north equatorial Africa, but some remain in southern Europe.

Polytypic, six subspecies recognised. All vagrants assigned to race have been nominate *scops* (Linnaeus, 1758) of France, Sardinia and Italy east to Volga, south to northern Greece, northern Turkey and Transcaucasia. Five other subspecies are found in Iberia, North Africa, Cyprus and the Middle East.

Irregular but quite frequent vagrant from southern Europe, mainly in the spring and summer, mostly to southern England and the Northern Isles.

Status: Prior to 1950 at least 55 birds are documented, 45 from Britain and 10 from Ireland (Keith Naylor *pers comm*). Since 1950 there have been 42 records, 37 from Britain and five in Ireland to the end of 2007.

Historical review: This is one of the relatively few species for which there were more records preceding 1950 than there have been since. The majority, 38, were recorded in the 19th Century, the first of which were two obtained at Audley End, Essex, c1821.

Many of the 19th Century records have vague dates (11 are undated). With one notable exception the dated records conform to patterns witnessed during the latter half of the 20th century (see Table 2). The outlier involves an extraordinary record of one apparently caught at Trevethoe, near Hayle, Cornwall, about 2nd January 1871. There are three reports of two birds together, including the first in Essex and a pair purportedly shot at Scone, Perth and Kinross, in May 1864 (Forrester *et al* 2007). There is also a record of two at Foulkes Mill, Co. Wexford, one of which was obtained on 31st May 1889.

A further 17 were documented between 1900 and 1948. What is evident from the pre-1950 records is a dearth of birds in northern England, the overwhelming majority coming from Ireland, southern England and East Anglia, plus overshoots to Scotland.

Five in the 1950s and four in the 1960s were all present for just one day; a two-day bird was present on the Calf of Man in late April 1968. Records were well dispersed, but the 'north or south' pattern exhibited by previous records remained (*see Table 3*).

Seven in the 1970s represented a slight increase in fortunes, though five were found dead or moribund. The two other records comprised a calling bird in Co. Fermanagh on 18th June 1974 and one on St. Mary's from 5th-14th April 1976.

Eight in the 1980s included a famous male at Dummer, Hampshire, from 12th May-14th July 1980. This bird was subsequently suggested to have been an escape (Clark and Eyre 1993), but remains part of the national archive for the time being at least.

The 1990s added a further eight records, including an obliging daytime bird at Morwenstow, Cornwall, from 9th-11th April 1995. The period 2000-2007 has added a further 10 birds, the best showing ever. The slight increase in records included an obliging bird at Porthgwarra from 24th-26th March 2002 and an even more entertaining male at Thrupp, Oxon, from 12th-30th June 2006 which returned to the same village from at least 15th May-5th June 2007. For those with long memories this individual mirrored the events experienced in Hampshire in 1980.

Table 2: Dated* monthly occurrence of pre-1950 Scops Owl records, by period and by region.

Region	Jan	Feb	Mar	Apr	May	June	July	Aug	Sept	Oct	Nov	Total
Scotland			5	5	1		1	1	1			**14**
N. England				1	1	1				1		**4**
S. England	1		1	2	3	4	1			2	2	**16**
Ireland			1	1	4	1	2				1	**10**
Totals	**1**	**0**	**2**	**8**	**13**	**7**	**4**	**1**	**1**	**3**	**4**	**44**

* 11 pre-1950 records were undated.

Table 3: Dated* monthly occurrence of post-1950 Scops Owl records by region.

Region	Jan	Feb	Mar	Apr	May	June	July	Aug	Sept	Oct	Nov	Total
Scotland				1	5	7	1			1	1	16
N. England				1		1						2
S. England			2	5	5	1		1	1	1		16
Ireland				2	1	1					1	5
Wales/Isle of Man				2								2
Totals	0	0	2	11	11	10	1	1	1	2	2	41

* one found 'long dead' excluded from totals.

There has been an average of 0.9 per annum since 1970. The recent increase in records has nudged the average up to 1.2 per annum over the period 1995-2007.

Where: The records since 1950 show an arrival pattern typical of a spring overshoot from the Mediterranean. Two areas dominate the archive: the Northern Isles and areas within close proximity of the south coast; few arrived in between. The open Shetland landscape has revealed eight birds, five of them between 2001 and 2006. Orkney has yielded five. Elsewhere in Scotland singles have reached the Outer Hebrides and Clyde and Argyll. One picked up in Sea area Forties was released in Northeast Scotland.

Southwest England and southern Ireland have fared well, with three each from Scilly, Cornwall, Hampshire and Co. Cork, and two in Devon. Elsewhere in England there were singles in Dorset, Pembrokeshire and the Isle of Man, and in Ireland singles in Co. Waterford and Co. Fermanagh. The sole record in the southeast involved one from Kent. Norfolk and Lincolnshire (one each) have contributed the only east coast records. Wiltshire, Nottinghamshire and Oxon have contributed inland records. How many more have sneaked through undetected?

When: There are no recent midwinter records, but birds have occurred in all other months. The bulk of sightings conform to a pattern of spring arrival between late March and early June. A few turned up in summer and a handful in autumn.

The earliest to be found was one caught aboard a fishing vessel about 2.4 km off Portland. It was released later at the Bird Observatory on 20th March 1990. One at Porthgwarra from 24th-

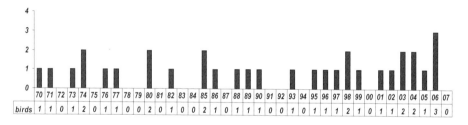

Figure 5: Annual numbers of Scops Owls, 1970-2007.

59

26th March 2002 was one of several early Iberian spring overshoots at the time and drew large numbers of observers to the county. A cluster of six records occurred between 2nd-7th April, all but one in the west and southwest. A further nine between 8th-20th May includes four from Shetland and a pulse of five between

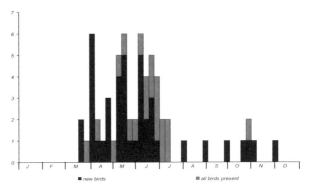

new birds   *all birds present*

Figure 6: Timing of British and Irish Eurasian Scops Owl records, 1950-2007.

4th-9th June (four of which were on Scottish offshore islands).

The six autumn records from late August show an easterly bias, possibly suggesting an origin from southeast Europe or southwest Asia. Two are from Orkney and singles from Kent and Norfolk. Birds in Co. Cork and Scilly at this season might also support such a trajectory. The latest to be recorded was found dead under wires at Holm, Orkney, on 27th November 1970.

Discussion: A propensity for overshooting north and northwest of its core range is illustrated by five records from Iceland up to 1975 (four between 20th April-30th June and one undated). There are 10 from Sweden (six in May-June, two in July, one in August and one in November). Two of the recent records from Denmark were in May and June. These north European sightings correlate well with sightings in the Northern Isles.

Always popular, this small owl is surprisingly rare here given the sizeable European breeding population. European population declines were reported between 1970 and 1990. Numbers have since remained stable, with increases in some countries but continued decline in others especially at the periphery of the range (BirdLife 2004). It seems unlikely that appearances will become more frequent, although the recent upturn in fortunes is doubtless appreciated. Whether this reflects improved detection rates or signs of a recovering European population benefiting from global warming is unclear.

Seventeen records of birds found moribund, injured or dead since the 1950s invite speculation on how many make a clandestine arrival and depart without mishap.

Snowy Owl *Bubo scandiacus*

(Linnaeus, 1758). Occasionally breeds in northern Scandinavia and Iceland, depending on availability of small mammals; casual breeding has occurred in Scotland and has been attempted in Ireland. Outside Europe, an erratic circumpolar breeder of the tundra and northern islands of Arctic Russia, Siberia, Alaska, Canada and northern Greenland. Most disperse southwards in winter, but some are resident, or nomadic, if food available.

Monotypic.

In 19th and first half of the 20th century an irregular winter visitor reaching mainly north and west Scotland. From 1960 annual occurrences, particularly in the Northern Isles, culminated in breeding on Fetlar from 1967-1975.

Status: Prior to 1950 there are 177 published records for Britain and 58 for Ireland. Since 1950 there have been a further 199 accepted records to the end of 2007 (181 from Britain and 18 from Ireland).

Historical review: Data seem to suggest periodic influxes, though disentangling inter-island movements is complex and establishing which birds represent fresh arrivals is not always clear-cut. Since 1950 obvious arrivals took place in the 1960s and 1970s. The first major influx involved a record 16 in 1965, quickly followed by 10 in 1967. The 1960s yielded at least 58 birds and temporary colonisation of Shetland commenced.

In the 1970s, 49 'new' birds included 12 in 1972, which until quite recently was the last time that a calendar year accumulated double figures. This magnificent owl became more of a rarity in the 1980s with just 18 recorded; the maximum number in any one year during the decade was just four in both 1980 and 1981. 1985 was the last year to draw a blank for new individuals, although three or perhaps four females were present throughout that year on Fetlar.

In the 1990s there was something of a return to form with 31 birds and a maximum of six in 1993. The millennium started slowly, but between 2000-2007 a new spate of arrivals produced 38 new arrivals, including 10 in 2005 and 13 in 2007 (the highest count for 35 years), all but two of which came from the new focus of Snowy Owl activity: the Outer Hebrides. There was a failed breeding attempt in Co. Donegal in 2001.

Where: The bulk of records come from Scotland, especially Shetland. There is some evidence that Snowy Owls bred on Shetland early in the 19th century (Pennington *et al* 2004). Events in the 1960s finally coupled the islands with the species. A small number of birds took up residence on the islands from 1963, culminating in the exciting discovery of a nest containing eggs on Fetlar in June 1967 (Tulloch 1968). This fine Arctic owl made nesting attempts on Fetlar in each year from 1967-1975 (1972 the only unsuccessful year). Robinson and Becker (1986) provided a thorough summary of these exciting events; see *Table 4*.

During the winter of 1975/76 the resident male died or disappeared. He was not replaced and all six males reared on the island died or emigrated. In early 1976 there were five females on Fetlar, reducing to just two in 1977 before a further increase to 1980-83 when four females were present. Infertile eggs were laid in each of these years. Two females laid in 1982. Infertile eggs were also laid between 1987-1990. The remaining female was last seen on Fetlar in May 1992 and it was assumed that a female seen at four locations in Shetland in summer 1993 was the last of Fetlar birds (Robinson and Becker 1986; Pennington *et al* 2004).

Away from Shetland sightings in Scotland are unusual. However, the Outer Hebrides has recently become a favoured location with records there since 1992 into double figures. Elsewhere in Scotland occurrences remain sporadic, though males spent the summers of 1979-1981 and 1984 in suitable habitat in the Cairngorms, a female was there intermittently during 1987-1991 and another male from 1996-98 (Forrester *et al* 2007).

Farther south Snowy Owl is an exceptional rarity, with just 19 records in England since 1950. The vast majority involves one-day birds, with a wide geographical spread: Suffolk (four), Kent (two), Scilly (three, one seen in Cornwall), Northumberland (two) and singles Devon, Hampshire (also seen on Isle of Wight), East Sussex, Essex, Lincolnshire (seen subsequently in Norfolk and East Yorkshire), North Yorkshire, Co. Durham and Cumbria. There are six one-day Welsh records over the same period: two each in Gwent and Anglesey and singles from Bardsey and Glamorgan.

In 2008 a juvenile male on the Isles of Scilly from 28th October-6th December – the first on the archipelago since 1972 – proved hugely popular, and even more so when it was relocated in Cornwall from 21st December through to at least April 2009.

The most popular of the English records was a 2nd-winter male in Lincolnshire. It was first found at Thornton Curtis on 13th December 1990 and then relocated to the Wainfleet and Friskney area from 24th December-18th March 1991. It crossed The Wash to Norfolk where it was encountered at Blakeney Point and Stiffkey on 23rd March, Burnham Overy on 24th and Burnham Norton on 25th. It briefly graced Spurn on 30th March. Possibly the same bird was present on the Northern Isles from late April onwards.

In 2001 a 1st-winter male was present in Suffolk at Felixstowe from 24th October-6th December and then Waldringfield from 8th-9th December. This bird was oiled-stained. It was linked to an arrival of several Snowy Owls on a ship off the coast of Canada that disembarked in the southern North Sea; several others were seen in the Netherlands and Belgium.

Irish records have shown a surge since 1992, presumably resulting from increased coverage in the northwest of the country, which culminated in a failed breeding attempt in Co. Donegal in 2001. Perhaps it is more than a coincidence that the majority of the recent records there has matched those in the Outer Hebrides?

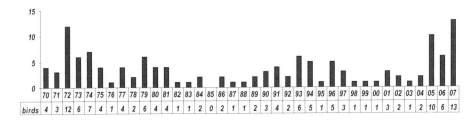

Figure 7: Annual numbers of newly arrived Snowy Owls, 1970-2007.

Discussion: Snowy Owl is a true nomad and the origin of birds reaching our shores has been disputed. Most are assumed to originate in arctic Eurasia and travel through Scandinavia. In late May 1992 an injured female on North Uist, Outer Hebrides, was taken into care where it died; it had been ringed at Hordaland in Norway. An adult male trapped on Fair Isle in June 1972 was found dead on Lewis in January 1975.

Year	1967	1968	1969	1970	1971	1972	1973	1974	1975	Total
Nests	1	1	1	1	1	1	2	2	2	12
Eggs Laid	7	6	6	5	5	4	5/3	5/1	6/3	56
Chicks	5	3	3	2	3	0	2/0	1/0	4/0	23

Some birds are of Nearctic origin. In 1989 one found on an oil tanker in mid-Atlantic was later released on Fetlar. In 1992 an adult female came aboard a Spanish fishing vessel about 320 km east of Newfoundland and was brought ashore at Aberdeen on 2nd July; it was released 3rd July and last seen 4th July. The popular 2001 Suffolk bird had hitched a ride from Canadian waters.

The last good irruption to southern Finland was during winter 1999/2000, when at least 90 birds were seen including 25 birds in Helsinki region (Rissanen *et al* 2001). Away from the far north of Europe records are rare. In the Netherlands there have been just 15 records up to and including 2008. Ten were seen between 1980-2008. In France there have been 17 records, the last of which was in 1996. All bar an adult male in the Ardennes (northern inland France) have come from costal départements from Brittany to the Belgian border. Birds have been recorded well into central Europe (for example 18 records from Slovakia). Others have penetrated to southeast Europe: a specimen from Hungary is dated 3rd November 1891. The wide-ranging vagrancy of this species is illustrated by birds on the Azores in January 1928 and on 17th February 2009.

Table 5: Seasonal distribution of post-1950 Snowy Owls in England, Wales and Ireland (by date of first arrival).

Region	Jan	Feb	Mar	Apr	May	June	July	Aug	Sept	Oct	Nov	Dec	Totals
England	2	1	5			1			1	4	3	1	18
Wales	1		2	1	1							1	6
Ireland	2	3		1	1	2	1	1	1	2	3	1	18
Totals	5	4	7	2	2	3	1	1	2	6	6	3	42

Northern Hawk Owl Surnia ulula

(Linnaeus, 1758). Nominate ulula breeds from northern Scandinavia east across northern Russia to Kamchatka and south to northeast Kazakhstan, Mongolia, northeast China and Sakhalin. Nearctic caparoch is found from Alaska east across Canada. A third race lives in central Asia and northern China. All populations are dispersive and irruptive in search of food, although most remain within the breeding range at times of plentiful supply.

Polytypic, three subspecies. Both the Nearctic race *caparoch* (P.L.S. Müller, 1776) – first record one caught on a ship off Cornwall in March 1830 – and *ulula* (Linnaeus, 1758) – first record shot near Amesbury, Wiltshire, pre-1876 – are presently on the *British List*. All historical records are under review. Compared with nominate *ulula*, *caparoch* is darker with a solid black patch on the hind neck and a broad black upper chest band. It also has broader bars on the underparts, sometimes tinged with rufous on flanks and lower belly. Undersides of the tail feathers are more contrastingly barred in *caparoch* than in *ulula* (Duncan and Duncan 1998).

Very rare vagrant, with only one record in the latter part of the 20th century and no records of the Nearctic race since 1868.

Status: Nine records, seven of them in the 19th century.

1830 Cornwall: Sea area Plymouth *caparoch* caught exhausted on board a collier a few miles from Looe, Cornwall, en-route to Waterford, Ireland, in March
1847 Somerset: Backwell Hill, Yatton *caparoch* shot 25th or 26th August
1860 Shetland: Skaw, Unst *ulula* shot in December
1863 Clyde: Near Maryhill *caparoch* shot some time before 29th December
1868 Clyde: Near Greenock *caparoch* shot some time before 20th November
Pre-1876 Wiltshire: Near Amesbury *ulula* shot, no date.
1898 Northeast Scotland: Gight adult female probably *ulula* shot on 21st November
1903 Northamptonshire: Orlingbury shot on 19th October
1983 Shetland: Near Lerwick, Mainland *ulula* 12th-13th September; same Bressay 20th-21st September

Discussion: Modern observers might raise an eyebrow at some of the older records, especially two of the Nearctic race in Clyde within the space of five years. There was even a third record rejected from Clyde in 1871 involving the same race and taxidermist. The area is close to ports where ships arrive from America, so perhaps the birds were ship assisted. More likely they are fraudulent. Likewise, the exceptionally early timing of the Somerset record suggests either a journey on a vessel bound for Bristol or a hoax. The earliest extralimital bird reported in the US during invasion years was 15th September (Duncan and Duncan 1998).

The recorded history of the first occurrence is well documented (Palmer 2000). The exhausted bird was captured by hand a few miles from Cornwall on a boat en-route to Waterford, Ireland, in March 1830. It must have been exhausted or confiding to allow hand capture. Voous (1988) documents a record of a 1st-year male *caparoch* that was taken on board a ship off Las Palmas in 1924, probably in October; the ship eventually docked at Rotterdam and the bird was taken to Rotterdam Zoo on 7th November 1924. It does not form part of the Spanish List.

Nearctic *caparoch* irrupts southward in some winters, usually from mid-October through late November, and rarely reaches Nebraska, Ohio, Pennsylvania and New Jersey. There are no irruptive movements documented for 1830 or 1924 (Duncan and Duncan 1998).

The farthest travelled ringing recovery in the US cited by Duncan and Duncan (1998) was 259 km from New York to Québec, although one ringed near Edmonton, Canada, in February 2000 was recovered dead that October at Dillingham, Alaska. This bird had apparently travelled 3,187 km (Bird Banding Office, Canadian Wildlife Service *pers comm*). The minimum distances involved if natural vagrancy was responsible for the Western Palearctic records of *caparoch* are exceptional: c4,700 km from Newfoundland to Canary Islands; and c3,700 km from Newfoundland to Looe. It would appear that unaided vagrancy of this race across the Atlantic is unlikely without some degree of ship transfer.

The remaining records assigned to subspecies are all nominate *ulula*. One from Gurnard's Head, Cornwall, on 14th August 1966 was accepted but is now considered not proven following review. A review of the veracity and racial attribution of the remaining eight records to 1903 is ongoing; some are likely fall by the wayside.

The sole appearance during the modern recording era was in Shetland in 1983. Many observers journeyed north to see it, but for many more the quest was foiled by its short stay. The bird was associated with an exceptional influx of the species into southern Scandinavia. Several hundred penetrated south to Denmark that winter.

Northern Hawk Owl is a rare vagrant in Europe away from Scandinavia. Three accepted birds in France are all from the 19th century. The Netherlands was luckier in the 20th century (5th October 1920, 2nd April 1995 and 30th-31st October 2005, all nominate *ulula*). The last coincided with the largest post-1983 influx into southern Finland. Such appearances on the near continent in recent years have heightened expectations of another in Britain. The next arrival here will no doubt draw a great crowd of admirers.

Tengmalm's Owl Aegolius funereus

(Linnaeus, 1758). Breeds discontinuously through southern and central Europe eastwards to the Ural and Caucasus Mountains. The European strongholds are from Fennoscandia eastwards across Russia. Nominate funereus occurs as closely as Belgium, France and The Netherlands; there are five other subspecies in central and eastern Asia to northeast Siberia, and another in North America. Dispersive at times of food shortage following high productivity, but mostly sedentary.

Ray Scally

Polytypic, seven subspecies. All records assigned to race are *funereus*, but Nearctic *richardsoni* – Boreal Owl – is a potential vagrant. The largest and brightest forms occur in northeast Siberia; smaller, darker birds live to the west and south. Nearctic *richardsoni* is among the darkest forms, and more similar to *funereus* than *magnus* and *sibiricus*, which are closer geographically.

Formerly a rare irruptive vagrant, typically from October to January on the Northern Isles and along the east coast. Few records since 1929.

Status: A total of 43 were recorded prior to 1929 (Keith Naylor *pers comm*). There have been just seven since.

Historical review: This sought-after owl was formerly a much more regular visitor to Britain. No fewer than 30 birds were found between 1812-1897, and 13 more between 1901-1929. The first was killed at Widdrington, Northumberland, in January 1812. Records were almost exclusively along the east coast. (see *Table 6*)

> *On January 31st, 1912, I received word that a small owl had been shot in the vicinity of the Seaton Burn, Northumberland. Accordingly, I immediately cycled over to the house of the owner at Seaton Sluice, and was pleased to find a specimen of Tengmalm's Owl sitting contentedly in a small wooden dove-cage. Mr. James Hall, in whose possession the bird was, informed me that he had "winged" it in Holywell Dene on December 11th, 1911, his setter having flushed it from a hawthorn bush in a small ravine. It was apparently much confused by the light, making no attempt to defend itself, and when brought home and placed on a table at first crouched down, and stared around with a somewhat bewildered expression. However, shortly afterwards, the news of its capture having spread, a party of the owner's friends collected and stood around gazing at the Owl, which had already so far lost its sense of fear that it devoured a Sparrow on the table.*
> **J. M. Charlton.** *British Birds 6: 8-10*

Ten are documented for Yorkshire, six from Northumberland, five in Norfolk and four in Shetland. Many more occurred along the east coast between Kent and Northeast Scotland. In the northwest there was a single in Cumbria. In the Midlands singletons came from Shropshire, Northants and Worcestershire.

The pattern of dated records to 1929 reflects our present expectations: 37 per cent in October-November, 33 per cent in December-February and 14 per cent in spring. The only anomaly is a single July record in Norfolk. This pattern reflects occasional irruptions reaching the east coast followed by secret wintering and a modest detection of a spring exodus.

Table 6: Dated* monthly occurrence of 1812-1929 Tengmalm's Owl records by region and 1959-86 records by month.

Region	Jan	Feb	Mar	Apr	May	June	July	Aug	Sept	Oct	Nov	Dec	Total
Scotland	2	2	1								1	1	**7**
N. England	2	1	1	1						5	4	2	**16**
E. Anglia	3			1		1				4			**9**
S. England	1		2									2	**5**
Totals	**8**	**3**	**4**	**2**	**0**	**0**	**1**	**0**	**0**	**9**	7	3	37
1959-86	1	0	1	0	2	0	0	0	0	1	1	1	7

* Six records were undated

Three in 1861 and four in winter 1872/73 would be marvelled at today, let alone the six located during October and November 1901 (followed by one more in the first week of 1902). This influx comprised two together in Suffolk and one in Norfolk on 30th October, followed by November birds in Shetland, North Yorkshire and Worcestershire. The January bird was in Northamptonshire. Seven in a season! How many would have been recorded given modern efforts?

Six more followed between 1903-1929. Yet since 1929 there have been just seven records, five of them from Orkney. Three have been in spring, suggesting birds returning from Britain to Scandinavia; two each have been in autumn and winter.

1959 Orkney: Cruan, Firth, Mainland 26th-27th December and 1st January 1960
1961 Orkney: Stromness, Mainland 1st May
1980 Orkney: Finstown, Mainland adult 13th to at least 20th October, trapped 14th
1980 Orkney: Finstown, Mainland adult trapped 18th November, later found dead
1981 Co. Durham: Fishburn dead (leg only) 10th January, ringed as nestling Greften, near Vang, Hedmark, Norway, 10th June 1980
1983 East Yorkshire: Spurn 6th-27th March, trapped 7th & 16th March; possibly since 28th January
1986 Orkney: Egilsay 31st May-1st June; presumed same Glims Holm on 25th June dead about 3-4 weeks

The decision to suppress the presence of the Spurn bird of March 1983 provoked more discussion regarding the withholding of information than any other rarity. Feelings still run high in some quarters. Conjecture that it had been present since 28th January was based on brief views of a small owl seen at dusk on that date, and the number of pellets found at its favoured roost site.

The discovery of a Tengmalm's Owl Aegolius funereus on Spurn Nature Reserve, Humberside, on 6th March 1983, and the Yorkshire Naturalists' Trust's subsequent decision to keep its presence secret, has generated considerable debate and on occasions somewhat hostile discussion. When the owl was discovered it was immediately obvious that a potential multitude of people would wish to see it. Its roost, however, was close to vital coastal services (lifeboat houses, coastguard and Humber Pilots facilities, an area closed to the public), and at the time Spurn was in its most precarious condition since 1960. February storms had seriously eroded the northern end of the peninsula, all but breaking through a flood bank, and high spring tides were expected in March which could have severed the peninsula. Throughout March, emergency works were undertaken to create further reinforcements against the sea, operations involving heavy cranes, bulldozers and lorries, working in a confined space; the sea of mud created was another problem. These works had to be carried out with the utmost urgency, since the implications of a breach of the peninsula are very serious: at the least isolating the lifeboat crew, pilots and coastguard, cutting off their water, telephones and electricity; at the worst affecting, to an uncertain extent, surrounding low-lying land and the patterns of silt deposition in the Humber estuary and shipping lanes.

During March, through the national media, the YNT was warning of the seriousness of the situation. It would, therefore, have been wholly irresponsible to have encouraged large numbers on to the reserve, particularly considering that emergency provisions for the evacuation of the warden in times of high risk had also been implemented for this period. The last thing the Trust could have coped with was the control of large numbers of people. All our efforts were involved in securing, for the short term, the future of the reserve and the ability of the coastal and emergency services to continue to function, a primary responsibility being to protect the interests of our tenants, especially the families of the lifeboat crew.

The decision to keep the owl's presence secret was not an act of elitism, as some have suggested, but was taken for the most extreme of practical reasons. Although I was a party to it, the ultimate decision was taken by the Executive Officer and by the President of the Trust, neither of whom knew the name of the species, merely the circumstances.
Ian Carstairs. British Birds 76: 416-417.

Discussion: Across most of its European breeding range populations are stable or increasing. The largest European populations are in Russia, Finland, Sweden, Norway and Belarus, with good numbers also present in the Baltic States. Across Fennoscandia Tengmalm's is among the commonest predatory species. This attractive owl is declining heavily in southern Finland and there are a reduced number of records at bird observatories (Lehikoinen *et al* 2008). It has presumably bred more or less annually on the island of Bornholm in easternmost Denmark since 1987. In 2007 a breeding pair was found in a forest in Jutland in western Denmark (at least 300 km west of the island of Bornholm (Sebastian Klein *pers comm*). Even closer to our shores there have been several breeding records in the Netherlands since 1971; two nests with young were found in 2008. It is an annual breeder in eastern Belgium, in some years in high numbers (140 pairs in 1996) (Arnoud van den Berg *pers comm*).

North European populations of nominate *funereus* are irruptive. Their main prey is voles, which in Fennoscandia show 3- to 4-year population cycles. Breeding success varies with the vole cycle (Korpimäki 1988). Most, if not all, British records presumably originated from these populations, a hypothesis supported not only by the easterly and northerly distribution of records but also the recovery of the leg of a Norwegian-ringed bird in Co. Durham in 1981.

Many of the British records sexed and aged have been females and young birds, as might be expected. A study of birds trapped during autumn in the Gulf of Bothnia found that juveniles were more frequent than adults, and the sex ratio of juvenile owls was significantly female-biased. Contrary to expectation adult males, as well as females, were also caught. Among juveniles, males migrated earlier than females (Hipkiss *et al* 2002). Both sexes tend to remain in the same localities if prey densities are high, but if vole densities crash then females move longer distances. Some move over 480 km between breeding sites in subsequent years (Wallin and Andersson 1981).

In 2008 a late autumn influx in Scandinavia peaked with 264 birds trapped and ringed at Falsterbo, Sweden. The majority were 2nd-calendar year birds (59 per cent), with 28 (per cent) birds of the year, 6 per cent 3rd-calendar year birds and 8% aged as older (Falsterbo Bird Observatory). As part of this movement at least 23 were recorded in Denmark. The recent rarity status of this highly desired owl has deprived British birders of sightings for nearly a quarter of a century: how much longer must we wait?

Red-necked Nightjar Caprimulgus ruficollis

Temminck, 1820. Breeds in parts of Portugal, Spain, Morocco, Algeria and Tunisia. Winters in West Africa, probably chiefly southern Mali.

Polytypic, two races. The British record was originally attributed to the race *desertorum* of north-east Morocco eastwards, but has recently been reassessed as a bird of the nominate race *ruficollis* of the non-desert areas of Iberia and the rest of Morocco; a much more expected overshoot. *C. r. desertorum* differs from nominate *ruficollis* in distinctly paler ground colour and narrower black marks. The inner primaries of *ruficollis* show a predominantly dark ground colour with narrow orange bands; *desertorum* shows approximately equal-width dark and orange bands. On *desertorum* the proximal portion of the inner primaries is predominantly orange; *ruficollis* shows the opposite pattern (*BWP*).

Exceptionally rare vagrant with just two records in northern Europe.

Status: One record.
1856 Northumberland: Killingworth 1st-winter, probably male, 5th October

The sole British occurrence was shot on 5th October 1856 near Killingworth, just north of Newcastle-on-Tyne. John Hancock (of Hancock Museum fame) bought the specimen from local game dealer Mr Pape the day after it was shot. No one knows if Pape shot the bird himself, but it was fresh-killed when Hancock received it. He commented that he could not determine the sex from dissection, but tentatively identified it as a male by the white coloration of the primary spots. Hancock waited six years before announcing the discovery in 1862 (*Transactions of Tyneside Naturalists' Field Club* 5: 84-85, *Zoologist* 20: 7936-7938, *Ibis* 1st series, IV: 39-40).

He gives his reasons for the delay as follows:
"I have delayed until now making this announcement; for I found, on comparison, that the bird in question differed slightly from a Hungarian specimen in my collection, and I was consequently anxious to see others before doing so. I have now had an opportunity of referring to a specimen in the British Museum, and find that it quite agrees with my bird. I have therefore no hesitation in stating that it is the true C. ruficollis of authors, and I have much pleasure in adding this fine species to the British list..."(Ibis 1st series, IV: 39-40)

The Hungarian specimen was a proverbial Red Herring, and nothing on its label says it was from Hungary. The Killingworth bird was a 1st-winter bird, with recently moulted adult head and body feathers, but juvenile flight feathers. Andrew Lassey verified that it was the expected age class for a natural vagrant and was at exactly the right stage of moult for the time of year it was found. A bird collected within its range and transported to Newcastle would not have been fresh, and would not have had time to undertake its post-juvenile head and body moult. John Hancock had an impeccable reputation and there is no reason to suspect fraud on his part (Melling 2009).

Discussion: One found dead in Denmark (4th October 1991) is the only other record from northern Europe. This bird was mounted and exhibited as a European Nightjar *C. europaeus* and its true identity was only discovered some time later (Christensen 1996). The coincidence in date of the British bird with the Danish record is striking. It is also interesting to note that the Danish record was found farther north than the Killingworth record further demonstrating the species' capability for autumn vagrancy well to the north of its breeding range.

Nine have occurred in France. The four recent records were from southeastern France: found dead in June 1997 and 28th May 2005, plus a singing male from 10th-15th June 2007 and a male from 12th-16th April 2008 in the Camargue. In the Mediterranean there are records from Italy (*desertorum* on Sicily in 1898 and 1946, plus two spring birds from 2008 yet to be accepted) and Malta (at least 12 occasions). Vagrants have also appeared in Madeira and the Canary Islands.

Given the sizeable Iberian population of 110,000 pairs (BirdLife International 2004), and its wintering quarters in West Africa there is potential for birds to overshoot in the right conditions. Spring arrivals in the Iberian Peninsula occur from mid-April, earlier than those for European Nightjar. British-ringed European Nightjars have been recovered in Morocco and Spain, but no European Nightjars ringed abroad have yet been recovered in Britain (Wernham *et al* 2002).

The difficulty of encountering nightjars during daylight hours complicates the detection of a vagrant Red-necked Nightjar. Several of the recent extralimital records involved fatalities; statistically this may offer the most likely circumstances for an upgrade to category A for this species.

Egyptian Nightjar Caprimulgus aegyptius

Lichtenstein, 1823. Breeds in North Africa, Middle East and southern Iran and Afghanistan north to southern Kazakhstan. Nominate aegyptius occasionally breeds or has bred in countries such as Israel and Jordan, although it is more likely to occur there as a passage migrant en route to its Asian breeding grounds.

Slightly smaller birds of the race saharae breed in central and eastern Morocco, northern and occasionally southern Algeria, central Tunisia and possibly also in northwest Libya. The wintering grounds of all populations lie in sub-Saharan Africa.

Polytypic, two races. The subspecies occurring in Britain is undetermined. Nominate *aegyptius* shows extensive grey vermiculation on pink-buff feathers of the upperparts, upperwing and chest, appearing greyish sandy pink in fresh plumage. North African *saharae* is slightly brighter pink-yellow or buff-yellow with fainter and less extensive grey vermiculation *(BWP)*.

Very rare vagrant overshoot in late spring.

Status: Two records.

1883 Nottinghamshire: Rainworth shot 23rd June
1984 Dorset: Portland 10th June

A monument was erected in Thieves Wood on the spot where the Nottinghamshire bird was obtained. This is probably the only permanent memorial to a rare bird in Britain. A stone now marks the place where it fell, and on it is inscribed:
This stone was placed here by J. Whitaker, of Rainworth Lodge, to mark the spot where the first British specimen of the Egyptian Nightjar was shot by A. Spinks, on June 23rd, 1883, this is only the second occurrence of this bird in Europe.

Spinks was a gamekeeper. His exceptional trophy was thrown away before its oddity was spotted and retrieved. J H Gurney identified it as an Egyptian Nightjar. Although the location and circumstances seem a little odd, nobody seems to have profited from the bird so fraud is unlikely (Tim Melling *pers comm*).

Discussion: The second British record from Dorset in 1984 endured a troublesome ascent on to category A of the *British list* (Walbridge 1999). The unique opportunity of the sole observer to enjoy the bird during three flight views will doubtless be viewed with envy by listers everywhere.

In addition to the scant British records, there are a number of occurrences elsewhere in Europe to give hope that this species might grace our shores again. Records have come from Malta (10 between March and May, the last in April 1978) and Italy (eight accepted from 1874 to 1991, mostly from Sicily in spring). More pertinent to observers here are those from Germany (22nd June 1875), Sweden (21st-22nd May 1972) and Denmark (female 29th May-18th June 1983).

Common Nighthawk *Chordeiles minor*
(JR Forster, 1771). Breeds throughout temperate North America, south to Panama and the Caribbean. Winters South America south to central Argentina. Some migrate over the west Atlantic, where it is regular on passage on Bermuda and Lesser Antilles.

Polytypic, nine subspecies. All British records assigned to race have been of the nominate race *minor*. The juvenile plumage of *minor* is the darkest among the subspecies and has the most extensive areas of black in the feathers of the dorsum, especially on the crown and wing coverts. It has generally heavier barring ventrally, including the undertail coverts (Dickerman 1990).

Rare vagrant, wind-blown across the North Atlantic from the temperate Nearctic; nearly all have been found in southwest England, mostly on the Isles of Scilly.

Status: There are 22 records, 21 from Britain and one from Ireland. Two in 2008 have yet to be accepted, but one involved a specimen and the other was photographed.

1927 Scilly: Tresco female shot 11th September
1955 Scilly: St. Agnes juvenile 28th September-5th October
1955 Scilly: St. Agnes female 28th September
1971 Scilly: St. Agnes immature 12th-13th October
1971 Nottinghamshire: Bulcote immature 18th and 21st October
1976 Scilly: St. Mary's female found dead 14th October
1976 Scilly: St. Mary's female found dead 25th October
1978 Orkney: Kirkwall Airport, Mainland juvenile trapped 12th September
1981 Scilly: St. Mary's 12th-14th October
1982 Scilly: St. Agnes immature 20th October-4th November
1983 Dorset: Studland 25th October
1984 Greater London: Barnes Common adult male found moribund 23rd October, died in care on 28th
1985 Cheshire & Wirral: Moreton 1st-winter taken into care 11th October; transported to Belize where released in good health about 25th October
1989 Scilly: Tresco juvenile 16th-22nd September
1998 Scilly: St. Agnes male 9th-13th September; found dead on 14th.
1998 Scilly: St. Mary's female or 1st-winter 12th-20th September
1999 Scilly: St. Agnes 22nd September
1999 Scilly: Bryher juvenile 23rd-30th October
1999 Co. Cork: Ballydonegan juvenile 24th October
1999 Ceredigion: Mwnt juvenile found dead 28th October
2008 Scilly: St. Mary's picked up dead 6th October*
2008 Cornwall: Church Cove flew in off the sea 7th October also seen 8th*

The Isles of Scilly is the premier location for the species in the Western Palearctic. Two on the archipelago in 1998 were the highlight of the year for many observers.

On the evening of 8th September 1998, a 'non birder' stepped out of the door of the Turk's Head public house on St. Agnes, Isles of Scilly, and noticed a 'nightjar' flying around, hawking insects. He reported it to resident birder Nigel Wheatley, who searched for it in vain the next morning. Fran Hicks, one of the farmers on the island, then found the bird on the ground at about 1.00pm, whilst doing the rounds of his fields. Fran had seen Common Nighthawk before and immediately realised the importance of the record and quickly informed other local birders. To the delight of many visitors, the bird stayed for several days, and most birders enjoyed excellent views of it, both in flight and on the ground. It was the first Common Nighthawk to be twitchable in Britain since the one on Tresco, Isles of Scilly, in September 1989.

Birders arrived on St. Agnes to see it on a daily basis, until it sadly died on the morning of 14th September. Early that morning, birders had watched it sitting on the ground, absolutely still – until it was realised that it was dead. This left them in something of a quandary: had it died before or while they were watching it? But, other than to the bird, it did not matter much; dead or alive, it moved little while perched!

Remarkably, a second Common Nighthawk was found on the adjacent island of St. Mary's on 12th September. It was first noticed near Green Farm by Bill Rogers as it flew quickly over him mobbed by Swallows, and he initially took it to be a Kestrel. But it was clearly a nighthawk, and he assumed it was the St. Agnes bird.

This second St. Mary's bird was then seen in flight around the northern end of the island almost every day until the last confirmed sighting on 20th September. It was soon clear that it lacked tail-spots, so it was not the St. Agnes bird. Fortunately, unlike Nightjar, this species regularly feeds during the daytime.

Will Wagstaff. Birding World 11:338-340.

Where: The Isles of Scilly has a virtual monopoly on British sightings with no fewer than 13 birds; the islands boast the only multiple arrivals. Records elsewhere have been moribund birds or seen only briefly. The eight records away from Scilly are well dispersed. Just one has been seen in each of Scotland (Orkney), Ireland (Co. Cork) and Wales (Ceredigion). Singletons from England have come from Cheshire, Nottinghamshire, Dorset, Greater London and Cornwall.

When: Eight have occurred in September. The earliest, a male on St. Agnes from 9th-13th September 1998, was found dead on 14th. September records also include four found between 11th-16th and three more between 22nd-28th. In October six have occurred between 11th-20th and six from 23rd-28th. The latest was a dead juvenile at Mwnt, Ceredigion, on 28th October 1999.

Discussion: Elsewhere in the Western Palearctic there have been a further 15 birds. These comprise 10 from the Azores to the end of 2008 (28th September-11th November), including two in November 2005 and four in late October 2007. Singles have reached the Faeroe Islands (found dead on 1st October 1955), Iceland (collected on 23rd October 1955), Canary Islands (found dead in December 1972), France (Ouessant 17th-28th September 1998) and Spain (Cádiz, found exhausted on 4th November 2005).

This migratory 'goatsucker' winters in South America. Nominate *minor* breeds in central and eastern Canada, and northeast USA. It takes a transoceanic route over the western Atlantic to its wintering grounds in South America. On migration it forms loose flocks. Movement is both diurnal and nocturnal. The main movement occurs in the second half of August and early September, somewhat earlier than a proportion of the sightings here.

The majority of European records have occurred following fast-moving Atlantic depressions. Multiple arrivals indicate the direct displacement of flocks from the transoceanic route. The two on Scilly in 1955 occurred at the same time as the birds in Iceland and the Faeroe Islands (the only records from the northern reaches of the Western Palearctic other than the sole Scottish bird). There are other instances of multiple arrivals, most noticeably four in 1999 following the Atlantic weather systems prompted by Hurricane Irene. Interestingly these, and those on the Azores in 2005, were associated with exceptional numbers of Chimney Swift *Chaetura pelagica*, a species that also migrates in large flocks and is similarly susceptible to mass displacement across the Atlantic.

Unlike Old World nightjars the diurnal displays put on by this dashing species doubtless facilitate the discovery of vagrants here and ensures that the sporadic performers are always well attended.

Chimney Swift *Chaetura pelagica*
(Linnaeus, 1758). Breeds southern Canada and throughout USA east of Rockies to Gulf of Mexico. Winters upper Amazon Basin, Peru and perhaps elsewhere in South America.

Monotypic.

Scarce but increasingly regular vagrant, wind-blown across the North Atlantic from the temperate Nearctic.

Status: Thirty-six birds recorded: 19 in Britain and 17 in Ireland.

Historical review: The first was at Porthgwarra from 21st-27th October 1982, and was joined by a second bird from 23rd-25th October. The thrid appearance followed quickly with one at Coldingham, Borders, on 5th November 1983; it took 17 years to be accepted, due mainly to the fact that it was seen by a single observer (Forrester *et al* 2007). There was also a delay in admitting the Porthgwarra records because of the difficulty of eliminating Vaux's Swift *C. vauxi*. Singletons in 1986 and 1987 appeared in Scilly and Cornwall before a well-watched and photographed 'long-stayer' was a surprise east coast discovery at St. Andrew's, Fife, from 8th-10th November 1991. This bird was observed (and photographed) clinging to buildings, a rare opportunity to see the species settled on this side of the Atlantic.

There was a hiatus of eight years before the next, but the wait was worth it. An exceptional influx of 12 in 1999 resulted from the fallout of Hurricane Irene hitting our shores on the 21st October. The deluge comprised seven in Ireland (the first for the country with three each in Co. Cork and Co. Wexford, plus another in Co. Wicklow). There were five in Britain – three in Cornwall and one each in Scilly and Devon – with the bulk of arrivals between 22nd-25th October.

Singletons followed in 2000 (a noteworthy bird passing Spurn on 6th August) and 2001. Hurricane Wilma in 2005 caused a staggering arrival of 16 birds between 29th October-9th November, including 10 in Ireland and six in Britain. Several more reports from Ireland remain undocumented. In Ireland eight zoomed over Co. Cork, three of these between Courtmacsherry and Broadstrand from 31st October-4th November, with four present on 1st November. Elsewhere in Ireland, singles graced the airspace of Co. Kerry and Co. Waterford. In Britain the six 'one-day' singletons were well dispersed following the main arrival, indicative of birds filtering east. Singles were seen in Scilly, Devon, Cheshire, Anglesey (the first for Wales), Northumberland and East Yorkshire: the last two most probably refer to the same bird passing south. Simultaneously the Azores hosted around 112 Chimney Swifts: a quite unprecedented arrival of these birds in the Western Palearctic and easily the largest ever arrival of a Nearctic landbird species.

Where: There is, as would be expected, a southwesterly bias to records: Co. Cork has hosted no fewer than 12 birds; there have been three in Co. Wexford; four have reached the Isles of Scilly; and six have visited Cornwall, twice including two together. East Yorkshire is the only other county with multiple records (two). Singletons have come from Fife, Borders, Northumberland, Cheshire, Anglesey, Co. Wicklow, Co. Kerry and Co. Waterford.

When: All records fall in late October and early November, with one notable exception. One through Spurn on 6th August 2000 is the outlier and was presumably a bird from the influx of the previous autumn that remained on this side of the Atlantic.

The next-earliest record was from near Truro, Cornwall, on 18th October 1987. Most have been in the last week of October. Four of the five east coast records fall in early November and presumably refer to birds filtering east. The latest record is from Faranfore, Co. Kerry, on 9th November 2005, the last of 11 early November records.

Discussion: *Chaetura* swifts present an identification conundrum: Chimney very closely resembles Vaux's Swift, a migrant breeder of the Pacific Northwest that winters in Mexico, Central America and Venezuela. All vagrant *Chaetura* swifts have been assigned to Chimney, which on probability is far more likely to turn up here as a vagrant. (See comments above regarding the delay in accepting the first British records.) The two have different calls but all records have been silent.

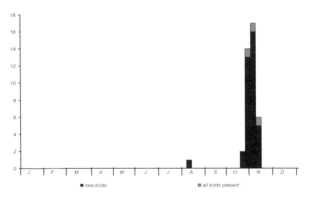

■ new birds   ■ all birds present

Figure 8: Timing of British and Irish Chimney Swift records, 1982-2007.

Three distinct flyways are suggested for the migratory routes, one of which is along the Atlantic Coastal Plain. Most cross directly over the Gulf of Mexico, passing over the Yucatan Peninsula before following the Atlantic coast of Central America and reaching Peruvian wintering grounds in early November. Results from ringing sites in the southeastern USA provide evidence of random movement during the autumn, including a number of fairly long-distance movements in a northwards direction rather then the expected continued southwards movement (Lowery 1943).

This Nearctic swift is also becoming more frequent elsewhere in the Western Palearctic. The first for Canary Islands was on 2nd October and 26th November 1997. In 1999, the first two for Spain were present on 23rd and 27th October, first for Portugal on 26th October and the first for Sweden from 6th-7th November and again 12th-13th November. The sole Norwegian record is the only spring record for the region, on 26th May 2000.

The extraordinary displacement of 2005 delivered a minimum of 112 to the Azores between 28th October-10th November, with peak flocks of 27 on Corvo and at least 30 on São Miguel. There had been just eight birds seen there previously and three since. In Spain, there were two on the mainland: Coruña on 30th October and Pontevedra same day. On 31st October, five were at Lanzarote (two remained to 11th November). A further record is under consideration from Fuerteventura on 1st November. The first for France also occurred during this period (30th October-2nd November).

In October 2005 Hurricane Wilma pushed more than 2,000 Chimney Swifts north from staging areas farther south in the US. At least 700 were found dead in the Maritimes after the event. Many thousands of birds reached Nova Scotia (Dinsmore 2006; Gauthier *et al* 2007). Events such as this may become more frequent as a result of climate change. The frequency, intensity and trajectories of hurricanes may alter and displacements of species that migrate in large flocks, such as Chimney Swifts, may become more frequent.

Summer 2007 yielded several claims of Chimney Swift in Britain, all of which were later found to be 'not proven'. Many observers watched an instructive, albeit elusive, bird over York, North Yorkshire, between 24th-25th July. Most observers left the site relieved that they had connected with an exceptional inland record of Chimney Swift. Subsequent photographic evidence illustrated findings to the contrary as the bird appeared to be an aberrant Common Swift *A. apus* missing its rectrices and outer primaries.

The York swift drew attention to a serious identification pitfall, and highlighted the need for rigorous documentation of any putative Chimney Swift outside expected periods. Had record shots not been obtained of this bird then the multi-observed nature of the record would doubtless have led to it being accepted with no query. Likewise, had the York bird flown quickly through a visible migration hotspot then it would also surely have passed through with little questioning. A cautionary tale from a bird without a tail.

White-throated Needletail *Hirundapus caudacutus*

(Latham, 1802). Two subspecies. Nominate caudacutus breeds from central Siberia east to Japan and Korea. Winters in Australasia. The Himalayan race nudipes breeds from northern Pakistan to southern China, though the wintering range is unclear.

Polytypic, two subspecies. All records involve nominate *caudacutus* (Latham, 1802). The Himalayan race *nudipes* differs from nominate *race* in uniform black crown, nape, and sides of head, glossed with dark greenish-blue and lacking white on the forehead and lores *(BWP)*.

Very rare vagrant, overshooting in late spring and summer from Siberia.

Status: There are 11 records, thought to involve eight birds: seven from Britain and one from Ireland.

1846 Essex: Great Horkesley 6th-8th July, when shot
1879 Hampshire: Near Ringwood shot 26th or 27th July

1964 Co. Cork: Cape Clear 20th June
1983 Orkney: South Ronaldsay 11th-12th June
1984 Shetland: Quendale, Mainland 25th May-6th June
1985 West Yorkshire: Fairburn Ings 27th May
1988 Orkney: Hoy 28th May-8th June
1991 Kent: Wierton Hill Reservoir 26th May
1991 Staffordshire: Blithfield Reservoir 1st June; presumed same as Kent
1991 Derbyshire: Near Belper 3rd June; presumed same as Kent
1991 Shetland: Noup of Noss 11th and 14th June; presumed same as Kent

Discussion: This is one of the most highly desired birds for visitors to eastern Asia and deservedly so as it is a breathtaking species to observe. All have been between May and late July. The four records in 1991 were all thought to involve one bird heading north, a testament not only to the mobility of the species but also the skills of the dedicated patch worker Don Taylor (who saw it first) and those who managed to relocate it over its journey. The cluster of records between 1983 and 1991 make it conceivable that a solitary bird was involved in all of them. A cruelly short stay in West Yorkshire in 1985 still ranks among 'my worst dip' for many observers. Extended stays in the Northern Isles in 1984 and 1988 unsurprisingly proved popular.

The island of Hoy hit the headlines on Wednesday 1st June when Keith Fairclough, RSPB warden of North Hoy, 'phoned Birdline with news of a Needle-tailed Swift Hirundapus caudacutus at the southernmost end of the island. The bird had first been seen at Melsetter by Stanley Thomson on 28th May, but then proved elusive for a day or two before it settled into a daily routine upon which (apart from the weekend when a close encounter with a Peregrine sent it scurrying) it could be relied until a week later, when it left in the late morning of Tuesday 7th June.

During the mornings and evenings the bird would feed almost exclusively around the garden of Melsetter house, careening back and forth along the edge of the surrounding sycamores at a breathtaking pace. It regularly disappeared for several hours in the middle of the day, and at night roosted clinging to the wall of the house itself.
Richard Millington. *Birding World 1: 200-202*

Elsewhere in Europe there are 12 records, nine of them from Scandinavia. There are four from Finland (21st May 1933, 21st April 1990, 10th May 1991, 8th May 2005), two from Norway (17th May 1968, 20th May 1995), plus two from Sweden (22nd-27th May 1994, 5th June 1998) and one from the Faeroe Islands (19th-20th June 2000). There is one from the Netherlands (22nd May 1996). South of British latitudes there are no European records in spring. Lost autumn individuals have appeared in Malta (mid-November 1971) and Spain (4th November 1990).

This 'Sibe' bucks the typical trend in that almost all British and European records have been in spring or summer. The nominate race breeds across much of northern China, Mongolia and Siberia, and winters in Australasia. It is extremely rare within the Urals/Western Siberia region, though locally common (Ryabitsev 2008). It seems likely that spring overshoots have continued on a westerly heading into northern Europe, a theory supported by the high incidence of

Scandinavian sightings. Some could represent lost individuals that have wintered with Common Swift *Apus apus* in Africa and were then picked up heading north, as may have been the case with the British bird(s) between 1983-1991.

This exciting swift, often touted as the fastest bird in the world in level flight, remains an exhilarating prospect for those who scrutinise summer skies in anticipation. The high proportion of inland records clearly offers hope for all.

Pallid Swift *Apus pallidus*

(Shelley, 1870). Locally common throughout Mediterranean basin from Iberia to Greece, but rare or absent from many regions. Recently discovered breeding in France on Atlantic coast. A colony was discovered in Switzerland in 1987. Outside Europe, breeds locally from Mauritania and Atlantic Islands across Africa and Middle East to Arabian peninsula and southern Iran. Most winter in the north-African tropics, but some remain in southern Europe.

Polytypic, three subspecies. Racial attribution of records is undetermined. Most have presumably been of the taxon *brehmorum* Hartert, 1901 found across much of European breeding range, Canary Islands, Madeira and coasts of North Africa. There is a record of *illyricus* Tschusi, 1907, of the eastern Adriatic, from Denmark in 1993 (Thorup 2001).

Rare but regular vagrant from European or North African populations, increasingly prone to influxes during the late autumn during the last decade. True status clouded by difficulty of identification.

Status: This problematic species is not only mercurial in its visitations but also most difficult to identify. The 70 accepted birds to the end of 2007 comprise 66 in Britain and four in Ireland. All but one have been since 1978.

Historical review: The first occurred long ago in 1913 when one was collected at St. John's Point, Co. Down, on 30th October.

There was a 65-year gap between the first and second records for Britain and Ireland. The second was well received at Stodmarsh, Kent, from 13th-21st May 1978. The next, in 1983, was also a spring bird in the south at Farlington Marshes, Hampshire, on 20th May. But, it was a warm southerly airflow, which coincided with deposits of Saharan dust, in early November 1984 that laid the foundations for the arrival patterns that we now expect of the species. Four were identified between 10th-14th November, including two together at Portland and singles in Kent and Pembrokeshire.

The next arrival was not until 1992 and involved a three-day late July bird off Flamborough Head that frustrated observers with its erratic appearances. Four more followed between 1993-1997. In 1993 there was a late-July bird in Norfolk and then an early-August bird in Dublin (which was moribund and died later the same day). A moribund adult male on Orkney on 26th October 1996 was thought most probably to be of the subspecies *brehmorum* (McGowan 2002). There was another Norfolk bird in late August 1997. A mini spring influx of three birds took place in 1998 with records spread between Kent, Scilly and Co. Louth (one in April and two in May) supplemented by a late-summer sighting in Derbyshire.

It was in 1999 that the conditions for occurrence hinted at some 15 years earlier were finally confirmed. Another mild southerly airflow in the late-autumn period delivered 12 birds to the east coast between Suffolk and Borders from 24th to 31st October. Another, tellingly a juvenile, was taken into care on 5th November. It was ringed and released 10 days later: the first Pallid Swift to be ringed in Britain (Preston 1999).

The vectors for autumn arrival were again conducive in 2001. Ten were accepted between 2nd-31st October, nearly all along the east coast between Shetland and Kent (except for one in Pembrokeshire). Earlier in the year there had been singles in late spring and summer, yielding an annual total of 12 birds. Fourteen more in 2004 arrived in a similar date pattern (12 between 15th October-4th November). These were more widespread geographically with birds mostly between Kent and Northeast Scotland and singles over Scilly, Cornwall and Dorset. Nine in 2005 included an arrival of eight more over the east coast between Norfolk and Northumberland between 30th October-5th November. No fewer than 42 birds have been accepted between 2001-2007, including a juvenile found moribund at New Romney, Kent, on 26th October 2006 (Walker 2006).

When and when: The nine spring records have a southwest and south bias, implying overshoots from Iberia or northwest Africa: three have been from Scilly and two from Kent, with singles from Co. Wicklow, Co. Louth, Hampshire and an inland bird in North Yorkshire. The earliest spring record was on Bryher – 25th-26th March 2002 – followed closely by one at Bray, Co. Wicklow – 27th-28th March 2006. Two others have come in late April and five in May.

Nine more records span the period mid-June to late August (four 12th June-3rd July, three 19th-23rd July and two 3rd-9th August). These singletons are likely to involve non-breeding birds moving around with Common Swifts. Such sightings are well spread: three come from Scilly plus singles in Cornwall and Co. Dublin; east coast birds have been seen in Norfolk, East Yorkshire and Shetland; and an inland bird in Derbyshire.

It is late autumn that dominates the archive for this confounding swift. Sandwiched between the last of the summer and the first cluster of October birds, one at Mundesley, Norfolk, on 28th August 1997 would appear to be exceptionally early. Four between late September and early October include singles from Dorset, Pembrokeshire, Shetland and Suffolk, with two in 2001.

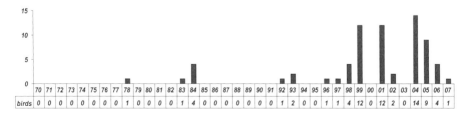

Figure 9: Annual numbers of Pallid Swifts, 1970-2007.

The overwhelming majority of late-autumn records are crammed into the period 15th-31st October. These are from the recent influxes. Nearly all are east coast birds: nine in Norfolk; four each in Kent and Suffolk; five in East Yorkshire; four in Northumberland; and two or less from eight counties between Orkney and Essex. Away from these areas there have been singles in Dorset, Scilly and Co. Down. The November records are slightly more dispersed: three in Dorset and singles in Cornwall and Pembrokeshire; four east coast birds from Norfolk to Kent, two of them from Norfolk. The latest record was from Stanpit Marsh, Dorset, on 22nd November 2002. The later birds exhibit a more southerly pattern of occurrence than earlier arrivals in October.

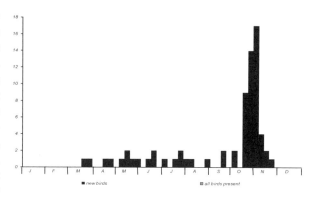

Figure 10: Timing of British and Irish Pallid Swift records, 1913-2007.

Discussion: The recent perception was that any pale swift seen here in late autumn off the back of warm southerly airflows would most probably be a Pallid. This notion is becoming increasingly sullied by the confusion potential of late juvenile Common Swifts of the nominate race and Asian Common Swift ssp. *pekinensis*. The pitfall of the young nominate birds has recently been brought into focus courtesy of well-watched and photographed birds at Filey in late September 2004 and Shetland in early August 2008, while *pekinensis* is starting to attract attention.

Asian Common Swift breeds much nearer than other vagrant Asian swifts that have occurred in Britain. It winters in southern Africa, presumably skirting Europe on the way south. This Asian taxon is not on the *British List* (Kehoe 2006), but has occasionally been suspected in Britain (for example, two records from the Isles of Scilly were tentatively suggested to be this taxon; Flood *et al* 2007). A confirmed record would presumably require a ringing recovery from the core range of *pekinensis* or molecular support, but this taxon is a very likely potential vagrant and could easily be mistaken for Pallid Swift. The possible late-autumn mix of aerial confusion highlights the likelihood that some previously accepted records of Pallid Swift might have been identified incorrectly. The challenge of eliminating *pekinensis* when claiming vagrant Pallid in the autumn has been vastly underestimated (Lewington 1999; Garner 2006). It seems likely that when a review of accepted Pallid Swift records takes place that some will fall short of acceptable. It may well be that only those that are well documented with good quality images will be considered safe enough to retain a place in the national archives.

In contrast to Britain, Pallid Swift remains rare elsewhere in northern Europe. The exception is Sweden, where 19 were recorded up to the end of 2006: the first in 1991 and the rest since 1999, including five each in 2004 and 2005; 14 have been in October and November. Elsewhere, countries such as the Netherlands (20th-21st October 2006) and Finland (two in late October 2004) have only recently logged their first. The seemingly disproportionate number of British records is driven by late autumn displacement from the Mediterranean.

A recent northerly range extension has been documented (Hagemeijer and Blair 1997). In France breeding first took place on Corsica in 1932, on the mainland in 1950, and from the 1990s at Biarritz on the Atlantic coast. In Switzerland a – possibly long-established – colony at Locarno was discovered as recently as 1987 *(BWP)*. Dierschke (2001) analysed records from north of the breeding areas in Europe and found that sightings in western Europe occurred from spring to late autumn, linked with the western breeding populations.

A few spurious records may well inflate the increase in sightings in Britain, but it is clear that more are occurring. It seems unlikely that this is purely as a result of observer ability and it may well be that range expansion has played a part. Weather conditions would appear to have been a principal component, with late southerly airflows prevalent in the last decade. Pallid Swift differs from Common Swift in being double brooded. Second clutches in Iberia are laid in late July and young fledge in early October *(BWP)*. Recent influxes have coincided with the fledging period of Pallid and at a time when most Common Swifts have already passed through southern Europe. The only bird ringed in Britain was aged as a juvenile, as was a moribund late-October bird in Kent.

The recent glut of records has perhaps lulled observers into expecting Pallid Swifts in late autumn, but clearly the season also offers up alternative taxa for elimination. The identification conflicts are unlikely to be resolved in the near future.

Pacific Swift Apus pacificus

(Latham, 1802). Breeds from Siberia eastwards to Kamchatka and Japan, south to Himalayas and northern southeast Asia. Nominate pacificus, breeds from Siberia to northern China and Japan. A long-distance migrant that winters in Indonesia, Malaysia, the Philippines, Melanesia, Australia, New Zealand and south to sub-Antarctic Macquarie Island.

Polytypic, four subspecies. The first British record was attributed to the highly migratory *pacificus* (Latham, 1802) and subsequent records have most likely been this subspecies. Three further races are all restricted in range: *kanoi* (Yamashina, 1942), southeast Tibet, southern China, and Taiwan; *leuconyx* (Blyth, 1845), outer Himalayas and hills of Assam; *cooki* (Harington, 1913), southeast Asia, east from eastern Burma, south of Kanoi *(BWP)*.

Julian R. Hough

Status: There are five British records of this long-distance migrant, one of which (in 2008) is still to be assessed.

1981 At sea: Sea area Humber Leman Bank, 53°06'N 02°12'E, about 45 km ENE of Happisburgh, Norfolk, first-year caught exhausted 19th June. Released Beccles, Suffolk, on the same day and seen in Shadingfield area, Suffolk, on 20th
1993 Norfolk: Cley 30th May
1995 Northamptonshire: Daventry Reservoir 16th July
2005 East Yorkshire: Spurn south on 1st July
2008 East Yorkshire: Kilnsea south on 22nd June and again over Spurn on 26th June*

Of these only the Norfolk bird in 1993 remained long enough to attract an audience. Conveniently found on a spring Bank Holiday several hundred birders were able to enjoy views of the bird, though many were less fortunate.

At 10.45am on Bank Holiday Sunday, 30 May 1993, I arrived back at Cley Coastguards' car park Norfolk, after an early morning sojourn on Blakeney Point to enjoy once again the Desert Warbler which had arrived there on the previous Thursday. Local birder Alan Brown was standing in the car park talking to Jackie and David Bridges about a strange swift he had just seen over the adjacent Cley reserve.

I joined them, and Alan explained that he had just been in the North Hide and had spent ten minutes watching an odd swift with a white rump. Mindful of the relative frequency of partial albino Common Swifts, he felt quite sure that it was no more than that, and Jackie and David set out for Blakeney Point and the Desert Warbler. I was anxious to see the bird, however, so, accompanied by Alan, I hurried down to the North Hide in case it was still there. It was indeed flying around over the North Scrape with Common Swifts. However, its very clearly demarcated white rump instantly suggested that this was more than just a Common Swift exhibiting partial albinism. It was certainly very close to the Common Swifts in size, but its wings were considerably more scythe-shaped and its tail was noticeably longer and more deeply forked. This was clearly very interesting indeed!

Without further delay, I called Richard Millington on my mobile phone and suggested that he hasten to the North Hide from the other side of the reserve, and Alan headed back to the car park to alert other birders.

There were really only two possibilities for the bird's identity; it had to be either a White-rumped Swift A. caffer or a Pacific Swift A. pacificus, although the former is not on the British list and the latter has not occurred naturally in Britain. It was very slightly larger than the Common Swifts, so that seemed to rule out the former, but Pacific Swift should show indistinct greyish feather-edgings to the body feathers. Although I knew that these greyish scaly markings were very difficult to see, they were not visible, even though the bird was often no more than 80 yards away in quite good light.

Despite having injured his foot rushing to the Desert Warbler three evenings previously, an anxious Richard Millington was soon in the hide, and other observers quickly began to gather. Within a minute or two, the bird obligingly did a fly-past just 25 yards in front of the hide, and it was clear that it did have faint grey scaling on both the upperbody and the underbody. It was a Pacific Swift!

All hell now let loose! After a quick mobile phone call to Hazel Millington, the news was on Birdline by 11.00am and we called everybody we could think of. Within minutes, birders from all over the country were converging on Cley Norfolk Naturalists' Trust Reserve!

The bird remained flying around the reserve with Common Swifts until about 4.10pm when it disappeared. By this time, several hundred birders had seen it, but other, less fortunate ones were still arriving from further afield well into the evening.

It was not seen again.

Steve Gantlett. Birding World 6:190-191.

Discussion: There are just three other European records, all from Sweden (6th July 1999, 30th July 2005, 19th August 2007).

The range extends northwest into the Urals/Western Siberia region where numbers are generally low, but very common locally on the Yenisey. Vagrants have been recorded west to the Urals (Ryabitsev 2008). Nominate *pacificus* is a long-distance migrant that, not surprisingly, shows similarities in arrival pattern with White-throated Needletail in Europe, with all records from late spring and early summer. The locations and timing suggests that they are spring overshoots.

Two have been observed passing through Spurn – the premier observation point for observing Common Swift movements in Britain – representing just reward for the effort of observers in counting Common Swifts there each year. Over 340,000 Common Swifts have passed through the peninsula since 1999. The joint-record day count for the site is 20,000 birds on 23rd June 1970 and 25th June 1989.

Little Swift Apus affinis

(JE Gray, 1830). Isolated population in northwest Africa increasing and expanding in Morocco and into Spain. Breeds locally and discontinuously in Middle East from Israel to southeast Iran and north along the Euphrates to southeast Turkey. Largely resident. Some Middle East populations migratory. Elsewhere, resident or dispersive throughout sub-Saharan Africa and Indian subcontinent to Sri Lanka.

Polytypic, six subspecies. The racial attribution of occurrences here is undetermined. Nominate *affinis* (Gray, 1832) of peninsular India south of Himalayas and coastal East Africa is closely similar to *galilejensis* (Antinori, 1855) of northwest Africa south to Niger, Chad, central Sudan, northern Somalia and Middle East east to Uzbekistan and western Pakistan. Nominate *affinis* has a slightly darker forehead and sides of crown, white rump slightly narrower, white throat patch slightly smaller, and under tail-coverts and undersurface of tail slightly darker. *A. a. aerobates* of West and central Africa slightly blacker on mantle, back, belly and wing-coverts than both nominate *affinis* and *galilejensis (BWP)*.

Rare vagrant from North Africa, formerly very rare but becoming increasingly regular with most records in spring.

Status: Formerly an extreme rarity but several accessible birds in recent years have eroded its elusive nature here. Twenty-three records: 22 in Britain and one in Ireland.

Historical review: The first was on Cape Clear on 12th June 1967.

On the evening of 12th June 1967, on Cape Clear Island, Co. Cork, I was sitting on top of a steep ridge overlooking Cummer, a narrow col situated between the north and south harbours. The soft evening light was directly behind me and it was perfectly calm. I was casually watching five or six Swallows Hirundo rustica, a House Martin Delichon urbica and five Swifts Apus apus which were hawking for insects through the col, when I spotted a swift with a gleaming white throat, contrasting with black under-parts. This bird passed several times about ten feet above and 30 yards away from me. Other features distinguishing it from the Swifts were its shorter wings and less deliberate, more 'fluttery' wing-beats. The bird then flew lower, passing 30 feet below me and 60 yards away, and I saw that it had a very marked, square, white rump and that its upper-parts were a glossier black than those of the Swifts. I watched it for about five minutes as it hawked back-and-forth sometimes below me and sometimes above me, but it apparently departed (along with the Swifts) while I was busy writing field-notes, for only the hirundines were present when I tried to relocate it.

J. T. R. Sharrock. *British Birds 61:160-162.*

The next were two birds in May 1981: Skewjack, Cornwall, on 16th May and Skokholm from 31st May-1st June. Five more in the 1980s brought the total for the decade to seven, but despite the increasing numbers of records, the typically brief stays or inaccessible locations ensured it remained a highly sought-after species.

A one-day bird on 1st November 1991 on Fair Isle did little to alleviate demand for an obliging individual. The thirst for an accessible record was finally sated in 1997 by a hugely popular performer on the Isle of Wight, at Bembridge and Foreland on 5th May and in the Brading Marsh area on 6th May. Two other brief spring birds in 1997 conformed to fly-over form, but three in a year constituted a new record. The following year equalled this tally. In 1998 two

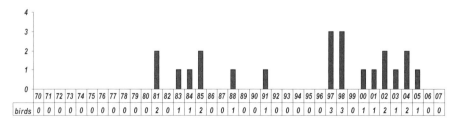

Figure 11 Annual numbers of Little Swifts, 1970-2007.

characteristically brief birds were located on the same day in Cornwall and Cleveland, but these did little more than enforce the perception of a transient visitant to our skies. However, one at Barton-upon-Humber, Lincolnshire, on 26th June lasted long enough for many admirers to divert attention away from the World Cup. Amazingly, after so many years of chasing shadows two had been collectable in as many years. The 1990s total of seven birds ended on par with the previous decade, but the shroud of elusiveness had been discarded.

A one-day bird in 2000 kicked off the Millennium, but a show stopper in 2001 at Netherfield Lagoons, Nottinghamshire, from 26th-29th May finally allowed those with less urgency the opportunity to see the species on British soil. Obligingly it roosted on a brick viaduct over the River Trent each evening and its showy performance and central location ensured it attracted large numbers of admirers during its extended stay (for a Little Swift). Eight have been found between 2000 and 2006, six of which were one-day birds. The last was again a crowd pleaser, this time at Cromer, Norfolk, from 12th-13th November 2005.

Following the review period two reports in 2008 both came from Yorkshire, where one passed south at Spurn on 26th June and one was well watched over Old Moor, South Yorkshire, for much of the day on 2nd July. Possibly the same bird was involved in both sightings.

Where: Records are well spread, though most have been in southern and eastern England. In the southwest, Cornwall accounts for three records, Dorset and Scilly two apiece and there is one from Devon. On the east coast, Lincolnshire has produced two birds and Norfolk, East Yorkshire and Cleveland each have a single. Hampshire and the Isle of Wight have one each. The two Welsh records were

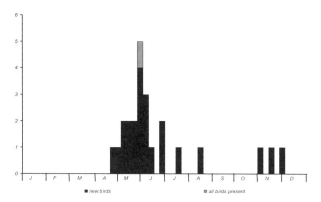

■ *new birds*   ■ *all birds present*

Figure 12: Timing of British and Irish Little Swift records, 1967-2007.

both in the south (Gwent and Pembrokeshire). The sole record for the northwest was from Seaforth, Liverpool. In Scotland two have reached Shetland and one Fife.

When: Brief stays are the norm for this mobile species. Spring has produced most records (17 birds). The earliest was on St. Mary's on 28th April 2003. Eleven in May and five to 12th June constitute a slight peak.

Two late June records both came from Lincolnshire. Single mid-July and mid-August records complete the quartet of summer records. Autumn birds are spread through November. The latest was at Studland, Dorset, on 26th November 1983.

Discussion: Little Swifts breed in most of Africa and southern Asia, where populations are mostly considered residents. The race *galilejensis* occurs from northwest Africa eastwards to Kashmir. It is partially migratory in northwest Africa. Some move south to the southern edge of the Sahara and may stray west to the Canary Islands and Madeira. Most, if not all, British records presumably relate to birds from this region and subspecies, as all others are sedentary.

The first ever breeding in Europe occurred when a pair presumably nested in Cadiz province, Spain, in 1996. Breeding was confirmed in 2000 in one location and from at least three locations from 2004 onwards, two of which were coastal and one inland, including a small colony. With a growing population in Spain it seems likely that more will visit Britain.

Extralimital records for northern Europe are rare, but include three from Sweden (9th June 1979, 16th August 1985 and 8th October 1988), two from the Netherlands (17th May 2001 and 20th November 2007), one from Guernsey, Channel Islands (22nd April 2000) and one found dead in Germany (9th November 2002). Interestingly, two of the four records from Italy have been since 2006.

Despite the recent increase in records Little Swift remains a devilishly difficult bird to get to grips with in Britain, being rare and typically brief and mobile. Those who have seen one of the obliging birds will be grateful for that opportunity, as predicting the location or date of the next long stayer would be little more than guesswork.

Belted Kingfisher *Megaceryle alcyon*

(Linnaeus, 1758). One of the most widespread landbirds in North America, breeds from southern Alaska and Labrador south to Mexico. Northern breeders winter south to Central America and the Caribbean, where common, and northern South America, where rare.

Monotypic.

Exceptionally rare visitor from the Nearctic, with most records in Ireland and the southwest.

Status: Six individuals of this stunning Nearctic vagrant: three each in Britain and Ireland.

1908 Cornwall: Sladesbridge female shot November

1978 Co. Mayo: Bunree River, near Ballina 1st-winter female 10th December-3rd February 1979, when shot

1979 Cornwall: Sladesbridge 1st-winter male 2nd October-June 1980; probably same Boscathnoe Reservoir adult male 23rd-29th August 1980

1980 Co. Down: Dundrum Bay female shot 12th October

1984 Co. Clare: Ballyvaughan female 28th October-early December

1985 Co. Tipperary: Near Killaloe female 6th February-8th March; same as 1984 individual

2005 Staffordshire: Tixall 1st-summer male 1st April; also in East Yorkshire and Northeast Scotland

2005 East Yorkshire: Eastrington Ponds 1st-summer male 2nd April; same as Staffordshire

2005 Northeast Scotland: Peterculter 1st-summer male 4th-8th April; same as Staffordshire

Discussion: Three birds elected to put in extended stays, particularly those in 1979-1980 and 1984-1985; the 2005 bird did not linger for so long, but was warmly welcomed by a new generation of birdwatchers.

The 2005 individual was discovered by a birder on his local patch on 1st April. The date of the find ensured initial skepticism, but this was superseded by panic once a photograph was placed on birdwatching websites and the veracity of the find was confirmed. Many saw the bird before darkness fell. For those unable or unwilling to travel immediately, anticipation turned to despair the following morning as it became apparent that the bird had departed overnight. Amazingly, later that morning it was relocated at a small nature reserve in East Yorkshire, some 135 km northeast of the Staffordshire sighting, but it showed only to its finders before vanishing again ahead of the descending crowds. Two days later it was found again, this time 385 km farther north in Northeast Scotland. This exciting trio of finds may represent the first instance of an American landbird being tracked heading northwards in the Western Palearctic.

Belted Kingfisher has been a special bird for our family since 2nd December 1979, when Ian (then 8 years old) was taken by his father, Roger, on his first 'twitch', to see the second for Britain, which spent that winter in Cornwall. Probably the most memorable thing of that day was the wonderful rattling call of the bird, which often assisted in its relocation.

It was therefore with incredulity that Roger heard that call again at Tixall in Staffordshire, just after midday on 1st April 2005, adjacent to the River Sow and the Shugborough Estate. "Surely not?" he thought as he recognised the call but, within a second, the bird came into view, with that familiar faltering, Jay-like flight. The bluish-grey and white appearance and long, dagger-like bill left him in no doubt – it was a Belted Kingfisher! It turned back towards the stunned observer and then perched, which allowed him to take one vital photograph with his digital camera, which, fortuitously, he had with him. Then the panic set in! He telephoned his wife Gill, to ask her to bring the camcorder to him, and then called Ian (at work in Aberdeen), whose advice was to the point – "get the news out!".

The bird flew closer, and perched above the canal before being startled by a narrowboat, whereupon it flew, calling loudly, towards Shugborough. After a frantic dash, Gill met Roger in a state of shock on the tow path, but the bird had flown. Then there were several tense phone calls, whilst parking arrangements were made for the impending invasion of birders. However, the greatest problem was that it was April Fool's Day and, clearly, it would be viewed as a hoax. Ian phoned Roger again: "You do know the date? You've got to get the photo on the internet fast". So it was over to Gill to play a vital role whilst Roger stayed on site to await the masses. She took the camera home and emailed the photograph to Ian, who uploaded it to Surfbirds and BirdGuides, with the note that this was "NOT an April Fool!".

With the urgent work done, Gill returned to the scene to help Roger in the search for the bird, which he was convinced would be found at the lake in the grounds of the Shugborough Estate. Fortunately, it was relocated there shortly after 5.00pm, and many birders managed to see this remarkable addition to the Staffordshire county list. Unfortunately, however, the bird was not to be found the next morning, much to the disappointment of

the assembled hordes of birders, but then news came through that the bird had been relocated at Eastrington Nature Reserve, some 85 miles to the northeast, near Goole in East
Yorkshire. Presumably, overnight or early that morning, the bird had followed the Trent
valley from Shugborough to the Humber Estuary. Neil and Caroline Smith, while walking their dogs as usual at Eastrington Ponds, spotted the unmistakable bird at 10.45am as
it flew and perched in plain view just 20 metres away raising its ragged crest as it called.
After 15 minutes of watching the water below, the bird swooped off its perch as a train
passed, and flew off into dense over at the north-eastern end of the reserve. Upon returning home, Neil realised the significance of the sighting and broadcast the news, but again
the masses were to be disappointed, as the bird was not seen again in Yorkshire. It seemed
as though the 'April Fool bird' had gained the last laugh.

 So it was with some considerable surprise that Ian took a phone call at 5.30pm on
Tuesday 5th April from Harry Scott, to say that he had received a report of a Belted Kingfisher on the River Dee near Peterculter in Aberdeenshire, and could we get a few birders out to the river to check it out? On arriving at the site after 6.00pm, Ian met Ewan
Weston, who had been looking for the bird all day. Apparently, it had been seen the evening before by Katy Landsman and her boyfriend, and its identity confirmed at 9.00am
the next morning by her mother, Joyce, when it was seen in the same areas by the island in
the river at the bottom of Pittengullies Brae. However, the bird had not been seen since.

 Just as he was starting to curse the ultimate 'one that got away', Ian noticed a distinctive blue and white blob perched on wires above the river – it was still present! Harry and
Ian immediately began phoning the news out, and about 30 local birders managed to see
the bird that evening, as it roosted on a low branch overhanging the river.

 The bird continued to delight many admirers (including television news crews) for the
next three days, and it was seen to go to roost in the same trees on the evening of Friday
8th April. Next morning, at dawn, the bird left its roost and was seen flying north over
the river at 5.50am, and off to the north. Unfortunately, it was not seen again, much to
the chagrin of the assembled throng of birders who had been unable to visit the site until
the weekend.

R. Broadbent, G. Broadbent & I. Broadbent. Birding World 18: 159-163.

Discussion: Elsewhere in the Western Palearctic there are six records from the Azores (March
1899, 21st October 1996, 2nd October-9th December 2001, 9th December 2005 to 22nd March
2006, 5th March 2007 and 17th October 2007), five from Iceland (late September 1901, 17th-
18th May 1998 and presumably same July to 15th September 1998, 18th-24th June 1998, found
exhausted on 24th February 2002 died next day and 10th-12th October 2003), and one from
the Netherlands (shot and collected on 17th December 1899; specimen lost in WWII but photograph survived).

 The range of dates for all Western Palearctic records varies. Of the 18, there is one September record from Iceland, seven have been found in October (three on the Azores, two in Ireland
and one in Britain) and four between November-December (one each on the Azores, Ireland,
Britain and the Netherlands). Southward autumn migration in North America takes place from

mid-September onwards and continues through November. Offshore ship records have been reported at distances greater than 645 km from the Atlantic coast (Scholander 1955). The dates of the autumn records suggest a transatlantic crossing off the back of autumnal Atlantic weather systems.

Six between February and June comprise two March birds from the Azores, one from Britain and three from Iceland. Perhaps these had crossed during the previous autumn and were picked up moving north in the spring after wintering farther south; certainly the 2005 British record seemed to be doing just that.

Blue-cheeked Bee-eater Merops persicus

Linnaeus, 1758. Nominate persicus breeds from Egypt and southeast Turkey east to Kazakhstan, northwest India and Afghanistan; winters mainly in tropical East Africa. The race chrysocerus is found discontinuously from Morocco to Algeria and from Senegal to Lake Chad; winters in sub-Saharan West Africa.

Polytypic, two subspecies. All records have been of nominate *persicus* Pallas, 1773. Adult *chrysocercus* shows less white on forehead than *persicus*, sometimes virtually none, pale blue supercilium and pale blue band on forehead narrower, and often no white below greenish-black ear coverts. Also, the tail streamers are longer *(BWP)*. Blue-cheeked forms a superspecies with Blue-tailed Bee-eater *M. philippinus* Linnaeus, 1766 from India and south-east Asia to New Guinea. These closely related species do not hybridise on breeding grounds where they meet in north-west India and are generally regarded as separate species (Fry *et al* 1992).

Very rare vagrant from southwest Asia and the Middle East.

Status: Nine records of eight birds, all in Britain.

1921 **Scilly:** St. Mary's adult shot 13th July
1951 **Scilly:** St. Agnes adult 22nd June
1982 **Cambridgeshire:** Peterborough adult 17th September
1987 **Devon:** Otter Estuary 30th June-2nd July
1989 **Cornwall:** Kennack Sands and Cadgwith 1st June
1989 **East Yorkshire:** Cowden 8th-10th July, possibly since 25th June; also in Lincolnshire
1989 **Lincolnshire:** Leverton Marsh 12th July; same as East Yorkshire individual
1989 **Kent:** Church Hougham 18th July
1997 **Shetland:** Bressay, Asta, Tingwall Valley and Lerwick area, Mainland 20th June 3rd July

The first two were on Scilly, the second of which highlights that rarities can appear from nowhere and sometimes at the strangest of times.

> *On June 22nd, 1951, having a visitor staying with me on St Agnes (Isles of Scilly), I went up the lane before breakfast to get the morning's milk. Something skimmed across the path which registered as strange…only a glimpse but it was definitely odd. I wondered if*

it could have been a Starling looking greener than usual, or a pale Swallow with a light sheen on it (the flight suggested something of the Swallow tribe). However, I had to go back and get breakfast for my guest, and leave the mystery for a moment.

While we were breakfasting, a neighbour came in to say that Mr Lewis Hicks had seen a strange and most wonderful bird. So we left breakfast standing and rushed out; collected Mr Hicks and went with him to the fields where he had seen it. (He told me afterwards that at first he could not believe his eyes, and went to fetch his wife to 'come and tell me if you see what I see!') Fairly soon we saw our quarry afar off, and presently it perched at some distance, but in good view. There was no doubt that it was a bee-eater; one knew it from pictures, and the curved bill and elongated tail-feathers could be clearly seen. It seemed to be returning fairly regularly to one spot on the telegraph wires, so I stalked gently up the lane to a position within 20 feet or so of where it came back to perch. It made frequent sallies after insects and brought them back to the wire to eat. ('Sitting there eating up my bees as fast as it can!' said Mr Hicks, who has hives…but we could not be sure that they were truly bees that it was catching.) Once, its prey escaped from its bill, and it did a lightning dive and turned to recapture it and bring it back to the wire.

About four hours after it was first seen it disappeared, and a party of bird-watchers, hastily summoned from St. Mary's by 'phone, spent the afternoon searching St. Agnes in vain.

Hilda M. Quick. British Birds 45:225

Discussion: This is one of the few mega rarities that is best sought out during the height of summer. All British records have been between 1st June and 18th July; the exception is an extraordinary single-observer sighting of an adult one foggy morning in a lorry park in central Peterborough in September. This pattern concurs with most records elsewhere in Europe, although several have been in the autumn.

As in Britain sightings of this attractive species elsewhere in northern Europe remain rare. In Scandinavia there are four from Sweden (2nd and 9th August 1961, 2nd July 1978, 27th May 1985 and 8th-11th June 1998), three from Finland (9th July 1991, 26th May 1996, 30th May 1998), three from Denmark (21st-22nd June 1989, 29th June 1993, 6th-12th July 1998) and one from Norway (22nd June 1998).

There are four from Germany (19th June 1993, 30th May 1996, 14th July 1997, 18th June 1998) and two from the Netherlands (30th September 1961 and 18th May 1998) and one of the French records is from Ouessant (15th October 1993). In Spain it is equally rare with just four records of six birds for the mainland and Balearics plus an old record in the Canary Islands.

Nominate *persicus* is a long-distance migrant, returning from its eastern Africa winter quarters from mid-March to early June, but mostly in April and early May. The British and north European records comprise individuals that are overshoots to northwest Europe. The closest breeding population to Britain is in northwest Africa and involves birds of the race *chrysocerus*. This relatively small population of short-distance migrants is less likely as a vagrant.

Three sightings in 1989 were considered to relate to different individuals and were thought to involve a small influx (one occurred in Denmark at the same time). The East Yorkshire bird

Table 7: Dated records of Blue-cheeked Bee-eater in northern Europe (n=18).

Month	May	June	July	August	September	October
Number	5	6	4	1	1	1

was exceptionally well attended, being the first-ever collectable individual. The Shetland bird of 1997 enticed more than 100 observers to the islands. Both birds performed well for their admirers, but the absence of a bird for over a decade has now reinstated this beautiful species to the ranks of 'most highly desired' for a new generation of birders.

European Roller *Coracias garrulus*

Linnaeus, 1758. Declining, yet widespread and numerous in northwest Africa and Spain. In eastern Europe occurs locally north to Estonia and east to the Ukraine, but nowhere common. Remains common from Turkey and southern Russia to southern Urals, southwest Siberia, southern Kazakhstan and western China. Winters locally in equatorial West Africa, but most in east Africa from Kenya to Zimbabwe. Race semenowi breeds Iran, Afghanistan and northern Pakistan, winters East Africa.

Polytypic, two subspecies. All records have been of nominate *garrulus* Linnaeus, 1758. The southeastern race *semenowi* is smaller in the western part of the range and clinally larger towards east. The colour of *semenowi* similar to nominate *garrulus*, but blue of head, neck, underparts and upperwing slightly paler, especially on the throat. It also sometimes more greenish and the bright violet-blue band on shorter lesser upperwing coverts of the adult is narrower *(BWP)*.

Rare vagrant from southern Europe, becoming increasingly rare and erratic in occurrence.

Status: Prior to 1950 there were 207 birds recorded: 198 in Britain and nine in Ireland (Keith Naylor *pers comm*). Since 1950 there have been 112 British and five Irish records.

Historical review: The first British record was killed near Crostwick, Norfolk, on 14th May 1664. European Roller is another one of the few species for which records prior to the modern recording era were more frequent than now. Unfortunately most of the early records on our shores met the same fate as other exotics and were shot by collectors.

> *On July 31st, 1907, a gamekeeper in St. Leonard's Forest observed one of these birds, a fine male, flying in the forest near Colgate. So conspicuous a bird can hardly escape notice, nor the desire of man to capture it, and the specimen in question met with the usual fate. I saw it in the flesh on August 2nd, and it has now been added to my collection.*
>
> *Formerly the Roller was a regular though scarce summer visitor to this part of England, but now its appearance is distinctly rare. As far as I can ascertain this is the only example that has been killed during the past ten years in West Sussex. The throat and crop contained several small beetles.*
>
> **J. G. Millais**. British Birds 1:189

Table 8: Monthly occurrence of dated* pre-1950 European Roller records in Britain & Ireland.

Month	Jan	Feb	Mar	Apr	May	June	July	Aug	Sept	Oct	Nov
Birds	1	2	0	5	24	30	12	8	33	28	3
%	0	1	0	2	12	14	6	4	16	14	1

* 29% (61 birds) were undated (one in winter, one in spring, four in summer, four in autumn and 51 with no date)

Table 9: Average numbers of European Roller by 5-year period since 1960 in Britain & Ireland.

Period	1960-1964	1965-1969	1970-1974	1975-1979	1980-1984	1985-1989	1990-1994	1995-1999	2000-2004	2005-2007
Average	1.4	3.4	2.8	3.8	2.6	1.0	2.0	0.6	0.8	2.3

Most old records are from east coast counties, for example: 23 in Norfolk, 16 in Suffolk, 15 in Northumberland and 11 Yorkshire. Along the south coast 11 came from Sussex. Three winter claims during this period are quite exceptional: one shot at Waxham, Norfolk, in February 1824; one near Sandon, Hertfordshire, on 6th February 1932; and one seen at Deptford, Greater London, on 31st January 1945. The two sight records struggle to convince, and the dates of the shot bird in Norfolk and one picked up dead at Eday, Orkney, in winter 1874 must raise questions regarding their provenance. A curious record, the provenance of which is also uncertain, involved a pair present near Skelton, Cleveland, on 26th June 1847, one of which was subsequently killed and found to be a female with eggs in the oviduct (Mather, 1986).

Close to one third (29 per cent) of old records are undated, but for those that were the majority were in May and June (26 per cent) and September and October (30 per cent); see Table 8. Seemingly spurious late winter records aside, the earliest from this period was from near Budleigh Salterton, Devon, on 11th April 1923, although two subsequent records have since pre-dated this sighting (see below). Two November records represent the latest-ever arrival dates for the species: one caught exhausted at Rainham Marsh, Gillingham, Kent, on 8th November 1888 and two seen at Westray, Orkney, on 10th November 1890.

Eighteen were recorded in the 1950s, including four each in 1956 and 1959. Numbers remained similar in the 1960s with 24 birds, including five in 1969 and six in 1968. One near Retford, Nottinghamshire, from 28th June-22nd July 1966 was considered to have possibly been a bird that escaped from Twycross Zoo, Leicestershire, in June of that year (even birds at the right time year can be suspect).

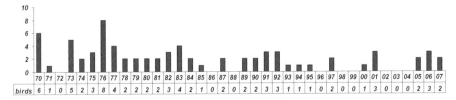

Figure 13: Annual numbers of European Rollers, 1970-2007.

The 1970s yielded the largest numbers to arrive here during recent times (33 birds). These included a record eight in 1976 (two 13th-16th June, four 1st-20th July, plus singles in August and October) and six in 1970 (all 19th May-28th June). This pinnacle in fortunes was short lived. Numbers declined in the 1980s, a decade that produced a modest 18 birds including a peak of four in 1983. The 1990s were worse still, with just 13 birds recorded. Peaks of just three in 1991 and 1992 reflected the poor showing.

As the Millennium commenced it appeared that this resplendent species had plunged to its nadir (see Table 9). Unexpectedly, 11 have been recorded since then, with no fewer than seven birds during 2005-2007 delivering the best period for this dazzling species since the early 1990s, including three each in 2001 and 2006. These records bucked the trend of recent times, but for how long will this renaissance continue?

Where: Since 1950 there appears to have been a shifting emphasis of records northwards, with just over a quarter coming from Scotland including six from Shetland and five from Highland. This, coupled with a strong east-coast bias, points towards birds originating in eastern Europe rather than spring overshoots from Iberia.

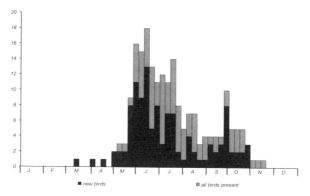

Figure 14: Timing of British and Irish European Roller records 1950-2007.

East Anglia has 15 per cent of records (including nine birds in Norfolk and six in Suffolk), a proportion matched by the southeast (including four each in Kent and Surrey). Surprisingly in light of the Scottish bias, northeast England has produced a mere eight per cent of records.

The southwest has accounted for 18 per cent of records (including four birds each in Dorset, Devon and Scilly); most were late spring arrivals suggesting birds of westerly origins. Wales has produced six per cent, the Northwest seven per cent, the Midlands four per cent and Ireland four per cent.

When: There is a strong late-spring and early-summer bias to records. The three earliest records all occurred during the 1950s: at Mungrisedale, Cumbria, on 15th March 1953; Yetminster, Dorset, on 7th April 1955; and picked up dead at Hook Head Lighthouse, Co. Wexford, on 18th April 1956. The westerly locations suggest that these may have involved birds of Iberian origin. The earliest in more recent times was near Arlesey, Bedfordshire, on 3rd May 1990.

Arrivals peak from late May to mid-June, with some to late July. Protracted stays at this time are indicative of lost wanderers. Late summer records to August are not unusual. Autumn records peak in the last week of September. Just nine have been found in October, the latest a

1st-winter on Skokholm, Pembrokeshire, on 26th October 2001. The second-latest find, at Great Holland, Essex, remained from 24th October-18th November 1968.

Discussion: The decline in Roller numbers has been apparent for 50 years across Europe, but against this backdrop the numbers occurring here showed little variation until the late 1980s (see table 9). Between 1965 and 1985 the species failed to make an appearance only in 1972. Since 1996, absences have been noted in six years.

The European Roller has two strongholds in Europe. The Iberian Peninsula supports modest numbers (up to 2,800 pairs). Farther east, key populations are found in Turkey (up to 60,000 pairs) and Russia (up to 20,000 pairs). This species declined across most of its European range between 1990-2000 (including the important populations in Turkey and Russia), a decline estimated at more than 30 per cent by BirdLife International (2004).

Since the 19th and early 20th centuries there has been a range contraction to the south and east, coupled with a serious population decline. The Polish population declined by more than half between 1970-1990. Similar decreases were observed in other central and eastern European countries (Hagemeijer and Blair 1997). The declines have been linked to agricultural intensification resulting in a loss of large insects and small reptiles, leading to poor chick survival. A study undertaken in southwest Spain (Avilés and Parejo, 2004) found that agricultural practices around nests affected chick mortality. Breeding success and egg productivity were also affected by farming activities.

Unless declines in core populations can be contained to some extent it seems likely that this resplendent species will become an increasingly rare and erratic visitor to our shores. The recent upturn in fortunes may well be a short-lived phenomenon.

Yellow-bellied Sapsucker Sphyrapicus varius
(Linnaeus, 1766). Breeds in much of North America, from southeast Alaska across southern Canada and into northeast USA. Winters in central and southern USA south to Panama and the West Indies.

Monotypic.

Exceptionally rare transatlantic vagrant from the Nearctic. Four Western Palearctic records.

Status: Two records in autumn.

1975 Scilly: Tresco 1st-winter male 26th September-6th October
1988 Co. Cork: Cape Clear 1st-winter female, trapped, 16th-19th October

The autumn of 1975 was exceptional for Nearctic vagrants. The second record was a welcome opportunity for those who missed out on the Scilly individual.

Sunday 16th October on Cape Clear was damp and misty with a light southeasterly breeze.
A new Red-breasted Flycatcher and Yellow-browed Warbler in Cotter's Garden boded

well and persuaded Willie McDowell, Anthony McGeehan, Dennis Weir and myself to spend the rest of the morning working Ballyieragh (the southern third of the island). At about midday, we were returning to the observatory when Anthony, Dennis and I decided to check a thicket about 100m off the High Road. Something small flitting about in the bushes (we never discovered what it was) attracted our attention and we surrounded the thicket to try and sort it out. Suddenly, Dennis said quietly but clearly, "there's a Yellow-bellied Sapsucker in here". After the initial shock, Anthony and I were quickly at his side, but the sapsucker had dropped below the level of the perimeter dry stone wall and was lost to view.

We split up and after five anxious minutes I saw it through a gap in the vegetation. A small woodpecker with speckled grey/brown upperparts, a scarlet crown and a large white longitudinal wing patch. I called the others over but it again disappeared from view. After a further 15 minutes, it reappeared and all of us plus Willie (who had by now returned wondering what had kept us) obtained more complete views.

It was now early afternoon and many of the birders on the island were leaving on the mailboat at 3.00 p.m. so it was essential to find everybody quickly. Willie's strategically placed note had the first birders running down the track towards us within 10 minutes. However, the noise they made on arrival disturbed the sapsucker which promptly flew out of the thicket over the heads of the newcomers. Fortunately, it flew straight into Cotter's Garden where it was quickly relocated and seen by almost everybody on the island. Soon it turned up in a mist net and was taken to the observatory where it was determined to be a first-year female.

The sapsucker was released back into Cotter's Garden. It stayed there for four days, during which time it was seen by nearly 100 British and Irish birders.
Nick Watmough. Birding World 1: 392-393.

Discussion: This short- to medium-distance nocturnal migrant leaves its the breeding grounds in the first half of September and arrives in the wintering areas in early October. It totally vacates its breeding range in winter (Walters *et al* 2002). Females migrate farther south than males and are more numerous in the southern part of the wintering range (Howell 1953).

This is a regular species along the Atlantic coasts of North America, reaching Florida in October. A proportion of the population undertakes a sea crossing to wintering grounds in the West Indies (Walters *et al* 2002). It has been recorded in Bermuda. There were four records on Greenland between 1845 and 1934 (Boertmann 1994).

Despite the apparent unlikelihood of a woodpecker crossing an ocean, Yellow-bellied Sapsucker is clearly susceptible to a transatlantic crossing under the right conditions. Durand (1963) recorded this species on board the RMS *Mauretania* on 8th October 1962, some 640 km east of New York on an eastbound crossing of the Atlantic, along with 10+ Northern Flickers *Colaptes auratus* and a Hairy Woodpecker *Picoides villosus*.

Two of the three additional Western Palearctic records are from Iceland. The first was a spring overshoot (found dead on 5th June 1961) and the other from 7th-13th October 2007. There is also a record of a 1st-year male on Corvo, Azores (11th October-3rd November 2008).

Eastern Phoebe *Sayornis phoebe*

(Latham, 1790). The extensive breeding area extends from the northern Canadian provinces into southeastern USA. The wintering range is confined largely to southeastern USA, with peak numbers along the Gulf Coast. The southern limit extends well into Mexico.

Monotypic.

Very rare transatlantic vagrant with one Western Palearctic sighting.

Status: One British record.
1987 Devon: Lundy 24th-25th April.

Discussion: A report from Slapton Ley, Devon, just two days previously (22nd April 1987) was initially thought to be the same individual. The description of the Slapton bird was deemed to be incompatible with Eastern Phoebe by *BOURC* and the record fell short of a first for Britain (McShane 1996). The Lundy bird was not described well enough to ascertain whether it was a different individual. The Slapton bird was described very well, but the description of the call was incompatible with Eastern Phoebe (call was not even mentioned in the Lundy description). Under *BOURC*'s watertight rules for a first, it was not possible to accept a claim where the description refers to something incompatible with a correct identification. Had the observer of the Slapton bird omitted this from the description the record would have been accepted. The *BBRC* had rejected the Slapton record before it was passed onto the *BOURC*.

Eastern Phoebe is a short-distance diurnal migrant that migrates early in spring. The first arrivals are back in breeding areas in March, although northern limits may not be reached until early May (Weeks and Harmon 1994). The routes of migratory movement for this nondescript flycatcher are not fully understood.

Clearly the species is capable of making sea crossings, as migrants have occurred on St. Pierre (off Newfoundland), Cuba and Bermuda. It seems possible that the Devon bird may have been a spring overshoot, though it could have been a northbound migrant that made landfall in Europe the previous autumn.

Alder Flycatcher *Empidonax alnorum*

Brewster, 1895. Breeds from Alaska across Canada to Newfoundland and south to northwest USA. Winters South America.

Monotypic. Formerly considered conspecific with Willow Flycatcher *E. traillii* and treated as a single species, Traill's Flycatcher, until 1973 (*AOU* 1973). Disentanglement of the two song forms of Traill's Flycatcher resulted in its taxonomic partition into Alder Flycatcher *Empidonax alnorum* Brewster, 1895 for the *fee-bee-o* singing populations and Willow Flycatcher *E. traillii* (Audubon, 1828) for the *fitz-bew* singing populations (Lowther 1999). The taxonomy and nomenclature of both species and the subspecies of Willow Flycatcher were reviewed by Browning (1993).

Very rare vagrant from North America, one record awaiting acceptance.

Status: One record, yet to accepted by the relevant committees.
2008 Cornwall: Nanjizal 1st-winter, 6th-9th October*.

As would be expected, this addition to the *British list* regardless of species pair at the time was hugely popular.

Discussion: All other Western Palearctic records of *Empidonax* flycatchers have come from Iceland. An Acadian Flycatcher *E. virescens* was found dead on 4th November 1967. A Least Flycatcher *E. minimus* flew into a house and was caught by hand in the late afternoon of 6th October 2003. It was taken to a local birder where it was studied and photographed before being released on the morning of the 7th October. Only three days later (10th October 2003) an Alder Flycatcher was found. It was later caught by mist net and following examination released unringed (Kolbeinsson 2003).

The biometrics of the Cornish bird and some of its plumage marks were indicative of Alder Flycatcher. Alder and Willow Flycatcher are notoriously difficult to separate during the autumn, even in the hand; Pyle (1997) proposed two discriminant formulae to separate the species by the shape of the wings. The statistics derived from the Nanjizal bird placed it within the range of Alder Flycatcher. The measurements for both formulae fell outside of the range of 95 per cent of Willow Flycatchers. A feather shed during the ringing process should enable DNA analysis (Wilson 2008). Those who journeyed to Cornwall for this cryptic tyrant wait eagerly to find out if the *BOU* find the evidence sufficiently watertight to eliminate Willow Flycatcher.

Calandra Lark Melanocorypha calandra

(Linnaeus, 1766). Abundant on steppe grasslands of Iberia and Morocco, uncommon and local in rest of Mediterranean basin and quite common in the Balkans and farther east to Jordan. To the east breeds in Ukraine, Turkey and southwest Russia to Kazakhstan and Afghanistan. European and south Asian populations resident or nomadic; north Asian populations disperse south of breeding range to the Persian Gulf coast of Iran.

Ray Scally

Polytypic, four subspecies. The racial attribution of occurrences here is undetermined. Nominate *calandra* from southern Europe is darkest race. Breeders from western and central Asia Minor are variable *(BWP)*.

Rare, but increasingly regular vagrant from southern Europe, North Africa or south-west Asia, almost exclusively on offshore islands and observatories.

Status: Fifteen records.

1961 **Dorset:** Portland 2nd April
1978 **Shetland:** Fair Isle 28th April
1985 **Scilly:** St. Mary's 26th-29th April
1994 **Outer Hebrides:** St. Kilda 21st September
1996 **Scilly:** St. Agnes 17th-18th April
1997 **Isle of Man:** Leanness 17th-18th May
1997 **Norfolk:** Scolt Head 19th May
1999 **Northumberland:** Farne Islands 28th April
1999 **Shetland:** Fair Isle 16th-17th May
2000 **Shetland:** Fair Isle 13th May
2002 **Orkney:** North Ronaldsay 10th-11th May
2004 **East Yorkshire:** Spurn 3rd October
2006 **Fife:** Isle of May 12th-17th May
2007 **Shetland:** Baltasound, Unst 12th May
2008 **Shetland:** Fair Isle 20th-22nd April*

Discussion: There is a clear bias towards spring overshoots, with 13 falling in spring (six in April with the earliest on the 2nd; seven in May) and just two in autumn (September and October).

Elsewhere in northern Europe most records come from Scandinavia, mainly in the past 25 years. There have been 10 records each from Norway and Finland and nine from Sweden. Compared with the British records, they are distributed more evenly with a good number of winter and midsummer records (see *Table 10*).

In Germany about 10 were recorded between 1802 and 1933 with the only recent record on 16th May 1996. There have been three from the Netherlands (10th October 1980, 16th May 1988, 27th May 2005) and singles from Estonia (31st December 2007 to 9th January 2008), Belgium (23rd November 1997, 12th May 2008) and northern France (29th April 2006). Luxembourg has one old record (21st March 1905).

In central Europe, 23 in Switzerland since 1993 (eight in 1993 and seven in 2008) constitute a major change in fortunes after five previous records since 1900. There have been six records in Austria since 1980. Clearly the recent increase in sightings is not restricted to countries at the periphery of northern Europe.

The recent surge of British records is difficult to interpret against a European backdrop of range contractions and population declines across much of its range between 1990 and 2000. The regular recent sightings are signs of an increasingly regular vagrant to our shores. That many have come from well-watched observatories supports the conclusion that the increase is genuine and not the product of increased observer effort. The British boost is mirrored by the analogous rise in extralimital records elsewhere to northern Europe.

Table 10: Monthly distribution of Calandra Lark records in Scandinavia.

Country	Jan	Feb	Mar	Apr	May	June	July	Aug	Sept	Oct	Nov	Dec
Finland			1	3	2					1	1	2
Sweden	2				2	1	1	2		1		
Norway				3	2	1	1			3		
Totals	2	0	1	6	6	2	2	2	0	5	1	2

It seems likely that the British birds in spring originate from populations farther east than the European populations. Southwestern and southern populations are mostly resident, while eastern populations migrate south in winter, rendering them more susceptible to overshoot on their return journey. The high number of Scandinavian records would support this hypothesis for the origin of British birds. Those in the southwest and west could have originated from Iberian populations, but this is nothing more than supposition as they too could also have been overshoots from the east.

One thing would appear to be clear from the British records: self-finders are pursuing a futile goal by scouring coastal fields on the mainland. The species' affinity with islands and observatories is clearly established.

Bimaculated Lark Melanocorypha bimaculata

(Ménétriés, 1832). Nominate bimaculata breeds from northeast Turkey to Iran, rufescens from southern Turkey to Syria and Iraq, and torquata from eastern Iran to western China. Some winter in the south of the breeding range: bimaculata migrates to Sudan, rufescens to Ethiopia and torquata to India.

Polytypic, three subspecies. The racial attribution of occurrences here is undetermined. Colour and size differences between subspecies are slight *(BWP)*.

Very rare vagrant from southwest Asia, the last in 1976.

Status: Three records. As with the last species all are from offshore islands, two in spring and one in autumn.

1962 Devon: Lundy 7th-11th May
1975 Scilly: St. Mary's 24th-27th October
1976 Shetland: Fair Isle 8th June

Discussion: Elsewhere in northern Europe there have been 12 records from Scandinavia: six have reached Sweden (24th May 1982, 10th-23rd December 1983, 27th-29th May 1996, 28th January-23rd March 1999, 17th June 2001, 24th November 2003); four Finland (12th-23rd January 1960, 26th-27th May 1996, 31st May-3rd June 1996, 17th December 2000-17th January 2001); and one each from Norway (1st August-1st October 2000) and Denmark (1st-4th January and

3rd-6th March 2006). There is one German record (6th July 1998). It was considered to be a possible escape and placed in category D.

The spring records from the far-flung outposts of Britain exhibit an arrival pattern similar to Calandra Lark. As with that species, dates are indicative of spring overshoots. In light of the records from Scandinavia since 1996 it is perhaps surprising that the last British record was more than 30 years ago. Scandinavia is arguably much better placed to receive overshoots returning from their wintering grounds compared to the more westerly position of Britain. Five of the Scandinavian records have been in December-January; presumably these birds wintered on the southern edge of their breeding zone and were displaced. The pattern correlates to some degree with that of the other highly prized central Asian larks: Black *M. yeltoniensis* and (to a lesser extent) White-winged *M. leucoptera*.

Bimaculated Lark is highly desired by a whole generation (or two) of listers. The Scilly individual is the only one to entertain an audience. The increased incidence of birds in northern Europe provides an air of expectation for another before long. There would be no doubting its popularity if it was to put in an extended stay. The Northern Isles would seem the most likely location.

White-winged Lark Melanocorypha leucoptera

(Pallas, 1811). Breeds from the lower course of the Volga to eastern Kazakhstan, China and Mongolia. Partially migratory with erratic winter movements, birds tending to move south and west to winter between the Black and Caspian Seas, reaching Iran. Southernmost birds are mainly resident.

Monotypic.

Very rare vagrant from Central Asia, both records in late autumn and early winter.

Status: Two accepted records.

1869 East Sussex: Near Brighton female caught 22nd November
1981 Norfolk: King's Lynn 1st-winter male 22nd and 24th October at the Beet Factory

Discussion: As with the previous species, several records from Sussex were dismissed as part of the Hastings rarities investigation (Nicholson and Ferguson-Lees 1962). Until quite recently there were five British records of seven individuals accepted for the period 1869 to 1981. Following review, only two records remain extant (Marr and Porter 1995; Anon 1996). For such a distinctive species, confusion possibilities would appear to be few. Snow Bunting *Plectrophenax nivalis* and partially leucistic Eurasian Skylark *Alauda arvensis* need eliminating, as does an escaped Mongolian Lark *M. mongolica* (Fisher 1988): there is a record of the last from near Schiphol airport in the Netherlands on 22nd March 1980; it is also known in the cage bird trade in Britain. Observers should beware of the pulse rate quickening too soon when confronted with a potential White-winged Lark.

This is another central Asian lark with a restricted range and reluctance for westwards vagrancy. Recent European records include six from Scandinavia, five of which have been since 1999: three from Finland (9th June 1971, 21st-23rd June 1999, 11th-13th April 2004), two Norway (24th-29th May 2001 and 19th July-14th August 2004), and one in Sweden (7th July 2002). Of these June and July have accounted for two records each and there were singles in April and May. Maybe midsummer is the time to seek out this mega? There are four recent records from Poland (21st October 1975, 4th August 1978, 30th March 1988, 12th May 1993), two of which lend further credence to the summer peak, as do two old records from Helgoland (2nd August 1881 and 2nd July 1886).

Occurrences from southeastern Europe, by way of contrast, appear to mostly involve winter birds. The most recent record from this region was one from Bulgaria (16th February 1997). There are nine accepted records from Italy in the 100 years since 1896, mostly winter occurrences: November 1869, two in October 1871, October 1885, January and March 1896, November 1926, undated bird in 1957 and 8th December 1996. Three from Greece (eight birds on 4th May 1959, 24th February 1963, 4th-18th June 1966) includes one winter record.

Formerly a sporadic, perhaps even regular, winter visitor to Turkey until the start of 20th century. A notable influx in February 1871 involved the collection of nine specimens. Two were shot on 14th October 1914. Four subsequent claims, none of which are documented, have been published (18th April 1965, at least 12 in late September 1979, three on 24th September 1981, 20th September 1986) (Kirwan et al 2008). It is often reported that the wintering area stretches west to Romania, but this is incorrect and is probably based on invasions in 1902-03 and 1907-08 (Lindroos and Tenovuo 2000). Geographically these countries lie much closer to regular wintering areas than do Scandinavia and Britain, and this is presumably reflected in more regular sightings.

The extralimital records of White-winged Lark share some similarities with Black Lark (Lindroos and Tenovuo 2002), but there appears to be a greater propensity for spring and summer occurrences from northern Europe, with late winter and early spring records less pronounced than for Black Lark. White-winged Lark forms post-breeding flocks in midsummer and move on, which correlates well with the majority of north European records. Most of the breeding range is vacated in winter. The wintering range extends south and (especially) southwest, with a propensity to wander west in winter (based on records from southern and central Europe). Return movement commences in March. Southern breeding areas are reoccupied by mid-April, though arrival in northern parts does not take place until late April or even early May (*BWP*); this again correlates with some northern European records, which presumably involve spring overshoots.

White-winged Lark is in a select group of rarities on the *British List* for which occurrences have never been observed other than by the finders, are both exceptionally rare and infrequent, and for which it seems likely that future sightings will remain at an absolute premium. Breeding across the dry grass steppes from Dagestan, the lower Volga River area, through central and northern Kazakhstan to about 80°E its range has decreased markedly in the last 100 years. Habitat destruction due to ploughing is one of the main threats to the White-winged Lark.

Against a backdrop of declines the upturn in sightings elsewhere in northern Europe appears anomalous, but does offer hope for the optimist that one could stray that bit farther west. Formerly perceived as lost late autumn migrants and occasional winter visitors, recent activity provides clear indications that a midsummer find may await some fortunate observer.

Black Lark *Melanocorypha yeltoniensis*

Forster, 1767. Largely resident on grassy steppes of central Asia from Lower Volga region of southeast Russia to northeast Kazakhstan. Outside breeding season, nomadic flocks remain within breeding range, some occasionally west to Ukraine.

Monotypic.

Very rare vagrant from central Asia, all males in spring.

Status: Three records of a most enigmatic species.

1984 **East Yorkshire:** Spurn male 27th April
2003 **Anglesey:** South Stack male 1st-8th June
2008 **Norfolk:** Winteron male 20th-21st April*

The Black Lark on Anglesey was one of the ornithological events of recent years, stimulating many a dormant twitcher to take to the road. Visitors were thrilled with the bird, the fantastic backdrop and a show-stopping performance. While many were enjoying views of this special bird it was evident that an earlier record from East Yorkshire nearly 20 years earlier was already in circulation with the BBRC. Coincidentally it was nearing acceptance at the time of the Anglesey bird's arrival. The Anglesey bird must surely have helped convince assessors that this enigmatic species really could reach Britain.

Following three days of light easterly winds with clear conditions at Spurn, East Yorkshire, the afternoon of 26th April 1984 brought a freshening northeasterly wind with associated low cloud and sea fret. On 27th April, a cold, light northeasterly breeze with thick sea fog persisted for most of the day, until the sun finally broke through at 19.00 hrs, leaving clear conditions for the remainder of the evening. A thin scattering of grounded migrants had been seen during the week, including small numbers of Northern Wheatears Oenanthe oenanthe, a few thrushes Turdus, one or two Black Redstarts Phoenicurus ochruros and a Whinchat Saxicola rubetra.

At 08.45 hrs on 27th April, Nick Bell (NAB) found an unusual passerine feeding on the short turf of the Parade Ground at the Point which he was unable to identify. He returned with B. R. Spence (BRS) and G. Thomas, who were equally puzzled by its identity. A single-panel mist-net was erected in order to attempt to catch and establish the bird's identity but, unfortunately, this proved unsuccessful. After flying over the net on the first attempt, the bird returned and, after a short wait, it was once again ushered towards the net. This time it flew above the net and kept going, flying over the nearby Helgoland trap, out across the Humber estuary and on towards Lincolnshire. It was never seen again.

The remarkable plumage and confiding nature of the bird led the observers to believe that it must be an escaped cagebird, albeit an unusual one. Although various books on cage-birds were consulted at the time, the observers were unable to put a name to the bird. The general size and gait reminded NAB of a cowbird Molothrus, although clearly not one with which he was familiar. Nonetheless, after considering the options available, the observers decided that it must be an escape from captivity, perhaps a species of cowbird from Central or South America, or one of the African weavers (Ploceidae).

The bird was described briefly in a short note in the Spurn Bird Observatory (SBO) log, but this made no mention of the only other birder to see the bird, Alex Cruickshanks (AC), who had watched the attempts to trap it from afar. To his credit, AC did suggest that the bird could be a Black Lark Melanocorypha yeltoniensis, a suggestion that did not find favour with the more experienced NAB and BRS at the time. Sadly, the bird was dismissed and quickly forgotten, with the details hidden away in the depths of the SBO log.

In the early 1990s, NAB saw the plate of Black Lark in Volume 5 of BWP, and this revived old memories of the mystery bird at Spurn. Unfortunately, he was unable to find his old notebook which contained his notes of the bird, and it was not until 1996 that this was discovered in the attic of his mother's house in Hull, East Yorkshire. The notebook included a description and sketch of the bird. At about the same time, NAB also became aware of a photograph of a Black Lark that had occurred recently in Sweden, and it was at this point that NAB and BRS realised that they had probably made an error. Regrettably, BRS had, by this time, discarded his old notebook with details of the bird. With only one source of evidence, the record was thought unlikely to find favour with BBRC, so, despite an initial rush of renewed enthusiasm, interest in the sighting was once again destined to languish.

Nothing further happened with the record until early in 2000. At that time, the old SBO logs were retained by John Cudworth, chairman of SBO, but following his retirement in 1999, they were transferred back to Spurn. Intrigued by NAB's tale, Dave Boyle, then SBO warden, looked back through the old logs and rediscovered the account of the mysterious 'escaped' passerine. He was surprised to discover that the nightly log for 27th April 1984 had been written by NAB himself, an action that he had long since forgotten about! With two primary sources of information now available, and with further encouragement from Spurn regulars, NAB once again retrieved his old notebook and this, coupled with the account in the SBO log, formed the basis of a formal submission to BBRC in spring 2000.

Lance Degnan and Ken Croft. British Birds 98: 306-313.

Discussion: Ever since Black Lark was removed from the *British List* as part of the Hastings rarities affair (Nicholson and Ferguson-Lees 1962), it held near-mythical status amongst birders. It was one of most sought-after birds for many, either as a hoped-for vagrant or on trips to breeding areas in central Asia. Almost the entire range lies in Kazakhstan, but it does extend west to the Volga and north into the steppes of the Urals/Western Siberia region. Movements, including winter, not infrequently take birds north of the breeding range to Ufa, Chelyabinsk, Omsk and Novosibirsk (Ryabitsev 2008).

Away from Britain there have been 10 records of 18 birds in Europe between 1958 and 2003. These include records from 1958 (male in Greece on 20th April), 1961 (one in Italy on 3rd May), 1963 (a flock of eight in Greece on 20th February), 1964 (two in Greece on 8th February), 1981 (male shot in the Czech Republic on 28th November), 1988 (one in Poland on 17th January), 1989 (males in Finland on 24th March and 8th April), 1993 (male in Sweden from 6th-7th May) and 1995 (female in Bulgaria on 25th May). Farther east, one was collected in Lebanon (31st October 1958) and a recent report of a male from Turkey (23rd May 2008). In Ukraine, 30 in mid-December 2008 were only the second record in the Crimea for over a century (Gantlett 2009).

Earlier records include Italy in autumn 1803, Austria in the late 19th century, Turkey (14th October 1914), Greece in autumn 1930, Malta in winter 1929, Romania in March 1897 and 1900, and Helgoland, Germany (27th April 1874 and 27th July 1892). The best period for vagrancy is in spring from March to early May.

The European records illustrate that this species and White-winged Lark – both short-distance migrants from the steppes – have different occurrence patterns to other eastern rarities. Typical eastern vagrants favour the northwest reaches of Europe (Lindroos and Tenovuo 2002), but the two larks make west or southwest movements. It would appear that Black Lark is, once displaced, a true nomad.

Outside the breeding season Black Lark forms single-species flocks. The females, unlike the sedentary males, move away from the breeding grounds. Such flocking behaviour could explain the very high proportion of males in the European records. Males are more susceptible to extreme weather conditions in the normal wintering areas, forcing them to disperse beyond their normal range during periods of inclement weather. Flocks of females typically disperse farther than adult males in winter, and are presumably less susceptible to adverse conditions and have a reduced tendency to extralimital vagrancy. Could the less-conspicuous females, undertaking their movements outside of the breeding season, be making undetected trips to Europe?

Early European records suggested a preference for southern latitudes, but most records since 1981 have been in northern Europe. This mystical lark appears to have overcome its reluctance to stray west. Britain, hosting two in five years, has become an unlikely and welcome hotspot for this enigmatic species.

Lesser Short-toed Lark Calandrella rufescens

(Vieillot, 1820). Breeds from Spain, the Canaries, and Morocco across North Africa to Turkey and the Middle East, and through the southern parts of the former Soviet Union to Manchuria. Largely sedentary in Spain, in North Africa it is resident to dispersive. Farther east it is more migratory. Breeders in the former Soviet Union winter mainly in the southern part of the breeding range. In the Far East, it moves southwards into northern parts of the Indian subcontinent and southern China.

Polytypic, seven subspecies. The racial attribution of the sole occurrence is undetermined. Variation between the subspecies is complex and involves colour (mainly ground colour and width of shaft-streaks of upperparts and chest, and amount of white in tail), size (as expressed in length of wing and tail) and shape of bill (heavy and conical or small and fine) (*BWP*).

Very rare vagrant, presumably originating from populations in central Asia.

Status: One British record.
1992 Dorset: Portland 2nd May

Discussion: In Ireland, four records involving a total of 42 birds between January 1956 and March 1958 were formerly accepted in Cos. Kerry, Wexford (twice) and Mayo (Anon 1960). Unsurprisingly this remarkable series of sightings attracted attention (Eigenhuus 1992; Smiddy 1993) and a subsequent review by the Irish Rare Birds Committee found them no longer acceptable.

Lesser Short-toed Lark is an exceptionally rare bird in northern Europe. There is an old record from Germany (26th May 1879). The small number of recent records includes seven in Scandinavia: three from Sweden (27th-28th April 1986, 9th-10th May 1991, 10th-14th February 2003); three from Finland (18th November 1962, 16th January-1st February 1975, 25th-26th September 2004); and one from Norway (7th-23rd November 1987). Elsewhere, there are two from Switzerland (28th-29th April 1989, 25th April 2000) and singles from Belgium (26th-27th April 2003) and Austria (7th April 1993).

The resident, or locally dispersive, movements of Lesser Short-toed Lark in the western part of its range reflects its extreme rarity here in contrast with the much more frequent Short-toed Lark *C. brachydactyla*, which accrued over 600 British records between 1958 and 2003 (Fraser and Rogers 2006). Lesser Short-toed Lark populations farther east are more migratory. It is likely that these are the source of north European overshoots, which would explain the greater incidence of Scandinavian birds compared to elsewhere in northern Europe. Records farther west could conceivably originate from the less mobile North African or Iberian populations. In northern Europe, the peak month for occurrence is in April (five records); two were in November. Singletons in January and February, May, July and September do little to aid the interpretation of these sporadic wanderings to northern Europe. The timing of the Dorset bird fits nicely with the peak early spring arrival period apparent for north European sightings.

Crested Lark *Galerida cristata*
(Linnaeus, 1758). Nominate cristata breeds from northern Spain and France to Denmark and east through central Europe to the Ukraine. Most are resident; some eastern populations are partial migrants.

Polytypic, 35 subspecies. Records attributable to race have all involved nominate *cristata* (Linnaeus, 1758). The European group of races comprises nine subspecies. Geographical variation between subspecies is marked and complex, mostly involving ground colour and intensity of streaking (*BWP*).

Rare vagrant, presumably originating from populations on the near Continent.

Status: There are 19 accepted records (21 birds), nine in the 19th century up to 1881 and 10 since 1947; the recent Kent bird awaits a formal acceptance.

Pre-1845 West Sussex: Littlehampton killed; no date
1846 Cornwall: Marazion Marsh two males shot 9th September
1850 Cornwall: Marazion Marsh shot 24th October
1863 West Sussex: Shoreham-by-Sea collected 20th October
1865 Cornwall: Falmouth shot in December
1879 East Sussex: Worthing shot in spring
1879 Kent: Dover Cliffs caught 22nd April
1880 Cornwall: Polbrean, The Lizard female shot 12th June
1881 East Sussex: Portslade-by-Sea adult male collected 10th October
1947 Greater London: Between Hammersmith Bridge and Chiswick Eyot two seen 8th March
1952 Shetland: Fair Isle 2nd November
1958 Devon: Exmouth 29th December to 3rd and 10th January 1959
1965 Cornwall: Marazion 4th April
1972 Somerset: Steart Point 8th April
1972 East Yorkshire: Tunstall 11th June
1975 Kent: Dungeness 28th September-1st October
1982 Caernarfonshire: Bardsey 5th-6th June
1996 Suffolk: Landguard 2nd and 9th October
2009 Kent: Dungeness 29th-30th April, probably since 27th

Most recent records have been brief. Those in Kent and Suffolk were the only collectable individuals. *As the morning of the 2nd October had been mildly productive at Landguard Point, Suffolk, with a good run of common migrants, Paul Holmes decided to have a quick look around Adastral Close before lunch, in the hope that there might be something of interest at this top housing estate site. Little was happening, however, and the thought of beans on toast was too much for him to bear.*

Driving back to the cottage at 12.30pm, he was surprised to see a lark run across the road in front of his car. With his binoculars still conveniently around his neck, he peered out of the smeary windscreen at the bird. Surprised by what he saw, he ran from the car to get better views. It was clearly a Crested Lark!

He was unable to believe his eyes, and immediately called Mark Grantham to come and make sure that he was not hallucinating. Nigel Odin had witnessed the incident from some distance away and, being intrigued by the abandoned car and frantic telephoning, hurried over to see what all the fuss was about.

The three of us had just a quick look at the bird and then watched as it took flight over the observatory and out of sight! The news was hastily released (initially as a Crested or Thekla Lark until better views could totally rule out the latter) and fortunately the bird was quickly relocated in an adjacent car park. Excellent views were obtained at close range and the identification was confirmed: it was Suffolk's first Crested Lark. Mike Marsh and Steve Piotrowski were quickly on the scene, and the identification was triple checked.

MG hastily returned to the observatory to close the nets in preparation for an afternoon of arriving birders and headless chicken activities. This was the first Crested Lark to be twitchable in Britain for 21 years! In the meantime, the lark had flown high to the north and was feared lost in the dockland wilderness. NO, MM and SP set out to search for it and PH returned to make one or two telephone calls.

On reaching the last net, MG was a little peeved to find a lone 'phyllosc' in the bottom shelf. It soon became apparent, however, that this was no 'Willow/Chiff'. Still reeling from post Crested Lark euphoria and fearing a case of rarity madness, he bagged the bird and returned to the ringing room for a calm assessment. Back in the observatory, PH was beckoned and the contents of the bag were revealed. This time, PH confirmed that MG was not hallucinating: it was indeed a Bonelli's Warbler. MG processed the bird and calmly pronounced that it was a Western Bonelli's Warbler. At this point, the lark was relocated at the car park at the north end of the reserve. The warbler was quickly released in the compound. Birdline was up-dated, and everyone migrated back and forth between the two sites.

The Crested Lark was watched well for about two and a half hours as it fed on and around the car park. It was very confiding – photographers had to back away from it as it came too close for them to focus on – and it even gave several snatches of song. About 150 birders managed to see it before it took flight to the north at about 4.30 pm. It appeared to land just a few hundred yards away on the edge of Felixstowe town, but, to the distress of later arriving birders, it could not be refound.

Amazingly, however, it was relocated exactly a week later by a visiting birdwatcher, John Walshe. He, and several members of the bird observatory team, watched it for 10 minutes as it fed in a spot less than 100 metres from where it was first found. Frustratingly, however, it then flew off south across the river into Essex, and it was not seen again.

Mark Grantham and Paul Holmes. *Birding World 9: 392-393*

Discussion: All 19th century records were discovered along the south coast and included four each in Sussex and Cornwall and one in Kent. Two were undated but there were singles in April, June and September, three in October and one in December. From the mid-19th century onwards Crested Lark underwent a marked range expansion on the Continent, culminating in the colonisation of southern Fennoscandia in the early 1900s (Hagemeijer and Blair 1997). The cluster of British records spanning this period doubtless stemmed from this change in fortune.

From the 1930s the range of Crested Lark began to contract, with both climatic reasons and agricultural intensification cited as possible causes (Hagemeijer and Blair 1997). Declines are ongoing with the large European breeding population experiencing further descent at the end of the 20th century: the last breeding in Norway took place in the 1970s, just a handful of pairs remained in Sweden in 1989 and sizeable populations in France and Germany have decreased, the latter by up to 79 per cent (Hagemeijer and Blair 1997; BirdLife International 2004). However, there are increasing numbers arriving in Finland: four in the 1970s, five in the 1980s, seven in the 1990s and 16 since 2000 (Aleksi Lehikoinen *pers comm*). Nearer to home, breeding numbers have dwindled in the Netherlands from 3,000-5,000 breeding pairs in 1973-77 (Hagemeijer and Blair 1997) to a handful of pairs in 2008 with a maximum of seven pairs at five localities (Arnoud van den Berg *pers comm*).

All nine 20th century records have been capricious with respect to both location and arrival date, with little hint of a pattern for future finders to follow. The wide spread of sightings during this period has come from Suffolk, Kent and Greater London in the southeast; the southwest has received birds in Cornwall, Devon and Somerset.

The first of the 20th century records was a rather unconvincing sight record of two together by the River Thames in London in 1947. One on Bardsey is the sole Welsh record and one on Fair Isle the only Scottish representative. Another wanderer reached East Yorkshire. Four have been found between March and April, two in June and singles in each month between September and December.

Despite its large breeding range and relative abundance on the Continent, this unexciting lark remains an exceptionally rare bird in Britain. This is even more apparent when only recent records are considered. Declining numbers in Europe combined with the sedentary nature of the species and its reluctance to traverse the English Channel and North Sea have presumably intensified its rarity on British soil in recent times. This is one species for which the rapid increase in observers has had no impact upon status.

Tree Swallow *Tachycineta bicolor*
(Vieillot, 1808). Breeds throughout North America below the tree line from Central Alaska to Labrador and south to the southern USA. Winters Gulf of Mexico and throughout Central America and Caribbean to northern South America.
Monotypic.

Very rare vagrant from the Nearctic.

Status: Two British records, the first also the first for the Western Palearctic.

1990 Scilly: St. Mary's 6th-10th June
2002 Shetland: Burrafirth, Unst 29th May

Discussion: Elsewhere in the Western Palearctic there have been seven on the Azores including five in 2005 (26th October-2nd November, three on 2nd November, one on 5th-6th November) and two in 2007 (19th and 23rd October). One claimed near Tata, Morocco, on 5th February 2005 was not accepted by the Moroccan Rare Birds Committee (Bergier *et al* 2006).

The first British record, on St. Mary's, was enjoyed by large numbers of birders during its extended stay. The second, on Unst, was observed and photographed with Barn Swallows *Hirundo rustica* and House Martins *Delichon urbicum* for an hour by a single observer.

The occurrence of both British birds in spring could imply a direct transatlantic crossing. The Shetland bird appeared after a period of successive weather systems crossing the North Atlantic.

The later date of the Scilly individual, at lower latitude, could involve a lingering northbound bird that had crossed the Atlantic during a previous autumn. Unlike other swallows, this species can subsist for extended periods on seeds and berries enabling them to remain in the USA as far north as Massachusetts in winter (Robertson *et al* 1992). Spring arrival dates in the USA are much earlier than other swallows, with the main passage from mid-March to early April. The British records are much later than might be expected of spring overshoots.

Departure to the winter quarters takes place from July-August. A proportion of the population migrates along the eastern seaboard of the USA to Florida and the Caribbean. It is a casual visitor to Bermuda and extralimital records have reached Greenland. Flocks are now encountered with regularity along the Caribbean coast of South America. This migration route renders them susceptible to the weather conditions that displace other Nearctic vagrants. Given the abundance of the species and the autumn migration route taken by some to their wintering areas, it is perhaps surprising that there are so few records when compared with the numbers of other aerial feeders such as Chimney Swift and Cliff Swallow that have been displaced across the Atlantic.

Purple Martin Progne subis

(Linnaeus, 1758). Nominate subis is widespread across much of eastern North America north to southern Canada, west to Montana and central Texas and south to the highlands of central Mexico. This form is replaced in southern Arizona and western Mexico by P. s. hesperia and to the west by P. s. arboricola, which breeds along the west coast of the USA and north to southern British Columbia. Winters in South America, south to the Amazon basin, northern Bolivia and Northern Argentina.

Polytypic, three subspecies. The racial attribution of the occurrence is undetermined, but probably involved nominate *subis* given its wider and more northern and eastern distribution. Adult males of all three races have identical plumage characters, although the western races may show less purple on the underparts. Compared with the nominate form, *hesperia* is slightly smaller and *arboricola* larger. Females of both *hesperia* and *arboricola* are paler than nominate *subis*, particularly on the underparts and the forehead, and show a grey-white collar that is typically paler than females of the nominate form and which can appear whitish on some individuals (Hill 2002).

Very rare vagrant from the Nearctic.

Status: One British record, although long claimed to occur.
2004 Outer Hebrides: Butt of Lewis and Europie, Lewis juvenile 5th-6th September

Discussion: Purple Martin is the northern representative of a group of closely related species whose systematics remains unclear. Genus Progne is widely distributed throughout Central and South America. At least eight different species have been described at various times. The difficulty in identifying species in the field, and the varying taxonomic treatments, have confused the status of each in regions of Central and South America where they overlap during at least part of the year. As a result the migratory routes and extralimital occurrence of Purple Martins are not known with certainty (Brown 1997).

Nominate *subis* has a northern and eastern distribution east of the Rocky Mountains and north to Alberta and Nova Scotia, thus it is the most likely form to occur here. Much of the eastern population migrates across the Gulf of Mexico to South America, though the early timing of the migration period presumably makes them less susceptible to the displacement by autumn weather systems that produce arrivals of other Nearctic vagrants.

There were at least four claims of Purple Martin in Britain and Ireland during the 19th century. In each case sufficient doubt surrounded the circumstances or identification to support the record. These include: one Kingstown, Dublin, in 1839 or 1840 (specimen held at the Museum of Science and Art, Dublin); an adult male and a juvenile male said to have been shot at Brent Reservoir, Greater London, in September 1842; one at West Colne Bridge, near Huddersfield, West Yorkshire, in 1854; and one from Colchester, Essex, 26th September 1878 (Coyle *et al* 2007). Of these, only the Dublin record is thought to have some credence.

The second for the Western Palearctic arrived on the Azores on the same day that the Lewis bird disappeared. These were the only accepted records for the Western Palearctic, but they have since been pre-dated by two birds on Corvo, Azores (28th September 1996). The early date of the Scottish and Azores birds in 2004 might at first appear at odds with most Nearctic landbirds. However, in the USA, autumn migration commences once the young fledge and takes place from late July to September with the main exodus in late August and early September.

Eurasian Crag Martin *Ptyonoprogne rupestris*

(Scopoli, 1769). Breeds northwest Africa and Iberian Peninsula northwards to southern Germany and east through Mediterranean and central Asia, north to Baikal region of southern Siberia, south to the Tibetan plateau and east to northeast China. South European population mostly resident but Asian populations migratory, wintering in northeast Africa, and northwest India to north-central China.

Monotypic.

Status: Eight records involving seven birds, the first as recently as 1988.

1988 Cornwall: Stithians Reservoir 22nd June
1988 East Sussex: Beachy Head 9th July
1989 Caernarfonshire: Llanfairfechan 3rd September
1995 East Sussex: Beachy Head 8th October
1999 Leicestershire: Swithland Reservoir 17th April; also in West Yorkshire
1999 West Yorkshire: Angler's and Pugney's Country Parks, Wakefield 18th April; presumed same as Leicestershire
1999 Orkney: Davey's Brig, Mainland 3rd May
2006 Surrey: Badshot Lea 22nd October

Discussion: The British records spread over a period of seven months, including singles in each month from April to July, September and October (two). Two occurred in both 1988 and 1999. One of the latter, a weekend bird in Leicestershire and the following day in West Yorkshire, allowed hundreds of birders to successfully hunt down this chunky martin.

Since 1980 this predominantly southern European species has extended its breeding range north. These include Bulgaria and Switzerland since 1980, Bavaria in 1981, Austria in 1982, Romania in 1986 and the former Yugoslavia in 1988 (Burton 1995). Between 1990 and 2000 populations were stable or increasing across much of its European range. The trend of the sizeable Spanish population (up to 100,000 pairs) during the period is unknown (BirdLife International 2004).

The flurry of records spanning just over a decade from the late 1980s was presumably associated with the increase in breeding numbers across Europe. Populations are mostly sedentary, but northern birds move as far south as sub-Saharan Africa; the increase in reports presumably represents overshoots from these areas.

In line with the European range expansion there have been a number of firsts across northern Europe. There have been two birds in Finland (9th-10th June 1988, 26th-27th May 2003) and one from Norway (27th June 2007) that is not yet submitted. The first two for Sweden also occurred recently (19th-22nd October 1996, 25th-29th October 2001). A further four Swedish records formed an exceptional influx with four on 20th-21st October 2006 and three of the same still on 22nd. The first six birds for the Netherlands also occurred during this period (three records each of two birds) as well as the most recent British sighting. There have been five in Denmark (8th May 1988, 8th-12th November 1990, 13th June 1997, 4th May 2000, 5th May 2000), three in Belgium (18th-24th April 1989, 4th-9th May 1994, 4th-6th November 2005) and three on the north coast of Germany (28th June 1991, 24th April 1992 and 30th May 1997).

Contrary to expectations at the time of the first few records, this species remains exceptionally rare. Surprisingly, in light of the sudden increase in sightings elsewhere in northern Europe, there has been just one bird since 1999. Presumably the northern European records are from the northern migratory populations, but the species appears to have a reluctance to cross large bodies of water, such as the North Sea.

Cliff Swallow *Petrochelidon pyrrhonota*

(Vieillot, 1817). Breeds throughout North America from western Alaska to Nova Scotia, USA (except the southeast) and south to southern Mexico. Winters South America from Brazil to Chile and Argentina.

Polytypic, five subspecies. Records are likely to be nominate *pyrrhonota* Vieillot, 1817, although none has been positively assigned to race. There is significant integration between races, which form a cline from north to south. Nominate *pyrrhonota* breeds on the eastern side of North America west to Manitoba and the Rocky Mountains, and from southwestern British Columbia south to northwest Baja California. Considering their distribution the other races are less likely vagrants (Turner and Rose 1989).

Very rare autumn vagrant from the Nearctic mostly to the southwest, but records include three on the south coast and even two east coast sightings.

Status: Ten records, nine in Britain and one in Ireland.

1983 Scilly: St. Agnes juvenile 10th October; same, St. Mary's 10th-27th October
1988 Cleveland: South Gare juvenile 23rd October
1995 Scilly: Tresco juvenile 4th-5th December
1995 Co. Kerry: Dunmore Head 16th November
1995 East Yorkshire: Spurn juvenile 22nd-23rd and 28th October
1996 West Sussex: Church Norton juvenile 1st October
2000 Hampshire: Titchfield Haven 1st October; possibly same as Dorset
2000 Dorset: The Verne, Portland 29th-30th September
2000 Scilly: St. Mary's 28th-30th September.
2001 Scilly: St. Agnes 1st-winter 26th October; same, St. Martin's 26th-27th October; same, St. Mary's, 28th-30th October

Discussion: Most records are from late September (earliest 28th) to late October. There are single November and early December records. A southwest bias is evident, which makes the east coast sightings in Cleveland and East Yorkshire all the more notable.

Vagrants have reached Wrangel Island, Siberia, and southern Greenland. Elsewhere in the Western Palearctic there are three birds from Iceland (collected on 12th October 1992 and two on a ship on 12th October 2004, one of which was collected) and one France (juvenile 30th September and 1st October 2000). Three birds have been found on the Azores (two 28th September-11th October 2001 and 5th November 2005), and singles in Tenerife, Canary Islands (juvenile 30th September 1991) and Madeira (27th October 2001).

Most migrant Cliff Swallows presumably follow the Central American isthmus between North and South America. It is not a rare bird along the eastern seaboard of the USA in autumn and is therefore prone to transatlantic vagrancy. Multiple records of three each in 1995 and 2000 recall the sporadic influxes of Chimney Swift *Chaetura pelagica*. Cliff Swallow is a diurnal migrant that moves in flocks; multiple arrivals are not unexpected.

The only species with which Cliff Swallow is likely to be confused is Cave Swallow *P. fulva* from southern North America and Mexico, the Greater Antilles, Ecuador and Peru. Cave Swallow is largely resident across much of its disjunct range, but its movements are poorly understood. Since the mid 1980s it has undertaken a dramatic range expansion in Texas and has also colonised south Florida (West 1995).

Historically, Cave Swallows have been observed in the Maritime Provinces of Canada as far back as 1968 and through the 1990s. Subsequently, the species has been observed annually in the northeast US. New York State's first autumn record was in November 1998, since when their appearance during late autumn has become predictable. During November 2005 there was an exceptional incursion of at least 1,000 to New York State (Spahn and Tetlow 2006).

The differences in distribution and migration strategies between the two species indicate that Cave Swallow would be an unexpected vagrant, but clearly it is an outside possibility for consideration in light of the small numbers reaching the east coast of the USA.

Blyth's Pipit *Anthus godlewskii*

(Taczanowski, 1876). Breeds from southern Transbaikalia and northern Mongolia to extreme northeast China and south to Tibet. Winters locally throughout Indian subcontinent and strays west to Israel and Arabia.

Monotypic.

Increasingly regular vagrant from Siberia. Probably overlooked previously.

Status: There have been 20 records involving 21 birds, 20 of them since 1988.

1882 **East Sussex:** Brighton caught 23rd October
1988 **Shetland:** Fair Isle 13th-23rd October
1990 **Cornwall:** Skewjack 1st-winter c22nd October-1st November
1993 **Scilly:** St. Mary's 1st-winter 20th-22nd October
1993 **Shetland:** Fair Isle 1st-winter 31st October-4th November, trapped 1st November
1994 **Suffolk:** Landguard 1st-winter 4th-10th November trapped 10th; taken by Common Kestrel *F. tinnunculus* on 10th
1994 **Kent:** South Swale NR 7th November-11th December
1996 **Norfolk:** Sheringham and Weybourne 1st-winter 14th-16th October
1998 **Dorset:** Portland two 22nd-24th November, one to 6th December, both trapped 22nd November
1999 **Norfolk:** Happisburgh 25th to at least 28th September
2002 **Nottinghamshire:** Gringley Carr 1st-winter 28th December-5th January 2003
2004 **Cornwall:** Land's End 1st-winter trapped 15th November to at least 20th December
2005 **Caernarfonshire:** Bardsey 1st-winter 16th–17th October
2005 **Scilly:** St. Mary's 1st-winter 23rd October
2006 **Shetland:** Sumburgh 1st-winter 12th October
2006 **Scilly:** St. Mary's 1st-winter 18th-27th October
2006 **Shetland:** Fair Isle 21st-24th October
2007 **Scilly:** Tresco 1st-winter 16th-23rd October
2007 **Shetland:** Sumburgh 17th-18th October
2007 **Shetland:** Fair Isle 1st-winter 27th October, trapped, died later

Where: Most have been on Shetland (six birds) and Scilly (four birds), plus four between Dorset and Cornwall in the southwest and five between Norfolk and East Sussex along the east coast (five birds). There is one Welsh record. The notable exception to coastal birds was an inland bird in Nottinghamshire, but given that Richard's Pipit *A. richardi* strays inland with some regularity this record might not be so unlikely as it first appeared. It has been suggested that inland Richard's might follow major river systems (Fraser and Rogers 2004). If so, perhaps Blyth's might behave similarly and further inland birds might be expected.

When: Blyth's is an early autumn migrant, leaving Mongolia in August and arriving on the winter quarters from early September. Surprisingly many of the European records occur significantly later than typical for Richard's Pipit (Alström and Mild 2003).

All have been in autumn or early winter. One in late September (25th, Happisburgh, Norfolk) was

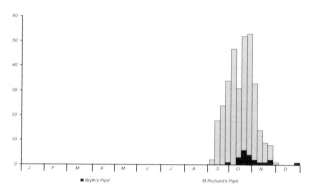

Figure 15: Timing of British Blyth's Pipit records 1882 -2007 with an overlay of autumn Richard's Pipits in Yorkshire (to 2005).

the earliest by some margin. Fourteen have been in October (13 between 12th and 27th). Five November birds are split between the early and latter part of the month. The inland bird was found in late December.

The arrival dates of Blyth's Pipits between Shetland and Scilly/Cornwall barely differ (median date 19th October versus 21st October), so there is no evidence of 'spill down' for this species from north to southwest. Arrivals may occur on a broad front so surely others in between are not being detected.

Discussion: Kenneth Williamson unearthed the first British Blyth's Pipit in 1963 when he was looking at specimens of Tawny Pipit *A. campestris* in the British Museum and discovered a specimen of Blyth's Pipit mislabeled as a Richard's Pipit *A. richardi*. Williamson said it was the most tatty skin that he ever inspected. The bird had been obtained at Brighton, Sussex, in October 1882 (Williamson 1977). It was not accepted in 1963 because it was thought to be an extremely unlikely vagrant to Britain. It was only when a bird was found in Finland in 1975 that the record was re-evaluated and accepted. This was published in the 10th *BOURC* Report (*BOURC* 1980) following rejection in the 8th Report (*BOURC* 1974) because "the majority of the committee was not convinced of the validity of the record".

Confusion reigned over the identity of a number of birds. Much of this was presumably due to this species being deemed an extraordinarily difficult bird to identify. Some of the confusion stemmed from reliance upon a supposedly diagnostic 'chep' call (for example, Kitson 1979a; Alström and Mild 1987, Alström and Mild 1988, Alström 1988; Heard 1990) which Blyth's rarely gives. Once observers learned to listen for a Yellow Wagtail-type call on a pale-lored pipit the identification clouds began to part.

Many observers did not appreciate quite how small Blyth's is, though perception of size is to be used cautiously on lone individuals. The species has now received ample attention in the literature for identification features to be highlighted (for example, Lewington et al 1991; Bradshaw 1994; Alström and Mild 2003). The first modern record to be accepted by the BOU was the Landguard bird of 1994 (BOURC 1996). Acceptance of the Skewjack record (1990) followed. The Fair Isle bird of 1988 was the fourth to be assessed by the BOU and became the earliest to be accepted on to Category A.

Some controversial birds have also been well discussed, most notoriously the 'Portland pipit' at Portland from 16th March to 3rd May 1989 (Grant 1989; Millington 1989). It was eventually considered to be a Richard's Pipit. Others have perplexed their observers by not conforming to expectations (Gray 1996; Page 1997). Such birds offer a cautionary note for observers who feel that this species is a clear-cut identification.

Nine more records made the grade between 1990 and 1999, most of which were trapped. Increasing confidence with the species has led to many subsequent birds being named in the field. Records become almost annual and it appeared that this enigmatic pipit was going to become a rarity of regular occurrence as the millennium approached. However, there was a stall in sightings at the turn of the century, with just one being found to 2004.

Since 2004 the impetus has returned. A record three in 2006 and another three in 2007 quickly followed singles in 2004 and 2005. Several more recent records are yet to be assessed. It now appears that this increasingly regular pipit is here to stay and is fast descending from a major rarity to an expected annual vagrant.

The change in status in Britain is in accordance with records from elsewhere in Europe. There are 19 from Finland (10 from 28th September-24th October and nine between 4th-23rd November), 12 from Sweden (five from 3rd-15th October, six between 2nd-20th Novemberand one in December) and seven from Norway. There are six records from the Netherlands (13th November 1983, 25th-28th October 1996, 24th November 2002, 30-31st October 2005, 14th November 2005, 1st January-7th February 2007). Five from France include a wintering bird in the southeast (Bouches-du-Rhône) from 16th January-25th February 1998 and four in October (between 14th-28th); three have been since 2002.

Belgium has five records (16th November 1986, 16th October 1998, 26th October 2004, 6th November-30th December 2005, 10th November 2005-21st January 2006). Denmark has three (22nd November-1st December 1998, 18th-26th November 1999, 16th October 2001), and there are two German records from Helgoland (25th September-2nd October 1996, 26th-30th October 2007).

Many European countries only recorded their first since the 1990s. Birds are also being detected farther south. The first for Spain – where Richard's Pipit winters regularly – was present from 26th November-12th December 2005. Two from Sicily, Italy, are to be submitted from winter 2007/08. The first for Turkey were both trapped in 2006 (one on Turkey's south coast on 19th September and another in Southeast Anatolia on an unknown date in autumn) (Kirwan 2009) and the first for Cyprus was seen from 4th-8th April 2007.

The archipelagos of Shetland and Scilly are emerging as the main sites in which to secure this pipit in Britain, but given the suspiciously close timing of arrivals in both areas many more may be passing undetected in between. Blyth's Pipit remains difficult to identify, but a greater awareness of what to look and listen for is leading to an increased rate of detection. Presumably this awareness is responsible for the accelerated rate of records rather than any change in distribution. Many fly-over birds may escape detection. The increase in records in northwest Europe suggests the species was overlooked in the past. Perhaps a number of old 1st-winter Tawny Pipit records (with which Blyth's was once considered conspecific), in late autumn actually involved this species. Blyth's has also been considered conspecific with Richard's Pipit (but not at the same time as Tawny) and with which on plumage it appears similar.

Olive-backed Pipit Anthus hodgsoni

(Richmond, 1907). European range restricted to northern Urals. Widespread across central and eastern Siberia to northern China, Kamchatka, Kuril Islands and Japan; winters widely across southern China, Taiwan and northern and central parts of Southeast Asia. A population in Himalayas and mountains of west-central China winters in Indian subcontinent.

Polytypic, two subspecies. Records are attributable to the widespread and westernmost subspecies *yunnanensis* Uchida and Kuroda, 1916. Nominate *hodgsoni* Richmond, 1907 from the Himalayas east to eastern China has more heavily streaked central crown and especially mantle and scapulars. Its chest streaks are bolder and extend farther down on to the belly and the flanks are more extensively streaked (*BWP*).

Increasingly regular migrant from Siberia mostly in October, with records noted from September to November. An increasing number of winter and spring records have been observed in recent years.

Status: This is now one of the most frequently occurring species still classed as a rarity, with a total of 318 records to the end of 2007 of which 311 are from Britain and seven in Ireland.

Historical review: The first was a spring bird trapped on Skokholm, Pembrokeshire, on 14th April 1948. In retrospect, the date was rather more astonishing than the species. It took 30 years for the find to gain official recognition as the first record of the species in Britain and Ireland (Conder 1979).

On 14th April 1948, Joan Keighley (now Mrs Joan Jenkins) and I caught a pipit Anthus in the Garden Trap on Skokholm, Dyfed. It was clearly none of the species known to us and, even after consulting The Handbook, we could not identify it. We recorded details in the bird observatory's log (now no longer available).

At the end of the season, I visited the British Museum (Natural History) and examined pipit skins. I decided that the bird had almost certainly been an Olive-backed Pipit A. hodgsoni, but hesitated to submit such an unusual record: not only a first for Britain and Ireland, but of a little-known species and at what seemed an extraordinary time of year for an Asiatic vagrant.

When thoroughly documented records of Olive-backed Pipits on Fair Isle, Shetland, in October 1964 and September 1965 were published in 1967 and I saw the accompanying photographs, I became certain that my Skokholm pipit was the same species. I sent the photograph and notes to I. J. Ferguson-Lees, who was then executive editor of British Birds, and the Rarities Committee was consulted. Its view was: 'Probably, but not sufficient evidence'.

Ten years later, in April 1977, Dr J. T. R. Sharrock was sorting through the British Birds files when he came upon the photograph, which he recognised as showing an Olive-backed Pipit. The only writing on the back, however, stated merely 'Skokholm pipit'. Knowing nothing of the previous history, but aware that no Olive-backed Pipit had ever been recorded on Skokholm, he wrote to Dr Christopher M. Perrins, who was likewise unaware of any such occurrence, and who commented that it must relate to a period before he was associated with Skokholm. At CMP's suggestion, therefore, JTRS wrote to me, and thus successfully tracked down the photograph's source. In the meantime, I had seen Olive-backed Pipits in India and no doubts remained in my mind; it was with enthusiasm, therefore, that I complied with JTRS's suggestion to resubmit the record, this time to the BOU Records Committee (the Rarities Committee now considers only those records since 1958; the BOURC considers all British 'firsts'). With a background of detailed notes on nine British records during 1964-76, and advice from Alan Kitson who has recently studied the species in Mongolia, assessment is far easier now than it was in 1967 when I first submitted the record (and incomparably easier than in 1948): the BOURC unhesitatingly accepted 'the Skokholm pipit' as an Olive-backed, which, thus, became the first for Britain and Ireland.

Peter Conder. British Birds 72:2-4.

The next was not until 1964, when one was on Fair Isle from 17th-19th October. Another was there on 19th September 1965.

The 1970s added a further 11 birds to the total and instigated a trend of annual occurrence from 1973 onward. The 11 included the first for Ireland, from Great Saltee Island, Co. Wexford on 21st October 1978. The distribution of birds was widespread with four from Scilly, two from Norfolk and one from Dorset. Over this period Fair Isle mustered just three indicating, for once, that a rare Sibe could turn up away from sacred ground.

The 1980s brought this attractive pipit to the consciousness of many birders courtesy of an inland bird in Bracknell, Berkshire, from 19th February to 15th April 1984. Frequenting gardens, this individual gave many their first experience of the species in Britain. The decade went on to amass a then quite exceptional 59 birds, including a widespread arrival of 20 in 1987 followed-up by 11 more in 1988.

The 1980s saw the status of this species rewritten within a short period of time. The trend had clearly turned. No longer was this creeping pipit an enigmatic Sibe, it was well on the radar of hunters and the 1990s were not to disappoint as the decade accumulated a staggering 139 birds.

The autumn of 1990 delivered a deluge of 46 birds (nine on Fair Isle alone). The brunt of the invasion extended from the Northern Isles and along the east coast, but nine also reached the Isles of Scilly and the influx was welcomed as far west as Co. Cork where three were found. Who knows how many more must have passed through without observer interception?

The unprecedented 1990 arrival raised expectations and double-figure counts in 1991 and 1992 consolidated the birds new-found status. Autumn 1993 came close to matching events of the previous three years, but fell short at a 'mere' 36 birds. This time just two reached each of Scilly and Cornwall and only one was found in Co. Cork. As is often the case, famine followed bounty and the period of 1994 to 1999 amassed a comparatively paltry total of 25 birds (just over four a year): thin pickings compared to what had gone before.

The new century brought a revival, including a creditable 18 in 2000 (12 in Shetland) that was just bettered by 19 in 2001. There were 18 again in 2003 (10 in Shetland). The seven years between 2000 and 2007 have amassed 106 birds at over 13 per annum. The average number of birds per annum between 1980 and 2007 was 10.9. The glory years of 1987 to 1993 produced 21 per annum.

Where: Just under half of all 'OBPs' have been detected in the barren Shetland landscape. Only 30 birds have been found elsewhere in Scotland: 18 in Orkney and five in Fife. The next-best area is the Isles of Scilly with 37 birds (12%), though the species remains rare in the rest of the southwest with a modest six in Dorset (all at Portland), five from Cornwall, and three from Devon.

Northeast England has fared well, producing 39 birds: 22 in Yorkshire, seven each in Lincolnshire and Northumberland, and three in Cleveland. Norfolk accounts for 17 birds, including one wintering inland at Lynford Arboretum from 2nd to at least 20th February 2002. Just three have graced Suffolk and only one Essex. The Southeast has yielded just 12 birds (five of them in Kent), but it is responsible for two inland records: Bracknell, Berkshire, from 19th February to 15th April 1984 and Woodford Green, London, from 23rd-26th October 1992.

In the far west the species remains rare. Of seven Irish birds six were in Co. Cork and one in Co. Wexford. Four Welsh birds were all on islands: three from Pembrokeshire and one on Bardsey. The only record from the northwest is one on the Calf of Man.

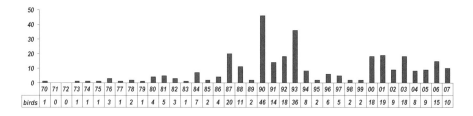

Figure 16: Annual numbers of Olive-backed Pipits, 1970-2007.

When: Six wintering birds have been detected in January and February: Berkshire (19th February-15th April 1984); Shetland (3rd January 1988); Co. Cork (23rd-24th January 1991); Essex (13th January-2nd April 1994); Devon (18th January-9th April 1997); and Norfolk (2nd to at least 20th February 2002). The secretive nature of the species probably obscures its true status here in winter.

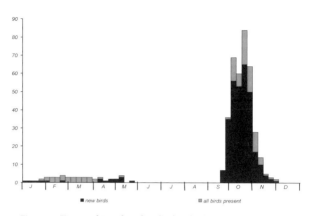

Figure 17: Timing of British and Irish Olive-backed Pipit records, 1948-2007.

Despite the rapid change in the autumn fortunes of this lovely pipit in recent years, spring records remain a rarity with just 11 birds detected. Eight of these have been since 1991, correlating with the winter records and the start of increasing numbers arriving in autumn. This pattern hints at an increasingly regular winter presence, be it in Britain or elsewhere in Western Europe. All bar one were on islands (if we take the tombolo at Portland to form an island): four in Shetland; two each in Pembrokeshire and Dorset; and singles on the Isle of May and the Calf of Man. The only mainland record was at Thorngumbald, East Yorkshire, on 22nd April 2007. Five have been between 12th-24th April (earliest Portland 12th April 1991) and six between 2nd and 22nd May (latest Isle of May 22nd-23rd May 1985).

It is autumn that accounts for the vast majority of records. The earliest ever was on Fair Isle on 18th September 2006 (pre-dating the next earliest there on 20th September 2001; the island accounts for four of the seven earliest records). The earliest on the mainland were at Holkham Meals, Norfolk, from 21st-23rd September 1996 and Flamborough Head from 22nd-23rd September 1994. The main thrust of arrivals takes place from late September, peaking in the third week of October and maintaining a high level of arrivals through to the end of the month. Interestingly, on Fair Isle, arrivals occurred much earlier in the 1990s than sightings prior to then, but since 2000 have reverted to later dates (see Table 11).

Table 11. Median arrival dates of autumn Olive-backed Pipits on Fair Isle.

Period	Pre 1990 (n=24)	1990-1999 (n=28)	2000-2007 (n=20)
Median Date	15th October	4th October	16th October

Numbers fall off rapidly in early November with just three after the 13th. Three late November records (20th-26th) include two on Shetland. The third and latest was at Hartlepool Docks,

Cleveland, on 26th November 2000 during a year in which a late autumn influx accounted for seven of the 25 November records for the species.

Discussion: The numbers of OBPs reaching us is clearly on the increase and it now seems a long time since this species was revered as an enigmatic Sibe. This most attractive pipit remains an excellent find in many parts of the country, but its days as a true rarity seem numbered if the present trend continues. For those who stood on the Berkshire housing estate in late winter 1984 it would be sad to see it go the way of those other popular Sibes Radde's Warbler *Phylloscopus schwarzi* and Dusky Warbler *P. fuscatus*, but there does seem an air of inevitability about its impending loss of status if the present magnitude of arrivals is maintained.

The reasons behind the increase are unclear, but given that the western part of the breeding range is imperfectly known it seems likely to be linked to a westerly range expansion or population increase. The variability in numbers recorded here presumably reflects this expansion of the breeding range, coupled with good or bad breeding success in any given year and subsequent population surges.

The British experience is not matched for magnitude elsewhere in Europe, although many countries have seen a similar escalation in numbers during recent times. Norway had a total of 57 to the end of 2006, Finland had 42 records (all but four since 1986) and Sweden 24 (all since 1988), but there are just five records from Denmark. Germany has seen more than 25 (most during October on Helgoland), the Netherlands 17 and Belgium two. Four have reached the Faeroe Islands and one has wandered to Iceland.

Interestingly there is a sizeable number of reports from the Mediterranean, with five from Malta, three records of four birds from Spain (including one on 13th December 2000 and two together from 16th-25th December 2001) and four from Cyprus (including one on 10th February 1989 and two together from 22nd-29th March 1998). Turkey (13th April 1992), Portugal, Italy and the Canary Islands have also scored with this species.

In the Middle East multiple records occur annually in Israel, especially in October-November. In recent years small numbers have wintered in the South Arava valley and Eilat (Yoav Perlman *pers comm*). In Kuwait it is a rare winter visitor in small numbers. There is a January record from Iran. It seems that sightings in Europe and farther east reflect an expanding reach of this endearing species.

Siberian vagrants were formerly considered to be reversed migrants to Europe, but growing evidence now points toward some Sibes occurring here not as accidental visitors, but as pseudo-vagrants that are performing annual migrations to presently undiscovered wintering grounds in western Europe or Africa (Gilroy and Lees 2003). The growing number of wintering and spring records of Olive-backed Pipit from Britain and elsewhere fits such a theory.

Pechora Pipit *Anthus gustavi*

Swinhoe, 1863. Breeds within narrow region of scrub-tundra and taiga of subarctic Eurasia, from Pechora region of northeast Russia across Siberia to Chukotskiy Peninsula and Kamchatka. Migrates through eastern China and Taiwan to wintering areas in Philippines, Northern

Borneo and Northern Sulawesi. An isolated race menzbieri breeds northeast China and Amur River region of southeast Russia.

Polytypic, two, sometimes three, subspecies. Records are attributable to nominate *gustavi* Swinhoe, 1863. *A. g. menzbieri* Shulpin, 1928 of northeast China and south-easternmost Russia is less rufous than *gustavi* with broader streaks on mantle and crown. It is also distinctly smaller (Alström *et al* 2003).

Rare migrant from Siberia, almost exclusively to Shetland in September and October. Rare elsewhere in Europe away from islands.

Status: There have been 81 records of this sometimes hellishly skulking pipit to the end of 2007; 79 in Britain and two in Ireland.

Historical review: The first was from Fair Isle on 23rd-24th September 1925 (when shot). As with several other Sibes this record was to fashion an association between species and island that has stood the test of time. The island also hosted birds in 1928 and in 1930 (two).

October 16th-18th 1952. This small, dark brown bird proved so great a skulker that neither Edward Skinner nor I were able to get a single satisfactory view of it during several encounters spread over the three days. It haunted the depths of two neighbouring cabbagepatches, going from one to the other each time it was disturbed. Our only brief view of it in the open was obtained late on the 16th and the suggestion of a buffish stripe down the mantle contrasting with the dark brown upper parts, plus the fact that whitish outer tail-feathers had already been seen when the bird flew, led us to suspect a Petchora Pipit. This provisional identification was confirmed on the following morning by James Wilson, who has seen and heard the species on previous occasions. The bird had the misfortune to fly into telephone wires on the 18th, and the skin is now in the Royal Scottish Museum. **Kenneth Williamson.** British Birds 46:210-212.

Four in the 1950s also came from Fair Isle (1952, 1953 and two in 1958) as did one of the two for the 1960s. It was 1966 that provided the first to break the mould, when a bird was trapped at Spurn on 26th September.

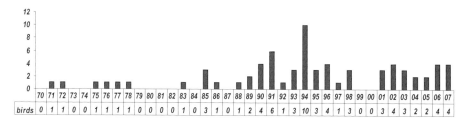

Figure 18: Annual numbers of Pechora Pipits, 1970-2007.

It was back to business as usual in the 1970s, with all six records coming from Shetland between 1971-78, including five on Fair Isle. One on Whalsay, Shetland, on 3rd October 1972 was only the second to be found away from Fair Isle.

There was then a five-year gap before a south coast bird at Portland on 27th September 1983. The respite from Fair Isle clutches was short-lived: six of eight records that decade frequented hallowed turf, the other exception being one from Orkney. Three from Fair Isle between 22nd-24th September 1985 represented a record annual total. Up to the end of 1989, no fewer than 20 of the first 24 Pechora Pipits had been found on Fair Isle.

One at Firkeel, Co. Cork, from 27th-28th September 1990 was the first record for Ireland. It started the run of records that was to change the status of this once great rarity forever. Only one record of four in 1990 was from Shetland. On 20th October birds arrived at Portland and Land's End. A new record of six followed in 1991. These included the only accepted spring record, on the Calf of Man, from 4th-9th May. Five made landfall on Shetland between 17th-22nd September that year.

During the 1990s records came from each year to 1998 and amassed a cumulative total of 35 birds. These included a monstrous 10 in 1994, with the first for Scilly and eight on Shetland. The other was a treat for some mainland birders: a one-day bird at Filey, North Yorkshire.

October 9th, 1994 dawned misty with light easterlies off a large high pressure extending into Siberia, so it was with some optimism that I set off to check the country park. By 08.00hrs I hadn't found any new arrivals as I stood somewhat disconsolately at the edge of the Totem Pole Field. With just a single observer it usually makes sense to walk along the edge of the field checking the scrub, but on this occasion I decided to zig zag across the field hoping to flush something.

Ten yards over the fence I flushed what was obviously going to be the first pipit of the day. It got up without calling, looked slightly bigger than a Meadow Pipit and as so often happens when one finds a rarity 'it just looked rare'. After ditching in long grass it again flushed with the same result, although this time it flew slightly further. Brief views revealed a very streaky pipit with the hint of pale braces, and after landing in shorter grass, a careful approach and a five minute wait brought obscured views on the ground.

The initial views gave the impression of a heavily streaked pipit with whitish tram-lines, pale lores and whitish underparts, heavily streaked across the breast. Unfortunately I could not get a clear view of the wings, but two off-white wing-bars were obviously present and at this stage there seemed to be no primary projection. In retrospect clearly the views were insufficient to see the primary projection (in fact at 8.8mm it should be easy to see in the field but only providing the bird isn't moving and is in full view!).

At this point the bird vanished and after a fruitless search I called Pete Dunn over, who was ringing in the top scrub. After ten minutes we eventually flushed the pipit from rank vegetation but were unable to see it on the ground. Over the next hour the pipit was seen briefly on the deck twice, both times by observers with experience of Pechora's on Fair Isle, all other sightings unfortunately related to flight views only. Pechora Pipit was discussed but with some features seemingly favouring Red-throated Pipit and the apparent lack of primary projection, all present eventually favoured the latter species. The pipit showed no signs of giving better views and, therefore, I decided to leave it for a while and hoped that it would settle and also that the light would improve.

As luck would have it over the next twenty minutes the sun broke through and with more birders arriving the pipit was pinned down more regularly. Whilst stood at the tip I received a CB message that eventually the bird had been seen well (but briefly) on the ground, and that several observers suspected that it was a Pechora Pipit, and that indeed it did appear to possess a primary projection. Hurrying back to the Totem Pole Field I met Mike Bayldon who had seen several Pechora's before and he was convinced that it was indeed a Pechora Pipit. However, to allay any doubts Pete Dunn trapped the pipit. I rushed up to the net took one look at the wing tips and as soon as I saw the primary projection, I can honestly say that I was the only person in the field who was absolutely gutted.

After processing, the bird was released back into the Totem Pole Field where it remained very elusive for the rest of the day. However, with up to 900 birders visiting the site before dusk – often forming 'viewing circles' around the bird's feeding area – most did get a glimpse of the bird on the ground. In the late afternoon it flew into the rose bushes at the eastern end of the tope scrub where it could be seen on the ground in the fading light. Just before dark it flew back to the Totem Pole Field to roost.

Craig Thomas. *Yorkshire Birding 3:120-122*

The species failed to appear in 1999 and 2000, but three were recorded in 2001. The period from that year to 2007 was slightly more subdued than the previous decade, though still accumulating 22 records. All but three of these were on Shetland, including three on Fair Isle in 2002. However, wider exploration of islands by visiting birders resulted in an increasing number of finds away from Fair Isle (for example, two on Foula in 2004). The second for Ireland was found in 2001 and only the second for Orkney (and North Ronaldsay) in 2005. By far the most well-watched bird of the decade was the latest ever and a first for Wales at Goodwick Moor, Pembrokeshire, from 19th-23rd November 2007.

The average number of records since 1980 is 2.3 per annum. The bounty of the 1990s produced 3.5 per annum in that decade.

Where: Pechora Pipits remain almost exclusively the preserve of Shetland, which accounts for a gigantic 81 per cent of sightings. An exceptional 51 per cent (41 birds) have been on Fair Isle, where the median arrival date since 1970 is the 24th September (n=32). With increased intensity of effort elsewhere in Shetland, Fair Isle's grip on records has slipped in recent years. But, the archipelago is still the place for self-finders to encounter the species.

Table 12. Median arrival dates of Pechora Pipits on Fair Isle.

Period	Pre 1970 (n=9)	1970-1979 (n=5)	1980-1989 (n=6)	1990-1999 (n=15)	2000-2007 (n=6)
Median Date	30th September	21st September	29th September	22nd September	30th September

There are three records from elsewhere in Scotland, two from Orkney (North Ronaldsay in 1989 and 2005) and one on the mainland in Northeast Scotland in 1993. Remarkably only two have been recorded from the east coast, both in Yorkshire (1967 and 1994). Yet the southwest has produced six records: one on Scilly (1994), two from Portland (1983 and 1990) and three in Cornwall (1990, 1995 and 1996). In Ireland records have come from Co. Cork (1990) and Co. Donegal (2001). There are singles from the Isle of Man (1991) and Wales (2007).

When: There is just one spring record (Calf of Man, 4th-9th May 1991), which is one of just three spring records for Europe (see Spain and Poland below). One from Minsmere, Suffolk, on 27th April 1975 is no longer considered acceptable.

Autumn records are condensed into a compact arrival period. The earliest

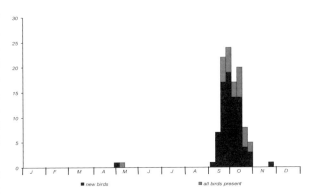

Figure 19: Timing of British and Irish Pechora Pipit records, 1950-2007.

was from Fair Isle on 8th September 1978, pre-dating the next earliest, also on the island, by five days from 13th-17th September 1995. The peak period for arrivals is the last two weeks of September (44% of records) but numbers drop only slightly in the first two weeks of October (35% of records). After mid October records become much more infrequent. There are four late birds between 15th-20th and three from 23rd-27th, including singles from Fair Isle on 27th October 1930 and Tresco from 27th-28th October 1994. These are followed over three weeks later by the showy bird in Pembrokeshire from 19th-23rd November 2007.

Five of the eight latest records come from the southwest from 20th October onwards. In southwest England and Ireland, excluding the late Welsh individual, the median arrival date is 13th October (n=8), nearly three weeks later than the expected arrival date for birds on Fair Isle. This pattern strongly suggests a trickle-down effect.

Discussion: The breeding range is thought to extend west to the Pechora, but, in reality, the species is found hardly at all west of the River Ob' where it breeds very locally and not annually (Sokolov 2003; Ryabitsev 2008). Sokolov (2006) also recorded Pechora Pipit on the west coast of Yamal Peninsula.

This species is exceptionally rare in Europe. There are single records from Finland (9th September 1972) and Sweden (5th-20th September 1991), but there have been 19 in Norway to the end of 2007 (29th September 1976, 21st September 1991, 1st October 1992, 10th October 1994, 7th September 1996, 3rd October 1996, no fewer than five records in the period 19th September-7th October 2002, 4th October 2003, 26th September 2004, 29th September 2004, 30th

September 2004, 1st-2nd October 2004, 22nd September 2006, 13th September 2007, 22nd September 2007).

Elsewhere there are singles from Iceland (collected on 9th October 1967) and Lithuania (4th November 1994), two from Poland (30th September 1983, 14th April 1985) and one each from France (16th September 1990) and Germany (3rd-4th October 2007). The only sighting from southern Europe was in Spain (29th April 1969), but as this record occurred prior to the formation of the Spanish Rarities Committee it has never been assessed and does not form part of the Spanish list.

The species can be remarkably elusive during its visits to our shores, being both skulking and reticent to call. Most stays appear to be of relatively short duration. It is tempting to speculate that others go undetected on the well-vegetated coast of the British mainland. Shetland holds a virtual monopoly on European sightings, which has led to suggestions that it is a classic reverse migrant (Cottridge and Vinicombe 1996). However, the few spring records in Western Europe suggest that this species may better be considered a potential pseudo-vagrant (Gilroy and Lees 2003). Considering how difficult the species is to find, the number of filter-down individuals that have been encountered elsewhere in Britain and Ireland suggests that others must turn up away from Shetland.

Buff-bellied Pipit *Anthus rubescens*

(Tunstall, 1771). North American race A. r. rubescens 'American Buff-bellied Pipit' breeds western Greenland, north and northwest Canada, and Alaska, winters west and south USA, Mexico and Central America. Asian race japonicus 'Siberian Buff-bellied Pipit' is a vagrant to the eastern regions of the Western Palearctic. It breeds northeast Siberia west to at least the Baikal region, winters northern Pakistan and northwest India to south and east China, South Korea and southern Japan.

Polytypic, three distinct subspecies recognised, sometimes four. All British and Irish records have been attributable to nominate *rubescens* 'American Buff-bellied Pipit' (Tunstall, 1771). Race *japonicus* 'Siberian Buff-bellied Pipit' Temminck and Schlegel, 1847 is an anticipated future vagrant. In winter *japonicus* is separated easily by its whiter underparts and larger, blacker and more clear-cut streaks reaching farther on to the underparts. The streaks often merge on the upper breast to form a necklace. Legs are typically pale (Alström *et al* 2003).

American Buff-bellied Pipit is a very rare vagrant from North America; all records between September and October, apart from two in November and one in December.

Status: Until 2007 there had been just seven records, but an unprecedented arrival in 2007 added a further 12. Four more were recorded in 2008*. Of the 23, 14 have been in Britain, eight in Ireland and one at sea.

1910 Outer Hebrides: St. Kilda immature male caught 30th September
1951 Co. Wexford: Great Saltee Island 8th-16th October, trapped 9th and 11th

1953 **Shetland:** Fair Isle immature 17th September
1967 **Co. Wicklow:** Near Newcastle 19th October
1988 **Scilly:** St. Mary's 9th-19th October
1996 **Scilly:** St. Agnes 1st-winter 30th September-2nd October; same, Tresco 7th-13th October; same, St. Mary's 23rd-28th October.
2005 **Lincolnshire:** Wyberton 1st-winter 5th-13th December
2007 **At sea:** Sea area Hebrides 200+ km NW of Outer Hebrides 19th-20th September, died on board ship
2007 **Shetland:** Fair Isle 1st-winter 23rd-25th September; presumed same, 1st-7th October
2007 **Scilly:** St. Mary's 25th September-2nd October
2007 **Scilly:** St. Mary's 27th September, presumed same Tresco 27th September-2nd October.
2007 **Co. Cork:** Lissagriffin one on 5th-6th October, two from 7th-21st October
2007 **Co. Clare:** Clahane Beach, Liscannor 7th-13th October
2007 **Oxfordshire:** Farmoor Reservoir 8th-10th October
2007 **Outer Hebrides:** Torlum, Benbecula 18th October
2007 **Co. Cork:** Ballycotton 31st October-10th November
2007 **Co. Wexford:** Carnsore Point 2nd November
2007 **Co. Cork:** Red Barn Beach, Youghal 25th November-21st March 2008
2008 **Outer Hebrides:** St. Kilda 19th-22nd September*
2008 **Scilly:** Bryher 3rd-7th October*
2008 **Orkney:** North Ronaldsay 3rd-13th October*
2008 **Outer Hebrides:** South Uist 1st-2nd November*

The events of autumn 2007 produced at least 12 birds: six in Ireland, five in Britain and one at sea. The pick of these records was an inland find at Farmoor Reservoir, Oxfordshire, from 8th-10th October.

Discussion: Until recently Water, Rock and Buff-bellied Pipits were considered to be a single species: Water Pipit *A. spinoletta*. In 1986 the *BOURC* proposed that the Water Pipit complex should treated as a superspecies comprising Rock Pipit *A. petrosus*, Water Pipit *A. spinoletta* and Buff-bellied Pipit *A. rubescens* (*BOURC* 1986; Alström and Mild 1987; Knox 1988a). American and Asian Buff-bellied Pipit are not yet treated as separate species, but Zink *et al* (1995) found substantial differentiation in mDNA suggesting that they may be considered candidates for a future split.

Reports of American Buff-bellied Pipit elsewhere in the Western Palearctic have multiplied rapidly. These include 17 from Iceland since 1977. Five between 20th-28th September 2008; one found dead 24th April, seven 16th-28th September, eight 2nd-21st October; latest 14th November. There is a single record from Norway (10th October 1997), two in Sweden (trapped, 3rd-6th December 2005, 22nd November 2008-23rd January 2009) and three from Helgoland, Germany (all in the 19th century: November 1851, May 1858 and September 1899).

During the 2007 influx singles also reached France (26th September-10th October 2007) and the Channel Islands (Jersey, 24th-27th November 2007). There have been six from the Azores

(29th October 2005, 2nd-3rd November 2005, 23rd January 2006, two on 18th January 2007, 1st November 2007).

Siberian Buff-bellied Pipit is yet to be recorded here. Elsewhere in the Western Palearctic it is a scarce but regular passage migrant and winter visitor to the Middle East, with arrivals from late October onwards and at least 60 during the winter of 1986/87 (Shirihai and Colston 1987; Shirihai 1996). There are just a handful of records in Europe: Italian Buff-bellied Pipits (13th November 1951, 26th October 1960) may refer to *japonicus*, though the 1951 bird could be *rubescens*; a record from Sweden (29th December 1996-14th January 1997); and a well-documented bird from Norway (12th January-8th April 2008) that is still to be assessed. There is also a recent record from Turkey (22nd November and 11th December 2008).

American Buff-bellied Pipits in eastern and central North America start leaving their breeding grounds in northern Canada by late August. It could be argued that this subtle pipit is surprisingly scarce in Britain and Ireland given that breeding birds occur as 'close' as western Greenland and that it is an increasingly regularly detected vagrant to Iceland. Williamson (1959) considered that there was strong evidence (based on plumage differences) that Meadow Pipits *A. pratensis* on west coast passage originated from southeast Greenland and Iceland. Further ringing proof of these movements is provided by three autumn English east coast controls. Meadow Pipits from such populations might be considered potential carriers for *rubescens* that have already been displaced eastwards.

Buff-bellied Pipit has doubtless been overlooked in the past. Observers with new interest because of the split started to find the species; increased familiarity has produced a surge in records that may well continue in years to come.

Yellow Wagtail Motacilla flava
Linnaeus, 1758

Yellow Wagtail is one of the most variable species in the Palearctic. Males in particular have distinctive plumages in different populations. There also appears to be extensive hybridisation wherever races/populations meet. Recent analysis of molecular data gives support for the idea that Yellow Wagtail should be separated into at least two species: Western Yellow Wagtail M. *flava* and Eastern Yellow Wagtail M. *tschutschensis*. Within Eastern Yellow Wagtail, races *taivana* and *macronyx* may be specifically distinct from *tschutschensis* as Chinese Yellow Wagtail (see for example: Ödeen and Alström 2001; Alström and Ödeen 2002; Voelker 2002; Pavlova *et al* 2003; and Ödeen and Björklund 2003).

(Western Yellow Wagtail breeds from the Atlantic coast of western Europe and North Africa to central Siberia. Subspecies reported from Britain and Ireland within this grouping include: *flavissima* of Britain and Ireland, the northern coast of France and bordering the North Sea to southern Norway; *flava* of western and central Europe north to southern Sweden and east to the Urals; *iberiae* of the Iberian peninsula, southern France and north-west Africa; *cinereocapilla* of Italy; *beema* of Russia and western Siberia; *thunbergi* from Fennoscandia across Siberia north of c60°N, possibly merging into *plexa*, which is synonymised with *thunbergi* by Alström *et al* (2003), but see below; *feldegg* of the Balkans through central Asia to western China. The various subspecies winter across much of sub-Saharan Africa and the Indian subcontinent.)

(Eastern Yellow Wagtail breeds largely to the east of Lake Baikal and comprises four or five sub-species that are well separated from Western Yellow Wagtail in both mitochondrial and nuclear DNA: *tschutschensis* (including *simillima* (Kamchatka), which Alström *et al* (2003) treat as a synonym of *tschutschensis*), breeds central and eastern Siberia to Alaska; *plexa* from Taimyr to Kolyma; *macronyx* from eastern Mongolia, northeast China and southeast Russia; and *taivana* from eastern Siberia and northern Japan; *leucocephala* from north-west Mongolia has not yet been analysed, but judging by its geography it is likely to belong with the Eastern group. These subspecies winter from India eastwards across southeast Asia and the Philippines, and south-east China.)

The two northernmost taxa (*thunbergi* and *plexa*) are perplexing. Alstrom *et al* (2003) consid-ered *plexa* to be indistinguishable on plumage from *thunbergi*, with which it was synonymised. These two taxa are phenotypically indistinguishable, but molecular analysis of these two appar-ently contiguous forms shows that they are widely separated and distinct taxa, with *thunbergi* belonging to Western Yellow Wagtail group and *plexa* to Eastern Yellow Wagtail (Pavlova *et al* 2003). Further analysis supported by larger datasets will doubtless lead to further advancement and perhaps revisions. It is currently unclear if these forms have contiguous distributions or even if there is a gap or an overlap. The difficulty is that molecular analysis is the only sure way of identifying one from the other.

Polytypic, a complex species with up to 18 taxa (Cramp *et al* 1988) though Alström at al (2003) recognised 13 forms assigned mainly on the basis of male plumage and geographic distribution. Only *flavissima* (Blyth, 1834), *flava* Linnaeus, 1758 and *thunbergi* Billberg, 1828 are considered to be regular visitors to Britain.

Several other races are regarded as vagrants. Black-headed Wagtail *feldegg* Michahelles, 1830, has been recorded on 13 occasions. Ashy-headed Wagtail *cinereocapilla* Savi, 1831 is on the *British List* (first record from Cornwall in 1860) and evidence indicates that it is a rare visitor to Britain. Other forms have also been accepted, but their identification and status is presently under review (Dudley *et al* 2006). For detailed overview of geographical variation see Alström *et al* (2003).

The *BBRC* recently clarified its position regarding Yellow Wagtail forms. Of the vagrant races only *feldegg* has traditionally been assessed by the *BBRC*, although several other forms are cur-rently on the *British List* and others are suspected of occurring. But, owing to individual vari-ation and racial intergrades at range boundaries, positive racial identification is sometimes extremely difficult or impossible (Kehoe 2006).

Most subspecies interbreed to varying degrees in areas of contact. British *flavissima* breeds with nominate *flava* along the coasts of the English Channel; hybrids are often termed 'Chan-nel wagtails' (Dubois 2001). In France there are four distinct intermediate populations (Dubois 2001; Dubois 2007) including: Channel wagtail; Middlewest wagtail intergrade *flava* x *iberiae*; Mediterranean wagtail hybrid *iberiae* x *cinereocapilla*; and eastern wagtail intergrade/hybrid *flava* x *cinereocapilla*.

Some Channel wagtails strongly resemble central Asian *beema*, making identification of vagrant *beema* exceptionally difficult. Pale individuals can resemble central Asian *leucocephala*, a seemingly unlikely vagrant. Together these forms may confound attempts to verify the occurrence of some vagrant forms. Both *beema* and *leucocephala* are presently on the *British List* based on old records that pre-date *BBRC*.

Western Yellow Wagtail Forms

Black-headed Wagtail Motacilla flava feldegg

Breeds Balkans and Greece east through Turkey to eastern Kazakhstan and Afghanistan and south to Iran. Western populations winter Nigeria to Uganda and south to Congo. Eastern populations winter northwest India.

Very rare spring overshoot from southeastern Europe.

Status: Thirteen accepted records (plus two in April 2009 in Norfolk and Northumberland).

1970 **Shetland:** Fair Isle male 7th-9th May
1983 **Norfolk:** Cley male 30th July-6th August
1984 **Lothian:** Skateraw male 28th April
1985 **Northumberland:** Cresswell Ponds male 2nd June
1985 **Suffolk:** Landguard male 30th June
1986 **Pembrokeshire:** Skomer male 7th May
1988 **Oxfordshire:** Brightwell-cum-Sotwell male 4th-5th June; same Roke 12th June-10th July
1998 **Caernarfonshire:** Conwy male 8th-9th May
1999 **Cornwall:** Camel Estuary male 11th-12th May
2003 **Scilly:** St. Mary's 1st-summer male 6th May
2004 **Lincolnshire:** Holbeach Marsh male 4th June
2005 **Devon:** West Charleton Marsh male 8th–18th June; same South Huish 8th July-2nd September
2006 **Somerset:** Minehead male 23rd-24th April

The criteria for acceptance of Black-headed Wagtail were outlined recently by the *BBRC*. *Observers of a potential pure feldegg should establish clearly the colour and extent of gloss on the crown and nape, and whether the bird shows any potential intergrade features. These would include the presence of any yellow, grey or white superciliary stripes and/ or the presence of any extensive white in the throat. A restricted amount of white in the malar region would not necessarily be incompatible with true feldegg, and many apparently pure feldegg appear to show a restricted area of white at the base of the bill.*
    The eastern form of 'Black-headed Wagtail' M. f. 'melanogrisea', typically shows a narrow white malar stripe, but also lacks a pale supercilium. Other supporting features which should be assessed carefully are the extent to which the blackish colour of the nape extends onto the mantle, the nature of the demarcation between the nape and the man-

tle, and the colour of the wing-bars. The wing-bars are frequently more yellow than on typical flava or thunbergi in fresh plumage, but become whiter than other forms owing to wear. This feature is particularly variable, however, and may be difficult to interpret in the field.
Adam Rowlands. British Birds 96: 291-296

Discussion: Many other birds have been suspected as *feldegg* but have fallen short because intergrades could not be eliminated conclusively (Rowlands 2003). Some confusing early literature misleadingly suggested that some *thunbergi* were near-identical to *feldegg* (van den Berg and Oreel 1985; Svensson 1988). The criteria outlined above have clarified the expectations of the *BBRC* for a successful claim in light of a well-observed bird that was ultimately found not to be acceptable (Rowlands 2003). Records dropped off in the 1990s after a good run in the 1980s, but seem to be returning to annual status as observers identify males with renewed confidence. The recent run of males raises the question whether female *feldegg*, some of which are distinctive, are escaping detection.

The Devon individual in 2005 was well watched over its long stay, easily eclipsing the length of residence by the Oxfordshire bird in 1988. All others have been brief. The east coast emphasis of earlier records has now been replaced with a southwest bias. Late April to early May appears to be the time to be out searching for this eye-catching form.

This taxon has been accorded species rank by some authors (Devillers 1980; Stepanyan 1990). That and its dramatic appearance makes a long-stayer an attractive proposition for those with an eye on future splits. However, molecular analysis indicates that the form is not separated genetically as much its appearance would suggest. Black-headed Wagtail is expanding its range in the Balkans and adjacent parts of south-east Europe. Perhaps the recent upturn in sightings is linked to range expansion as well as a better understanding among observers as to what might constitute a pure *feldegg*.

As with all potential vagrant Yellow Wagtail forms, call is a key variable that observers should try to describe in detail or even better record. The sole Netherlands record (10th May 1988) has recently been rejected because the call was not heard let alone sound-recorded (Arnoud van den Berg *pers comm*).

Ashy-headed Wagtail M. f. cinereocapilla

Form *cinereocapilla* Savi, 1831 from Italy is presently on the *British List* (first record Cornwall in 1860). A review of Scottish records found none were documented sufficiently well (Forrester *et al* 2004). Two pairs of *cinereocapilla* reportedly bred in Belfast, Co. Antrim, in 1956 (Ennis and Dick 1959). There is also a record from Rathlin Island, Co. Antrim, on 1st May 1985 and also one from Copeland Island, Co. Down, on 9th May 1998 and for the following two days at Belfast Harbour Estate.

Evidence suggests that *cinereocapilla* is a rare visitor, but the extent to which intergrades (such as between *flava* and *thunbergi* and their backcrosses) produce birds that suggest *cinereocapilla* is unclear. Of possibly greater concern is the issue of *iberiae* x *cinereocapilla* intergrades; some birds may be acceptable only as *iberiae/cinereocapilla*. Claims of this taxon will be treated informally by the *BBRC* until diagnosability issues are investigated further (Alström *et al* 2003; Nieuwstraten 2004; Kehoe 2006).

Spanish Wagtail M. f. iberiae

This taxon is not on the *British List*, but seems likely to occur. One extremely detailed claim (North Warren, Suffolk, 25th April 1998) included high-quality illustrations, but lacked voice details and photographs. It has been assessed by both *BBRC* and *BOURC* and was found not to be acceptable because of the problem of eliminating near-identical hybrids. A well-documented claim from Conwy RSPB in late April 2008 is still to be assessed. A first record will require all key characters to be covered, including detailed recording of the voice. Intergrades are a potential problem and some apparent *flava* can also show quite extensive white on the throat and could be an *iberiae* pitfall (Alström and Mild 2003; Winters 2006).

Other forms

Two central Asian forms: White-headed *leucocephala*, which was not sampled in the recent genetic work; and Sykes's Wagtail *beema* are on the *British List* based on old records that pre-date *BBRC*.

The status of these races and Eastern Yellow Wagtail is under review by *BOURC* (Dudley *et al* 2006). It is thought likely that most birds in Britain resembling *beema* (of which there are many, including some that are a close match to actual specimens of *beema*) and *leucocephala* (of which there are few) represent variants of *flava* or intergrades such as Channel wagtail. The occasional birds resembling the yellow-headed *lutea* (not on the *British List*) are most likely to be variant or old male *flavissima* (Kehoe 2006).

Eastern Yellow Wagtail Forms

Eastern Yellow Wagtail M. f. tschutschensis

This form is on the *British List* based on two specimen records of what were formerly called *simillima* Hartert, 1905, both from Fair Isle: 1st-winter female collected on 9th October 1909 and a 1st-winter male collected on 25th September 1912. These birds are a good match with *simillima*, but the extreme variation exhibited by all Yellow Wagtails make it necessary to look at DNA to confirm these two specimens as the Eastern taxon. Genetic material from these birds is being analysed to try and assess their taxonomic attribution.

Some genuinely monochrome birds in Britain in autumn may be of east Asian origin and a number of recent autumn claims in late September and early October have coincided with vagrants from Siberia and beyond. Some individuals of *beema* (part of the Western species complex) breeding in the Volga steppes and wintering in East Africa can appear similarly monochrome, as can *feldegg* (Alström *et al* 2003) and even some breeding birds in Britain (Ian Wallace *pers comm*). Such variation renders racial assessment problematical in the absence of genetic material. The buzzy Citrine Wagtail-like calls of this eastern group may aid identification, but other races (for example, *feldegg*) also have buzzy calls and complicate the usefulness of this feature.

In conclusion, tackling identification of any of these forms is not to be undertaken lightly. Many or most will likely remain as possible/probable until firm criteria emerge. Adult males are potentially solvable, but even these are problematical on current knowledge. The conclusion coming out of recent work on these complex birds is that there is benefit in obtaining voice data for individuals suspected of being Iberiae or *cinereocapilla* and for all suspected vagrants as much documentation as possible should be gathered.

White Wagtail Motacilla alba

(Linnaeus, 1758). Breeds across and beyond the Palearctic from southeast Greenland and Iceland to western Alaska and south to the Himalayas. The nominate race, alba, occurs across the Western Palearctic, but is replaced in Britain and Ireland by yarrellii and in northwest Africa by subpersonata. Further races occur in the Eastern Palearctic. The most widespread form, ocularis, is replaced by lugens from Kamchatka through Japan to eastern China. In the south of the region migratory races personata, baicalensis, alboides and leucopsis occur.

Polytypic: eight to 11 subspecies recognised; Alström *et al* (2003) list nine. White Wagtails were included in molecular analyses by Odeen and Alström (2001), Alström and Odeen (2002), Voelker (2002b) and Pavlova *et al* (2005). From these analyses there was some evidence that the south Asian races *personata*, *alboides* and *leucopsis* formed a separate clade, though this was not supported by the nuclear sequences. White Wagtail *alba* and Pied Wagtail *M. a. yarrelli* Gould, 1837 were the only races proven to occur here, until a well-documented claim of Amur Wagtail *leucopsis* Gould, 1838 was accepted recently.

Amur Wagtail Motacilla alba leucopsis

Breeds in central and eastern China, Amurland and Ussuriland, Korea and southwest Japan. Winters from northern India to Southeast Asia.

Status: One record. Several other birds with characteristics of other eastern forms have been reported, but have been inadequately documented.

2005 Co. Durham: Seaham male 5th-6th April

Discussion: This record is one of the farthest-east passerine vagrants to have been recorded here. The escape potential needs to be investigated thoroughly for such an amazing sighting.

The systematics of *alba* wagtails have been treated variably over the years. A recent overhaul by Alström *et al* (2003) suggested that the *alba* complex comprises nine species under the phylogenetic species concept, as had been proposed by Sangster *et al* in 1999: White Wagtail *alba* (including *dukhunensis* Indian Pied Wagtail); Pied Wagtail *yarrellii*; East Siberian Wagtail *ocularis*; Black-backed Wagtail *lugens*; Himalayan Wagtail *alboides*; Masked Wagtail *personata*; Baikal Wagtail *baicalensis*; Amur Wagtail *leucopsis*; and Moroccan Wagtail *subpersonata*.

A female Amur Wagtail *leucopsis* was in Norway from 1st-2nd November 2008. Masked Wagtail *personata* has been recorded from Cyprus (1st-winter male 22nd September 1966), Norway (15th November 2003 to 9th April 2004) and Sweden (29th April 2006). There have also been recent European records of Moroccan Wagtail *subpersonata* in Portugal (male 13th-14th July 1995), Corsica (male captured and photographed 15th May 1997), Gibraltar (late July to 6th August 2006) and Spain (24th March 2007 and 25th March 2009).

Several other forms seem likely candidates for westward vagrancy and with greater awareness of these vagrants more forms may be identified. Using the classification adopted by Alström *et al* (2003) only four of the nine have not been recorded in Europe: Himalayan *alboides*; Black-backed *lugens*; East Siberian *ocularis*; and Baikal *baicalensis* (Addinall 2005). Given that all are long-distance migrants except *alboides* their chances of turning up here are high. The question remains whether observers are in a position to identify these birds in anything other than adult breeding plumage?

Citrine Wagtail Motacilla citreola

Pallas, 1776. Nominate race breeds in northern Russia, from East Kola and Kanin peninsula across northern Siberia to Taimyr peninsula and south to central Siberia. Small numbers now breed regularly in Belarus, the Baltic countries and occasionally southern Finland; otherwise from Ukraine and southern Russia, east across Kazakhstan and Mongolia to northern China. Black-backed race calcarata breeds south/central Asia to Tibetan plateau. Winters throughout Indian subcontinent, southern China and southeast Asia to peninsular Thailand.

Julian R. Hough

Polytypic, two well defined subspecies. The racial attribution of occurrences here is undetermined, but adult males have lacked a black back so *calcarata* is unlikely. Geographical variation is marked in males in summer plumage, but slight in females and non-existent in juvenile, 1st-winter and winter plumage. Male *calcarata* Hodgson, 1836 is separated by its all-black upperparts and black nape that reaches the central mantle. The head and underparts are a deeper yellow. Nominate male *citreola* Pallas, 1776 has greyer upperparts and paler yellow head and underparts. A third subspecies *werae* (Buturlin, 1907), breeds from southern Russia, Kazakhstan and northwest China. It is treated as a separate subspecies by most authors, but Alström *et al* (2003) contest that differences in plumage between *werae* and populations of *citreola* are too inconsistent to uphold the former taxon. Birds breeding in the Baltic, eastern Europe and eastern Turkey are generally assumed to belong to *werae* (Alström *et al* 2003).

133

Increasingly regular rare migrant from Siberia and eastern Europe. The majority involves 1st-winter birds on Shetland in late August and September, but spring sightings throughout Britain and Ireland are becoming more frequent.

Status: The total of 213 records to the end of 2007 comprises 195 in Britain and 18 in Ireland.

Historical review: The first occurred as recently as 1954, when a 1st-winter was caught in the Gully Trap on Fair Isle. It stayed from 20th to 25th September and was followed quickly by another there from 1st to 5th October.

> *That night a new party arrived on "The Good Shepherd", consisting of W. Conn, W. J. Eggeling, I. J. Ferguson-Lees, G. Mountfort, D. I. M. Wallace and W. J. Wallace. They were all out searching for the bird early next morning, but were unable to find it because one of the hostel staff had caught it in the Observatory "Helgoland" and put it in the laboratory pending my return from the traps! All were able to examine the bird, and after it had been photographed by Guy Mountfort it was set free.*
> **Kenneth Williamson.** British Birds 48:26-29.

Subsequent records also came from Fair Isle in: 1960 (October), 1961, 1962 and 1964 (all in September). One at Minsmere from 17th October to 14th November 1964 broke the isle's monopoly on British sightings. The Minsmere bird, as well as the first on Fair Isle, apparently lacked the full cheek surround now considered to be diagnostic of the species, but racial variation in this character remains unchecked (Ian Wallace *pers comm*). The 1960s produced 12 birds in all. The last blank return for occurrences was as long ago as 1965. The first for Ireland was at Ballycotton, Co. Cork, from 15th to 16th October 1968.

The 1970s fared marginally better with a total of 15 birds and each year of the decade producing one or two birds. The first record outside of autumn occurred in 1976: an unprecedented record of a male feeding four young wagtails at Walton-on-the-Naze, Essex, from 4th to 24th July. The female was not observed, but a mixed pairing with a Yellow Wagtail *M. flava* was considered likely and the nestlings were suspected of being hybrid Citrine x Yellow Wagtails (Cox and Inskipp 1978), even though hybridisation with *flavissima* is thought likely to be extremely rare (Dubois 2007).

> *On 4th July 1976, Mr and Mrs H. Huggins discovered an unfamiliar wagtail carrying food to a nest at an Essex coastal locality; they considered that it resembled a male Citrine Wagtail Motacilla citreola. At their invitation, SC went to look at it on 7th; he agreed with the identification and also readily located the nest, situated in an area of common saltmarsh-grass Puccinellia maritima on a small muddy peninsula in a saline lagoon. The nest contained four young wagtails, about one week old. SC watched the birds on eight more occasions up to 24th July, for a total of at least 16 hours; the fledged young were observed being fed by the adult male on 18th and 24th. On the last date, TI – who had been invited in view of his special interest in wagtails – also had excellent views of the male and the young. At the request of the landowner, and to safeguard a rare breeding bird, it was decided not to circulate widely details of the locality at the time.*

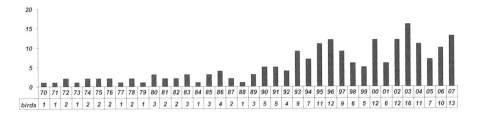

Figure 20: Annual numbers of Citrine Wagtails, 1970-2007.

At no stage was an adult female seen definitely associated with the nest or young, although, on 11th July, a female Yellow Wagtail M. flava (at least two pairs of the race M. f. flavissima were regular in the vicinity) alighted close to the nest, inspected the area, flew off, returned a few minutes later, and then was apparently chased away by the male Citrine. **Simon Cox and Tim Inskipp.** British Birds 71: 209-213.

The total for the 1980s was a new record of 24 birds, including a new record tally of four in 1986. With the notable exception above, all birds so far had been autumn vagrants. Despite the increase in numbers through the decade, Shetland went five years from 1980 without a single one as the geographical spread of records widened.

In 1990 two May birds at different locations in Kent (Sandwich Bay from 8th to 10th May and Bough Beech Reservoir on 29th May) constituted the first-ever spring overshoots. Annual tallies increased rapidly with 11 in 1995 and 12 in 1996, and the decade total was a whopping 73 birds. The surge in sightings was coincident with the start of regular breeding in the Baltic countries.

The period 2000-2007 continued in a similar vein with 87 birds recorded, including in 2003 a new record count of 16. The increase was associated with many more spring records. Since 1990 there has been an average of 8.9 birds per annum and double-figure tallies in eight years. The average since the turn of the millennium has increased to 10.9 birds per annum.

Where: The most likely area in which to encounter this striking wagtail in Britain is unquestionably Shetland, where 45 per cent of all records has occurred (96 birds, all bar two in autumn). There were seven during 1954 to 1969, 12 in the 1970s, five in the 1980s, 29 in the 1990s and 43 between 2000 and 2007. No fewer than 54 (56 per cent) of Shetland records have come from Fair Isle; up until 1988, 21 of the 24 Shetland birds were on that island. Since the 1990s, increased coverage of the rest of the archipelago resulted in Fair Isle losing its monopoly. The total of 15 found on the Fair Isle in the 1990s amounted to 52 per cent of all Shetland records. Since the millennium the 18 on Fair Isle accrued for only 42 per cent of Shetland records.

The next-best area is the English coast between Hampshire and the southwest with 41 birds (19 per cent), seven of them in spring. There have been 18 in Scilly and 14 in Cornwall, five in Hampshire, two in Dorset and singles in Devon and Avon. In Hampshire there is an inland record of a 1st-summer male at Fleet Pond from 15th to 16th May 1993.

Scotland away from Shetland has amassed 22 records (10 per cent), just two of which were in spring. Eleven of these are from Orkney (nine on North Ronaldsay), five from Lothian, four in the Outer Hebrides and singles in Fife, Highland and Northeast Scotland. The paucity of records from the east coast of Scotland seems surprising, but the theme continues along the English east coast. Between Northumberland and Kent there are only 30 records (14 per cent), with seven each in Northumberland and Norfolk, four in Suffolk, three in Yorkshire, two each in Cleveland, Essex, London and Kent and a single in Lincolnshire. In London there are inland records from Beddington sewage farm, a juvenile to 1st-winter from 24th to 28th August 1993 and King George V Reservoir on 22nd August 1994.

The 18 Irish records (eight per cent) have all bar two been in autumn. There are two from Co. Donegal, one each from Co. Mayo and Co. Clare, three from Co. Kerry, seven from Co. Cork and four from Co. Wexford. Surprisingly, there was just one Welsh record over the review period, from Skomer, Pembrokeshire, in 2000 and one in northwest England (Lancashire, May 1997), yet three have been found in the Midlands: spring birds from Leicestershire (1st-summer male at Eyebrook Reservoir on 18th May 1991) and Warwickshire (female at Brandon Marsh on 18th May 1997) and an autumn bird from Northamptonshire (Pitsford Reservoir from 15th to 18th November 1996).

When: Spring occurrences are a recent phenomenon and are correlated with the recent general upsurge in records. Of the 23 spring birds, nine have been along the east coast between Kent and Lothian, an area that fares extremely badly for autumn birds. Elsewhere, seven were in the south-west, three on the Northern Isles, two each in the Midlands and southern Ireland

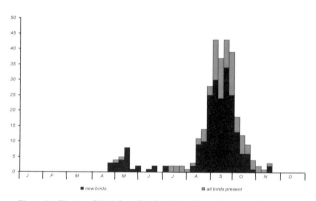

Figure 21: Timing of British and Irish Citrine Wagtail records 1954-2007.

and one in northwest England. The first occurred in 1990, when two were in Kent in May, followed by two more May birds in 1991. Others occurred in 1993, 1995 (two), 1996 (two), 1997 (three), 2001 (two), 2002, 2004 (three), 2005 (two), 2006 and 2007 (two).

The earliest record was a male at Salthouse, Norfolk, on 24th April 2004. There are just three other April arrivals: two from Fair Isle (26th April 1996 and 29th April 2005) and Bryher on 30th April 2005. Two follow in early May, but the peak period for spring arrivals takes place between 8th and 18th May (11 birds), after which there are two late May birds and one in early June. Three in late June were all on the east coast (Orkney, Lothian and North Yorkshire). In addition there are two midsummer records: the 1976 Essex bird and a long-staying 1st-summer male at Farlington Marshes, Hampshire, from 6th July to 6th September 1996.

It is in autumn that birders most expect to encounter this species, a season that accounts for 88 per cent of all records. The earliest was a 1st-winter trapped on Lundy, Devon, on 6th August 1998, followed by one at a more expected location, Fair Isle, from 10th to 13th August 2002. A further 33 (16 per cent of all records) have been found by the end of August. September is the best time to search, as the month accounts for 57 per cent of all records, with a discernible peak from 17th to 23rd. October arrivals diminish quickly; only seven arrivals occurred after mid-month and just one in the last week of October (Fair Isle, 28th October to 1st November 1997). The two latest records come from inland locations: Pitsford Reservoir, Northamptonshire, from 15th to 18th November 1996 and Welney, Norfolk, 16th to 17th November 1980.

With only 15 birds dated after 9th October (six in the 1960s, two in the 1970s, three in the 1980s and four in the 1990s), it seems possible that some late birds were variant juvenile Yellow Wagtail M. *flava* or of one of the monochrome Eastern Yellow Wagtails. Awareness of this potential pitfall came into focus only recently. Despite the surge in Citrine Wagtail numbers reaching our shores, birds later than the first week of October remain exceptional.

The trend in autumn is firmly for earlier arrivals. Fair Isle's dominance of records and constant coverage makes it a good indicator of changing occurrence patterns as shown in *Table 13*. Away from Fair Isle, the 18 autumn records in the southwest since 1992 have a median arrival date of 5th September, showing that birds arrive at the extremes of the country at roughly the same time rather than filter southwards.

Table 13. Arrival dates of autumn* Citrine Wagtails on Fair Isle.

Period	Pre 1990 (n=21)	1990s (n=14)	2000-2007 (n=17)
Median Date	16th September	6th September	2nd September

Two spring records from the island are excluded

Discussion: The change in status for this appealing wagtail has been rapid, given that the first birds came little over half a century ago. The vector for this onslaught on the archives is almost certainly the westward extension of the breeding range. At the end of the 1920s, the western limit of the breeding range ran through the territories of Ryazan, Tambov, Penza and Gorky regions and Tartary. By the beginning of the 1950s, the species had penetrated the Moscow region and has since extended farther west. In Ukraine it was first recorded breeding in 1976 (Wilson 1979) and there are now 8,300 to 13,800 pairs.

From 1976 Citrine Wagtails paired with Yellow Wagtails have been found nesting in Estonia, Finland, Sweden (with *thunbergi*) and England (*flavissima*); in the Middle East hybrids with *feldegg* have been reported (Shirihai 1990). The first breeding in Belarus occurred in 1982 and that population now stands at 1,000 to 2,500 pairs. In Lithuania it first bred in 1986, with two pairs in 1988-89 and six in 1992; now 50 to 100 pairs breed. In Finland the first breeding by a pure pair took place in 1991. It has become a scarce but regular breeding bird in southern Finland during the 2000s, with breeding confirmed in four locations in 2007 (Keskitalo *et al* 2008). Breeding commenced in Latvia in 1993 and the population now stands at 5 to 20 pairs. The first breeding in Poland (at least four pairs) was in 1994. The dozen or more pairs in 1995 rose to 25 to 50 pairs presently (Burton 1995, BirdLife international).

In Germany breeding has taken place twice. The first (unsuccessful) breeding was in 1996. In 2005, a pair in Niedersachsen became the most westerly breeders yet recorded. Breeding in Switzerland occurred for the first time in 1997 and in Austria in 2007 (Burton 1995, BirdLife International 2004). A bird hatched in the Gdańsk region was controlled the following year in northeast Germany (Hampe *et al* 1996).

In line with the increase in breeding pairs and British migrants there is a corresponding rise in records through western Europe. There are 18 records from the Netherlands between 1984 and 2006 and eight in Belgium between 1993 and 2007. At least nine reached Iceland up to the end of 2006 and two to the Faeroe Islands. There are 16 records from Switzerland between 1980 and 2008, including the first breeding in 1997. In Austria there were 47 records between 1980 and 2006, and breeding took place in 2007. The first for Hungary was as recently as 1989, but there have now been 38 records. Greece has 29 records. There are 28 records from Spain (26 in autumn and two in spring) including two on the Canary Islands, one of which wintered (20th November 2005 to 18th January 2006). Another wintered on mainland Spain from 14th December 2006 to 19th March 2007. Two have reached the Cape Verde Islands (including one from October 2007 to March 2008).

Clearly the trend is for increased numbers of birds and earlier arrivals coinciding with the westward expansion of the breeding range. Might another breeding attempt, this time involving a pure pair, not be too far away?

Cedar Waxwing *Bombycilla cedrorum*

Vieillot, 1808. Breeds from southeast Alaska and British Columbia across Canada to Newfoundland and south to northwest California, northern Georgia and western South Carolina. Winter range variable, but occurs from southern Canada, south to Central America and the Caribbean.

Monotypic.

Very rare vagrant from North America.

Status: Two records.

1985 **Shetland:** Noss adult 25th-26th June
1996 **Nottinghamshire:** Nottingham 1st-winter 20th February-18th March

The Nottinghamshire bird of 1996 occurred during something of a purple patch locally (the first British Redhead *Aythya americana* was discovered shortly after). The Nottingham bird consorted with a minimum of 600 Bohemian Waxwings *B. garrulus* during its stay. Needless to say, it proved a popular attraction.

For several weeks from late January 1996, the number of Waxwings in the city of Nottingham had been building up from the more normal groups of 20 to 40 birds to amazing flocks of up to 400. I had been to see them on several occasions and, on the morning of Tuesday 20th February, I counted a flock of 160 in the Sherwood district of the city, near to where I work.

I went home for lunch and dipped into the National Geographic Society field guide to the birds of North America (1983). On the Waxwing page I saw the Cedar Waxwing (*Bombycilla cedrorum*), and idly thought, "wouldn't it be nice to find one of those...." With so many Waxwings in the country and the past autumn's exceptional number of American vagrants, maybe the idea was not so far fetched? I noted the field characters and went back to work. Things were quiet and so, being self-employed, I decided to finish work at 3.15pm and relocate the waxwings.

Within 15 minutes I had found them, in a quiet street, feeding on Hawthorn berries in the hedge next to a nursing home. I parked my car across the street and watched them through the open window. By now, I had found counting them a little dull, so I started looking for white undertail-coverts. The effect was electrifying! How to pep up your birding in one easy move; the first bird I looked at had white undertail coverts! My first reaction was "Oh no! why me?" I quickly scanned the nearby birds to see if there were any more and , as I did not find any, looked back to this individual to note further details.

The bird looked much like the others, but it was obviously separated by its undertail coverts, which were white rather than rust coloured like those of all its neighbours. The lowest part of it belly also looked different: a paler, yellowish-buff. The wings were also very different, appearing almost completely plain. As I started to look at the face mask, the bird moved out of sight and I could not immediately refind it, but I had noticed that it looked rather different around its face mask to the other birds.

Now I was panicking. If nobody else saw it, who would believe me? But I had not seen the white above the mask properly, the fourth field mark I had noted in the book. Should I look for this first, to put it beyond doubt in my own mind, or get somebody else to see it and corroborate my report? I decided to go for reinforcements, I phoned around to find a birder who was not at work, and eventually got hold of Eric Birkinshaw. We both got back to the waxwings at about 3.45pm, not long before the time that they normally go to roost. Very quickly, I spotted the white undertail coverts, high above us in a beech tree. I now noticed for the first time that the bird was appreciably smaller than the other waxwings. I pointed it out to Eric, just before it dived down into some bushes behind a wall, but before we could relocate it, the whole flock took off and flew away into the distance, presumably to roost. This was very unsatisfactory, but at least Eric had been able to confirm that it had white undertail-coverts.

That evening, I was kept busy on the telephone as people checked my story; I got very little sleep that night. Needless to say, I was relieved in the morning, when Mark Dennis phone to say "It's confirmed. Well done, mate! Thirty people have just had crippling views of it"

Peter Smith. Birding World 9: 70-73.

Discussion: The first record, from Noss, was originally placed in Category D by the *BOURC* as a bird of uncertain provenance. On the first circulation, nine of 10 members voted for Category D, perhaps influenced by the presence of a probable escapee seen in Oxfordshire on 12th July 1985 and found dead the following day (McKay 2000). The discovery of the Nottinghamshire bird in 1996, among one of the largest irruptions ever into Britain of Bohemian Waxwings, led to a status upgrade of the Noss bird to Category A and thus a British first (McKay 2000).

The only additional Western Palearctic sightings come from Iceland (one from April to July 1989 was originally identified as a Bohemian Waxwing but was reconsidered in 1995 and found to have been Cedar and another on 8th October 2003). Of interest is the observation by Durand (1963) of five birds in the Atlantic 640 km out from New York, on 8th to 9th October 1963.

There are two spring migratory surges in eastern and central USA, one from February through March and another from May to early June. Brugger *et al* (1994) reported that both first-year and adult Cedar Waxwings were at their maximum mean distance from nesting areas in February and that there was no evidence for differences in migration distances to wintering areas between immature and mature birds. They also found that some Cedar Waxwings move about steadily through the winter. Although the return migration occurs mostly in March, movement of Cedar Waxwings may continue throughout the spring; in some years they do not depart Central America until mid-May and movement along the coastal states of eastern Canada is noted regularly in June (Witmer *et al* 1997). Cedar Waxwings are infamously late breeders in North America (Lea 1942), with the peak of breeding occurring from mid-June to July, even extending into September (Leck and Cantor 1979).

The species' propensity for late movements ties in well with the idea that the Shetland bird was a spring overshoot. The Nottingham bird may well have made landfall in northwest Europe during the previous autumn and joined with Bohemian Waxwings. But, there were unprecedented numbers of Bohemian Waxwings in Newfoundland that winter. Single flocks of 6,000 birds were recorded and perhaps 20,000 or more birds were within a 25 km radius of St John's in early January. The influx surpassed all previous records on the east coast of North America (Smith 1996; Brown and Grice 2005). During the time of this influx, large numbers of Bohemian Waxwings were recorded in Ireland and northwest Britain; some birds in Britain may have originated in the Nearctic and brought the Cedar Waxwing with them.

The highly dispersive nature of this species is demonstrated by records as far south as northern South America. It is a regular winter visitor to Bermuda, with sporadic influxes (Amos 1991). Now that birders are alerted to this species as a possibility, perhaps we can expect more of these attractive birds in due course.

White-throated Dipper *Cinclus cinclus*
Linnaeus, 1758. Largely resident from northwest Africa, western Europe and Scandinavia, east to Mongolia and western China.

Polytypic, 10 to 12 subspecies recognised, four of which are in Europe. Two races breed in Britain: endemic *gularis* British Dipper (Latham 1801) over most of the British breeding range; and *hibernicus* Hartert 1910 Irish Dipper in western Scotland and Ireland. Nominate *cinclus*

Linnaeus 1758 Black-bellied Dipper occurs as a rare migrant and *aquaticus* Bechstein 1803 (from central Europe) has been suspected but is not on the *British List*.

Black-bellied Dipper Cinclus cinclus cinclus

Breeds Scandinavia, Baltic countries and western Russia. Outside the breeding season resident or dispersive to the south and west of breeding range.

Scarce migrant from northern Europe mostly in early spring and autumn. Occasionally winters. Status uncertain but a decline in recent records imply it is rarer than previously.

Status: The difficulty of identification probably clouds its true status here but there are around 210 records (Keith Naylor *pers comm*), around half of them from Norfolk (mostly wintering birds) and a quarter from Shetland (mostly passage birds). There is one record for Ireland.

Away from Shetland many birds have been recorded wintering, especially in Norfolk, so it seems likely that Black-bellied Dipper occurs as a wintering bird in small numbers. They probably occur throughout eastern Britain, but are less likely to be noticed in areas with resident British Dippers. In flat East Anglia with few fast-flowing water bodies, British birds are generally absent, so migrants will attract attention. Similarly the Northern Isles do not have residents and so migrants are more conspicuous.

Historical review: The first two British records both came from Suffolk: Leathes Ham, Lowestoft in 1849 and Boulge Hall Park in 1850. A further 11 records followed between 1873-1898, nearly all of which from east coast counties: Yorkshire (three), Lincolnshire (three), Norfolk (two) and Suffolk (one). There were also singletons from Shetland and an inland bird from Nottinghamshire.

A further six were documented for Norfolk between 1909 and 1920 and in 1931. One shot in East Yorkshire was the only other record from this period. Singles on Shetland in 1934 and 1936 were the last of pre-1950 records, giving a total of 23 birds.

The 1950s added a further 12, all in Norfolk and Shetland except one in Essex and an exceptional record for Ireland, on the River Tolka, Dublin, from 12th January to 3rd February 1956. The 1960s yielded 42 birds including nine in 1966. Norfolk and Shetland produced the majority, and east coast counties added to the tally. Away from these areas a bird on the River Rib, near Bengeo, Hertfordshire, from 29th December 1962 to 17th February 1963 was notable, as was a bird in Dorset early in 1965 that frequented the River Stour in January and was relocated on the River Allen at Witchampton from January into February.

The 1970s saw around 56 birds, including a record 13 in 1976, 12 of which were in Norfolk. During the decade Norfolk was once again the favoured area, along with Shetland. One at South Cerney Gravel Pit, Gloucestershire, on 17th December 1972 was a rare sighting from western England and one frequented Lemsford Springs, Hertfordshire, from 20th November 1974 to 21st February 1975, an unusual sighting for southern England.

Fewer were recorded during the 1980s, with just 33 birds documented. Not surprisingly the east coast Norfolk-Shetland bias prevailed. A noteworthy duo visited Lemsford Springs in 1981 (one from 25th January to 28th March and another from 8th to 27th March; could one have a returning bird from 1974-75?). One at Skerton Weir, Lancaster, from 20th to 27th November 1998 remains an unequalled sighting from the northwest.

A modest 20 were reported in the 1990s, nine from Shetland. One at Stornoway, Lewis, on 10th November 1990 was quite exceptional. There have been just 16 since 2000, all but four from Shetland. A popular bird wintered on the River Glaven, Norfolk, from November 2008 to March 2009.

This form has only recently been added to the list considered by the BBRC (Hudson *et al* 2008) and the criterion for acceptance outlined:

> Only birds that completely lack any hint of chestnut on the breast are likely to be acceptable, although a limited/narrow brown band is not unusual in this race, while some nominate cinclus show a narrow chestnut band at the breast-belly interface.

> Birds showing chestnut on the underparts probably cannot be distinguished from darker individuals of the British forms C. c. gularis and C. c. hibernicus, or from C. c. aquaticus from central Europe, which has occasionally been suspected here.

That Black-bellied Dippers cross the North Sea is unequivocal. An adult ringed in southwest Sweden on 4th March 1985 was recaptured in Fife on 3rd and 5th April 1987, a Norwegian bird ringed as a chick on 31st May 1993 was hit by a car at Malton, North Yorkshire, on 28th October 1993 and another chick ringed in Norway in May 2004 wintered on Mainland Shetland in 2005-2006 (caught on 2nd February 2006). It is interesting to note that two of these birds were not inexperienced 1st-winter birds when they were first located in Britain. Another on Fair Isle from 7th to 14th April 1998 also bore a foreign ring, but managed to evade capture during its stay.

A Black-bellied Dipper present at Burnham Market, Norfolk, from October 1990 was ringed there on 16th February 1991. It was subsequently controlled at Belstead, Suffolk, on 14th November 1991, where it had been present from 7th and remained to 25th March 1992. This suggests that Scandinavian birds may return to Britain in successive winters, but not always to the same site (unless it had summered elsewhere in Britain). If returning birds adopt new wintering sites in subsequent winters the actual number of birds reported from areas such as Norfolk may be less than the records imply.

Where: The majority of records come from Norfolk, where there have been around 100 birds, although the interconnecting river systems of central and eastern Norfolk make it difficult to decide how many birds may have been involved over the course of a winter. There have been 52 records from Shetland, 15 from Lincolnshire and 11 from Yorkshire (Wilson and Slack 1996; Taylor *et al* 1999; Pennington *et al* 2004). Elsewhere records would appear to be exceptionally rare though the difficulty of identification may cloud the issue. Previously it was assumed that any dipper with reduced chestnut on the belly was a Black-bellied, which is now known to not be the case. A review of all records may result in a number no longer making the grade.

When: The peak of sightings between late March and early April through Shetland presumably suggests birds on return passage from wintering sites in Britain though to Scandinavia, where northern populations are partly migratory. The late autumn peak presumably represents the passage of birds from Scandinavia heading through to British wintering locations.

Records from Norfolk have become extremely rare with just five since 1988, suggesting that far fewer birds are now arriving here than was formerly the case. Alternatively, this situation might reflect a better understanding of the identification criteria with some of out of range birds being correctly identified as variant British Dippers and are not assumed to be Black-bellied.

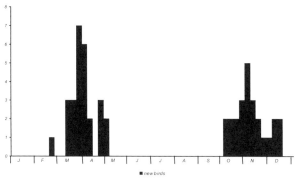

Figure 22: Timing of Black-bellied Dipper records in Shetland, 1934-2007 (first-date only).

It would appear that British and Irish Dippers are relatively sedentary, a fact confirmed by the geographical distribution of ring recoveries, for which the median distance for 220 recoveries was just 2 km and 95 per cent were within 35 km (Wernham et al 2002). There are very few movements of birds greater than 50 km (Mark Grantham pers comm).

Another possibility for the recent downturn in sightings may be due to less emigration resulting from warmer winters. Almost annual records from Shetland suggests that this may not be the case, though could these be an artifact of increased coverage on the islands?

Discussion: Attribution of birds as this form has undoubtedly been oversimplified in the past. Some older records away from expected locations may have involved gularis with reduced amounts of chestnut on the belly. It also appears that some birds within the range of hibernicus in western Scotland may lack chestnut, and there is some variation in the amount of chestnut on the belly in British birds (Tyler and Ormerod 1994; Forrester et al 2007).

A recent study by Hourlay et al (2008) found a complex phylogeographic structure with at least five distinct lineages for birds from the Western Palearctic region, which supports previous findings that there is a weak genetic differentiation between West European populations (Lauga et al 2005). Thus, the wide variation between individuals and forms is complex and it may be that variation within the European taxa is clinal with dark-bellied birds in cooler and wetter climates and chestnut-bellied birds in warmer and drier areas (BWP).

Clearly many dippers with little or no chestnut in the belly in eastern Britain are most likely to be nominate cinclus, as are passage birds on Shetland or wintering birds in areas such as East Anglia that are not frequented by residents. Records committees may have to accept that even birds that seem 'good' may well evade a confident assignment under close scrutiny. Even in typical locations caution should be exercised: three chestnut-bellied birds on Fair Isle in late spring and at least 10 in Norfolk have shown chestnut on the underparts and were thought to belong to either gularis or aquaticus (Taylor et al 1999; Forrester et al 2007).

Despite the fact that some chestnut-bellied birds in eastern Britain have been suspected aquaticus, the confirmation of this race without ringing proof, or perhaps DNA evidence, would be nothing more than circumstantial against the backdrop of variation within and between populations. It is also worth bearing in mind that East Anglia is much nearer to British breeding areas than it is to central European *aquaticus*. Although very helpful, neither geography nor habitat is entirely foolproof for any of the taxa, but the addition of Black-bellied to the *BBRC* list will doubtless ensure that dippers are studied more intimately than before.

Northern Mockingbird *Mimus polyglottos*

(Linnaeus, 1758). Breeds in southern Canada south across the USA, extending south through Baja California and Mexico to Oaxaca and Veracruz. Range includes Bahama Islands and Greater Antilles. Winter range is similar, though northern populations may be partially migratory.

Polytypic, two subspecies. The racial attribution of occurrences here is undetermined, but most probably involved *polyglottos* (Linnaeus, 1758) of eastern North America south to east Texas and Florida.

Very rare vagrant from North America.

Status: Two records.

1982 Cornwall: Saltash 30th August
1988 Essex: Horsey Island 17th-23rd May

Discussion: The largely sedentary Northern Mockingbird seems an unlikely vagrant, but it is reportedly partly migratory in the northern part of its range and there has been a recent northward expansion along the east coast of the US into southern Canada. Some ringed individuals have travelled up to 800 km (Cooke 1946) and vagrants have reached Bermuda and Honduras. The species was not documented by Durand (1972) on his North Atlantic voyages in the 1960s and was categorised in the least-likely vagrant group (along with White-crowned Sparrow *Zonotrichia leucophrys*, Bay-breasted Warbler *Dendroica castanea*, Wilson's Warbler *Wilsonia pusilla* and Swamp Sparrow *Melospiza georgiana*) by Robbins (1980) in his paper that assessed statistically the likely future American landbirds to reach Europe.

Additional Western Palearctic records include one in the Netherlands from 16th-23rd October 1988 and a male holding territory on Gran Canaria, Canary Islands, from November 2004 intermittently until February 2006 (when it may have been captured and taken into captivity). The origin of this bird was unknown.

Four Northern Mockingbirds were formally accepted by the *BBRC*. After assessing the relative likelihood of vagrancy and escape from captivity in each case, the *BOURC* assigned two to Category A, one to Category D (Worm's Head, Glamorgan, 24th July-11th August 1978) and one Category E (Blakeney Point, 20th-28th August 1971).

The Essex and Glamorgan birds seem likely to have been ship-assisted given the proximity to nearby ports. The Glamorgan bird was deemed too early for a true migrant and was assigned to Category D. The Essex bird was placed in Category A due to the acceptable spring timing and the greater likelihood of ship-assistance (given the frequent occurrence of other species of Nearctic passerine in the area). The treatment of the two records appears inconsistent under present guidelines.

The Cornwall bird had a strong supporting cast of Nearctic vagrants, occurring the day after a Black-billed Cuckoo *Coccyzus erythrophthalmus* in the Isles of Scilly, two days before a Black-and-white Warbler *Mniotilta varia* in Cornwall and six days before a Tennessee Warbler *Vermivora peregrina* in Orkney. As such, it probably has the best vagrancy credentials of any of the European records. Its arrival provides one of the earliest autumn dates for a Nearctic landbird.

Brown Thrasher Toxostoma rufum

(Linnaeus, 1758). Breeds southern Canada south to central Texas, Gulf Coast and southern Florida. Northern birds winter in southern USA within and slightly to the west of the range of residents.

Polytypic, two subspecies. The sole record was attributed to nominate *rufum* (Linnaeus, 1758).

Very rare vagrant from North America.

Status: One record.
1966 Dorset: Durlston Head 18th November-5th February 1967

Discussion: Predominantly a short-distance partial migrant. There has been a shallow but steady decrease in numbers since the 1960s (Sauer *et al* 2008). In winter it moves away from the north of its range into the southern part of the breeding range in southeastern USA. Ringing recoveries suggest that wintering areas are located directly south of breeding sites (Cavitt and Haas 2000).

Durand (1972) recorded the species during his North Atlantic voyages during 1961-65, including a probable on 30th September 1963 1,500 km from New York and three birds the following day 350 km from New York. A record from Bermuda is reported to have been ship-aided (Amos 1991). The only other record from the Western Palearctic was from Helgoland, Germany, in autumn 1836. Robbins (1980) gave it a very low prediction for vagrancy.

The Dorset individual was on a headland close to busy shipping lanes and almost certainly arrived on a ship. No one knows if it could make such a crossing without human interference, but despite being a seemingly unlikely vagrant it has stayed on Category A. The bird had a particular liking for Holm Oak *Quercus ilex* acorns, which it hammered fiercely and loudly (Ian Wallace *pers comm*).

Grey Catbird Dumetella carolinensis

(Linnaeus, 1766). Breeds southern Canada from British Columbia to Nova Scotia, south through central and eastern USA to Texas and Georgia. Winters southern USA through Central America to Panama.

Monotypic.

Very rare vagrant from North America.

Status: Two records.

1986 Co. Cork: Cape Clear 4th November
2001 Anglesey: South Stack 4th-6th October

The Anglesey bird led a good many observers a not so merry song and dance.
The morning of 4th October 2001 dawned cloudy, and with a southwesterly wind blow-
ing, much as it had done for the previous week, and I began the day seawatching from the
steps of the South Stack lighthouse at the western tip of Holy Island, Anglesey. After 30
minutes, during which time I had seen only a single Manx Shearwater Puffinus puffinus,
my enthusiasm for seawatching had waned and my thoughts turned to passerine migrants
– this was October after all – and I made my way inland a few hundred metres to the
'Plantation', a small stand of conifers interspersed with a few hawthorns Crataegus and
elders Sambucus. This is an area which I check regularly each spring and autumn, and in
the past I have discovered Barred Warbler Sylvia nisoria, Pallas's Leaf Warbler Phyllosco-
pus proregulus and Red-breasted Flycatcher Ficedula parva here.
    While walking up from the car park, slowly working my way to the waterworks com-
pound adjacent to the Plantation, I failed to see a single bird. Finally, around 09.00 hrs,
I glimpsed a movement in the hawthorns at the top end of the Plantation. In hope rather
than expectation, I focused my telescope on the bird and saw that it was a Goldcrest Regu-
lus regulus, but while I was watching it, another bird emerged from some ivy Hedera and
appeared in the same field of view. It looked almost entirely blue-grey, except for the black
crown and eye, and a long black tail. I stood there, quite stunned, as I realised that it could
only be a Grey Catbird Dumetella carolinensis! Agonisingly, it then began to play cat-
and-mouse with me, disappearing into the vegetation but then quickly emerging again.
It continued to show on and off like this for five minutes or so, as it hopped about through
the bushes. Unfortunately, a Eurasian Sparrowhawk Accipiter nisus then flashed past and
behind it, not particularly close but sufficient for the catbird to dive deep into the cover of
an elder bush. I waited a further ten minutes for it to reappear, but there was no sign.
    At that point, I decided that I had better get to a telephone and put the news out.
Having done so, I returned to the site, where five members of the South Stack RSPB staff
had already raced across from their office and seen the bird in the elder. As they watched,
it had flown across the adjacent field and into dense cover. Other birders soon began to
arrive but the bird proved to be extremely elusive. Remarkably, given that the catbird
was the first American passerine recorded on Anglesey, a Red-eyed Vireo Vireo olivaceus
was found that afternoon, but was of little consolation to most! Although still present
the following day, the Grey Catbird remained extremely unco-operative, spending much
of the morning skulking in an area of thick gorse Ulex, being glimpsed just occasionally

before diving back into cover, and calling infrequently. During the afternoon, it gradually
became more obliging and was eventually seen by numerous birders.
Ken Croft. British Birds 97: 630-632.

Discussion: A medium-distance migrant. Wintering range includes the Atlantic coast from New England to Mexico and the Caribbean. Migrants pass through Bermuda in April and mid September to early November (Amos 1991).

There have been four other Western Palearctic records most or all presumably ship-assisted. There are singles from Germany (Helgoland, shot 28th October 1840), Channel Islands (Jersey mid October 1975 onwards, trapped and apparently held captive for several weeks, but not identified until December 1975), Canary Islands (Tenerife 4th November 1999) and Belgium (1st-winter 15th-16th December 2006). Another German record on the mainland at Leopoldshagen on 2nd May 1908 is no longer accepted.

Durand (1963) recorded two birds at sea: 640 km east of New York on 8th-9th September 1962 and six hours out from New York on 1st October 1963. In October 1998 one took up residence on the liner *Queen Elizabeth II*, cruising from New York to Europe. It remained on board while the ship was in port at Southampton on 21st October and was last seen off Malta on 29th. This passenger does not qualify as a British record under the present ruling as it was provided with food. An additional stowaway on the same voyage came to light in the Mediterranean: a Song Sparrow *Melospiza melodia*.

Alpine Accentor *Prunella collaris*

(Scopoli, 1769). Breeds discontinuously in mountains of northwest Africa, Spain and central Europe north to southern Germany, southern Poland and through Balkan countries to Greece. In Asia, breeds in all major ranges from central Turkey across central and south Asia to the Himalayas and east to Japan and Taiwan. In winter, most European birds descend below snowline near breeding areas but some disperse to lowlands.

Polytypic, nine subspecies. Records are attributable to nominate *collaris* (Scopoli, 1769) of northwest Africa and western and central Europe east to the Carpathian mountains. *P. c. subalpina* (C L Brehm, 1831) of southeast Europe and also southern Asia Minor is similar to *collaris*, but the upperparts are paler grey less tinged olive and buff. The chest is paler and flanks have more narrowly streaked with paler rufous; throat-patch less heavily barred, chin often almost uniform white (*BWP*).

Rare vagrant from southern and central Europe, mainly in late April and May.

Status: There were 20 records involving 26 birds between 1817 and 1932 and a further 16 records from 1955 onwards; all have been in Britain, with no records from Ireland.

Historical review: The pre-1950 records include a number of seemingly implausible sightings, including three together in Hampshire and two together in Cambridgeshire on two occasions.

Of the 21 dated birds for this period, 11 were between November and January, seven between March and June and three between August and October: an arrival pattern quite dissimilar to the spring bias evident in modern records.

There are 16 records between 1955 and 2004, 14 of them since 1975.

1955 East Sussex: Telscombe Cliffs 24th April
1959 Shetland: Fair Isle 27th-28th June
1975 Kent: Ramsgate 7th May
1976 Kent: Dungeness 8th May
1977 Scilly: St. Mary's 30th October-9th November
1978 Dorset: Portland 8th-30th April, trapped 11th
1978 Norfolk: Sheringham 30th April-4th May, trapped 1st May
1990 Isle of Wight: The Needles 27th May-6th June
1990 Cornwall: Rough Tor 4th November
1993 Devon: Lundy 8th May
1994 Lincolnshire: Saltfleetby 14th-18th November
1997 Pembrokeshire: Strumble Head 30th October
2000 Kent: St Margaret's 6th May
2000 Suffolk: Corton 13th May
2002 Suffolk: Minsmere 16th-19th March
2004 Norfolk: Overstrand 20th April

Where: The records between 1955 and 2004 include nine along the south coast from Scilly to Kent, with birds having graced the latter county on three occasions including sightings a day apart in consecutive years. There is one Welsh record from mainland Pembrokeshire and one Scottish record from Fair Isle. Along the east coast five have been found between Lincolnshire and Suffolk. That only one has strayed farther north than Lincolnshire is surprising.

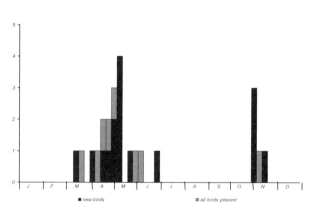

Figure 23: *Timing of British Alpine Accentor records, 1955-2007.*

When: Twelve have arrived in spring, with the earliest on 16th March and four recorded in April. Sightings peak in early May, with five between 6th and 13th. There is one late May record and another in late June. The four autumn sightings fall between 30th October and 14th November.

Table 14: Records of Alpine Accentor in Northern Europe* since 1950.

Country	Apr	May	June	October	November	Total
Norway			1			1
Sweden	4	4	1	2		11
Finland	1	3				4
Denmark	6	5	1			12
Poland	4					4
Netherlands	3	3				6
Belgium	1	3				4
France	1			1	4	6
Other**		3				3
Totals	**20**	**21**	**3**	**3**	**4**	**51**

*All records involve birds away from breeding locations.
** Includes Estonia, Channel Islands and Luxembourg.

Discussion: Young birds and females abandon the breeding sites more readily than adult males. Males depart their lowland wintering quarters from early March. Upland areas are reoccupied fully from mid-March onwards, mostly during April; females arrive back in alpine zones in early May (Hagemeijer and Blair 1997). It seems probable that the spring peak relates to birds overshooting their breeding areas. The small cluster of autumn records presumably represents birds displaced during their descent to lowland wintering areas.

An analysis of extralimital records in northwest Europe between 1940 and 1986 showed that spring records were most numerous (van IJzendoorn and Nuiver 1987), a pattern reflected by modern British records. Analysis of north European records since 1950 (see *Table 14*) shows that the overwhelming majority of extralimital records have been in April (39 per cent) and May (41 per cent); German records are excluded from the table as small numbers reach the northern regions of Germany annually in spring (Peter Barthel *pers comm*)

In the face of climate change we might expect this species to be less regular as winter in mountainous areas ameliorates and altitudinal migration becomes less necessary. But, some recent autumn records in northwest Europe are from before the winter really begins, so it is not solely cold weather that forces birds to wander.

Rufous-tailed Scrub Robin Cercotrichas galactotes

James Gilroy

(Temminck, 1820). Nominate galactotes breeds in south-west Europe, North Africa, Sinai, Israel, Jordan, and southern Syria; syriacus in southeast Europe, west and central Turkey, and the Levant south to Lebanon. Beyond the Western Palearctic breeds from eastern Arabia north-east to Kazakhstan and western Pakistan, and in sub-Saharan Africa from Mauritania and Sénégambia east to Somalia. Eurasian and North African populations are migratory, wintering in the northern Afrotropics.

149

Polytypic, five subspecies. Records that have been positively identified to race have been attributed to nominate *galactotes* (Temminck, 1820), which differs from *syriacus* (Hemprich and Ehrenberg, 1833) by having rufous-brown upperparts similar in colour to the rump and tail as well as cleaner underparts. Further structural differences are evident with *syriacus* having a more pointed wing and shorter tail (*BWP*).

Exceptionally rare vagrant from Iberia and North Africa, mainly in autumn.

Status: 11 records, eight in Britain and three in Ireland.

1854 **East Sussex:** Plumpton Bosthill male shot 16th September
1859 **Devon:** Near Start Point shot 25th September
1876 **Co. Cork:** Old Head of Kinsale shot 10th September
1876 **Devon:** Slapton immature shot 12th October
1951 **Kent:** Dungeness 12th September
1951 **Co. Wexford:** Great Saltee Island 22nd September-4th October
1959 **Devon:** Gammon Head 20th October
1963 **Lincolnshire:** Skegness adult male 2nd-9th September, trapped 4th
1968 **Co. Cork:** Cape Clear 20th April
1972 **East Yorkshire:** Flamborough Head 5th-6th October
1980 **Devon:** Prawle Point 9th August

The week-long stay of the bird in Lincolnshire is the envy of birders today.
One of my weekend trips in autumn 1963 was typical of that period and illustrates the relatively leisurely and laid back character of early 1960's birding. It is permanently etched in my memory largely because it involved a major 'dip' – odd how such disappointments are often easier to recall in vivid detail rather than one's successes.

The trip had been arranged well in advance for the weekend of 7th/8th September with Martin Coath (still an active birder), Dave Putman and his brother John. Our transport was to be the Putmans' mother's car, a much prized and immaculate Ford Anglia, which had been loaned to us for several previous trips. It was generally held that early September was the peak time on the east coast and our plan was to visit Cley in the hope that there would be anticyclonic conditions over Scandinavia with easterly winds across the North Sea to produce a good fall of migrants. In the event, the weather forecast for the weekend was indifferent with slack westerly conditions but, despite this, we headed to north Norfolk, quietly confident that we would see much of interest and, in any case, we were unaware of any rarities to attract us elsewhere.

Cley was very quiet on that particular Saturday but, there on the famous East Bank, we met Richard Richardson, the local birding expert par excellence. Richard mentioned that he had received a postcard a few days earlier from a friend staying on a weeks family holiday at Butlin's Holiday Camp at Skegness. He casually added the more interesting news that, on the previous Saturday afternoon, the lucky camper had found a Rufous

Bush chat and, at the time of writing, on the Tuesday, the bird was still around the camp sewage farm. Such 'hot' news seemed too good to ignore and even Richard, who seldom ventured out of Norfolk, was surprisingly keen to visit Butlin's.

Five of us crammed into the Ford Anglia and we set off along the tortuous route to Skegness. We arrived at Butlin's in the early afternoon and a uniformed commission-aire, seemingly on duty to repulse non-guests, enquired if we had come to 'clock on'. He appeared both startled and unimpressed by our story that we wished to visit the camp sewage farm for an hour or two to see a bird. After some while, we established that the finder of the Rufous Bush Chat had already departed but, despite considerable effort, we were unable to obtain permission to enter to the camp grounds. However, we did glean the information that the sewage farm could be viewed from a public footpath and, after seeking directions from various bemused holiday makers, found the spot suitably secure behind a high chain-link fence. There were no birdwatchers to be seen and a few min-utes observation revealed that the tiny sewage farm had minimal cover and was devoid of birds. Walking along the path peering into the camp through the fortifications revealed only two Wheatears and some puzzled campers. After two hours Richard pronounced that the bird had clearly gone and he was late for tea, so we retraced the long, tedious journey back to Cley.

Highlights of this weekend included a Barred Warbler, a Pomarine Skua and a Red-necked Grebe, as well as a nude octogenarian bather plunge-diving into the Cley surf. We missed a Purple Heron at Minsmere and a rumoured Two-barred Crossbill at Cley, but still managed to amass 90 species. The 550 miles clocked in the Ford Anglia repre-sented a major excursion, bearing in mind that the journey from London to Cornwall, for example, in these pre-motorway days usually took over nine hours each way.

Sadly, there is a postscript to this account. In mid October, I heard from Richard Richardson that the Rufous Bush Chat had still been at Butlin's on the afternoon of our visit. Indeed, it was still there two days later and a local press photographer had even pho-tographed the bird whilst it sat on the fence surrounding the sewage farm. Presumably, at the time of our visit, the Rufous Bush Chat was somewhere around the chalets or enjoy-ing the camp facilities. Much to my chagrin, I have still to see the species in Britain. Per-haps there will be one next September, but I suspect that there may well be more than fiver birders looking for it on a Saturday afternoon and that news of it will not have been cir-culated by postcard!

Ron Johns. Birding World 4: 315-317.

Discussion: Just one of the 11 birds was in spring and the rest in autumn: one August; six Sep-tember; and three October. Most have been on southern coasts, with singles in Kent and East Sussex and four in Devon plus three in southern Ireland. East coast records are restricted to East Yorkshire and Lincolnshire. Although three have been seen since the Lincolnshire bird, only the Yorkshire bird of 1972 provided a multiple viewing. The last was a single-observer record as along ago as 1980.

Just five others have occurred in northern Europe during recent times. These include the first for Finland (27th May 1995), three from Germany (6th October 1995, 1st September 1997, 9th June 1998), and one from the Channel Islands (Jersey 7th June 1998).

The decline of the key Spanish population in recent decades was quantified at between 50 and 79 per cent (BirdLife International 2004). Underlying reasons are linked to biocides (herbicides and pesticides) or the effects of Sahel droughts (Hagemeijer and Blair 1997). With what would appear to have always been a rare vagrant to the north of its core range, declines of such magnitude in the European stronghold offer little prospect of a change of fortunes. This species must surely rank among the most wanted for most keen listers.

Eurasian and North African populations are migratory, wintering in the northern Afrotropics, including the Sahel. The exodus from breeding areas occurs from September to early October, which correlates nicely with the bulk of British and Irish records. In spring it arrives in southern Iberia in late April and early May, but not until late May does it reach central Spain and some take until early June to get to Portugal (*BWP*). The sole spring record presumably represents a spring overshoot; the lack of other records at this season is surprising in light of other recent records from the Channel Islands, Germany and even Finland.

Anyone wishing to unearth the next bird would do well to lurk in south Devon as the area has accounted for no fewer than four British records, a quite disproportionate share of the spoils.

Rufous-tailed Robin *Luscinia sibilans*

(Swinhoe 1863). Breeds southern Siberia from northern Sakhalin and Russian maritime provinces bordering the Sea of Okhotsk, west to the Altai Mountains and upper Yenisey River, north to 62°N in Yakutia and south to mountains in northeast China. Winters from China south of Yangtze River to northern Indochina and Thailand.

Monotypic.

Exceptionally rare vagrant from Siberia, with two Western Palearctic records.

Status: One record for Britain was the first for the Western Palearctic.
2004 Shetland: Fair Isle 1st-winter on 23rd October.

This species shares a similar range and migratory route with its congener the Siberian Blue Robin *L. cyane*, breeding as far west as the lower Yenisey valley and wintering in southern China and southeast Asia. Its occurrence here has long been anticipated (for example, Wallace 1980) and there could be no more a fitting location than Fair Isle in which to proclaim its presence on the Western Palearctic list.

With a light breeze blowing from the northeast, conditions seemed promising for vagrant birds on Fair Isle, on 23rd October 2004, and it was with my usual optimism that I headed out on the morning census to cover the northern part of the island. Other than a few thrushes and some cracking 'Northern Bullfinches' Pyrrhula p. pyrrhula, however, I had not seen a lot to shout about by the time I reached the top of Ward Hill at about 11.00 hrs.

Meanwhile, Mike Wood (one of the Directors of Fair Isle Bird Observatory) was ambling south along the road from the Observatory with his wife, Angela, and two young daughters (Emily and Kate), when he spotted a bird resembling a juvenile Robin *Erithacus rubecula* hopping along the roadside by Bull's Park. The only person in sight was Mark Newell, and Mike went over to ask him if this was possible in late autumn. Mark replied 'No!' and they returned to look for the bird.

As I descended from the top of the hill towards Lower Station, my mobile phone rang and an out-of-breath (but still running) Alan Bull (my Assistant Warden) was shouting down the phone: 'Mark has just described to me what sounds like a Veery (*Catharus fuscescens*) at Bull's Park – well a *Catharus* thrush anyway! I'm on my way to check it out but there's no point you coming down yet. I'll keep you informed!' 'Okay! Thanks!' I replied and immediately started down the hill. I did not care whether it was a Veery or not; any *Catharus* thrush would be a lifer and definitely worth running for. By the time I arrived, it had been identified as a Veery, and was showing well behind an old gate leaning against the dry-stone dyke. As I looked, I could see a small bird with a heavily mottled breast, a rufous tail contrasting with cold, olive-brown upperparts and pink legs. At this stage, I did not think it seemed quite right for Veery and thought it looked more like a Hermit Thrush *C. guttatus*, except that the breast pattern was more like that expected of a Veery. As I had not (and still haven't!) seen any of the *Catharus* thrushes, I thought it best to just take in the features of this bird. It was feeding close to the base of the dry-stone dyke and would periodically disappear into it for several minutes at a time. Discussion about its identity continued among the dozen or so people present, and although only one person had previously seen both species, his opinion strongly favoured Hermit Thrush. No other species were even considered at this time!

Regrettably, with my mind firmly fixed on a *Catharus*, I did not consider that it could be anything else and I tentatively put the news out that we had found a possible Hermit Thrush. The debate continued until lunchtime, when we returned to the Observatory for lunch and to check more references. It was over lunch that Nick Dymond casually mentioned that it 'looked a bit like a Rufous-tailed Robin (*Luscinia sibilans*)', but added that 'it couldn't be that 'cos they are small, the jizz wasn't right and besides they are from southeast Asia'. Pandemonium ensued as references for Rufous-tailed Robin were sought, and shortly afterwards I was staring with incredulity at a picture I had found on the internet. Alan Bull came in with a similar picture in a copy of Birding World. We looked at each other in disbelief. That was it! Incredibly, it looked like we had another first for the Western Palearctic!

We were confident that this was our bird, but thought we had better see it in the flesh again to be absolutely certain. Mark, who had stayed behind to keep tabs on the bird, had not seen it since we left for lunch but, fortunately, a walk along the wall soon relocated it and doubly confirmed our suspicions. The next few moments will live with all of us forever – a feeling of relief that the bird was still here, then stunned shock as the realisation sank in that it really, truly was a Rufous-tailed Robin, followed by cheers and other such signs of elation! Now that the identification had been confirmed, I phoned out the *updated news. This included a distorted telephone call to Paul Harvey, who was aboard the Cyfish with the Shetland crowd heading our way – not knowing if they were coming to see a Veery or a Hermit Thrush!*

Once all had soaked in the moment, I decided that it could be easily trapped and that it should be examined in the hand to be absolutely sure of the identification and to check for signs of captivity. A net was erected next to the wall and the bird was gently coaxed into it. A glance at the underwing to check that we were not making some horrendous faux pas showed it to be plain buffish-white. In the hand I was amazed at how small it felt – even smaller than a Robin! Back in the ringing room, measurements and a brief description were taken. It was in good condition with no feather, claw or bill damage and was aged as a 1st-winter based on the retained juvenile greater coverts, which were brown with obvious small deep-buff tips (although the innermost two had been moulted, and were more olive and lacked buff tips), and distinctly pointed tail-feathers.

After ringing, it was photographed and then released back at the same site, where it remained until dusk. As it had been feeding voraciously throughout the day and was in good condition when examined, it was no surprise that it had departed by the morning, following a clear night.

Deryk Shaw. British Birds 99: 236-241.

Discussion: Also alluringly known as Swinhoe's Robin or Whistling Nightingale, the Fair Isle bird was savoured by just 20 or so indebted observers. The arrival of this unfamiliar species doubtless caught many by surprise. Many Shetland birders chose initially not to travel to Fair Isle because they had seen Veery before. The delayed re-identification meant that few managed to see the bird. To the disappointment of those who journeyed north hoping to share the occasion it departed during a clear night. However, somewhat amazingly, the second Western Palearctic record came just over a year later beside a sewage-farm at Białystok, eastern Poland, on 30th December 2005. It remained on the following day and was seen and photographed by a handful of fortunate observers.

The breeding range of this enigmatic 'Sibe' extends west beyond the Yenisey to the watershed with the Ob' river basin and the environs of Tomsk (Ryabitsev 2008). It overlaps with those of many species that occur regularly in Western Europe in late autumn. With two in as many years, this species is now clearly etched on the radar of finders. Events of recent years have produced a cast of extreme far-eastern vagrants, so could these two European records be the forerunners of others to come? The reason for the clear upsurge in rarities of this magnitude is unknown, but there appears to be an apparent tendency in some 'Sibes' to extend or reoccupy their range to the west-north-west. It will be interesting to see if the tendency continues.

What other unexpected species will be delivered to our migration hotspots in the coming years?

Thrush Nightingale Luscinia luscinia

(Linnaeus, 1758). Widespread in eastern Europe with a dramatic population increase in 20th century, particularly from the 1960s. Range still expanding northwest into western Norway; locally abundant in southern Scandinavia and Baltic countries. Ranges from Denmark southeast to Romania, Ukraine and through temperate Russia to southern Siberia. Winters east Africa from southern Kenya to Zimbabwe.

Monotypic.

Rare migrant from Eastern Europe and Scandinavia, predominantly a spring drift migrant to Shetland and along the east coast.

Status: There have been 171 records – 168 in Britain and three in Ireland – between 1911 and 2007.

Historical records: The first was a male shot on Fair Isle on 15th May 1911. The next was not until 1957, when one was trapped on the same island on 10th May, a scenario that was repeated on 15th-17th May 1958.

There were no further records until seven years later, when two were trapped, again on Fair Isle, in the space of three days in late May 1965. That year also delivered the inaugural autumn records, both of which were trapped at Low Hauxley, Northumberland, one on 26th-27th September and the other on 2nd October. Four in a year was unprecedented. Further singletons were recorded in 1967 and 1969, bringing the cumulative total for the decade to an exceptional six. Little did observers at the time realise that these were the precursors to an imminent change in status for this subtle species.

Seven in 1970 kick-started the decade, including an impressive run of four on Fair Isle between 8th-14th May and two on the Isle of May between 9th-17th May. One trapped on St. Kilda from

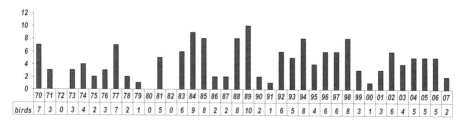

Figure 24: Annual numbers of Thrush Nightingales, 1970-2007.

29th to 30th May 1975 was the first to be recorded away from the east coast. It was soon followed by the first for Wales on 20th September 1976 when a 1st-winter was found dead on Bardsey. The record 1976 tally of seven was equalled in 1977, and it was noteworthy that three of these were autumn finds. No less than 32 had been logged by the close of

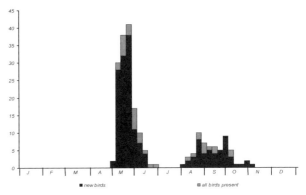

■ new birds   ■ all birds present

Figure 25: Timing of British and Irish Thrush Nightingale records, 1911-2007.

the decade, including seven autumn birds, and there was just one empty year (1972).

Sandwiched between non-appearances in 1980 and 1982, there was a creditable five in 1981. The 1980s continued in a similar vein to the preceding decade, delivering an aggregate of 49 birds and the commencement of continuous annual arrivals from 1983. A new record total of nine in 1984 included an unprecedented mainland run of three birds in the Spurn area between 23rd-30th May, two of which were trapped and aged as 1st-summer birds. Even this trio was outshone by an inland bird in song on the Ouse Washes, Cambridgeshire, from 2nd-3rd June. Eight more in each of 1985 and 1988 came close to establishing a new level of occurrence, but it was the unequalled 10 in 1989 that saw out the decade in style. During the 1989 arrival Fair Isle weighed-in once again with four birds trapped during 21st-25th May. Later in the year the first Irish record graced Cape Clear from 29th October to 1st November (the first of three autumn records to come from that far southwest island). The decade amassed 50 birds, nine of them in autumn, at an average of five a year.

The 1990s followed suit with a further 49 birds, including eight in both 1994 and 1998. The 1994 records were dispersed between Dorset and Shetland. The epicentre of arrival in 1998 was once again Fair Isle, where four were found between 12th May-3rd June; three 1st-summers and an adult. During the 1990s no fewer than 15 birds were recorded in the autumn. The most obliging of these was an extremely popular 1st-winter that spent an extended stay at Landguard Point from 27th August to 15th September 1995. Atypically showy, it allowed several hundred birdwatchers a fantastic opportunity to observe the species at close range performing out in the open for extended periods in stark contrast to the usual skulking glimpses.

Between 2000 and 2007 numbers were slightly lower at 31 birds. All bar one in East Sussex were on the east coast between Norfolk and Shetland. Just under half (15 birds) were found in the autumn, a much greater proportion of records for the season than was witnessed previously. The bulk of these (10 birds) were on Shetland, possibly a result of the greater coverage of the islands during autumn since the turn of the millennium.

An overall average of 4.3 per annum has been recorded since 1970, with a peak of five per annum during 1980-1999.

Where: Not surprisingly there is a strong east coast bias to records. Shetland leads the way with 74 birds, 51 of which were on Fair Isle (43 per cent of the national total). On Shetland five were pre-1970 before numbers leveled off, with 19 in the 1970s, 18 in the 1980s, 17 in the 1990s and 15 since 2000. Elsewhere in Scotland only Fife has recorded double figures, with 17 birds: 16 from the Isle of May and one from Crail. A further 15 birds have landed elsewhere in Scotland, seven of them on Orkney.

In northeast England, 29 coastal birds have been found between Northumberland (seven records) and Lincolnshire (five records), plus 10 in East Yorkshire and five in Cleveland. Coastal East Anglia has produced 15 birds: 10 in Norfolk and five in Suffolk. In southeast England, Kent has produced six and East Sussex three. It remains very rare in the southwest with two spring birds in Dorset and September birds on Scilly and Lundy.

Records from the west and northwest are very rare. All three Irish records have been in October from Cape Clear. The sole Welsh record was on Bardsey Island in September, the same month of the only Lancashire occurrence. One on the Isle of Man and the only record from the Outer Hebrides were both in May.

The inland bird on the Ouse Washes in 1984 was unprecedented. However, following the review period another inland adult was trapped on Salisbury Plain, Wiltshire, on 9th August 2008*; raising the intriguing possibility is that it had spent the summer unnoticed at an inland location in southern England.

When: The species, also known as Sprosser, is predominantly a spring drift migrant. Numbers vary from year-to-year in accordance with prevailing weather conditions. In some years, several birds may appear in a short period as illustrated by seven between 20th-26th May 1989. Despite range increases during recent decades, there appears to be no evidence of a trend towards earlier arrivals (see *Table 15*).

Sprossers are significantly more numerous here in spring than autumn, with 71 per cent of all records coming at that time and nearly 60 per cent falling in the period 7th to 27th May. Many are singing males, implying that a fair number of silent individuals of this skulking species must pass undetected. The two earliest records have both been on 6th-8th May: Isle of May in 1978 and Bexhill-on-Sea, East Sussex, in 2001. There are 16 early June records, the latest of which was trapped at Sumburgh on 17th June 1994.

Just 29 per cent (49 birds) have been found in autumn, most of them since 1990 and with a surge of 13 birds in the years 2004 to 2007. It seems likely that many more make a clandestine passage at this season as most records feature birds caught in mist nets. The earliest autumn bird was one at Hillswick, Shetland, from 31st July to 7th August 1970. Slight peaks come in late August and again in late September. There are just eight October records: three from Ireland; two from Shetland; and others from Orkney, Northumberland, Norfolk and Kent. The latest of all was on Cape Clear from 29th October to 1st November 1989.

The origin of most of our birds is likely to be Scandinavia, a fact supported by a first-year at Beachy Head on 26th August 1984 that had been ringed at Molen, Vestfold, Norway 12 days previously. It had travelled 1,099 km and remained until the 1st September, being controlled on 26th, 27th August and 1st September.

Table 15: Median arrival dates of spring Thrush Nightingales on Fair Isle.

Period	1970-1979 (n=13)	1980-1989 (n=11)	1990-2007 (n=15)
Median Date	21st May	21st May	25th May

Discussion: During the 18th century, breeding numbers were high from central Sweden to Poland and on the Austrian Danube. By the late 1800s, the population and range began to contract. In the 1920s these trends began to reverse, at least in northwest Europe. The 1960s and 1980s saw a marked increase over the entire northern and western part of the distribution. The population in Finland increased from 200 pairs in the 1950s to 8,000 pairs in the 1970s and up to 20,000 pairs by the late 1990s. Similarly the Norwegian population rose from 200-500 pairs in the 1980s to 300-1,000 pairs at present. The population increased in eastern Germany in early 1990s and numbers rose in Poland from the late 1970s. The first breeding for the Netherlands took place in 1995 and was attempted for a second time in 2005. Some marginal populations experienced declines between 1990 and 2000. This may be reflected in the slight downturn in fortunes in recent years.

Thrush Nightingale is sympatric with Common Nightingale *L. megarhynchos* in a narrow zone from Denmark to the Balkans. Hybridisation, proven through ringing, occurs frequently (Hagemeijer and Blair 1997). Hybrids are not proven to occur in Britain, but it seems probable that if and when Thrush Nightingale attempts to breed here that mixed pairings are likely. Given the absence of a plumage character totally diagnostic of either species, it seems likely that such birds would be virtually impossible to identify in the field. Mixed wing-structure features might be evident on trapped birds (King 1996).

Away from Shetland, Sprossers are very rare birds. The bumper numbers recorded on Fair Isle (and to a lesser extent the Isle of May) are a testament to the lack of cover and intensive trapping there. Consideration of this artifact suggests that many more must be passing through undetected along the well vegetated east coast, especially in autumn. Observers away from favoured locations usually have to rely on the assistance of an eager male in song or an opportunistic extraction from a mist net.

Common Nightingale *Luscinia megarhynchos*
C.L. Brehm, 1831. Nominate megarhynchos breeds across most of Europe and North Africa east to central Turkey and Levant. Winters sub-Saharan Africa from Gambia to Uganda.

Polytypic, three subspecies generally recognised. Nominate *megarhynchos* C.L. Brehm, 1831 occurs as a scarce and declining breeding species in southeastern England and is a scarce passage migrant in Ireland. There have been three accepted records of Eastern Nightingale *golzii* Cabanis 1873, which breeds from the Aral Sea and eastern Turkmeniya east to Mongolia. Eastern Nightingale *golzii* differs markedly from *megarhynchos* with cold brownish-grey upperparts contrasting with a long rufous tail. The chest is sandy buff and underparts whitish. It has a distinct fore-supercilium, dark cheek patch and clear pale edges to the tertials and pale tips to the greater coverts, forming a distinct wing-bar (*BWP*). Caucasian Nightingale *africana* (Fischer and Reichenow, 1884) breeds in eastern Turkey, the Caucasus, Iraq, northwest and north Iran and southern Turkmenistan. It winters in East Africa. It is intermediate between *megarhynchos* and *golzii* and is also a potential vagrant. It differs from *megarhynchos* in being duller (not darker) brownish-grey on the upperparts and has a longer tail (although the tail is shorter tail than *golzii*) (Duquet 2006).

Eastern Nightingale *Luscinia megarhynchos golzii*
This distinctive form breeds widely in central Asia from the Aral Sea eastwards through Kazakhstan to western China. It winters in coastal East Africa from Kenya to Tanzania.

Very rare vagrant from Central Asia with all records in October; possibly overlooked.

Status: Three records.

1971 **Shetland:** Fair Isle found dead 30th October
1987 **Scilly:** St. Agnes 23rd October
1991 **East Yorkshire:** Spurn 6th-14th October, trapped 6th

Discussion: In addition to the three British records there are two others from northwest Europe: Sweden (taken into care on 18th October 1964) and Iceland (collected on 5th November 1980).
    Clearly all October nightingales are worth a second look, but this striking form instantly appeals as something different, even briefly recalling Rufous-tailed Scrub Robin *Cercotrichas galactotes* (Bradshaw 1996; Duquet 2006).
    Eastern Nightingale is mooted as a potential split, being not only different in plumage but also long-tailed with a different song (Duquet 2006). Those who were fortunate enough to see either the Scilly or Yorkshire bird, even if they did not appreciate the rarity of what they were looking at, might at some point thank their opportune timing.

Siberian Rubythroat *Luscinia calliope*
(Pallas, 1776). Occurs throughout Siberia from Ob River east to Anadyr and Kamchatka, with small numbers to the European foothills of the Urals in the west. Southern limit reaches northern Mongolia, Ussuriland, northeast Hokkaido and northeast China, with isolated population on the eastern slopes of Tibetan plateau. Winters from Nepal east through the Himalayan foothills to northeast India, Burma and northern Indochina to central Thailand, southern China and Taiwan.

Monotypic.

Very rare vagrant from Siberia, with a noticeable increase in records in northwest Europe during the present decade.

Status: seven records.

1975 **Shetland:** Fair Isle 1st-winter male trapped 9th to 11th October
1997 **Dorset:** Osmington Mills 1st-winter male 19th October
2001 **Shetland:** Bixter, Mainland male, probably 1st-winter, freshly dead on road 25th October
2003 **Shetland:** Fair Isle 1st-winter female 17th-19th October
2005 **Shetland:** Fair Isle 1st-winter female 23rd-27th October, when trapped
2006 **Co. Durham:** Fulwell 1st-winter 26th-28th October
2007 **Shetland:** Foula male 5th October

Discussion: For 22 years after its first record in 1975 this stunning species epitomised the most precious jewel of rarity value, boasting both beauty and elusiveness. The second, in 1997, although on the mainland and collectable by many observers within quick-twitching range, did little to banish the mythical allure of this stunning chat. Little over 10 years on, there have now been a total of seven records and the near-annual sightings since 2001 have slowly begun to erode the mystique. Surely a mainland bird available to all-comers is just around the corner to banish the woes for those unable to travel to Dorset on that autumn afternoon in 1997 or who happened to be in Co. Durham nine years later?

Elsewhere in Europe in the past 20 years there has been a correlated increase in records, mostly dated in October. There have been three in Finland (15th October 1991, 12th October 2000 and found dead in a Pygmy Owl *Glaucidium passerinum* nestbox on 27th October 2005) and two each from Denmark (20th October 1985 and 29th October-2nd November 1995) and Norway (6th October 2005 and 1st-winter female trapped and ringed on 1st November 2008). Singles records have come from Estonia (25th May 1974), Germany (5th-12th November 1995) and Malta (ringed on 25th January 2004). Elsewhere in the Western Palearctic a record from Egypt is no longer accepted.

Turn back the clock, however, and it appears that the recent vagrancy of the Siberian Rubythroat across Europe was first signalled 125 years ago. Four 19th century records from France are placed in category D, but seven still accepted records come from Italy (1883, December 1886, October 1889, December 1898, March 1903 and December 1906) and Iceland (8th November 1943). Why the species went missing for lengthy periods between records is not worth conjecture, but the remarkable recent capture of a January bird on Malta indicates that the species can cope with winter in Europe, perhaps explaining the three early winter and one spring bird in the old records.

The Red-flanked Bluetail *Tarsiger cyanurus* had long been the mega of choice for self finders, but its increasing numbers have toppled it from the podium to be replaced by this rapture-inducing fellow 'Sibe'. Identification problems are negligible for a ruby-throated bird, but fortuitous finders need to eliminate the superiorly attired White-tailed (or Himalayan) Rubythroat *L. pectoralis* hiding its tail. The latter was formerly traded as a cage-bird, as indeed was Siberian Rubythroat. Although the legal importation of birds into the European Union has now been banned, both species may still occur as part of small scale smuggling (Tim Melling *pers comm*).

Could it be that Siberian Rubythroat, off the back of a surreptitious westward range expansion or an increasing population size, is to become an annual vagrant in small numbers? Its rarity on tick lists might be diminished by a run of birds in recent years, but few observers would be disappointed by the appearance of one of these gems in front of them on their coastal patch; especially those who drooled longingly at Asian field guides throughout their birding apprenticeship.

Siberian Blue Robin *Luscinia cyane*

(Pallas, 1776). Breeds southern Siberia from Russian Altai east to Sakhalin and south to Mongolia, northeast China and northern Japan. The range extends west to the Tomsk region. Generally scarce, but locally common in the Urals/Western Siberia region. Winters throughout southeast Asia.

Polytypic, two subspecies. The racial attribution of occurrences is undetermined but most records probably involve nominate *cyane* (Pallas, 1776) of southern Siberia from Altai to Sea of Okhotsk. The subspecies *bochaiensis* (Shulpin, 1928) is found from the Amur basin of eastern Siberia, to northeast China, Korea and Japan (*BWP*).

Extremely rare vagrant from Siberia, with both records and additional European ones, in October.

Status: Two records, in consecutive Octobers.

2000 Suffolk: Minsmere female or 1st-winter 23rd October
2001 Orkney: North Ronaldsay 1st-winter male 2nd October

Discussion: The first European record came from the Channel Islands (Sark on 27th October 1975) shortly after the first British Siberian Rubythroat. The first British record was preceded by a matter of days by the second for the Western Palearctic in northeast Spain (1st-winter trapped at the Ebro Delta on 18th October 2000).
The Minsmere bird, seen briefly by very few observers and supported with little evidence, struggled on *BOURC* circulation due to unease over the lack of images so favoured in the new digital era. There were also a few discrepancies between the descriptions and as the light was failing, there was concern that required details might not have been noted. In addition, there was a worry that the diagnosis was talked 'up' when a news service was phoned for help with identification. The diagnostic tail quivering only 'registered' after the observers had been told about this feature.
Disquiet was also expressed over the number of Siberian Blue Robins in the bird trade at the time; the species is known to be kept relatively commonly in captivity. Three European records in two years might have seemed anomalous to some, as a tight cluster of records can also indicate a period when it was in frequent trade.
Similarly, the Mugimaki Flycatcher *Ficedula mugimaki* record in East Yorkshire in 1991 coincided with the only time in recent years that it was in trade in Western Europe.

However, the circumstances of the record were felt to be sufficiently indicative of the bird being a natural vagrant rather than an escapee, a decision doubtless supported by the presence of a 1st-winter bird in Spain five days earlier and another on North Ronaldsay a year later (Foster 2006).

Expected as a visitor to our shores since 1975, it is perhaps surprising that it took so long to occur. Unfortunately neither of the British records chose to linger for long. However, the recent run of extreme rare far-eastern vagrants raises expectations for additional records in due course.

Red-flanked Bluetail *Tarsiger cyanurus*

(Pallas, 1773). A small population persists in northeast Finland. Main range in cool temperate forests of northern Eurasia from eastern Russia and Siberia to Kamchatka, northern Japan and northeast China. Winters southern China, Taiwan and southern Japan, through southeast Asia to northern peninsula Thailand.

Polytypic, two subspecies. European records are attributable to nominate *cyanurus* (Pallas, 1773) of northern Eurasia. Extralimital *rufilatus* (Hodgson, 1845) occurs from Afghanistan to north-central China. The subspecies differ diagnostically in size, plumage and vocalisations (Martens and Eck 1995) and may form two monotypic species (Knox *et al* 2008): Red-flanked Bluetail *Tarsiger cyanurus* and Himalayan Bluetail *T. rufilatus*. Distinct features include the longer tail and tarsus and more rounded wing-tip of *rufilatus*. Adult male *rufilatus* has brighter blue upperparts, paler blue supercilium (usually without white to the front) and narrower and more pure white throat patch than *cyanurus*. Females and immature males of two forms similar, but *rufilatus* averages whiter throat and belly (*BWP*)

Rare but increasingly regular vagrant from northeast Europe or Siberia, mostly in autumn.

Status: Formerly one of the most eagerly sought after Sibes at the time of the first twitchable mainland bird in Dorset in 1993, since when it has almost been annual. There have been 48 records, all of which have been in Britain; one in 2007 was the first for Wales.

Historical review: Although no longer the great rarity it once was, this beautiful species can still entice an approving crowd. This is yet another dream Sibe that has made the descent rapidly from a species that most could only ever fantasise about to one that is firmly on the radar for self-finders.

The first British record described above was an adult male at North Cotes, Lincolnshire, seen on about 21st September 1903. Since then that county has developed an enviable affinity with the species that is difficult to explain given the shortage of finds of this species elsewhere along the English east coast.

The next was a 1st-winter shot on Whalsay, Shetland, on 7th October 1947. Others followed from Kent in 1956 (Sandwich Bay, killed in a Helgoland trap on 28th October) and Northumberland in 1960 (Hartley, on 16th October).

In the 1970s the species first began to develop toward its current status in Britain. Four during the decade included the first ever spring record (male on Fetlar, Shetland, from 31st May to 1st June 1971). Birds in consecutive years in Fife (1975 and 1976) and the second for Lincolnshire in 1978 hinted at an end to its extreme rarity. In contrast the 1980s saw few: Fair Isle delivered September birds in 1981 and 1984, and Lincolnshire bagged its third in 1988. By then the total

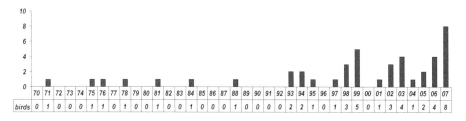

	70	71	72	73	74	75	76	77	78	79	80	81	82	83	84	85	86	87	88	89	90	91	92	93	94	95	96	97	98	99	00	01	02	03	04	05	06	07
birds	0	1	0	0	0	1	1	0	1	0	0	1	0	0	1	0	0	0	1	0	0	0	0	2	2	1	0	1	3	5	0	1	3	4	4	1	2	8

Figure 26: Annual numbers of Red-flanked Bluetails, 1970-2007.

stood at 11 records, but none had lingered and the species became regarded as a notorious short-stayer, a situation was about to change dramatically.

After a five-year gap, birds came in 1993 and began to redefine the occurrence parameters for this once-fleeting rarity. True to previous form, one on Fair Isle was an inaccessible one-day bird that served only to further tantalise mainland observers. Things changed dramatically with a 1st-winter at Winspit, Dorset, from 30th October to 8th November. It broke the mould regarding both location and duration of stay: its accessibility and 10-day performance ensured that several thousand birders could pay homage to one of their most-wanted species. At the time, few observers could have imagined how they would soon have plenty of chances for a repeat view of a British bluetail.

On Saturday 30th October 1993, Mike and Tanya Langman, their young son Adam and I went for a walk along Winspit valley in Dorset.

The weather was cold and grey, and any hope we may have had of finding late autumn migrants soon vanished in the gloom of a lifeless landscape. To be quite honest, I think we were all pretty soon looking forward to getting back to the pub for lunch.

At least some vestige of hope was kept alive by the very occasional Blackcap, although soon even the two birders across the valley appeared to have given up and were heading home. Having reached the end of the footpath at the cliffs, we turned round, intent on taking the quickest route back to the car, warmth and food.

Then, at 1.00pm, as I walked past an ivy-clad tree, I noticed a 'robin', with its back turned to me, sitting in the ivy. I turned away, but there was something about its jizz which made me look again. It turned round, revealing a lack of red breast, and then promptly darted away in the dense foliage. As it disappeared, I thought I saw a fleeting flash of blue at the base of the tail. My mind raced, and I shouted "I've got a Red-flanked Bluetail, Mike!" Indeed, I exclaimed loudly enough for the two birders across the valley to hear this proclamation. Mike replied indignantly, and with a touch of sarcasm, "You're joking!". The two birders apparently dismissed Mike and I as either pranksters or simply novice birders. I felt embarrassed at having blurted out my presumption, especially as the bird had now disappeared and, were it to reappear, it would surely present its red breast and brown tail for all to see. "Yeah, only joking perhaps, but let's just check this." Then Mike cried "B-Hell!" as the bird re-appeared and gave us both clear views of its cobalt blue rump and tail.

The next few moments of intense concentration, watching the bird flitting low in the undergrowth, served to confirm my presumption, and also eliminate all thought of Siberian Blue Robin, or worries about escaped 'blue flycatcher' species. After just a few moments more, we had seen orange-red flanks, plain face and pale eye-ring. I looked for those two other birders and, with binoculars raised, I recognised Peter Gamage. I yelled, "Peter, we HAVE got a Red-flanked Bluetail!"

I do not recall whether he and his companion, Graham Stephenson, ran, fell, jumped or flew across the valley, but within seconds their two white faces appeared on the footpath. The colour soon rushed back into their cheeks, however, as they feasted their eyes on the bluetail.

A group of four other birdwatchers plus one couple ambled down the path and were promptly beckoned to join our excited huddle. We had simply gone for a short 'family' walk, so I had no notebooks and no telephone numbers with me, but at about 2.10pm I ran up the hill to the village telephone box and I dialed the only three numbers Mike and I could recall. I returned down the valley, camera in hand, and took some very shaky photographs in the gloom of dusk. We left the site at 3.25pm, having lost the bird in one of the private gardens. Despite this, we were confident that the bird was still in the same area, as it had been following a circuit, frequently disappearing from view only to return a little later.

We spoke to some of the local villagers, as well as with the owners of the gardens frequented by the bluetail, and we confidently released the news of our exciting discovery during the evening.

Not surprisingly, by dawn the next morning (a Sunday), a large crowd of birders had assembled at the site. Fortunately, the bird was still present and proved very confiding, but it is to the credit of those who helped to organise the crowds and car parking, and the patience of those queuing for the limited viewing position, that everyone managed to see the bird well that day. Many more birders followed over the ensuing week and weekend, and hardly any left disappointed (although the bird could be elusive on occasions). It was last seen on 8th November.

Neil Morris. Birding World 6: 431-434.

The 1990s delivered no fewer than 14 bluetails, including an unprecedented three in 1998, eclipsed smartly by five in 1999 including four arrivals in two days between Shetland and Cornwall. The most remarkable record of the decade was the bird seen by fewest observers; a quite exceptional inland 1st-winter extracted from a mist-net near Loughborough, Leicestershire and Rutland, on 19th October 1997; yet another reminder of just how many vagrants must go unfound.

Since 2000, the spiraling trend of occurrence has continued apace. No fewer than 23 were logged between 2000 and 2007. A record eight in 2007 included the first for Wales, a fine male on Bardsey on 1st October. The species has missed only two years since the iconic Dorset bird of 1993 and has become an expected feature of the autumn rarity scene.

Four additional birds in late September 2008* have taken the total past 50, including the first records for North Ronaldsay and the Isles of Scilly.

Where: There is a Shetland and east coast bias to records. Shetland has hosted no fewer than 15 birds and nine since 2002. Elsewhere in Scotland, three have been recorded in Fife and one each in Northeast Scotland and Borders. Three have graced Northumberland and singles Co. Durham and Cleveland. Two reached East Yorkshire in 2007. Lincolnshire has a disproportionate showing of five. There have been four in Norfolk and three in Suffolk.

Along the south coast two have made landfall in Kent and there was the well-watched bird in Dorset. In the southwest two were seen in Devon in 2005 and two have reached Cornwall, but the species took until 2008 to impact upon the avifauna of Scilly (record pending).

Caernarfonshire had the sole Welsh record (only the second to be found away from the east coast and the southwest). The inland bird in Leicestershire is unique.

When: Two of the four spring records were in 2007. The earliest, at Easington, East Yorkshire, on 31st March, was later found dead. One was on Out Skerries a few days later from 2nd to 3rd April. In 1995 a male was on Holy Island, Northumberland, on 23rd April. The latest spring bird by some margin was a male on Fetlar, Shetland, from 31st May to 1st June 1971. It is tempting to suggest that the early dates of the spring birds in 1995 and 2007 imply that these individuals had wintered somewhere in Western Europe. The date of the first spring bird in 1971 is the only one that fits the pattern of a typical spring overshoot bound for Fenno-Scandia.

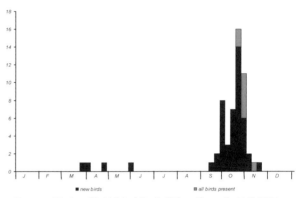

Figure 27: Timing of British Red-flanked Bluetail records, 1903-2007.

Red-flanked Bluetail is traditionally an autumn vagrant. The earliest was a 1st-winter on Fair Isle on 16th September 1993. Next were two on 19th and 21st and then eight between 25th and 29th September, all on the east coast. There is a lull in early October before the pace quickens in the second week of the month. Perhaps October birds are from farther east in the range than the September forerunners? Around one third of all records fall between 15th and 20th October before an abrupt decline to two records in November: one at Cot Valley, Cornwall, on 3rd November 2007 and the latest at Gibraltar Point from 15th to 16th November 2002.

Discussion: In Europe Red-flanked Bluetail breeds only in Russia and Finland. During the first part of the 20th century the range reached the western slopes of the Ural mountains. Buturlin and Dement'ev (1941) commented that it was one of a group of eastern species that were continually expanding their range westwards. The first records for the Kola Peninsula, in August 1937, involved a shot family party that had probably been nesting in the area, representing the then westernmost confirmed breeding record (Mikkola 1973). In Finland, it was first recorded 1949 and apparently spread west with some 40 singing males in the 1950s. Appearances declined almost to nothing in the 1960s, but 27 were found during 1969-1973. In the 1990s only 10 birds were found annually in northern Finland (Hagemeijer and Blair 1997). Numbers have increased since and 68 singing males were present in 2007. Its relative scarcity as a breeding bird in Finland suggests that the upsurge in British sightings reflects the species' status slightly farther afield. Perhaps the population has increased more rapidly in an eastern area that is as yet undetected?

Away from Finland this remains a scarce species in Western Europe. In Sweden there are 31 records, all bar one since 1973 and including a breeding pair in 1996. There are just 14 records from Norway to the end of 2007. There are three records from Estonia (May 1977, 24th April-22nd June 1980, 13th July 1995) and three from Poland (30th October 1995, 25th September 2001, 16th October 2007). The first for Lithuania was very recent (10th November 2008). The first for Denmark was in 1976, with 10 subsequent records all from 1994 onward. Dates include one each in May, June and September (28th), four between 2nd-6th October and the latest on 19th October. Three trapped and photographed in 2008 included one on 26th May and two in October. The dozen or so from Germany were supplemented by three more in October 2008. Exceptional birds include singing males on Helgoland on 6th-7th June 1998 and in Mecklenburg-Vorpommern from 2nd-10th July 2001 and a wintering bird from 27th January-5th February 2006. Some German birds are considered to be possible escapes (Peter Barthel *pers comm*).

There are eight from the Netherlands (16th October 1967, 29th September 1985, 11th-16th October 1999, 5th November 2001, 3rd January to 23 February 2007, 6th October 2008, 29th October to 8th November 2008 and 30th October 2008), three from Belgium (25th-27th September 2001, 15th October 2005, 17th October 2006), one from the Channel Islands (Sark 31st October-2nd November 1976), five from France (four Ouessant and one Pas-de-Calais: 27th October 1993, 20th to 23rd October 2002, 24th October to 2nd November 2003, 23rd to 27th October 2005, 16th to 27th November 2007) and one from Czech Republic (1st October 2007).

Intriguingly there are two from Spain (Llobregat Delta, Barcelona on 17th November 1998 and Valencia 16th November 2005 to 26th January 2006, when found dead) and nine from Italy including three winter records (2nd December 1997, 26th-29th December 2001, 23rd January 2005). The first for Israel was found wintering in the north from 1st January 1996. One in UAE from 7th December 2008 to at least 20th February 2009 was the first record for Arabia.

Contrary to the show-stopping performances of some recent individuals in Britain, Red-flanked Bluetail is usually a skulker. British individuals are probably drift migrants. Records from farther south in Europe and the increasing incidence of winter finds suggests purposeful filtering through in the autumn. Wintering birds in Western Europe are presumably the source of early spring east coast records in Britain. A future wintering record in Britain would not be unexpected in light of such records from Germany and the Netherlands, plus a number from the Mediterranean.

White-throated Robin *Irania gutturalis*
(Guérin-Méneville, 1843). Breeds across much of south-central Asia from southern Turkey east through northern Iraq and central Iran to western Afghanistan and the western Tien Shan mountains in eastern Kazakhstan; also recently recorded breeding on the fringes of the Turkish Black Sea. Winters mostly in southern Kenya and Tanzania.

Monotypic.

Status: Two records.

1983 Isle of Man: Calf of Man male 22nd June
1990 Pembrokeshire: Skokholm female 27th-30th May

Both birds were inaccessible for the masses and the listing fraternity awaits its chance to add this striking chat to the tick bag. The four-day stay of the Skokholm bird would have proved a popular draw but for news of its presence being withheld due to the fragile nature of the island, which holds huge numbers of burrow-breeding Manx Shearwaters *Puffinus puffinus*.

On 26th May 1990, I joined a small group of would-be wildlife artists, led by Peter Partington, at the jetty at Martinshaven, Pembrokeshire, to take the short ferry ride to the island of Skokholm. Shortly after breakfast on 27th May, our first morning on the island, I emerged from the dining hall and walked towards the cottage, which has a small walled garden and one of the few significant trees on Skokholm. As I did so, a small bird dropped from the tree onto the ground.

As passerines are not particularly numerous on the island, I immediately looked at it through my binoculars, and was surprised to see a bird that I could not put a name to. It was the size of a small thrush (Turdidae) and was hopping on the ground with an upright stance and cocked tail. Its size, large dark eye and plain underparts gave it an appearance reminiscent of a Common Nightingale Luscinia megarhynchos. It was clearly a chat or small thrush species with which I was totally unfamiliar and it definitely looked like something very interesting! I attracted the attention of some of the other members of the group who had appeared after breakfast, including Michael Betts, the Warden, and Jack Donovan. We were all perplexed by its appearance and uncertain what species it was. Fortunately, a copy of BWP was close at hand, and careful inspection revealed that we were looking at a female White-throated Robin!

Although a few other observers came to see the bird, the news was not released by the Warden owing to the fragile nature of the island, which is honeycombed with the burrows of tens of thousands of Manx Shearwaters Puffinus puffinus. The White-throated Robin remained on the island for four days and was last seen on 30th May. The weather at the time of the bird's arrival was warm, dry and settled, and a few other migrants recorded on the island during this time included a male Golden Oriole Oriolus oriolus.

After finding the robin, I had the honour of being asked to immortalise it on the toilet wall, alongside paintings of other island rarities as portrayed by their finders! According to Jo Harvard, the present Warden, these paintings remain a fascinating (and humorous) record of the key Skokholm rarities.

David Thelwell. British Birds 99:361-364.

Discussion: Both records conform to expectations regarding a spring overshoot and perhaps a dispersing male for the late June sighting.

Elsewhere in northern Europe there are 15 records. These include seven from Sweden (11th June-14th July 1971, 14th May 1977, 10th May 1981, 10th-19th May 1986, 16th-21st May 1989, 9th August 1995, 20th-21st May 2000) and two from Norway (15th May 1981 and 17th August 1989). There are four from the Netherlands (3rd-4th November 1986, 2nd June 1995, 30th August 2003, 31st October 2005 until found dead on 2nd November) and Switzerland (25th May 2000). The Dutch record in 2005 was of the less well-known pale morph. One trapped in Belgium (25th August 2007) is still in circulation. Seven of the spring birds have been in May and two in June.

Four of the autumn records have been in August and singles in October and November. Six of the Scandinavian birds have been in May and one in June, with both autumn birds in August.

In autumn, migrants cross the Arabian peninsula to Africa through Eritrea and Ethiopia, from where they move south. Return passage commences in April and is believed to follow a similar route. Arrival on the breeding grounds in Turkey from mid-April onwards is supplemented by continuing arrivals into May. Birds breeding in the western part of the range are likely to migrate out of Africa towards the northeast before reorientating to the north-west. It is presumably during such movements that overshooting takes place, accounting for the northern European records in May and June, the majority of which are from Scandinavia. The bird's rarity farther west would appear to be a result of the northerly orientation of spring movements leading to overshooting of the western part of the breeding range.

Most leave the breeding grounds by the end of August. Lost migrants after this time presumably account for the autumn northwest European records. A slightly different trajectory at this season could account for the bulk of the records coming from the Low Countries. Three of the four Dutch records and the sole Belgian record occurred between August and November.

Black Redstart *Phoenicurus ochruros*

(S G Gmelin, 1774). Breeds from southern Europe, Middle East and Caucasus through Iran to eastern Tibet. Winters southern Europe, northern Africa, Middle East, Arabia and central and west Indian plains and southern Asia.

Polytypic, seven subspecies recognised by *BWP* but only five by *HBW*. European Black Redstart *gibraltariensis* (J F Gmelin 1789) occupies Europe and northwest Africa. Caucasus Black Redstart *ochruros* (S G Gmelin, 1774) lives in Asia Minor, Caucasus and Northwest Iran. Levantine Black Redstart *semirufus* (Hemprich and Ehrenberg, 1833) is found from Syria to Jordan and Israel. Tian Shan Black Redstart *phoenicuroides* (Horsfield and Moore, 1854) breeds in the mountains of central Asia, notably Altai and Tien Shan. Both *phoenicuroides* and *rufiventris* intergrade and birds from Western China (Southern Xinjiang) *xerophilus* Stegmann, 1928 are considered by some authors to be probably intermediate between them (*HBW*). Unlike the other subspecies, *phoenicuroides* is a long-distance migrant to wintering grounds in the central and west Indian plains, southern Iran, Arabia, Somalia and Ethiopia. Pamir Black Redstart *rufiventris* (Vieillot, 1818) occurs from Turkmenistan through the Himalayas to central China and winters in southern Asia and possibly the Middle East and Egypt.

Black Redstart can be separated into three major groups based on morphology, biogeography and mtDNA cytochrome b sequence data (Cramp *et al* 1988; Ertan 2006). These include central and eastern Asian forms comprising *phoenicuroides*, *rufiventris* and the little-known Chinese Black Redstart *xerophilus*; eastern European/western Asian forms comprising *ochruros* and *semirufus*; and Western European forms including *gibraltariensis* and *aterrimus*. Separation of Black Redstart subspecies depends on size, extent of black and grey on upperparts, extent of red and black on underparts and underwing, plus the presence of white on forehead and the amount of white on wing (*BWP*).

Eastern Black Redstart Phoenicurus ochruros ochruros/phoenicuroides/rufiventris

Status: A recent review of all British records of 'eastern' Black Redstarts (races *ochruros, phoenicuroides* and *rufiventris*) concluded that it was impossible to eliminate the possibility that past claims of these forms referred to hybrids between Black Redstart and Common Redstart *P. phoenicurus* (*BOURC* 2002). Suspected hybrids of this pairing do not appear to be unusual (for example: Andersson 1988; Ukkonen 1992; Lindholm 2001).

Four records were removed from the *British List*: immature or hybrid Bryher, Scilly, on 13th October 1975; adult male Saltfleetby, Lincolnshire, 15th to 17th October 1978; first-year male Dungeness, Kent, 7th to 8th November 1981, trapped on 7th; and male Donna Nook, Lincolnshire, 21st October 1988. If images exist of the trapped bird at Dungeness, these may be worthy of a re-examination to assess the wing structure (see below).

Discussion: Recent records of Eastern Black Redstart of race *phoenicuroides* accepted elsewhere in northern Europe include 1st-winter males in the Netherlands from 21st-23rd October 2003 and Vazon Bay, Guernsey, Channel Islands, from 28th-31st October 2003. The latter was trapped, but unfortunately a feather collected for DNA analysis was subsequently lost. These two are the first for Western Europe to be fully acceptable (Steijn 2005). Three records from Sweden are attributed to one of the eastern races: male on 15th November 1986, 1st-winter female trapped, and present from 23rd -27th October 2000 (DNA analysis suggests it to belong to either *phoenicuroides* or *rufiventris; Steijn 2005)* and 20th October 2005. A number of other claims are now considered to be hybrids, which occur annually in Germany (Peter Barthel *pers comm)*.

The rarity of *phoenicuroides* in western Europe is surprising considering that it is a long-distance migrant from its central Asian breeding grounds to wintering areas in central India west to northeast Africa. Departure of *phoenicuroides* from the breeding areas commences during late August and early September and arrival at the wintering areas is from October onwards. Its breeding and wintering distribution is comparable to species that occur regularly in northwest Europe, including Desert Wheatear *Oenanthe deserti* and Isabelline Shrike *Lanius isabellinus*. Of the three red-bellied subspecies *phoenicuroides* seems the most likely to deviate to western Europe. Most could be overlooked given that immature birds would not be eye-catching and are difficult to identify.

Assessing the sex of Black Redstarts is not easy, especially for immature birds. Immature males show two different plumages: most (88 per cent) resemble females (the 'carei' plumage') but some (12 per cent) resemble adult males (the 'paradoxus' plumage); the latter plumage form is also found in the eastern subspecies (Nicolai *et al* 1996; Steijn 2005).

Both of the accepted European records were 1st-winters in less troublesome 'paradoxus' plumage; those in 'carei' plumage are much harder to identify due to their resemblance to female *gibraltariensis* (see Steijn 2005 for differences). Anyone faced with a potential Eastern Black Redstart should be aware of the pitfall of Black Redstart and Common Redstart hybrids. The most useful difference between Eastern Black Redstart and hybrid Black x Common Redstart is the wing structure. Steijn (2005) proposed differences based upon the emarginations on p3-6 for *phoenicuroides* and a ratio for spacing between p5-6 and p6-7.

Those who have encountered Eastern Black Redstarts on trips abroad will no doubt find it difficult to believe that the striking rufous-bellied birds they saw were the same species as birds in Western Europe.

Ertan (2006) concluded that the western races (*gibraltariensis*, *ochrurus* and *semirufus*) form a monophyletic group and that *phoenicuroides* and *rufiventris* were the most divergent subspecies. Further, *phoenicuroides* and *rufiventris* appear to be more closely related to Hodgson's Redstart *P. hodgsoni* than to Western Black Redstart (Ertan 2002; Ertan 2006).

Steijn (2005) suggested there might be merit in treating rufous-bellied *phoenicuroides* and *rufiventris* as a separate species (Eastern Black Redstart). The taxonomic status of intermediate *ochruros* is not clear. The next British suspect will no doubt be studied closely by observers now better prepared to tackle the identification of these complex stunning birds.

Common Redstart Phoenicurus phoenicurus

(Linnaeus, 1758). Breeds from northwest Africa and Europe east to Lake Baikal and south to the Balkan countries and Ukraine. Nominate phoenicurus found across the range, except in the southeast where it is replaced by samamisicus from Turkey, Crimea to Turkmenistan, and Iran; possible zone of intergradation in Greece and southern Balkans. Wintering grounds are in Sahel zones of Africa with some in southern Arabia.

Polytypic, two subspecies. Ehrenberg's Redstart *samamisicus* (Hablizl, 1783) is a vagrant of uncertain status. A review of all past records, including those previously not accepted, is in progress and a number of new claims are being assessed. The *BBRC* published a detailed set of criteria by which *samamisicus* can be separated from nominate *phoenicurus* (Small 2009). The conclusion was that adult male *samamisicus* should be separable easily from adult male *phoenicurus*, even individuals that show conspicuous pale fringes to the tertials and inner secondaries. In *samamisicus* the white edges extend from the tertials and secondaries on to the primaries and form an eye-catching wing panel that contrasts with a blacker mantle and scapulars. The ear coverts also have more extensive black than *phoenicurus*. First-winter male *samamisicus* is far more difficult to separate as it matches adult male *phoenicurus* closely.

Attention to correct ageing is important. Adult *phoenicurus* showing a white wing panel on the tertials and inner secondaries may be identifiable from first-winter *samamisicus* only by loral colour: black in *phoenicurus* and fringed paler in *samamisicus*.

Genetic differences between the two forms were highlighted by Ertan (2006) who found a 2.3 per cent divergence between *samamisicus* and *phoenicurus* (marginally below that of the recently re-split Green Warbler *Phylloscopus nitidus* and Greenish Warbler *P. trochiloides*). There also appear to be differences in the calls and songs of *samamisicus* and *phoenicurus*, and males are morphologically diagnosable.

Ehrenberg's Redstart Phoenicurus phoenicurus samamisicus

Breeds from central Turkey, east to the Crimea and Caucasus Mountains and into northern Iran. Winters in northeast Africa.

Status: Two birds identified as this form are presently accepted (Small 2009). The status of this sometimes striking form in Britain is poorly known as many claims have not been submitted to the *BBRC* because of a lack of confidence in the criteria for identifying anything other than adult males.

1975 Norfolk: Heacham male 26th October
1989 Norfolk: Holkham Meals 1st-winter male 12th September

Previously accepted records of males from Fife Ness on 23rd September 1976 and from near Keynsham, Avon, on 22nd-23rd September 1989 were found to be unacceptable during the early part of the review of records for the form (Fraser *et al* 2007). Four additional records are under review by the *BBRC* from Scilly (St. Agnes on 22nd October 1994 and Tresco from 10th-12th October 2004); Shetland (Grutness from 23rd-26th September 2000) and Cleveland (Hartlepool on 25th September 2001) (Small 2009). Numerous others have been considered to involve this form. Forrester *et al* (2007) added a further eight in Scotland from 1948-2000 (one in early September, five in late September and two in early November) that they felt should be reviewed. These birds were often associated with drift conditions. Putative claims from elsewhere have also been reported in recent years. How many fall within the variation of male *phoenicurus* is uncertain, but the two accepted records and others will be re-examined by *BBRC* under criteria set by Small (2009).

Discussion: There are accepted records for Sweden (male 25th-26th September 1988, female 20th October 1993, male 21st September 1996, female 18th October 1996) and Finland (29th April 2004) and a claim is pending from the Netherlands (10th September 2005). There are several records from France (the first was a male caught in the Camargue on 5th April 1959) and one from Italy. A record from the Balearic Islands remains under review.

It seems likely that many or most British claims will lack the detail for acceptance, even though some have been well photographed. All are autumn records. Young birds are the most likely age class to wander in autumn and, given that only adults are unequivocally diagnosable, many may be unidentifiable under present criteria. Might it take a spring overshoot to stake an unequivocal claim on the British List?

Moussier's Redstart Phoenicurus moussieri
(Olph-Galliard, 1852). Breeds in Morocco, Algeria and Tunisia. Mostly resident, but makes altitudinal and short-distance movements outside the breeding season.

Monotypic.

Exceptionally rare vagrant from northwest Africa; one British record the only sighting in northern Europe.

Status: One record.
1988 Pembrokeshire: Dinas Head male 24th April

Discussion: As a short distance or altitudinal migrant this would appear to be an unlikely over-shoot from its North African breeding areas. However, it is a regular straggler to Mediterranean regions, where 10 have reached Malta, four Italy (plus two yet to be submitted), three Spain (7th April 1985, 26th April 2000, 19th-22nd October 2008), two Greece (30th March 1988, 18th September 1994) and one Portugal (11th November 2006-14th January 2007). A female in northern France on 14th May 1993 is now rejected following review (Marc Duquet *pers comm*).

The species' propensity for straying from its breeding range to countries in the Mediterranean offers some hope that this may not be the one-off that it perhaps appeared at the time.

Common Stonechat *Saxicola torquatus*

Linnaeus, 1766. Breeds across Africa and Eurasia. In Eurasia occurs from northwest Africa and Iberia to the far east of Siberia.

Polytypic, 24 subspecies recognized; taxonomy confusing. Relevant subspecies for European birdwatchers comprise *hibernans* (Hartert, 1910) of Britain, Ireland, western Brittany and west coast of Iberian peninsula; *rubicola* (Linnaeus, 1766) of northwest Africa and Europe (except for range of *hibernans* and to the south of range of *maurus*) and Turkey to Caucasus; *maurus* (Pallas, 1773) from eastern Finland to Mongolia and Pakistan; *stejnegeri* (Parrot, 1908) of east Siberia to Korea and Japan; *variegatus* (S G Gmelin, 1774) to the east of the Caucasus to Iran; *armenicus* Stegmann, 1935 of southeast Turkey to southwest Iran. Birds from west and south of Eurasia are generally resident and those from the north and east are migratory (*BWP*).

The taxonomy of the Common Stonechat complex has come under much scrutiny in recent years by several authors (for example: Wittmann *et al* 1995; Urquhart 2002; Wink *et al* 2002). Wink *et al* (2002) concluded that European *rubicola*, Siberian *maurus* and African *torquatus* had diverged substantially, perhaps allowing the complex to be divided into three separate species. However, their study did not encompass all forms within the range. Siberian Stonechat includes both *maurus* and the easternmost form *stejnegeri*, which some authorities consider to be synonymous with *maurus* as differences between them are very slight. It is unclear whether *stejnegeri* can be diagnosed with confidence; the possibility of *stejnegeri* x *maurus* intergrades could never be eliminated, particularly in first-year birds, thus records tend to be referred to as Siberian Stonechat *maurus/stejnegeri*.

Urquhart (2002) also includes *variegatus* within Siberian Stonechat on the basis that it shares many morphological features, although it shares more features with White-tailed Stonechat *S. leucurus* from Pakistan and the northern part of the Indian subcontinent to Myanmar. Despite the shared features of *variegatus* and *leucurus*, these taxa are separated by the more *maurus*-like *armenicus*. Larger and more comprehensive DNA studies will hopefully yield further insights into the complex relationships of these forms.

There is increasingly strong evidence for elevation of Siberian Stonechat to full species status, as adopted already by some national committees. Such a treatment is not yet endorsed by the *BOURC*, who want to make a decision based on information from throughout the range, including *variegatus* (which has occurred in Britain).

Siberian Stonechat *Saxicola torquatus maurus/stejnegeri*

Breeds widely across northern Asia from north Urals south to the Caspian Sea, Mongolia and northern China east to Kolyma basin, Okhotsk coast and north Japan. Winters from northern Indian subcontinent to southern China and southeast Asia.

Status: There have been 331 accepted records to the end of 2007, 325 in Britain and six in Ireland.

Historical review: The first was a female obtained on the Isle of May on 10th October 1913. The second was not seen until 1960, when a female or 1st-winter was at Hartlepool, Cleveland, on 26th October. Other autumn birds followed from Fair Isle in October 1961, November 1964 and October 1965.

The sixth bird was the first in spring, from 6th-7th May 1972 at Cley. Records have been annual since 1974, presumably due to increased observer awareness and better understanding of the identification criteria following Robertson's 1977 article in *British Birds*. The 1970s produced 35 birds, including 10 in 1978, and the first for Ireland on Great Saltee Island, Co. Wexford, from 12th-19th October 1977.

The 1980s produced 87 birds, with a sudden annual peak of 25 in 1987. The 1990s yielded a whopping 151, with 32 in 1991 establishing the single-year record and 25 in 1993.

After 17 in 2001, numbers have been lower, with just 54 aggregated to the end of 2007. An average of 15 per annum during the 1990s has more than halved to 6.8 per annum between 2000 and 2007, a most puzzling decline given the trends in most sympatric taxa.

Where: Records have a predictable bias towards the Northern Isles and east coast, though it remains surprisingly rare in some east coast counties that would appear to be geographically well-positioned to receive birds. Shetland weighs in with 28 per cent of all records. Elsewhere in Scotland, 34 birds include 16 on Orkney and nine on the mainland of Northeast Scotland.

Northeast England accounts for 24 per cent of all records, including 28 birds in East Yorkshire, 21 in Northumberland and 12 in North Yorkshire. Norfolk boasts 11 per cent of national records, but southeast England musters just three per cent (eight in Kent). The southwest has attracted 12 per cent of all records, including Scilly with 16 birds, Dorset 12 and Cornwall 11. Away from these areas records are rare. There are nine Welsh records (four in Caernarfonshire,

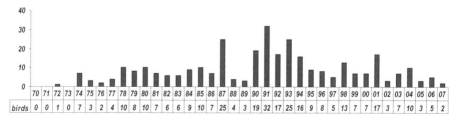

Figure 28: Annual numbers of Siberian Stonechats, 1970-2007.

174

three in Pembrokeshire and two on Anglesey), six in Ireland (five in Co. Cork and one Co. Wexford) and two in northwest England (one each on the Isle of Man and Lancashire).

A number of birds have penetrated well inland, with two reaching each of Wiltshire (Chirton Down, 21st November 1987 to 7th February 1988 and Whiteparish on 13th October 2003) and Cambridgeshire (Ouse Washes 27th to 30th November 1974 and 24th November 1979), and singles from Berkshire (Brimpton Gravel-pits, 31st October to 2nd November 1986, when trapped) and West Yorkshire (Wintersett Reservoir, 19th to 24th October 1976, trapped 22nd). Many of the inland records were from a period when Siberian Stonechat was still quite a rarity. Given its present status could it be possible that inland birds are now being missed among increasing numbers of Common Stonechats?

When: The *BBRC* recently reviewed all spring records of Siberian Stonechat (Rogers *et al* 2004). It was apparent that identification problems were caused by some *rubicola* showing a glaring white rump and whiter breast, causing them to look almost exactly like a Siberian Stonechat (for example, Corso 2001; Walker 2001). The review confirmed that spring records are indeed rare, with just 20 records (14 males, four 1st-summer males and two females). Seven were from Norfolk, six from Yorkshire and three from Shetland, plus singles in Lincolnshire, Cleveland and Northeast Scotland; one on Scilly is the only spring find away from the east coast.

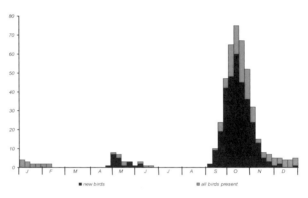

Figure 29: Timing of British and Irish Siberian Stonechat records, 1913-2007.

The earliest record was a 1st-summer male on Fair Isle from 28th-30th April 1990. The spring peak falls in the first two weeks of May (60 per cent of spring records). Records extend through to early June, the two latest were part of an exceptional east coast arrival of migrants in spring 1992 when Norfolk hosted a male at Holme from 7th to 8th June and a male at Cromer on 8th June. A number of spring records have been associated with falls: three in May 1978 between East Yorkshire and Lincolnshire; two in May 1985 with singles in East Yorkshire and Cleveland; three in May and June 1982, all in Norfolk; and two in early May 1993. The most notable record was a male in territory at Forest of Birse, Banchory, Northeast Scotland, from 2nd-24th June 1993. Siberian Stonechat bred in eastern Finland in 1992 (three pairs) and 1997 (a pair raised five young).

Autumn is well established as the time to seek out these sometimes subtle birds. The earliest records have both been from Norfolk, with a male at Cromer from 4th to 5th September 1994 preceding another male at Blakeney Point, from 9th to 13th September 1995. The earliest date from the Northern Isles is 10th September (North Ronaldsay in 1995 and Fair Isle in 2002). Records increase from mid September.

The peak period begins in late September, reaches a high during 8th-14th October, and continues to the end of October. Multiple records are rare, but at least five were on Fair Isle on 29th September 1987 and on five occasions two have been found together (three times on Shetland and once in Northumberland and East Yorkshire). After mid November records decrease markedly, although new birds continue to be found through to early December. A bird found in late December presumably constitutes the latest newly discovered wintering bird, and a handful have remained through to early February.

Discussion: In most of Europe this taxon is much rarer than in Britain. In Finland there are more than 175 records and tens of records where the subspecies was not identified. In contrast, there were just 34 from the Netherlands during 1973-2006 and eight in Belgium (six between 1993 and 1998 and two since). There is evidence of filtering south, but it is very rare in the Mediterranean region (Corso 2001), though the difficulty of separation from *rubicola* may mask its true status. There are four accepted records for Spain (2nd March 1997, 6th April 1997, 13th September 1997, 30th-31st March 1998), one from southern France (Camargue, 4th November 1996) and at least eight from Italy (the first of which was in October 1988). There are also three accepted records for Cyprus (December 1971, January 1991, September 2000).

In Britain and Ireland records in the south and west tend to occur later, suggesting that these birds originally made landfall farther north and east. For example, the median arrival date of autumn birds is 10th October for Fair Isle, 11th October for Norfolk, 14th October on the Isles of Scilly and 18th October for Portland. The recent fall-off in records is mystifying. Increasing numbers of observers during the period of reduced numbers of records make the decrease clearly real. Perhaps this situation indicates a range contraction or population fall?

Caucasian Stonechat *Saxicola torquatus variegatus*

Breeds west and north Caspian Sea from Azerbaijan and Georgia, north to lower reaches of Volga River, and northeast to delta of Ural River, Kazakhstan. Migrates southwest through east Iraq and throughout Arabian Peninsula west to Israel. Most winter northeast Africa from north and east Sudan, northern Ethiopia and Eritrea, occasionally north to Israel; small numbers remain southeast Transcaucasus.

Status: Three records.

1985 Cornwall: Porthgwarra 1st to 4th October
1993 Suffolk: Landguard 11th September
2006 Shetland: Virkie, Mainland 1st-summer male 7th May

Discussion: There have been just three other north European records. One was seen in Norway (June 1983) and one Denmark (18th-19th May 2003). A 2008 record from Sweden (27th May) is yet to be assessed. Elsewhere one, dead for about a month, was found in Italy (1st January 1996), one was in Greece (ringed in western Greece on 4th December 2006) and there are nine records from Cyprus (25th March 1999, 4th April 2003, 4th March 2004, 23rd March 2004, two on 24th March 2004, 13th February 2005, 17th April 2005 and 23rd April 2007).

Urquhart (2003) treats *variegatus* as a race of Siberian Stonechat, based on shared morphological characters, although it also shares many plumage features with White-tailed Stonechat S. *leucura*; a third form *armenicus* is found between the ranges of the two. The relationship between *variegatus, armenicus* and *leucura* will, hopefully, be clarified through DNA studies.

Isabelline Wheatear *Oenanthe isabellina*

(Temminck, 1829). Small European population restricted to eastern Greece, Bulgaria, Romania, Ukraine and southwest Russia. In Asia, breeds widely across arid grasslands from Turkey through Kazakhstan, Mongolia and northern China, south to Iran and northern Pakistan. Winters from northern Sahel zone to East Africa and throughout Middle East from Arabian Peninsula to south Iran, Pakistan and northwest India.

Monotypic.

Very rare vagrant from southeast Europe or Asia, all but one in autumn.

Status: A total of 27 records was recorded by the end of 2007, 26 in Britain and one in Ireland.

Historical review: The first was a first-year female shot at Allonby, Cumbria, on 11th November 1887. It was to be 90 years before the next and the only accepted spring bird at Winterton, Norfolk, on 28th May 1977. This re-acquaintance rapidly led to the discovery in 1979 of a long-staying first-winter at Girdle Ness, Northeast Scotland, from 17th October to 10th November and trapped on the 23rd. There was yet another in 1980, a probable adult at Bamburgh, Northumberland, from 16th-20th September. The renaissance then stalled until the next in 1988, on Scilly in early October. Observers (and records committees) began to get the measure of this subtle wheatear, aided by early tentative steps towards establishing the criteria by which to identify the species (for example, Kitson 1979b; Tye and Tye 1983; Clement 1987).

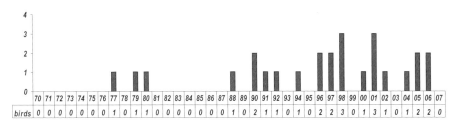

Figure 30: Annual numbers of Isabelline Wheatears, 1970-2007.

Two were recorded in October 1990, a long-staying bird into early November on Scilly and one at Kilnsea, East Yorkshire. Greater appreciation of the identification features led to the decade accumulating 12 individuals of what has always been a birder's bird in the self-finding fraternity. The period included the first for Ireland at Mizen Head, Co. Cork, from 10th-17th October 1992 and a record three in 1998 (late September birds on Shetland and North Yorkshire and an early October bird in Suffolk). The detection points for the decade were evenly segregated into an east/ west split. In the west, two reached within a matter of days the extremes of Wales (in 1997) as well as two in Scilly and singles in Cornwall and Co. Cork. In the southeast singles were found in Kent and Suffolk, and in the east two appeared in both Yorkshire and Shetland

By the turn of the century records had become virtually annual. Three more followed in 2001 to equal the best ever showing (September birds in Shetland, Norfolk and Suffolk). Increased observer familiarity had led to no fewer than 10 of these pallid waifs being scrutinised between 2000 and 2007, with birds arriving at almost two per annum. Finds over this period showed an east coast bias: two each in Norfolk and Suffolk, plus two from Shetland and one from Orkney and a single in North Yorkshire. In the west, only North Wales offered birds: singles on Anglesey and Bardsey.

Where: As might be expected an east coast prevalence is evident, with four on Shetland and three each in Yorkshire, Norfolk and Suffolk, and singles in Northumberland, Orkney, North-east Scotland and Kent. The Isles of Scilly notched up three between 1988 and 1991 and four have reached the offshore islands of Wales, two of them on Bardsey and singles on Skokholm and Anglesey. Other records have come from Co. Cork, Cumbria and Cornwall. The subtlety of identification and the uneven temporal distribution of records imply that many others have probably gone undetected in between. Any pallid-toned wheatear with white underwings and much black in the tail is clearly worthy of closer inspection. A large pale Greenland Wheatear *leucorhoa* is the only possible bogey.

When: All but one of the records was in autumn. The sole spring record was from Winterton, Norfolk, on 28th May 1977.
Autumn records fall in the period mid-September to early November. The earliest was on Fetlar from 14th to 15th September 2001, with the next earliest at Bamburgh, Northumberland, from 16th-20th September 1980. A late September and early October peak is evident (just under half of all records have been between 20th September and 6th October), followed by another mid-October

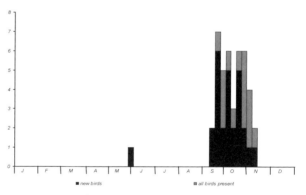

Figure 31: Timing of British and Irish Isabelline Wheatear records, 1887-2007.

■ new birds   ■ all birds present

peak. A third of all records have been between 10th and 23rd October, just two occurring after this: Church Cove, The Lizard, Cornwall on 29th October 1996 and the first record, in Cumbria, on 11th November 1887.

Discussion: Elsewhere in northern Europe sightings have increased in line with those in Britain. There have been 29 in Scandinavia with 13 from Finland (five between 13th April-8th May, two 6th-21st June and six between 5th September-13th October) and 11 from Sweden (one in early April, two in May and one in June and seven between 9th October-3rd November), plus five from Norway (29th September-18th October 1977, 14th November-2nd December 1999, 16th October 2001, 16th October 2003, plus a well documented record from November 2008 that is yet to be submitted) and two from Denmark (21st September 1989, 13th October 2005).

In northern Europe a further 17 birds have been recorded. Sightings have come from Poland (two: 29th May 1986, 3rd August 1997), Germany (three from Helgoland: 14th-17th October 1999, 28th September 2000, 15th October 2006), the Netherlands (five: 21st October-8th November 1996, 22nd-23rd September 2000, 14th-25th October 2000, 18th November 2005, 31st August-4th September 2006), Belgium (two: 13th October 2003, 18th-22nd October 2005) and northern France (five: 27th September 1970, 31st May-1st June 1988, 16th October 1993, 25th October 1997, 14th October 2006).

Table 16: Isabelline Wheatear records in northern Europe (excluding Britain and Ireland)

	April	May	June	July	Aug	Sep	Oct	Nov
Fennoscandia	5	3	3			6	11	3
N. Europe		2			2	3	9	1
Totals	**5**	**5**	**3**	**0**	**2**	**9**	**19**	**4**

The east Fenno-Scandian records show a markedly different occurrence pattern to the British and Irish records. Half of the records from Finland and Sweden fall in the spring period (a stark contrast with Britain). The dearth of spring records farther west in northern Europe is not that surprising given that departure from the north African wintering areas is on a north or north-easterly heading with little expectation for a northwesterly orientation. The Fenno-Scandian birds are presumably overshooting their return to southeast European breeding grounds on a northerly bearing. Spring passage is regular in the eastern Mediterranean and occasional spring falls occur. An amazing 83 were recorded 1st-24th March 1996 in southeast Sicily, representing the largest invasion in Europe outside of the limited breeding areas.

Based on the arrival dates for autumn birds in Britain it seems probable that most originate from central Asia, rather than the closer southeastern European or Middle Eastern popula-tions. In central Asia autumn departure is protracted with stragglers present in the Gobi until mid October and in Kazakhstan until early November; movement in Turkey is most conspic-uous during August to September (*BWP*). Central Asian populations, through necessity, have to undertake much longer migrations than the populations in southeast Europe and the Mid-dle East. Inexperienced young birds from Central Asia would appear to be more prone to dis-placement into northwestern Europe. Comparable species from southeast Europe (including

Masked Shrike *Lanius nubicus*, Cretzschmar's Bunting *Emberiza caesia* and Olive-tree Warbler *Hippolais olivetorum*) are incredibly rare in Britain during the autumn compared with central Asian species (including Isabelline Shrike *Lanius isabellinus*, Paddyfield Warbler *Acrocephalus agricola* and Hume's Leaf Warbler *Phylloscopus humei*).

The European population is at the western end of the breeding range. Since 1960 the breeding range has extended westwards, perhaps as a result of drier summers. This occurred first into southern Ukraine (where there are now up to 7,000 pairs), Thrace, northeast Greece and some of the Aegean Islands and then into Bulgaria (now c4,000 pairs) and Romania (now c700 pairs) (Hagemeijer and Blair 1997; BirdLife International 2004). It seems probable that this westward expansion has contributed to more birds occurring in northwest Europe, especially during the spring. The difficulty of identification surely clouded the species' early status in Britain and Ireland. Although still not an easy claim, birders now seem to have got the measure of what to look for. Doubtless plenty more will follow as this species continues to make the progression from exceptional to expected rarity.

Pied Wheatear *Oenanthe pleschanka*

(Lepechin, 1770). European range is centred on Black Sea, reaching eastern Romania and Bulgaria. To the east, small numbers in south and east Ukraine, occurs widely across southern Russia, southern Siberia, Kazakhstan and Mongolia to northern China, east to Gulf of Bohai. Winters in northeast and East Africa and southwest Arabian Peninsula.

Monotypic.

Rare vagrant from southeast Europe or Asia, typically in late autumn.

Status: The total stands at 56 records to the end of 2007, 53 in Britain and three in Ireland.

Historical review: There are just two records prior to the modern recording period. The first was a 1st-winter male shot on the Isle of May on 19th October 1909. This individual was of the white-throated morph '*vittata*' (Stoddart 2008). It is unclear what such birds are. There are suggestions that '*vittata*' might have originated through ancient hybridisation between Pied and Eastern Black-eared Wheatears. However, hybrids between the two species do not produce *vittata*-like offspring and '*vittata*' occurs way out of the range where Black-eared Wheatear

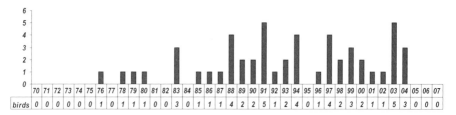

	70	71	72	73	74	75	76	77	78	79	80	81	82	83	84	85	86	87	88	89	90	91	92	93	94	95	96	97	98	99	00	01	02	03	04	05	06	07
birds	0	0	0	0	0	0	1	0	1	1	1	0	0	3	0	1	1	1	4	2	2	5	1	2	4	0	1	4	2	3	2	1	1	5	3	0	0	0

Figure 32: Annual numbers of Pied Wheatear, 1970-2006.

occurs. It, therefore, seems likely to be simply a pale-throated morph, just as its sister species Black-eared Wheatear has both black and white throated forms in both western *hispanica* and eastern *melanoleuca*. There are not thought to be any other records of this morph in northwest Europe (Hinchon 2008; Stoddart 2008).

The second was in 1916 (Swona, Orkney, on 1st November) and the third in 1954 (Portland, 17th-19th October). The next was also the first for Wales, a female on Skokholm from 27th-29th October 1968.

After an eight year absence the late 1970s produced three singletons. The first, in 1976, was a male that made a protracted stay at the mouth of the River Don, Northeast Scotland, from 26th September to 7th October. Others followed in 1978 and 1979. The first in spring was a 1st-summer male at Winterton, Norfolk, on 28th May 1978. The first of the next decade was also the first for Ireland, at Knockadoon Head, Co. Cork, with a 1st-winter male from 8th-16th November 1980.

The status of this attractive wheatear changed for good from the mid-1980s as virtually an annual arrival pattern commenced. The first multiple arrival took place in 1983 and included three well dispersed birds, in Norfolk, Devon and Co. Wexford. A total of 13 birds were observed during the decade, including four in 1988, three of which arrived on 16th and 17th October between Norfolk and Orkney and one that made landfall in western Ireland. All three Irish birds were recorded between 1980 and 1988, a period when the species was still something of an extreme rarity.

During the 1990s the increase in records intensified and 24 wanderers were delivered to our shores. They included a record five in 1991, beginning with a 1st-summer male tracked along the Yorkshire coastline from Spurn to Scarborough over the period 20th-23rd June. It was followed by a protracted autumn influx of four birds between early October and early November. In 1994 four more late autumn birds were logged, including an almost simultaneous arrival of three individuals in Kent, Suffolk and Orkney during 22nd-24th October. Four more in 1997 made another concentrated landfall over a nine-day October period, three of them on the east coast between Norfolk and Northumberland and a rare west coast bird in Lancashire.

The 12 records between 2000 and 2007 represent a downturn in sightings from the previous decade, although five in 2003 equalled the record annual total. This influx came in a protracted arrival spanning a month. The first was on Scilly, followed by one in Northeast Scotland and a 1st-winter male and female on the Norfolk coast. All four arrived between 15th and 31st October. Another graced North Ronaldsay in mid-November.

The unpredictability of occurrences is illustrated by a blank year in 1995 and, somewhat surprisingly, three consecutive blank returns from 2005 to 2007 (some records are still being considered in light of the difficulty of separating autumn Eastern Black eared and Pied Wheatears). Despite the recent glitch in annual sightings this cold-toned wheatear is now almost an expected annual rarity with an average of just over two records per annum since the mid-1980s.

Following the review period a female was found at Reighton Sands, North Yorkshire, from 8th-15th November 2008 and relocated at nearby Bempton Cliffs from 16th-18th.

Where: As might be expected, there is a significant east coast bias. The coastline between Kent and the Northern Isles accounts for more than 75 per cent of records. Ten have been seen in Norfolk and six each in Orkney and Shetland, plus records from most counties in between with, for example, four each in Northeast Scotland, Northumberland and East Yorkshire.

Along the south coast it is much rarer, with just seven birds found between Sussex and Scilly, and two each in Scilly and Dorset. Two have reached Wales, both on the Pembrokeshire islands, and three have strayed west to Ireland. The sole record from the northwest was a male at Seaforth, Lancashire, on 27th October 1997.

When: Predominantly a rarity of the late autumn period with only two found in spring and one in midsummer. A first-summer male was at Winterton, Norfolk, on 28th May 1978. Another first-summer male was trapped at Spurn on 20th June 1991 and relocated farther north at Scarborough from 22nd-23rd June 1991. The mid-

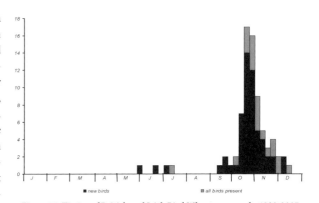

Figure 33: Timing of British and Irish Pied Wheatear records, 1909-2007.

summer individual was a fine male at East Newhaven, East Sussex, from 7th-9th July 1990.

Autumn records range from mid-September onwards. The earliest of four records from that month was a 1st-winter male at Winterton, Norfolk, from 13th-14th September 1989, followed by a male at Toab, Shetland, on 17th September 2000. A further eight have occurred in the first half of October, but the bulk of arrivals (52 per cent) have been between 15th and 31st October. Ten have been found in November and there have been two stragglers in December: a female at Paignton, Devon, from 4th to 6th December 1983 and a 1st-winter female at Tynemouth, Northumberland, from 8th to 14th December 1998.

Discussion: There are some signs that the range is spreading slowly in southeast Europe. The Bulgarian (up to 500 pairs) and Romanian populations (up to 750 pairs) are increasing and small numbers have started to breed in Croatia (Hagemeijer and Blair 1997; BirdLife International 2004).

Records from northern Europe are increasing in line with those in Britain. The first of 26 birds from Sweden was in 1975 and the first of 31 birds from Finland was in 1979. There have been double-figure numbers from Germany (15, 11 from Helgoland), the Netherlands (12) and Norway (11). In contrast there have been just seven from Denmark, five from France, four in Poland and one from Estonia.

Spring records are rare in Britain, but just under half of the German and Danish records have been in spring. It is tempting to speculate that the spring birds farther east involve spring over

Table 17: Pied Wheatear records in Northern Europe

	Apr	May	June	July	Aug	Sep	Oct	Nov	Totals
Finland		1				1	25	4	**30**
Sweden*		2				1	17	5	**26**
Norway		1					4	6	**11**
Germany		1	6				7	1	**15**
Netherlands		1	1				8	2	**12**
Denmark		2	1				4		**7**
France							3	2	**5**
Elsewhere**			1			3	2		**6**
Totals	**0**	**8**	**9**	**0**	**0**	**5**	**70**	**20**	**112**

table excludes January record.

** includes four records from Poland and singles from Estonia and Austria.

shoots from southeast European breeding grounds with Britain lying too far west of the over-shot trajectory. In autumn there have been just five birds in September, the bulk (63 per cent) coming in October. The sole winter record for northern Europe outside of Britain was a 1st-winter male in Halland, Sweden, from 1st-13th January 2004, which followed an influx of five into the country during the previous autumn.

Range extension might account for some of the increase in regularity of this species in northwest Europe, but it seems likely that most birds encountered have originated in central Asia. There is a high degree of association between the years of peak influxes (1991, 1997 and 2003) and numbers reported over the past 20 years with those of Desert Wheatear.

Around two-thirds of records involve males. Most aged birds were 1st-winter individuals. The separation of Pied and Eastern Black-eared Wheatear is notoriously difficult (Ullman 1994), with particular pitfalls concerning 1st-autumn individuals. Despite advances in identification, some individuals may not be identified with certainty. Indeterminate birds have been recorded in the Netherlands and Scandinavia. There have been similarly troublesome birds in Britain.

In some years records can be few and far between or even absent. This pattern of boom or bust is typical for species from central Asia. Late autumn easterlies seem to be a key component of such arrivals. Presumably, the occasional gluts are when such vectors coincide with good breeding years.

Migration from the breeding grounds takes place on a broad front. Eastern populations take a southwest orientation to the African wintering grounds. Departure from breeding grounds takes place from August and is a protracted affair, with many still in central Asian in late October. Multiple late arrivals in northwest Europe presumably involve birds prompted to make a departure en masse, perhaps due to a sudden weather stimulus. The northeast orientation of spring movement accounts for the paucity of spring records; not surprisingly two of the three records during this period have involved inexperienced 1st-summer birds.

Black-eared Wheatear *Oenanthe hispanica*

(Linnaeus, 1758). Western Black-eared Wheatear hispanica breeds in Iberia and southern France, northern Italy, Slovenia, Croatia and from Morocco to Libya. It winters in sub-Saharan Africa from Senegal east to Mali. Eastern Black-eared Wheatear melanoleuca breeds from southern Italy to Greece and in southwest Asia from Turkey to southern Caucasus, south to Israel and southwest Iran. It winters from Mali east to Ethiopia.

Polytypic, two subspecies. Previous authorities (Dementiev and Gladkov 1954) treated Black-eared as conspecific with Pied Wheatear; more recently they have been regarded as separate species. Records include both races (most unassigned): six attributed to nominate *hispanica* (Linnaeus, 1758) and three to *melanoleuca* (Güldenstädt, 1775). Western *hispanica* typically has warmer, sandier plumage than Eastern *melanoleuca*, but differences are not always marked. White-throated and black-throated forms occur in both subspecies, the proportion of the latter increasing from west to east (*BWP*)

Very rare vagrant, mostly as a spring overshoot, from southern Europe and southwest Asia.

Status: The total of 59 records comprises 55 in Britain and four in Ireland and includes 12 prior to 1950.

Historical review: The first was a male at Radcliffe Reservoir, near Bury, Lancashire, around 8th May 1875. That four of the 11 additional records between then and 1949 originated from Lancashire and North Merseyside (including two together near Lytham from 18th-22nd April 1940) appears highly anomalous. Other early records include two in Yorkshire and singles from Shetland, Outer Hebrides, Devon and Fife. The first for Ireland was 1st-summer male at Tuskar Rock Lighthouse, Co. Wexford, on 16th May 1916.

The first of the modern recording era was a male at King's Weir, near Eynsham, Oxfordshire, from 16th-18th April 1953. Eleven years passed until the next when, in 1964, Fair Isle bagged two in a year with a female on 19th May and a 1st-winter female on 27th September. Both were trapped. These were the first of five during 1964-1969, which included a well-watched male at

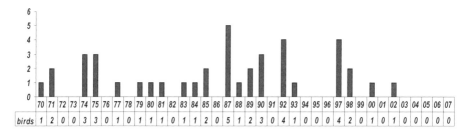

	70	71	72	73	74	75	76	77	78	79	80	81	82	83	84	85	86	87	88	89	90	91	92	93	94	95	96	97	98	99	00	01	02	03	04	05	06	07
birds	1	2	0	0	3	3	0	1	0	1	1	1	0	1	1	2	0	5	1	2	3	0	4	1	0	0	0	4	2	0	1	0	1	0	0	0	0	0

Figure 34: Annual numbers of Black-eared Wheatears, 1970-2007.

Salthouse, Norfolk, that began a two-week stay on 30th August 1965 and gave many birders their first opportunity to see the species in Britain.

Fortunes for this striking wheatear improved during the 1970s, perhaps allied to greater numbers of observers. Of the 11 recorded, the first was a male on Bardsey on 18th April 1970 (the first for Wales). Three in 1974 included the unique presence in modern times of two males together at Dungeness on 21st May 1974; one remained to 29th. Three more followed in 1975, when two spring birds visited the north Norfolk coast. Singles in 1977 and 1979 at opposite ends of the country – Scilly and Shetland – saw out the decade.

The momentum continued during the 1980s and 14 more birds were located. A record five in 1987 included an early May bird in Co. Wexford, followed by three on the south coast between East Sussex and Cornwall over a two-week period in late May and early June. A September record from Scilly completed the stats for a record year. Most records during the decade involved short stays or inaccessible birds, but a 1st-summer male lingered on the Isle of May from 2nd to 23rd May 1980.

The pattern in the 1990s was a more hit and miss, though once again 14 were added to the statistics. Hits of four apiece came in 1992 and 1997. Both years yielded well-dispersed records typically involving one-day birds, though one on Fair Isle in 1997 put in a two-month stay from July to early September. The most widely cherished individual of the decade repeated events from nearly 30 years earlier. A 1st-winter male frequented Stiffkey, Norfolk, from 24th October-1st November 1993 and attracted many birders as one of the few collectable individuals of an elusive species.

After the last of the decade in 1998, the modern run of occurrences nearly stopped. Just two more have arrived: one for two days in Dorset in 2000 and an exceptionally well watched male at Nanquidno, Cornwall, from 23rd March-1st April 2002. This was a cracking Western Black-eared Wheatear, one of a host of Iberian overshoots to the southwest at that time.

The 46 birds over the period since 1964 have arrived at an average of roughly one per year, though it is not unusual for several years to elapse without any at all.

Where: The geographical spread of records is as expected for a species that is predominantly a spring overshoot, with most along the south coast. Since 1964, 46 per cent of records have been between Kent (four) and Scilly (six); Cornwall and Dorset have four each, Devon has two and Hampshire and the Isle of Wight one apiece.

In the west since 1964, three have occurred in Wales (two on Bardsey and one in Pembrokeshire) and three in Ireland (two in Co. Wexford and one in Co. Cork). Just one has been found in the northwest during recent times, on the Isle of Man.

In Scotland over the same period two have reached the Outer Hebrides, six Shetland (four Fair Isle and two Out Skerries) and one each Highland and Fife. Along the east coast, four have reached Norfolk (two in 1975) and two Essex. Some 39 years after the Oxfordshire bird of 1953, a male at Chearsley, Buckinghamshire, on 25th April 1992 provided only the second inland record during recent times.

When: Predominantly a spring overshoot (80 per cent of records), this species shows a remarkably long span of arrival dates over at least four months. The earliest was a male at Nanquidno, Cornwall, from 23rd March-1st April 2002. The only other March record was a male seen at Burnage, near Didsbury, Greater Manchester, on 29th March 1915.

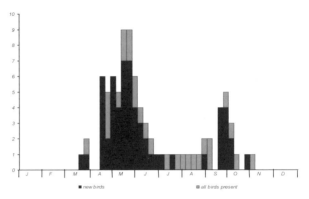

Figure 35: Timing of British and Irish Black-eared Wheatear records, 1875-2007.

All eight April records have occurred between 16th and 25th. May has the peak, particularly the last two weeks (17 records). Four more have been found in the first 10 days of June, three between 12th and 18th June and one late bird on 25th June. Two July records included a long-staying 1st-summer male on Fair Isle from 17th July to 8th September 1997.

Autumn records are very much in the minority, with just 20 per cent occurring at this time. The earliest was a male at Salthouse, Norfolk, from 30th August to 14th September 1965. The next eight were all found between 18th and 30th September. Three have occurred in October: two during the first week of the month and the widely appreciated 1st-winter male at Stiffkey, Norfolk, from 24th October-1st November 1993.

Racial attribution: The *BBRC* has attempted to assess subspecific identification. Nine male birds have been confirmed: three *melanoleuca* and six *hispanica*.

The three Eastern Black-eared Wheatear *melanoleuca* records, all 1st-summers:
1975 **Dorset:** Portland 14th June
1980 **Fife:** Isle of May 2nd-23rd May
1985 **Dorset:** Portland 27th-28th May

The six Western Black-eared Wheatear *hispanica* records, all 1st-years:
1965 **Norfolk:** Salthouse 30th August-14th September
1970 **Caernarfonshire:** Bardsey 18th April
1987 **Hampshire:** Farlington Marshes 5th June
1992 **Kent:** Denge Marsh 16th May
1993 **Norfolk:** Stiffkey 24th October-1st November
2002 **Cornwall:** Nanquidno 23rd March-1st April

BBRC has outlined features to look for when ascribing racial provenance for those fortunate enough to stumble across a Black-eared Wheatear. These included careful ageing and assessment of the precise details of colour on the mantle and crown. For males the exact size and shape of the black head markings is important: on Eastern the black feathering of the lores extends clearly above the culmen and the black of the mask extends above the eye. On Western the black mask reaches to the upper edge of the eye and the bill base, but does not extend on to the forehead; for dark-throated birds the lower edge of the throat patch extends to the upper breast. Females and immatures are the most difficult to deal with, and spring birds, especially some 1st-summers, are more difficult than those in fresh autumn plumage.

Most spring records presumably relate to Western Black-eared Wheatear *hispanica* arriving as direct overshoots from western parts of the range. This is at odds with the seasonal balance of the British records with all three *melanoleuca* having been in spring. European records also show *melanoleuca* can be expected in spring. In the Netherlands there have been three *melanoleuca* (two in early May and one in early June), two *hispanica* (May and one from 30th October-28th November 2004) and six unattributed. In Switzerland there are 25 records, 11 attributed to *hispanica* (29th March-21st June), three to *melanoleuca* (end of March-16th May) and 11 unattributed. In Sweden there are 11 records, one attributed to *hispanica* (November), five to *melanoleuca* (four 12th May-2nd June and one late September) and five unattributed.

The wintering grounds of *hispanica* are in Africa south of c.18°N, mainly north Senegal, southwest Mauritania and Mali, also southwest Niger and northern Nigeria. The route taken by easternmost populations of *hispanica* to the wintering grounds is unclear. A southwesterly heading from the easternmost breeding grounds would take birds across the Adriatic and Mediterranean Seas, over the North African coast to their wintering areas. However, the direct route would bring more birds to western Sicily and Pantelleria than are recorded and it is possible that a route to the west of the direct line is used, perhaps via northern Italy and southern France. Westernmost *melanoleuca* migrate only slightly west of south to reach wintering regions in Africa east of 5°E, though most occur in Chad, east through Sudan to Ethiopia and Eritrea (Azzopardi 2006). Although spring records comprise mostly *hispanica*, there appear to be vectors for both races to reach northwest Europe in both spring and autumn.

Discussion: Dementiev and Gladkov (1954) treated Black-eared Wheatear and Pied Wheatear as conspecific. More recently they have been regarded as separate species. The challenge of separating autumn Eastern Black-eared from Northern and Pied Wheatear remains as problematic as ever; a well photographed bird in north Devon in 2007 has still not been positively identified.

The European breeding population is very large, but has undergone a substantial decline across the western part of the range. This has been particularly evident in Spain, the likely source of many of the birds reaching our shores. The range has also contracted south in southern France and central Italy. In the east both the range and population were considered more stable with a slight increase in Romania (Hagemeijer and Blair 1997). The western declines presumably account for the recent fall-off in sightings, though occasional influxes are apparent after weather conditions suitable for overshooting.

More than three-quarters of all sexed birds have been males, suggesting, as with other wheatears, that females and immatures slip through undetected. Stays tend to be short and birds can be difficult to catch up with, but the periodic long-stay birds have been regular enough to satiate pent-up demand.

Desert Wheatear *Oenanthe deserti*

Julian R. Hough

(Temminck, 1825). Breeds widely but discontinuously across arid and desert regions of North Africa from Morocco to Middle East, north to the south Caucasus, and across central Asia from central Iran and northern Pakistan to Mongolia and northern China. Some African birds are resident, but many winter in Sahara and Sahel region from Mauritania east to Ethiopia and Somalia. Asian breeders winter from Arabian Peninsula to northwest India.

Polytypic, four subspecies but Svensson (1992) and *HBW* merge *atrogularis* with *deserti* and consider only three races to be valid. British records have been attributable to three subspecies.

There is one record of *deserti* (Temminck, 1825), of the Levant, a male shot on Fair Isle on 6th October 1928. There have been four records of *homochroa* (Tristram, 1859), of North Africa, the first of which was a female shot at Spurn on 17th October 1885. There has been one record of *atrogularis* (Blyth, 1847), from the Caspian Sea and Iran east through Kazakhstan and Afghanistan to Tien Shan, Altai, and Mongolia, a male killed at Pentland Skerries, Orkney on 2nd June 1906. No other records have been attributed to race. Differences between the subspecies are largely clinal with only western race *homochroa* and eastern *oreophila* distinct, others variably intermediate (*BWP*). Two subspecies, *homochroa* and *deserti*, were included in the study by Aliabadian *et al* (2007) and found to be indistinguishable in DNA sequence.

Rare but increasingly regular late autumn vagrant from central Asia and North Africa.

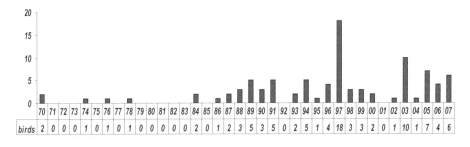

	70	71	72	73	74	75	76	77	78	79	80	81	82	83	84	85	86	87	88	89	90	91	92	93	94	95	96	97	98	99	00	01	02	03	04	05	06	07
birds	2	0	0	0	1	0	1	0	1	0	0	0	0	0	2	0	1	2	3	5	3	5	0	2	5	1	4	18	3	3	2	0	1	10	1	7	4	6

Figure 36: Annual numbers of Desert Wheatears, 1970-2007.

Status: A total of 108 records had been recorded by the end of 2007, 104 in Britain and four in Ireland.

Historical review: There were nine prior to 1950, the first a male at Gartmorn Dam, Forth, on 26th October 1880. Five of the eight that followed were in Scotland, two in East Yorkshire and one in Norfolk. Seven fell in the expected months of October and November, plus one in December. There was a spring bird in 1906, a male killed at Pentland Skerries Lighthouse, Orkney, on the very late date of 2nd June. Remarkably, a bird attempted to winter at Gorple Reservoir, West Yorkshire, from 12th November 1949 to 22nd January 1950.

The week commencing November 7th, 1949, was a period of high winds and rain which developed into a full gale by the morning of November 12th. On this date a reservoir keeper in the Halifax area reported that on November 9th he had picked up the half-eaten remains of a Little Auk (Alle alle) and had also seen "a wheatear." The latter bird, he said, had been about the reservoir embankment for several days and showed no inclination to continue its migration. I visited the reservoir on the following day expecting to see perhaps a late Greenland Wheatear, but failed to see the bird, although within an hour of my departure the keeper saw it again.

No further reports were received, and it was not until December 18th, when A. M. and P. C. A., both junior members of the Halifax Zoological Group (now the Ornithological section of Halifax Scientific Society), visited the reservoirs that they learned from the keeper that the wheatear was still present. They soon located it on the reservoir embankment feeding on insects, but unfortunately it was standing with its back towards them and took flight while they were still some distance away. However, they noticed through their field-glasses that the rear plumage pattern consisted of a buff-white rump and an almost completely black tail, instead of the normal pure white rump and black tail with half white outer tail-feathers of the Common Wheatear. While these two boys were having lunch in the keeper's house, and unknown to them, R. C. (also of the junior H.Z.G.) arrived and located the bird himself. He was fortunate enough to see the bird at close range and at once noticed its black face and throat and also the tail pattern and light markings on the closed wing. He then met the others, and together they took very full notes and made numerous sketches throughout the remainder of the daylight hours. That same evening they consulted The Handbook and came to the conclusion that the bird was a Desert-Wheatear, then brought the sketches and notes for me to see. These were sufficiently accurate for me to say conclusively that the bird was indeed one of the forms of Desert-Wheatear, so the following day (December 19th) A. M. and I visited the reservoir and had excellent views of the bird from a few feet distance and often up to the focusing limit of the x 9 binoculars.

A few days later Mr. R. S. R. Fitter was able to add his confirmation to the record on December 24th. The bird was seen between Christmas and New Year by several members of the Yorkshire Naturalists' Union, and also by P. A. D. Hollom, but while the bird was definitely established as a Desert-Wheatear, it was obviously impossible to determine the subspecies without first procuring the bird. Two methods of doing this were open to choice

- shooting or trapping; but, as we are against the former practice on the grounds that it would have deprived our visitors and all other ornithologists of the unique opportunity of seeing a Desert-Wheatear alive in this country, we trapped it.

The bird was carefully examined, photographed, plumage details sketched and notes taken. It was then ringed (No. F3270) and released. This took place on January 9th, 1950, and the bird was seen, and its ring noted, on January 15th, and on several later dates by the keeper and his wife. It was last seen on January 22nd by myself when the reservoir was partly frozen over.

George R. Edwards. British Birds 43:179-183.

The 1950s added two further records, in December and January, Co. Durham in 1955 and Essex in 1958. The 1960s produced four more between 1960 and 1966, three of them on the south coast between East Sussex and Hampshire and one in East Yorkshire. Two of the records came in April and two on the south coast in October and November.

The species remained something of a rarity throughout the 1970s, which added just five more including two one-day east coast birds in autumn 1970 and one from Lincolnshire (the earliest accepted British record). Autumn singletons from East Yorkshire in 1974 and Norfolk in 1978 put in stays of several days, as did another rare spring bird on Scilly in 1976.

With just 11 records between 1955 and 1978 expectations were not high for a change in status. A six-year wait ensued until the first of the 1980s, but the pause in arrivals was ended by males at Porthgwarra from 17th-20th November 1984 and Freswick, Caithness, from 26th December 1984 to 13th January 1985. Both performed well and alleviated the prior famine for a burgeoning generation of listers. These two were followed by another obliging bird in Cumbria in November 1986 and signalled a change in fortunes for this appealing wheatear. Two in 1987 included a 1st-winter male at Landguard, Suffolk, from 20th-24th October (trapped on 23rd) that re-located to East Prawle, Devon, from 26th-30th October. This individual illustrates how intent on migration some of the rarities reaching our shores still are. Without the ring, might it have been treated as two individuals?

Three in 1988 was a short-lived record as the following year delivered a new record of five. Times had clearly changed. The decade had added 13 birds to the rapidly increasing aggregate total. Not for the first time a major rarity at the beginning of the decade had demoted itself to expected fare by the end. Since 1984, the species has failed to appear in only three of the following 22 years.

The 1990s saw the pace pick-up further. Three in 1990 were followed by five autumn birds in both 1991 and 1994. The species really made an impact in 1997, courtesy of a record-breaking influx of 18, including one in the spring. The 17 autumn occurrences began with two October sightings, followed by three more between the 7th and 9th November. The bulk of the arrival took place from late November, with 12 birds located between 20th November and 12th December. Western outposts picked up the bulk of sightings: six from Gloucestershire to Cornwall, one in Pembrokeshire and another in Co. Wexford. The east coast of Scotland produced three between Northeast Scotland and Lothian. Two reached Northumberland and Kent. Singles were in East Sussex and Norfolk. Dispersal had clearly taken place on a broad front, and a number of sightings elsewhere in northwest Europe added to the deluge of sightings. The 1990s ended with whopping 41 records.

The period 2000 to 2006 has added a further 25 records, including another double-figure autumn influx in 2003 involving 10 birds. The first in Devon on 26th October was followed by a gap until an arrival of three on 10th November and five between 11th and 16th. A later bird followed at the end of the month. Once again the influx occurred on a broad front, comprising two in both Kent and Norfolk, three in Scotland and singles in West Sussex, Devon and Pembrokeshire.

Just one occurred in 2004, but an arrival of seven followed in 2005, largely restricted to the east coast between Kent and Northumberland. Four more in 2006 included two on Scilly and singles in West Sussex and Northeast Scotland. Six in 2007 included an exceptional early March inland male in Greater Manchester. That autumn provided a further five: two along the east coast, one in Cornwall and singles in north Wales and inland in Cheshire.

The 20 years between 1987-2007 have produced nearly 80 per cent of the national total, at an average of just over four per annum.

Where: There is a distinct south and southeast bias to records. Norfolk accounts for 14 records and Kent 11. Seven have made landfall in Cornwall. Just three have reached the Isles of Scilly, two of them in 2006. Most east coast counties have attracted multiple records, though north of Norfolk only Shetland (seven) has produced more than a handful. A number of birds have been recorded inland, a male at Barton Gravel-pits, Staffordshire, from 23rd to 30th November 1996 being particularly noteworthy. Two inland birds in 2007 included a 1st-winter male at Irlam Moss, Greater Manchester, on 8th-9th March 2007 and a male in Crewe, Cheshire, from 12th-14th December. Just five have pushed west to Wales and four to Ireland. The northwest has hosted six: two in Lancashire and singles on the Isle of Man, Cumbria, Cheshire and Greater Manchester.

When: Spring birds have been encountered with some regularity. Most are presumably reorientating wintering birds rather than fresh overshoots. Six have been found in March, three of them between 5th-11th (Studland Heath, Dorset, 5th-6th in 1997, Irlam Moss, Greater Manchester, on 8th-9th in 2007 and Carnsore Point, Co. Wexford, from 11th-21st in 1990). Five more have followed in April, the latest on the 17th (Beachy Head male from 17th-21st

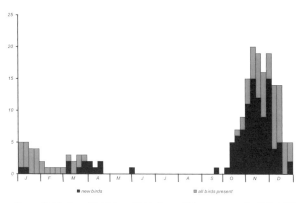

Figure 37: Timing of British and Irish Desert Wheatear records, 1880-2007.

April 1966). Just one has been found in the late spring period, a male killed at Pentland Skerries Lighthouse, Orkney, on 2nd June 1906.

The earliest autumn record is a 1st-winter male at Donna Nook on 23rd September 1970, the only September record of the species. Early October records are also scarce, just six arriving prior to mid-month and 17 more in the second part of October. November is the best month to find one of these engaging birds, fuelling the view that any wheatear species found during the month carries an increased probability of being a rarity. Twelve more have been found in December and two more in January, a similar late arrival pattern to that shown by Pied Wheatear. A number of birds have lingered for part of the winter and the two January finds, presumably newly discovered wintering birds.

Discussion: Although some of the increase must be observer-related, particularly during influxes when observers seek out potentially suitable areas for wheatears spurred on by reports on the news services, this attractive wheatear is now considered one of our more regularly occurring rarities. Where sex has been specified around 60 per cent have been males and 30 per cent females. Although blank years still occur there does appear to have been a marked change in the numbers reaching our shores, a trend shared with elsewhere in northern Europe. There have been 42 from Sweden (nine between March-May, the rest in autumn; most since 1985 with three assigned to *atrogularis* and the remainder thought likely to be this race), 15 from Finland (two in March and single in May, 11 October-November, one in December), 19 from Norway and 10 from Denmark (one in March, the rest late October to early December). There were 20 records from the Netherlands during 1970-2006, a dozen from Germany and four each from Estonia and Belgium, plus two from Poland and one from each of Switzerland and Hungary. Farther south there have been 19 accepted records in Spain (24 birds in total; half of them from the Canaries), 24 birds from Italy and 12 from Greece.

The majority of records presumably involve inexperienced 1st-winter birds, but the ageing of some, especially females, is difficult. Ageing can be made through the fringing on the primary coverts showing contrast between new and old greater coverts, a feature that is not always easy to observe.

Where the influxes come from is difficult to ascertain. Desert Wheatear is predominantly a late October and, especially, November rarity with a penchant for late arrivals shared with other central Asian species such as Pied Wheatear and Isabelline Shrike. The association with Pied Wheatear influxes would imply that the origin of many of our birds is also central Asia, involving *atrogularis* Eastern Desert Wheatear, which migrates to northeast Africa, Arabia and India. The North African race *homochroa* Western Desert Wheatear, which winters in the western Sahara south to Chad, has also been recorded. The largely sedentary *deserti* Levant Desert Wheatear, from Syria to northeast Egypt and Sinai and northern Saudi Arabia, has also strayed here on at least one occasion.

Migratory populations of Desert Wheatears are thought to leave their breeding grounds in September. Why late autumn arrival in Britain is favoured remains unclear. It is possible that many birds remain in central Asia until late autumn and dispersal is triggered by the onset of inclement weather conditions, thus explaining the late arrival dates here and elsewhere in

Europe. Periodic influxes presumably occur when dispersal is rapid and involves birds picking up weather vectors that allow them to reach northwest Europe.

White-crowned Black Wheatear *Oenanthe leucopyga*

(C.L. Brehm, 1855). Breeds from Morocco discontinuously eastwards to the Middle East and Arabia. Largely sedentary. Some individuals or populations may make short-distance movements in winter.

Polytypic, three subspecies. The racial attribution of the single British occurrence is undetermined.

Very rare vagrant from North Africa or the Middle East.

Status: One record.
1982 Suffolk: Kessingland male 1st or 2nd-5th June

Discussion: This attractive wheatear is largely sedentary throughout its range, though some may make short-distance movements in winter. There are relatively few records of vagrants away from breeding areas. The only other record from northern Europe was from Germany (9th-13th May 1986) and placed in category D as a possible escape.

The problem of record assessment was highlighted by a recent report from Lancashire and Merseyside. One was watched and videoed for about 20 minutes at Fazackerley on 17th October 2002. A couple of days later a man approached the finder and asked if he had seen a *"black bird with some white on it, because his mate had lost it"*. The location, circumstances and age (but perhaps not date) all point to this bird being of possible captive origin, but had the record been submitted to *BBRC* the identification would probably be accepted and it may well have been one that "got through" were it not for the chance encounter (Tim Melling *pers comm*).

Elsewhere, there is one from southern France (21st April 1884), two together in Spain (28th May 1977) and one from Portugal (25th March 2001). The first for the Canary Islands (10th January 2005) almost coincided with the first for the Cape Verde Islands (16th January 2005). In the eastern Mediterranean there are single records from Malta (18th April 1872), Greece (15th April 1993) and Croatia (1st-4th August 2001) and three from Turkey (12th August 1993, 9th March 1996, 2nd March 2005). There have been seven on Cyprus since 1970 (one in February, three in March, two in April and one in May).

In light of the Suffolk record a review of all the (then) accepted British and Irish records of Black Wheatear *Oenanthe leucura* was carried out by the *BOU*. This resulted in all records being rejected, partly due to possible confusion with White-crowned Black Wheatear. Black Wheatear was removed from relevant lists (*BOU* 1993). These included birds from Fair Isle (male, 28th to 30th September 1912, 19th October 1953) as well as one at Altrincham, Cheshire, on 1st August 1943 and one at Dungeness on 17th October 1954. There is also a Black Wheatear or White-crowned Black Wheatear *O. leucura/leucopyga* record from Co. Donegal at Portnoo on 10th June 1964.

Those observers fortunate enough to see the Suffolk bird can consider themselves lucky to have witnessed a rarity for which the prospects of a second-chance in the near future appear slim in comparison to some of the other 'blockers' on the *British List*.

Rufous-tailed Rock Thrush Monticola saxatilis
(Linnaeus, 1766). Widespread across the southern Palearctic from Iberia north to the southern Alps and east through Mediterranean basin to mountains of central Asia, Lake Baikal region, Russia, Mongolia and across China from Pamirs to northern Tibetan plateau and east to mountains in northeast provinces. All populations are migratory, wintering across northern sub-Saharan Africa.

Monotypic.

Rare vagrant from Europe, the majority as spring overshoots to the south coast.

Status: With just 28 records, 26 in Britain and two in Ireland, this dazzling thrush remains a rare and erratic visitor to our shores.

Historical review: The first accepted record was an adult male shot at Therfield, Hertfordshire, on 19th May 1843. This was followed by others in Orkney (17th May 1910), Shetland (8th November 1931), Kent (23rd June 1933) and Shetland (16th October 1936).
  Three of the first five records fell in the period May to June. The two Shetland records came in October and November.
  The first of 23 modern records was a female on St. Kilda on 17th June 1962, which was followed by a male at Eddystone Lighthouse, Devon, from 30th to 31st May 1963. Another followed in 1968 (April) and two more in 1969 (both in May).
  In the 1970s, two more reached Britain in June 1970 (Fair Isle) and May 1979 (St. Mary's) and there was also a first Irish record at Clogher Head, Co. Louth, from 20th-21st May 1974.
  The bulk of modern-day records was found during the 1980s (10 records and absences noted in only two years). At the time this species seemed an almost expected annual visitor, misleading the birding fraternity into the expectancy of a never-ending supply of records. There were double records in 1983, 1984 and 1989. The spread in locations during the decade was

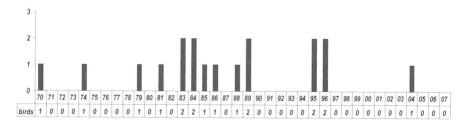

Figure 38: Annual numbers of Rock Thrushes, 1970-2007.

geographically even. There were six in the west from Dorset to Scilly and two in Wales. Four graced the east coast between Kent and East Yorkshire. One was inland in Hertfordshire. A male in consecutive years at Portland in 1988 and 1989 is interesting. The male in 1988 remained for just over a week in late April and that in 1989 was just a one-day bird: could the same male have been involved?

Just when birders were getting accustomed to annual arrivals there followed a six-year gap in occurrences until 1995 provided two, including a well-received spring male in Norfolk. In 1996 two more followed, one an obliging autumn bird on Scilly. So it appeared that the blip in records was finally over and it would be back to business as usual. An eight-year gap ensued before the next, a late afternoon and early evening male in Devon in 2004.

Thus it remains a tantalisingly rare bird and an obvious gap in the lists of a new generation of birders unable to travel to Devon for the evening performance in 2004. With just five in nearly 20 years there would appear to be little prospect of a dramatic change in status.

Where: Just over 40 per cent of records have occurred along the south coast counties: Scilly (four), Devon (three), Dorset and Kent (two apiece) and West Sussex (one). Two have penetrated inland to Hertfordshire, the second of which was a male at Graveley on 8th May 1983. To the west, singles have reached Co. Clare and Co. Louth in Ireland. The two Welsh records were in Ceredigion and on Anglesey. East coast records are scarce, with three in Norfolk and two in Yorkshire, one of which was well inland at Sutton Bank on 17th May 1969. Five have overshot to Scotland, one each for the Outer Hebrides and Orkney and three to Shetland (all on Fair Isle).

When: Spring is prime time to encounter this rare species, with 79 per cent (22 records) between mid-April and late June. Three April records constitute the forerunners of the protracted spread of spring arrival dates, the earliest a male at Portland, Dorset, from 16th-24th April 1988. May is the best month, accounting for half of all records, five 3rd-9th May and six 17th-20th May, plus three more in the last week of the month. Five June records include four from 17th-30th, the latest of which was a 1st-year male trapped on Fair Isle on 30th June 1970.

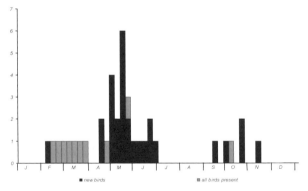

Figure 39: Timing of British and Irish Rufous-tailed Rock Thrush records, 1843-2007.

The five autumn records comprise two each on Shetland and Scilly and one in Co. Clare. Two are dated in September, the earliest at Kilbaha, Co. Clare, from 14th-16th September 1995 and two in October. The latest record was on Fair Isle on 8th November 1931.

One record well outside the arrival parameters for this species involves a male at Minster, Kent, from 5th February to 1st April 1983. In addition to the winter date, the extended length of stay is also most unusual for a British record and the possibility of an escapee is difficult to discount for this individual.

Few birds have remained long enough to attract a crowd. Males at Portland from 16th-24th April 1988 and at Hunstanton, Norfolk, from 22nd-25th May 1995 buck the trend. The latter fell foul of a Eurasian Sparrowhawk attack on the evening of its first day, but it was only slightly injured, kept overnight and released the following morning to the delight of the many observers who travelled to view it. A bird on Bryher from 28th September-2nd October 1996 was also well appreciated, but the events of the year before relegated it to a mere year tick for most viewers.

Discussion: In the 19th Century, the breeding range extended as far north as Belgium. In the 20th Centruy there was a long-term decline in population and a southwards retreat in range. The decline of the European breeding population continued between 1970 and 1990 and there was a further small drop between 1990 and 2000. There have been local recoveries in some parts of the range and it has recently started breeding again in the south of Germany (Hagemeijer and Blair 1997; BirdLife International 2004).

Despite breeding as close as southern France and central Switzerland and its status as a long-distance migrant this species remains a rare overshoot to northwest and northern Europe. Since 1950, records include five to Sweden (22nd November 1954, 8th-9th October 1988, 29th June-31st August 1998, 11th-14th May 1999, 25th May 2005), four to Finland (21st October-3rd November 1991, 28th July-3rd September 1992, 26th-27th June 1994, 10th June 2002) and three to Norway (26th April 1969, 3rd June 1974, 13th October 1974). There are also six from Denmark (1st June-15th September 1971, 9th June 1974, 17th-18th May 1977, 9th-11th May 1983, 5th May 1984, 19th May 1996), one from Estonia (5th-6th October 1999) and nine in the Netherlands (22nd-23rd April 1951, 12th-13th May 1994, 17th May 1998, 4th May 1999, 25th April 2000, 27th April 2000, 4th May 2000, 8th May 2006, 9th May 2008). At least 15 have been recorded from Helgoland, but as it has always been a commonly kept species in captivity in Germany the occurrence pattern may be distorted (Peter Barthel *pers comm*).

Rarely observed on migration this nocturnal migrant appears less prone to vagrancy than similar species from the region. It is highly migratory. Birds breeding in northeast China travel at least 8,000 km to African wintering areas. There is some evidence of site-fidelity on the wintering grounds with one individual returning to the same site in northern Tanzania for three winters (Clement and Hathway 2000).

Breeding birds of the west Mediterranean migrate south or east of south in the autumn, perhaps with a similar return pattern in spring explaining the more westerly distribution of north European records. More northerly breeding sites are occupied from mid-April onwards coinciding with the bulk of overshoots in late April and May. Against a backdrop of range contraction and declines, it may be some time before the fortunes of this rare vagrant recover from the recent paucity of occurrences.

Blue Rock Thrush Monticola solitarius

Ray Scally

(Linnaeus, 1758). Resident or dispersive throughout Mediterranean basin from northwest Africa and Iberia to northern Italy and east to Greece and Turkey. Other races through mountains of central and southwest Asia to Himalayas, east China, Taiwan and Japan. Winters within or to south of breeding range.

Polytypic, five subspecies. The racial attribution of occurrences here is undetermined, though the first most probably belonged to *longirostris* (Blyth, 1847) from eastern Turkey to northern Pakistan.

Very rare overshoot from central Asia and probably southern Europe.

Status: Six records.

1985 **Argyll:** Skerryvore Lighthouse, SSW of Tiree 1st-summer male 4th-8th June when found dead
1987 **Caernarfonshire:** Moel-y-gest male 4th June
1999 **Scilly:** St. Mary's male 14th-15th October
1999 **Cornwall:** Cot Valley male 25th October; possibly same as Scilly
2000 **Cornwall:** Geevor, Pendeen 1st-summer female 14th-18th May
2007 **Radnorshire:** Elan Valley male 11th April

Discussion: Elsewhere in northern Europe this remains a rare overshoot, with three from Sweden (21st-25th April 1981, 6th June 2007, 13th May 2008), two Finland (2nd-3rd June 1995, 5th April 1997) and one from the Netherlands (20th September 2003). There are also two from Germany where the species is not yet admitted to category A (8th June 1962, 10th June 2007); it is unlikely that these records will be admitted to category A as the species is not rare in captivity in Germany (Peter Barthel *pers comm*).

During 1991-1992 the *BOURC* undertook a reappraisal of all existing records for which identification was not under scrutiny for an assessment of the likelihood of natural vagrancy versus escape from captivity. Two birds were placed on Category E. Two others (Argyll and

Caernarfonshire) were placed on Category A (Hume 1995). Those on Category E were on North Ronaldsay (29th August-6th September 1966) and at Rye, East Sussex (10th August 1977). An additional bird also considered to be an escape occurred after this review: Hemel Hempstead, Hertfordshire (17th-27th August 1996). All three of records occurred at time inconsistent with expectations for natural vagrancy; the first two coincided with the period when birds were being imported from India, probably of the smaller and darker race *pandoo*.

The first British record was considered to be most probably of the race *longirostris*, which breeds from eastern Turkey to northern Pakistan and winters from northeast Africa north to eastern Israel and eastwards to northwest India. This assumption is based upon biometrics, but it was not assigned to race by the *BOURC*. Examination of feather mites also indicated a wild origin for this individual (Hume 1995). Subsequent records have not allowed such close examination. Assigning individuals to race is unsafe in the field because of variations in plumage: longirostris from central Asian origin and nominate *solitarius* from southern Europe seem the most likely. The presence of three birds in the southwest within a short space of time in 1999 and 2000 is exceptional for a species so rare away from its core range.

White's Thrush *Zoothera dauma*
(Latham, 1790). Palearctic race Z. d. aurea widespread in central and southern Siberia from Yenisey River to Ussuriland south to northern Mongolia, extreme northeast China, Korean peninsula and Japan. A small (isolated?) population extends west to the foothills of European Urals. Winters widely throughout southern China, Taiwan and southern Japan to Indochina and central Thailand. Nominate race resident or altitudinal migrant in Himalayas, southwest China and Taiwan.

Polytypic, seven subspecies. Records are attributable to the migratory race *aurea* (Holandre, 1825).

Rare vagrant from Siberia with modern records almost exclusively in September and October and on the Northern Isles.

Status: This irresistible *Zoothera* is always sought-after. Prior to 1950 there were 28 records; 25 in Britain and three in Ireland. In 1950-2007 there were a further 40 records; 39 in Britain and one in Ireland.

Historical review: The first British record was shot at Heron Court, Christchurch, Dorset on 24th January 1828 (formerly Hampshire). This was followed by the first for Ireland, one shot near Bandon, Co. Cork, in early December 1842.

White's Thrush was named after Gilbert White, the eighteenth century author of the *Natural History of Selborne*. Gilbert White had many achievements during his life but finding a White's Thrush wasn't among them. In fact, the first British record was shot in Hampshire, 35 years after White's death. It was named in honour of Gilbert White because he had lived and worked in Hampshire.

It had already been collected and described from its native Asia in 1790, where it had been given the name Golden Mountain Thrush. The British decided to stick with the eponymous White's Thrush (Tim Melling *pers comm*).

Table 18: Monthly occurrence of pre-1950 White's Thrush records.

Month	Jan	Feb	Mar	Apr	May	Sept	Oct	Nov	Dec	Total
Pre 1950	11					1	4	5	5	**28***
1950-1970	2				1		2	3		**8**
Post 1970	2	1		2		9	17		1	**32**
Total	**15**	**1**		**2**	**1**	**10**	**23**	**8**	**6**	**68***

Two records were undated.

The main period for the pre-1950 records is from December to February (57 per cent), no fewer than 11 being obtained in January. Just one was from September, from East Sussex, where it was reportedly present for two or three weeks before being found dead at the end of the month. Four were in October and five in November: three of the nine autumn records came from Fair Isle; two from Yorkshire; and singles from Northeast Scotland, Northumberland, Norfolk and Warwickshire.

Analysis of the older records (see Table 18) results in a pattern at odds with present vagrancy expectations, initially implying that some of these records might not have been as genuine as they first appear. However, it is worth remembering the change in hunting tactic between the two periods. All 19th century records occurred in the epoch of collection or bush shooting. The first field observation to be accepted was seen on Holy Island on 2nd November 1914. Of all pre-1950 records only it and one other involved individuals seen in the field; 93 per cent were shot or found dead. The modern tactic of field searching at observatories or coastal patches produces a strong bias towards autumn arrivals. More doubtless pass through and even winter undetected in well vegetated areas.

The 1950s and 1960s accumulated four apiece, with a southwest bias to the records. Four were found between Gloucestershire and Scilly: two of them in October, one in early November and one in early January. Two Scottish records included one from Fair Isle in early November and another from Perth and Kinross shot in early January. The remaining records were a coastal bird in Co. Durham in early November and a questionable Cheshire individual in early May.

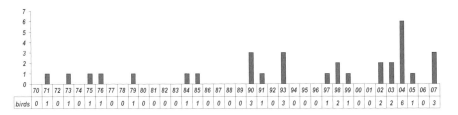

Figure 40: Annual numbers of White's Thrushes, 1970-2007.

Five followed in the 1970s, three of them coming from Shetland where there were records in February, September and October. The two others comprised a January bird from Clyde and a December bird from North Yorkshire. The 1980s mustered just two records, one apiece in 1984 (Devon in January) and 1985 (Shetland in October).

The 40-year period between 1950 and the end of the 1980s delivered just 15 of these prized thrushes, 10 of them elusive one-day birds and four two-day birds. The only bird present for an extended period was the first of the 1950s, on Lundy, from 15th October-8th November 1952. The birds recorded over this 40-year period had a widespread distribution. In Scotland five were from Shetland and singles from Perth and Kinross, Clyde and Borders. In England two reached Devon and singles from Co. Durham, Cheshire, Gloucestershire, North Yorkshire and Cornwall.

The detection dates also make interesting reading. The autumn records conform to modern expectations: the September record was on Shetland as were two of those in October and one in November. The other two October birds were in Devon and Gloucestershire and the November birds in Co. Durham and Scilly. Those outside of the expected autumn period included one in December (North Yorkshire), four found in January (Perth and Kinross, Clyde, Cornwall and Devon) with singles in February (Shetland) and quite exceptionally May (Cheshire).

Three Scottish birds in 1990 heralded a change in fortune for the species, followed in 1991 by the first truly twitchable bird at Brora, Highland, from 27th-29th September. In 1993 three more were logged, including a spring bird at Copeland Bird Observatory Island, Co. Down, from 16th-20th April which enticed about 180 birders to make the journey. The next was in 1997, followed by two more in 1998. The decade was to amass 11 records, almost as many as in the previous four decades, culminating in one in 1999 which famously shared part of its stay on Scilly with a Siberian Thrush *Z. sibirica*, a unique coincidence on British soil.

The trend set in the 1990s has continued. Two in 2002 and a total of 14 were accrued between then and 2007 included an unprecedented arrival of six in 2004 between 1st-11th October. Five of these were in Scotland, but much more widely appreciated was a one-day bird in East Yorkshire: the first collectable individual on the English mainland ever, enticed large numbers of visitors to Easington on 10th October.

Roy Taylor, Steve Addinall and myself had arrived at Spurn at dawn on Sunday 10th October 2004. We spent a couple of hours searching the point, then drove to Kilnsea for the first organised flush of the Olive-backed Pipit. Straight afterwards, the Kilnsea area was teeming with birdwatchers, so we opted to try the Easington area to try and find something ourselves, arriving at about 10am. This turned out to be one of our better decisions!

The trees and hedges were full of Goldcrests, and we were scouring the flocks looking for something rarer. I was a few yards ahead of the others as we approached Easington Cemetery. I was checking a bramble covered ditch under a hawthorn hedge when I flushed a bird that was obviously a thrush. The first things I noticed were large white corners to the tail, and that fact that it was big. It flew just a short distance and perched, largely obscured on a low hawthorn branch about five metres away. I managed to get it in the binoculars momentarily and saw that the olive-coloured upperparts were covered with small black crescents and the gleaming white underparts were similarly scaled. I didn't hesitate for a second in screaming "White's Thrush!! I've found a (expletive deleted) White's Thrush!!!" but it then flew into deep cover, probably encouraged to do so by my screams.

Roy and Steve arrived quickly, but there was no sign of it. A minute or two later I found it again, partly obscured in the hawthorn. I quickly confirmed the large size, scaled plumage, white underparts, and also noted a rather busy wing pattern. Unfortunately, I was too panicky to be able to give coherent directions, and the others failed to see it. It then disappeared for several minutes. It is interesting the kind of things that race through your mind in those minutes when nobody else has managed to corroborate your sighting. "Will anyone believe it? Did I get enough to clinch the identification? Could I have been hallucinating?" By this time, Roy had entered the field and was walking up the hedge away from the road. I remember the great relief when he called "It's here!" Seconds later it flew right across the field in front of us and we saw the stunning zebra crossing underwing pattern. There was no doubt. It was a White's Thrush.

Roy immediately radioed the news out and we stood back, careful not to flush the bird again. Within minutes cars started to arrive, but there was no sign of the bird in the small hedge we had seen it fly into. Steve checked the spot where we had originally found it and it was back there. It then did a carbon copy of its earlier moves, first sneaking up the hedge, then flying across the field into the sparse hedge. The big difference this time was that it had an appreciative crowd of spectators. Again, it didn't stay long in the sparse hedge, and flew fast and low to the long hedge next to Willow Cottage, where it remained until dusk. During the course of the day it was flushed several times to allow the growing crowds to see it. It occasionally perched in view for a few seconds, but typically it flew straight back into the same thick hedge and disappeared.

It was seen by well over a thousand people during the day and was not present next day. The small number of well-organised flushes enabled everyone to get good, albeit brief views, while the disturbance to the bird was kept to a minimum. The fact that it was not present the next day is probably a reflection of how well it had fed. Given the typically brief views and complicated plumage, it was quite difficult to compile a description. However, perseverance allowed us to compile the following.

Tim Melling. Birding World 17:432-434

The 25 birds since 1990 have exhibited quite a different occurrence pattern compared to those between 1952 and 1985. No less than 13 have been on Shetland, three have been on the Outer Hebrides and two on Orkney. Further island birds involve singles on Scilly and Co. Down. Three on mainland Scotland included dead birds in Borders and Lothian and one in Highland. Two from East Yorkshire are the only records from the English mainland, one of which was dead. Two have been in April, eight in September and 15 in October.

Short stays are the norm. One on Lewis, Outer Hebrides, remained from 14th to at least 26th October 1998, attracting admirers from near and far. At the time this represented the longest stay of any individual since the Lundy bird of 1952, but it was exceeded by one on St. Agnes from 6th October to 9th November 1999.

A further seven were recorded following the review period in autumn 2008. These included three on Fair Isle, including two birds on the island at the same time on 1st October. Singles graced Inishbofin, Co. Galway, a well watched bird was at Kergord, Shetland, another slightly inland at Parkhill Forest, Aberdeenshire. One trapped on the Isle of May in early June 2009 was unprecedented. These records are subject to acceptance.

Where: Since 1950, 18 birds have been found on Shetland cementing the archipelago as the place for self-finders to focus their efforts and fulfill their dreams. In contrast, just two have been found on Orkney and three have penetrated west to the Outer Hebrides. In England, two have filtered down to Scilly and Devon and three have been found in Yorkshire. Ten other counties have been graced on a single occasion.

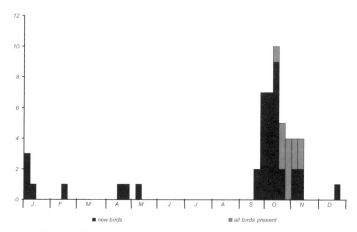

Figure 41: Timing of British and Irish White's Thrush records, 1950-2007.

When: Six winter records since 1950 were all recorded prior to 1984, including one in December in North Yorkshire, and four in early January and one in early February. Four spring records include two April birds: a 1st-winter on Copeland from 16th-20th April 1993 and at Wester Quarff, Shetland, on 28th April 2005. Other records comprise one at Weaverham, Cheshire, on 7th May 1964 and, outside the review period, one trapped on the Isle of May on 2nd June 2009.

Autumn is the expected arrival time, and of records since 1950 nine have been found in the last week of September; all in Scotland and six on Shetland. The earliest, on St. Kilda, Outer Hebrides, on 21st September 1993, was almost matched by one at Sumburgh, Shetland, on 22nd September 1990. A further seven fall in the first week of October (six on Shetland and one on Scilly) and eight between 10th and 14th (seven in Scotland and one in East Yorkshire). Three between the 15th and 20th includes birds in East Yorkshire, Devon and Orkney. The remaining October record was from Gloucestershire on 30th. Three early November records all fall between 3rd and 7th and all occurred between 1958 and 1965 (one on Shetland and one from Scilly, but the latest from South Shields, Co. Durham, on 7th November 1959).

Discussion: White's Thrush is locally common in the montane forests of the Urals and in east western Siberia. It is perhaps completely absent from large parts of western Siberia, so that the area of distribution in the Urals is thus isolated from the rest of the range in central and eastern Siberia. It has also bred just west of the Urals, indicating isolated breeding there (Ryabitsev 2008). A singing bird was noted in the Kirov region at 58°04'N, 48°29'E, on 20th May 1999, but not subsequently (Kondrukhova 2003). The first record for Udmurtia was of a bird shot south of Izhevsk in late August 2000 and a fledgling, not able to fly well and with tail feathers not fully developed was seen near Bol'shoy Bilib, eastern Udmurtia (57°21'N, 54°00'E) on 5th July 2007. This record is proof of successful breeding in Udmurtia and suggests that the species is extending its range westwards (Pyatak 2007).

European records suggest a westerly orientation, rather than a northwesterly heading towards Scandinavia, where it is rare. There are just four records from Finland (three between 29th September-5th October and a singing male on 15th June 2009), seven from Sweden (18th April 2003, 10th September 1966, four 3rd-23rd October and one November) and nine from Norway (2nd June 1979, three in September, earliest 28th in both 1978 and 2004, and five October). Denmark has six records (one each in January and March, two each in April and October). Three have reached Iceland (14th October 1939, 9th October 1982, 5th-9th November 1982) and two the Faeroes (male found dead on 9th November 1938, one found dead October 1974).

There have been c38 from Germany, nine from Poland (singles in February and May, two in September, one October, two each in November and December), three from Austria (all pre 1847), and one from Latvia (November). There have been 18 in the Netherlands (one in March, two in April and the rest in autumn, including three in September (earliest 8th September 1974), 11 in October and one in December), 15 in Belgium (10 pre-1907 records, four October and one November record since) and 16 records in France (one in 18th century and seven in 19th century, plus three more to 1932: one February and two October with five October records between 1987 and 2006).

From southern Europe there have been c33 from Italy including eight since 1950 (26th October 1966, 8th November 1973, 12th October 1974, October 1982, 6th November 1999, 9th December 2001, 24th November 2003, 3rd-4th March 2006), two from Greece (autumn 1954, 3rd February 1965), plus singles from Spain (shot on Mallorca, January 1912), Slovenia (19th November 1973) and Romania (September 1981).

All recent British records in September, and around four-fifths of all October records, have been from Scotland. Perhaps Scotland represents the most direct point of landfall and later records, of which there have been very few recently, represent birds that have subsequently spilled down to the south and west.

Although the events of 2008 may eventually better it, the influx of 2004 was unprecedented. It occurred on a broad front with records from a number of European countries. Such sporadic influxes are becoming characteristic of Siberian vagrants, which have occasional bumper years, presumably following a good breeding season and resulting from optimum weather conditions for carrying inexperienced birds westwards. Sustained periods of good breeding success coupled with a westward range expansion presumably lead to influxes to northwestern Europe.

Siberian Thrush _Zoothera sibirica_

(Pallas, 1776). Breeds central and eastern Siberia from Yenisey and Lena Rivers south to northeast Mongolia and east to northeast China, Amurland, Sakhalin and northern Japan. Winters central Burma, Indochina and Thailand south to Singapore, Sumatra and Java.

Polytypic, two subspecies. The racial identity of occurrences is undetermined. A recent paper has recommended that this species be reassigned to the genus Geokichla (Voelker and Outlaw 2008).

Ian Wallace

Exceptionally rare but increasing vagrant from Siberia.

Status: Eleven records, two of which were well photographed and are subject to acceptance at a national level.

1954 Fife: Isle of May adult male 1st-4th October, trapped 2nd October
1977 Norfolk: Great Yarmouth male 25th December
1984 Orkney: South Ronaldsay male 13th November
1985 Co. Cork: Cape Clear 1st-winter female 18th October
1992 Orkney: North Ronaldsay 1st-winter female 1st-8th October, trapped 1st October
1994 Norfolk: Burnham Overy 1st-winter male 18th September
1999 Scilly: St. Agnes 1st-winter male 5th-8th October
2004 Co. Clare: Loop Head adult male 31st October
2007 Shetland: Foula 1st-winter male 28th September
2008 Shetland: Fair Isle 1st-winter male 25th September*
2009 Norfolk: Natural Surroundings NR 1st-winter male 4th March*

One on North Ronaldsay in 1992 was keenly sought throughout its week-long stay and attracted large numbers of observers to the island.

East to southeast winds from 26th September 1992 brought a good passage of migrants through North Ronaldsay, Orkney, including some record day-totals such as 23 Yellow-browed Warblers and 11 Richard's Pipits. Also recorded at this exceptional time were an Olive-backed Pipit and a Bonelli's Warbler on 27th, a Little Bunting on 28th and a peak of 532 Bramblings on 30th.

On 1st October, early morning rain shed a new pulse of migrants, including 400 Robins, 38 Redstarts, 19 Ring Ouzels, 230 Song Thrushes, 500 Redwings and 376 Goldcrests. The day had started early for Alison Duncan, Lyn Wells and myself; we had been alerted *in the small hours by the principal lighthouse-keeper to an arrival of birds at the tower. But, having caught eight Redwings, it was clear that conditions were deteriorating, with increasing wind and rain making further attractions unlikely, although casualties by this time had included 40 Redwings and eight Jack Snipe, and we returned to our beds anticipating a good day to come.*

As soon as conditions allowed, at about 10am, we set the mist-nets at Holland House, and Alison and Lyn spent a busy day ringing total of 201 birds – the busiest day of the year. From about 6pm, visitors to the observatory began gathering at the ringing station to see what was coming in to roost and to swap notes on the day's sightings. By about 6.30pm, everybody was present and the general 'buzz' was that although it had been an excellent day with plenty of migrants (including Little Bunting, five Richard's Pipits, four Yellow-browed Warblers, two Red-breasted Flycatchers and a Great Grey Shrike), there was a slight air of disappointment that nothing more spectacular had been found. Perhaps we had just been spoiled by the quality birds (Pallas's Grasshopper Warbler and Yellow-browed Bunting) of the previous week…..

Ringing was getting very busy with a large roost of Bramblings and smaller numbers of thrushes, Robins and Blackcaps coming into the nets. In fading light, I was extracting the birds while Alison and Lyn were ringing them. Having systematically cleared the nets, I returned to the bottom net to start again, walking past more recently emptied nets already containing new birds, and noted a thrush in one of them. At a distance, I assumed it to be a Song Thrush or Redwing, but when I worked my way back to this bird, I immediately noticed something odd about the head. I checked the underwing pattern and looked for spots on the outer tail feathers. Yes, it was a Siberian Thrush!

Knowing the excitement that it would cause (I was already experiencing it), I decided to carry the bird the short distance to the ringing station so that all present could get a brief view before it was put in a bag. I shouted "Siberian Thrush", starting a panic rush, and then added "in my hand" – and everyone froze in disbelief.

Because of the number of birds waiting to be processed, the thrush was put to one side. Martin Gray and Peter Donnelly arrived in a cloud of burning rubber after a 'phone call to them, and with their help the rest of the birds were dealt with, the nets furled and we all adjourned with the bird to the observatory. Here PD ringed it, and MG scribed the in-hand description dictated by AD.

Kevin Woodbridge. Birding World 5: 377-379

Discussion: In Russia it is common in the middle and northern taiga on the Yenisey and its range extends to the wooded tundra. In the west, the range limit roughly follows the western edge of the Yenisey basin with vagrants recorded west to the River Taz (80–82°E) (Ryabitsev 2008). As with White's Thrush, the main push for European records would appear to be on a westerly rather than northwesterly orientation, but this species is much rarer.

An exceptionally rare species in Scandinavia, there is one record from Sweden (1st-year male 11th September 1990) and seven from Norway (8th October 1905, 29th September 1953,

29th October 1955, 4th October 1966, 11th October 1980, 30th September 1984, 4th September 1986).

There are about 10 old records from Germany and a similar number from Poland (but these include accepted flocks of 18 on 23rd January 1976, 17 on 20th March 1978 and a flock of 25 on 27th October 1996; other records include October 1826, 25th August 1851, 25th September 1851, 10th October 1877, 25th September 1881, 1st October 1961, 16th October 1962, 17th October 1975). There are two old records from the Netherlands (1st-winter males in September 1853 and on 1st October 1856) and three old records from Belgium (1st-year birds in September 1877, 18th October 1901, late October 1912). Just two of six from France are recent (1847, c1859, 1861, 1870, 7th January 1982, 15th October 2008). In central Europe there are two from Austria (1928/29, 26th December 1962) and one from Switzerland (8th December 1978).

Southern Europe has just three early 20th century records from Italy (27th October 1908, 13th October 1910, 11th October 1930) and one from Malta (trapped in October 1912).

The first British record, an adult male with a deformed bill, although occurring in the same period as Britain's first Citrine Wagtails seems more likely to have been an escape, given the bill deformity on an adult bird. Siberian Thrush breeds farther north and east than White's Thrush, so the fact that half of British records come from the south and southwest is odd given that most major rarities from Siberia in Britain have a more northerly occurrence pattern. However, this bias accords with the markedly southerly distribution of European Siberian Thrush records. Birds migrate south from their breeding range in eastern Siberia or southwest from the east of the range across Mongolia and China to wintering areas in southeast Asia. Departure from the breeding grounds takes place from early September, though some remain until mid October. Given the predominantly southerly orientation of the route to wintering areas it is surprising that such a large number have reached Western Europe, and those that have most probably originated from the western part of the range.

This species remains a high quarry for self-finders: that the first for Shetland was as recent as 2007 highlights just how difficult finding your own will prove will be. The origin of the Norfolk bird in 2009 was the subject of much debate regarding its credentials. A notorious skulker, mainland birds will prove devilishly difficult to find in cover and it is not surprising that the majority have been found in exposed locations offering little in the way of concealment.

Varied Thrush *Ixoreus naevius*

(JF Gmelin, 1789). Breeding extends from Alaska, western Canada and western USA to California where a small breeding population was discovered recently in San Mateo Mountains. This attractive thrush is a short-distance partial migrant wintering mostly within the breeding range, and east to Montana and south to Baja California.

Polytypic, two subspecies. The validity of *meruloides* (Swainson, 1832) as a separate taxon is often questioned, though other races have been proposed (Clement and Hathway 2000). The racial attribution of the British record is undetermined. Formerly treated as a *Zoothera* (for example, Clement and Hathway 2000), phylogenetic analyses of mitochondrial DNA sequences indicate that Varied Thrush does not group with African and Asian *Zoothera* thrushes, but is part of a New World radiation of thrushes (Klicka *et al* 2005), resulting in Varied Thrush being reinstated in a monotypic genus by the *BOU* (Sangster *et al* 2007).

Very rare vagrant from western North America.

Status: One record.
1982 Cornwall: Nanquidno 1st-winter 14th-23rd November

Discussion: For those who were not actively birding at the time, this must represent one of the most highly desired birds on the *British list*, a classic blocker. The Cornish bird lacked orange pigmentation (it exhibited schizochroism, literally a split of colour, where one pigment is affected but the remainder is not). Birds in such plumage are exceptionally rare. There is just one known museum specimen and a small number of field observations (Madge *et al* 1990; Clement and Hathway 2000). It is astonishing that the sole British record was of a bird in such plumage. However, as this 1st-winter individual occurred at the end of a very good autumn for Nearctic land birds, it was felt unlikely to have been an escape. The species was duly admitted to the *British list* with little opposition (Madge *et al* 1990), to the chagrin of those who found the combination of an odd plumage and a west North American origin a disincentive to journey down to see the bird itself. The only other Western Palearctic record was from Iceland (3rd-8th May 2004). The only other vagrant outside the Americas was a female on Wrangel Island, Russia, on 11th June 1983.

Occasional irruptive movements take place. Periodically, large numbers venture outside the normal winter range with peaks every two to five years (Keith 1968). Records have come from all but a handful of Canadian and US States. The Icelandic bird, coupled with the number of records from eastern Northern America, makes this Pacific Coast species less of an unlikely overshoot than first appeared.

Wood Thrush *Hylocichla mustelina*
(JF Gmelin, 1789). Breeds in southeast Canada and eastern USA. Winters in Middle America mainly along the Atlantic slope south from southeast Mexico to the western half of Panama, and on the Pacific slope from Oaxaca, Mexico, south to western Panama. Occasional to very rare in the southeastern USA, western Caribbean, northwest Mexico and northern South America.

Monotypic.

Very rare vagrant from eastern North America.

Status: One record.
1987 Scilly: St. Agnes 1st-winter 7th October

Discussion: The Scilly individual was the third for the Western Palearctic, after one on the Azores in 1899 and one from Iceland (male either found dead or collected on 23rd October 1967).

The Scilly individual was the forerunner of an amazing eight-day procession of Nearctic thrushes between 7th-15th October 1987. All the small North American thrushes arrived except for the most regular, Grey-cheeked Thrush *Catharus minimus*. The Wood Thrush was followed by an initially misidentified Veery *C. fuscescens* on Lundy on 10th, a Swainson's Thrush *C. ustulatus* in Cot Valley, Cornwall, on 11th (then one on St. Mary's on 12th and another on Lundy on 15th) and finally a Hermit Thrush *C. guttatus* on St Agnes on 15th. The arrival of some could have been facilitated by the gathering Hurricane Floyd which attained full status late on 12th October, but all pre-dated the famous Great Storm which hit southern Britain overnight on 15th-16th.

The Wood Thrush population has decreased significantly over much of its range since the late 1970s (Sauer *et al* 2008). Departure from northern breeding areas takes place from mid-August to mid-September. There is no evidence of trans-Gulf flights to more southern areas and this species is less susceptible to the weather systems responsible for displacing a number of other Nearctic landbirds.

Hermit Thrush *Catharus guttatus*

(Pallas, 1811). Breeding range extensive and complex; includes much of western USA, Canada, and northeastern USA, with an isolated population in northwest Baja California, Mexico.

Polytypic, eight subspecies. The racial attribution of occurrences is undetermined.

Very rare vagrant from North America.

Status: Eight records.

1975 **Shetland:** Fair Isle 2nd June
1984 **Scilly:** St. Mary's 28th October
1987 **Scilly:** St. Agnes 15th-16th October
1993 **Scilly:** Tresco 11th October, presumed same 15th-18th October
1994 **Essex:** Chipping Ongar taken into care exhausted 28th October, released 2nd November, last noted 3rd November (record now withdrawn)
1995 **Shetland:** Fair Isle 19th October
1998 **Shetland:** Fetlar 30th April-1st May
1998 **Co. Cork:** Galley Head 25th-26th October
2006 **Co. Cork:** Cape Clear 19th-20th October

Discussion: Unusually for a *Catharus* thrush, two records involve spring occurrences. There is also a spring record from Sweden and three in western Greenland. Spring individuals are presumably birds that have overshot their breeding grounds. This was the only species of *Catharus* recorded by Durand (1963) at sea and involved a bird photographed on 8th October 1962 over 640 km east of New York. It remained to the 9th and took a liking to the safety net over the First-class swimming pool.

Autumn records predominate in the Western Palearctic. There are 10 autumn records from Iceland (eight mid October-early November, singles 26th September and early December) and one from the Azores (Corvo on 12th October 2008). One from Sweden (27th April 1978) is the only spring record away from Shetland. Additionally, there are four 19th century records from Germany (December 1825, around 1828, October 1836 and autumn 1851). Koblik *et al* (2006) give two records of Hermit Thrush from northeast Russia: Wrangel Island and Chukotka.

The first British record was a one-day bird photographed on Fair Isle in late spring. The distribution of British and Irish records conforms to expectations. The other three of the first four came from Scilly and two each of the next five were in Co. Cork and Shetland. The last two suggest that some birds make landfall farther north than was previously the case, perhaps reflecting the differing track of Atlantic depressions. The Essex record stands out as an anomaly and, although supported by photographs and accepted by *BBRC*, it attracted considerable suspicion, culminating in the finder admitting recently that the report was a hoax.

Hermit Thrush is increasing in northern parts of its breeding range (Sauer *et al* 2008), which bodes well for more occurrences in Europe.

Swainson's Thrush *Catharus ustulatus*

(Nuttall, 1840). Breeds across south Alaska and Canada to south Labrador and Newfoundland, generally south of range of Grey-cheeked Thrush C. minimus, south to northern California, New Mexico, Great Lakes and West Virginia. Migrates across eastern USA to winter from Mexico south to northwest Argentina.

Polytypic, four to six subspecies. The racial attribution of occurrences is undetermined, though the first picked up freshly dead was suspected to be *swainsoni* (Tschudi, 1845) of Newfoundland and northeast USA west to eastern Alberta. Geographical variation is slight, mainly involving the colour tone of the upperparts. Eastern race *swainsoni* is rufous-olive on upperparts with birds in the west warmer brown above (*BWP*; Clement and Hathway 2000)

Rare vagrant from North America, typically earlier than Grey-cheeked Thrush with most in early October, predominantly to the southwest.

Status: There have been 30 records to the end of 2008, 25 in Britain and five in Ireland; the latter includes one in 2008 that was photographed but is yet to be accepted.

Historical review: The first was the only spring bird, one picked up freshly dead at Blackrock Lighthouse, Co. Mayo, on 26th May 1956. The next constituted the first British record, on Skokholm from 14th-19th October 1967, followed by the second Irish record on Cape Clear from 14th-16th October 1968.

The next was in 1976, but this bird was a surprise extraction from a Sandwich Bay mist net on 27th October. The first-ever multiple arrival took place between 20th-23rd October 1979 with three (all in the southwest, with two on Scilly and one in Cornwall) representing an exceptional influx at the time.

The 1980s added a further six birds, all but one in the southwest. One at Scatness, Shetland, from 25th-29th October 1980 was the first for Scotland. There were singles on Scilly in 1983 and 1984 and three October birds in 1987 were found in Devon, Cornwall and Scilly. The total of six for the decade, although up on the 1970s, was easily outnumbered by the 20 Grey-cheeked Thrushes recorded during the 1980s.

The best decade for the species took place in the 1990s, including a record four in 1990 with singles on Shetland and Co. Cork and two on Scilly. Single birds put in appearances in five further years, again in far-flung locations from Scilly to Orkney and the Outer Hebrides to Co. Cork

Between 2000 and 2008 a further eight have occurred, keeping up the average of just under one per annum for the last two decades. These birds are split between Shetland (four) and Scilly (three), plus one from Co. Cork.

Where: Not surprisingly it is the southwest that has accounted for the bulk of records. The Isles of Scilly leads the way with 11 records between 1979 and 2004 (nine of these were between 1979 and 1991). There have been two each in Cornwall and Devon, one in Pembrokeshire, four in Co. Cork and one in Co. Mayo. Away from the southwest, Shetland has weighed in with six records (four since 2000). Other Scottish records have been from Orkney and the Outer Hebrides.

The sole record for the mainland east coast was one trapped at Sandwich Bay on 27th October 1976. It was the only record that year and occurred shortly after an invasion of Grey-cheeked Thrushes to the southwest and as part of a major arrival of Nearctic landbirds.

Interestingly, over half the records for the last 10 years came from Scotland, two from Ireland and just three from Scilly, perhaps a testament to the more northerly trajectory of Atlantic weather systems during recent autumns?

When: The spring record in Co. Mayo on 26th May 1956 is one of only a handful of spring *Catharus* records in Britain and Ireland and presumably represents a spring overshoot.

The autumn arrival dates reflects that for most Nearctic passerines, being totally dependent upon the prevalent conditions at the time. The earliest autumn

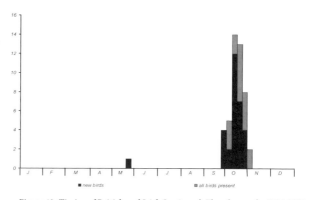

■ new birds   ■ all birds present

Figure 42: Timing of British and Irish Swainson's Thrush records, 1956-2008.

record was on Unst, Shetland, from 27th-30th September 2003. Three additional birds have been found between 28th and 30th of the month. The peak period for occurrences is the middle of October, 57 per cent (17) of records coming between 6th and 17th and 27 per cent (eight) between 20th and 27th October. The latest was the Sandwich Bay individual of 1976, which was trapped on 27th October.

Discussion: Records from elsewhere in Europe have an exceptionally broad geographic spread, but aside from those in Italy all have been in northern Europe. There have been four in Iceland (14th October 1978, 27th September-5th October 1995, 9th October 1996, 30th September-6th October 2005), four in Norway (20th September 1974, 28th September 1996, 30th September-6th October 1997, 12th-15th October 1999), one in Sweden (12th-13th November 1995) and two in Finland (19th October 1974, 1st November 1981).

Farther south there are four from Belgium (6th October 1847, 15th October 1885, 28th October 1896, October 1906) and France (one in the 19th century, 17th February 1979, 16th-18th October 1995, 12th-14th October 2000), two from Germany (undated record from 1866, 2nd October 1869) and three from Italy (all in the 19th century: autumn 1843, autumn 1878 and 12th December 1929). There is also an old record obtained near Kharkov, Ukraine, in 1893, published as 24th October by Dement'ev & Gladkov (1954) and 10th November by Stepanyan (2003). There is also a vagrant record in the Russian Far East – Ussuriland on 14th October 1978 (Stepanyan 1990). A 19th century record from Austria is no longer accepted.

The transatlantic vagrancy patterns of the two most frequent *Catharus* thrushes to Britain and Ireland are difficult to disentangle. Swainson's is the earlier migrant through North America, but there would appear to be little or no difference here in arrival dates and no correlation between peak years for the two species. Grey-cheeked was the commoner in the 1970s and 1980s, but Swainson's caught up to level pegging in the 1990s at a time when there was a dearth of Grey-cheeked. The new century has so far exposed Swainson's to be marginally the rarer of the two.

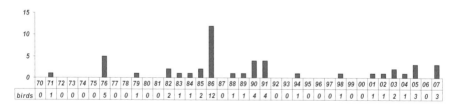

Figure 43: Annual numbers of Grey-cheeked Thrushes (top) and Swainson's Thrushes (bottom), 1970-2007.

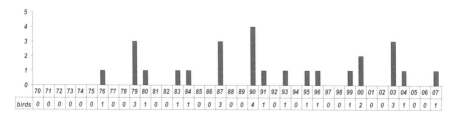

211

That the bulk of records have come from offshore islands is not surprising and presumably reflects the easier circumstances of detecting skulking *Catharus* thrushes in such locations. Doubtless others pass through undetected on the mainland as the mere handful of records testifies. Of the birds that have been aged, all have been 1st-winters, as would be expected from such vagrants.

Grey-cheeked Thrush *Catharus minimus*

(Lafresnaye, 1848). Breeds in the taiga forests and tundra shrubs of extreme northeast Siberia east throughout Alaska and northern Canada to Labrador and Newfoundland. Migrates across eastern USA to winter in northern South America.

Polytypic, two subspecies are weakly differentiated and some authorities treat the species as monotypic. The racial attribution of the vagrants to Europe is undetermined. Grey-cheeked Thrush was formerly considered conspecific with Bicknell's Thrush *C. bicknelli*. Their treatment as separate species was initially based on a study by Ouellet (1993). The split was adopted by the *AOU* in 1995. Using mt DNA sequences, Outlaw *et al* (2003), Klicka *et al* (2005) and Winker and Pruett (2006) identified a sub-clade of *Catharus* thrushes that comprised Veery, Grey-cheeked and Bicknell's Thrushes. The differences between these three, although not high, support the case for treating Bicknell's Thrush as a distinct species confirming Ouellet (1993).

Morphological differences between the two are subtle. Bicknell's is smaller than Grey-cheeked, with warmer olive brown upperparts and a larger area of colour on the lower mandible. However, nominate *minimus* Grey-cheeked Thrush (Newfoundland and southern Labrador south to St Pierre and Miquelon Islands) can also have some warmth to the upperpart colouration. The song differs too, a stronger reason for splitting than mtDNA, but of little help for an autumn vagrant.

Rare autumn vagrant from North America, typically later than Swainson's Thrush.

Status: There have been a total of 55 birds to the end of 2007; 48 in Britain and seven in Ireland (one of which, in 2007, is yet to be accepted).

Historical review: The first was on Fair Isle from 5th-6th October 1953. The next was trapped on the same island on 29th October 1958. Five well-dispersed records followed in the 1960s: two from Bardsey (including the first for Wales trapped on 10th October 1961 and died later); Scottish birds from the Outer Hebrides and Moray and Nairn; and the first English record dead at Horden, Co. Durham, on 17th October 1968. The decade included the first multiple arrivals with two each in 1965 and 1968.

Seven in the 1970s were the first to be accessible for birdwatchers. All bar one came from the southwest. Another from Bardsey in 1971 was the third for the Observatory. In 1976 the first multiple arrival of an American thrush in Europe came courtesy of a fall of birds in the southwest and, surprisingly, also the first Grey-cheeked Thrushes away from north Wales or northern Britain. These birds were the first to be collectable by the masses and the mini-invasion was received gratefully by the birding fraternity of the time. No fewer than five birds arrived between 14th and 21st October, four of them on Scilly (three graced St. Mary's and one Tresco). The other put in a five-day stay in Cornwall. A single from Scilly in 1979 saw out the decade. The perception of this transatlantic vagrant as an extreme rarity had been eroded for good.

The 1980s started innocuously enough in 1982, with the first Scottish record for 17 years on Shetland and the first-ever Irish record (Cape Clear on 19th October). Singles from Scilly in 1983 and 1984 conformed to revised expectations for points of landfall, as did two more in the south-west in 1985. The status of this transatlantic vagrant was transformed, temporarily at least, by events of 1986. On 19th October a vigorous depression crossed England, bringing strong wester-lies to the south, and although too late to catch most American migrants, the depression carried the biggest-ever fall of Grey-cheeked Thrushes so far witnessed in Europe. Three on Scilly on 20th were the vanguards. By the end of the week the islands had accommodated no fewer than eight birds, including one taken by a cat and another drowned by a wave. Additionally, three made landfall in Cornwall from 21st to 26th. One reached Lundy off north Devon on 26th and become the longest stayer of any when it remained to 2nd November (another found dead on Lundy on 27th was initially claimed as a Swainson's Thrush; the record has not been published by the *BBRC*). The arrival was centred on the southwest. The 12 birds recorded were perhaps only a fraction of those that had been carried across the Atlantic in the first place, as many were clearly exhausted. The fall involved few other American landbirds, with just a Red-eyed Vireo *Vireo olivaceus* turning up at the same time. Singles in 1988 and 1989 drew little attention; the status of this skulking *Catharus* had become that of an annual vagrant.

The 1990s started off in much the same way, courtesy of a mini-influx of four birds in 1990. Records were spread through the month and included one found dead at Slimbridge, Glouces-tershire, and others from Co. Cork and two on Scilly. Four more in 1991 including three from Scilly kept up the momentum of records. Surprisingly, the remainder of the decade produced just two more: a one-day bird on Orkney in 1994 and a dead bird in Cornwall in 1998. Despite a decade total of 10 birds, Grey-cheeked Thrush had suddenly become very rare again.

Fortunately for a modern generation of birders, the period 2000 to 2007 produced a fur-ther 11. Records have been nearly annual since 2001. Several obliging birds in Scilly and Co. Cork were augmented by an inland find in Norfolk extracted from a mist-net in 2004 and a long-staying bird in Hertfordshire in November 2005.

The autumn of 2008, a good one for Nearctic vagrants, delivered further records that are excluded from the analysis. These included the first for Dorset, trapped at Portland, and a more predictable record from St. Agnes, Scilly.

Where: In keeping with *Catharus* records in general, it is the southwest that hosted most arrivals over the years, although records from the last decade have been more dispersed.

Scilly has accounted for 43 per cent (23) of all records, often providing multiple falls: five in 1979, eight in 1986 and three in 1991. Since 1991 it has graced the archipelago in just three years, with singles each year from 2002-2004 (and another in 2008). Six have occurred in Cornwall (three in 1986, but just one since in 1998) and six in Co. Cork, with one in Co. Clare the only other Irish record. On the Scottish islands, five have made landfall on Shetland and two each on the Outer Hebrides and Orkney. All three Welsh records have come from Bardsey (1961, 1968 and 1971) and both Devon records from Lundy.

One found dead at Slimbridge, Gloucestershire, on 14th October 1990 was noteworthy, but its inland location has since been trumped by one trapped farther inland in Norfolk, at Croxton, on 10th November 2004 and eclipsed by an obliging bird at Northaw Great Wood, Hertfordshire, from 13th-25th November 2005. Two other east coast records include a 1st-winter male found dying at Lossiemouth, Moray and Nairn, on 26th November 1965 and one found dead at Horden, Co. Durham, on 17th October 1968. Most of these birds had much later detection dates than most, implying that they made their way to northwest Europe some time prior to their discovery. Perhaps those on the east coast made landfall farther north before crossing the North Sea?

Occurrence pattern of records: All records have fallen in the autumn, the earliest on St. Agnes from 22nd-26th September 1991. Two others have made landfall in September, both on Shetland since 2003: Foula, from 27th-30th September 2003 and Fair Isle on 30th September 2007. October accounts for nearly nine in 10 of all records, 31 per cent

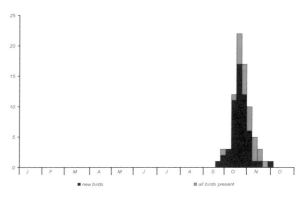

Figure 44: Timing of British and Irish Grey-cheeked Thrush records, 1953-2007.

(17) arriving between the 15th-21st. Three November records comprise the records from Norfolk (10th), Hertfordshire (13th to 25th) and Moray and Nairn (26th).

On the Isles of Scilly this species is generally later than Swainson's Thrush with a median detection date of 20th October (n=23) compared to 13th October (n=11) for Swainson's.

Discussion: The small number of records from elsewhere in the Western Palearctic do not show the geographical spread and penetration exhibited by Swainson's Thrush. Three have reached Iceland (found dead late September 1964, collected on 30th October 1983 and 17th October 2005) and one Norway (28th and 30th October 1973). There are four from France (20th October 1974, 22nd-25th October 1986, 12th-14th October 2002, 7th October 2008) and one from Germany (18th October 1937).

Three have been found on the Azores (22nd October 2002, 1st November 2005 and 21st-27th October 2007) and one in Italy (November 1901).

Peak passage through eastern USA is in late September and early October. Migration across the Caribbean is thought to be accomplished by a single flight, as it is only a vagrant or accidental on most islands. In Britain and Ireland, it is the commoner of the species pair with Swainson's and more prone to multiple arrivals, but as with that species there can be consecutive years when it fails to make an appearance. In keeping with other *Catharus* thrushes there is a tendency for records from well-watched islands; elsewhere birds have tended to betray their presence through stumbling into mist-nets or being found dead. This suggests that many others sneak through in between.

Bicknell's Thrush Catharus bicknelli
Ridgway, 1882. Breeds from northern Gulf of St. Lawrence and easternmost Nova Scotia southwest to the Catskill Mountains of New York State. Probably numbers no more than 50,000 individuals across the fragmented breeding range. In winter the range is even more restricted as it occurs on just four islands in the Greater Antilles, where specimen and field-survey data indicate that the bulk of wintering population resides in the Dominican Republic.

In light of the split from Grey-cheeked Thrush by the AOU there was a need to re-examine all records of Grey-cheeked Thrush. Three British records had formerly been considered to have possibly been Bicknell's Thrushes. A specimen from Bardsey on 10th October 1961 was formerly identified as such, but the bird had a 100-mm wing and a dull lower mandible which are consistent with Grey-cheeked (Knox 1996). A well-photographed bird on Tresco, on 20th October 1986 was considered to show features compatible with *bicknelli* as did another on St. Mary's from 19th-25th October 1990. Both records from Scilly were felt equally likely to have been extreme examples of Grey-cheeked Thrush.

The review of records (Knox 1996) concluded that there was no evidence to admit *bicknelli* to the *British List* on the basis of any of the records considered. Clearly Bicknell's is a possible vagrant to our shores. A first would require impeccable documentation of its biometrics and plumage details.

Veery Catharus fuscescens
(Stephens, 1817). Breeds Canada from southern British Columbia east to southwest Newfound-land, south through warm temperate USA and east of Rocky Mountains south to northern Arizona and Georgia. Winters south, central and southeastern Brazil.

Polytypic, between three and six subspecies (depending upon authority). Racial attribution of occurrences undetermined.

Very rare vagrant from North America.

Status: Seven records.

1970 Cornwall: Porthgwarra 1st-winter trapped 6th October
1987 Devon: Lundy 1st-winter 10th October-11th November, trapped 10th, 20th, 23rd and 30th October
1995 Outer Hebrides: Newton, North Uist 20th-28th October
1997 Devon: Lundy trapped 14th May
1999 Cornwall: St Levan 1st-winter 13th October

2002 Orkney: North Ronaldsay 30th September-5th October, trapped 30th
2005 Shetland: Northdale, Unst 1st-winter trapped 22nd September, later killed by a cat

The 1987 individual on Lundy is the most famous of the septet, but lightning clearly does strike twice.

On my way down to catch the boat back home after a week's birding on Lundy, Devon, in October 1987, I was told that the visiting ringers had caught a Thrush Nightingale in Millcombe Valley. With about ten other birdwatchers, I rushed down to have a look at the bird before it was processed and ringed. We were shown the bird in the hand and, despite a few enquiries about breast spots and ground colour, we were emphatically told that it was a Thrush Nightingale, and we accepted the verdict. A few minutes later, I caught the boat back to the mainland. I had just missed Britain's second Veery; some ten days later I received a telephone call to tell me that the bird had been caught again by other ringers, Brian and Les Tollitt, and re-identified. This incident taught me the valuable lesson of not accepting what others told me and always trying to be satisfied with my own thought on a bird's identification.

Almost ten years later, during my regular spring trip to the island, I was asleep in the old lighthouse on 14 May 1997, when I heard Rob Marshall shouting "Richard, wake up …. Steve's got a Veery in Millcombe!" Was I dreaming? No, it was a nightmare!

It was 5.00am. I rushed down to Millcombe and met Steve Wing and Mary Gade who duly showed me the bird in the hand. Of all the bird illustrations and photographs I had looked at in the last ten years, I must have studied the ones of Veery the most, as indeed had Steve who had also been with me ten years ago.

The underwings showed the classic American thrush pattern, and the grey lores, brown malar stripe, diffuse brown spotting on the breast and pale grey flanks all indicated that, yes, it was a Veery. The upperparts were not as I expected, however; they were not particularly rufous and appeared more of a warm olive brown. We put this down to a combination of early morning light and it being a worn spring adult. Steve processed the bird, and it was photographed and released. It was Britain's fourth Veery. As soon as the island's radio telephone came on line at 8.00am, I called Richard Millington at Birdline and the news was quickly broadcast.

After release, the bird was not seen in the field for four hours, but then it showed, intermittently, for about half an hour, feeding on the paths in the small area of woodland in the valley, and occasionally perching in the trees. Its upperparts now seemed much warmer in tone, but still were not the classic rufous as illustrated in the field guides. The bird remained quite elusive for the rest of the day, but it did show again a couple of times and, in the early evening, was successfully twitched by several birders from the mainland. The next morning there had been a general emptying out of birds from the island; all of the previous day's 50 Spotted Flycatchers had gone, and so had the Veery.

It was not seen again.

Richard Campey. Birding World 10: 184.

Discussion: The sole spring record in May was notable. Veery, along with Grey-cheeked Thrush, can undertake a longer transoceanic route when moving from wintering grounds to breeding areas in spring, as they use less fat than other thrush species because of their more efficient wing designs (Yong and Moore 1994).

Other Veerys have occurred at a more expected time with two in September and four in October. Surprisingly, it is so far an absentee from the avifauna of the Isles of Scilly. Both September records came from the Northern Isles. In Scotland, Nearctic landbird vagrants tend to occur earlier on average than those in the southwest. These presumably involve birds which move from more northerly parts of the breeding range earlier in the season that are susceptible to displacement in their autumn migration at more northerly latitudes.

The only record from elsewhere in Europe was Sweden (trapped 26th September 1978). The slightly more southerly breeding range than its congeners renders it less susceptible to displacement. Populations have been in decline over much of its range since the 1960s (Sauer *et al* 2008). Much of the autumn migration occurs over the Atlantic. It is an early migrant, with passage commencing mid-August and continuing into September. The timing of its migration is presumably the main cause of its rarity on this side of the Atlantic.

Eyebrowed Thrush *Turdus obscurus*

(JF Gmelin, 1789). Breeds in Siberia from Yenisey river east to Sea of Okhotsk and Kamchatka, south to Lake Baikal, northern Mongolia and Amurland. The western range limit is approximately a line running from Tomsk to the River Taz, and the species is found north to the wooded tundra. Migrates across much of China to winter in Taiwan, Indochina and Thailand south to Singapore, Sumatra, Philippines and northern Borneo.

Julian R. Hough

Monotypic.

Very rare vagrant from Siberia, mostly in autumn. More regular during the 1980s and early 1990s; exceptionally rare of late.

Status: 18 records, all of which have been in Britain

1964 **Northamptonshire:** Oundle 1st-winter 5th October
1964 **Outer Hebrides:** North Rona 1st-winter 16th October
1964 **Scilly:** St. Agnes 5th December
1978 **Clyde:** Lochwinnoch male 22nd October
1981 **East Yorkshire:** Aldbrough 16th-23rd April
1981 **Northeast Scotland:** Newburgh male 27th May

1984 Orkney: Evie, Mainland 1st-winter 25th-26th September
1984 Scilly: St. Mary's male 20th October
1987 Shetland: Fair Isle 1st-winter 7th-15th October, trapped 7th
1987 Scilly: St. Mary's 1st-winter 12th October
1987 Scilly: St. Agnes 1st-winter 27th October
1990 Scilly: Tresco female or 1st-winter male 21st October
1991 Scilly: St. Mary's 1st-winter 12th-13th October; presumed same, Tresco 15th-16th October
and St. Mary's 18th October
1992 Shetland: Fair Isle 1st-winter 4th October
1993 Scilly: St. Mary's 1st-winter 7th-14th October; presumed same, St. Agnes 15th-16th October
1995 Angus and Dundee: Auchmithie male 28th-30th May
1999 Caernarfonshire: Bardsey 1st-winter trapped 12th October
2001 Outer Hebrides: St. Kilda 1st-winter, 1st-2nd October

Discussion: This attractive thrush displays a curious pattern of records. Three in late 1964 represented a spectacular arrival for a species never before recorded on these shores, especially so the first, which remains one of only two inland sightings. The autumn records are all in keeping with a typical Sibe in terms of date and location, showing the

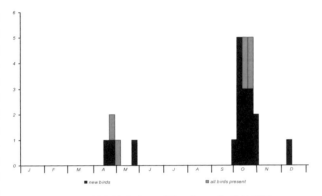

Figure 45: Timing of British Eyebrowed Thrush records, 1964-2007.

usual association with offshore islands. There is some suggestion of a spill-down in regional arrival dates, with birds in the north occurring earlier than those in the southwest. The five autumn records from Scottish offshore islands have a median arrival date of 4th October; the six autumn records from Scilly have a median arrival date of 16th October. It is perhaps surprising that there have been no more winter records, but all three spring records presumably relate to birds that had wintered in Europe rather than spring overshoots.

As with other rare thrushes the occurrence pattern elsewhere in Europe is dominated by records in central Europe. There are 12 records from Scandinavia, including five from Finland (11th-12th June 1978, 1st-14th December 1984, 1st-5th November 1995, 4th November 2000, 6th October 2002), four from Norway (2nd November 1961, 29th December 1978, 3rd October 1981, 12th September 2001) and three from Sweden (28th December 1989, 28th September 1998, singing male on 15th May 1999).

The bulk of records are from central Europe, with 16 from Germany (all but three prior to 1900; 5th October 1994, 9th October 1994, 31st October 1995), c12 from France (seven or eight in the 19th century, 13th January 1962, 25th-29th October 1986, 21st March 1990, 9th-23rd

218

April 1995, 27th October 2001, 5th October 2003), eight from Poland (six specimens in 19th century, 13th April 1995, 11th November 2001), eight from the Netherlands (27th October 1843, two 24th-26th April 1977, 30th September-4th October 1988, 5th-7th May 1989, 18th October 1992, 21st October 2000, 25th September 2001), nine from Belgium (three 19th century, October 1904, 20th October 1918, 2nd November 1932, 24th November 1990, 29th November 1995, 1st October 2004) and one from the Czech Republic (15th March 1980).

In southern Europe there are 28 from Italy, 14 of which have been since 1950 (two in April, one September (27th), eight in October and singles in November and December plus one undated). There are five from Malta (26th October 1966, 1st October 1975, 31st October 1990, January 1999 (kept as a cage bird and still alive in February 2006) and October 1999 or 2000; the latter was found as a specimen hence the vague date. There is one record from Portugal (shot on 28th October 1991). The first for Morocco was photographed on 17th December 2008. Farther east there are two from Israel (17th October 1996, 4th-6th November 2007).

This fine thrush has become surprisingly rare of late, an almost unique pattern for a Sibe. Those who saw the lingering birds on the Isles of Scilly will be thankful, as will those who saw the last male in spring 1995 in Angus and Dundee. Against a backdrop of increasing numbers of other Sibes, its decline in the 2000s is difficult to explain given the continuing arrivals of birds to other European countries.

Dusky Thrush *Turdus eunomus*
Temminck, 1831. Breeds farther north than Naumann's Thrush, from north-central Siberia east to Kamchatka. Winters from Japan south to Taiwan and to southern China and Myanmar.

Monotypic. The proposal to separate Dusky and Naumann's Thrushes *T. naumanni* to full species status is a recent one (Knox *et al* 2008). There is a limited zone of hybridisation between these two closely related taxa. As yet there is no published analysis of their molecular data and they were not included in the analysis by Voelker *et al* (2007). Birds with intermediate phenotypes exist in museum collections, but are extremely rare. Apparent hybrids in the past may have been confused with the variation shown by Naumann's Thrush. There appears to be little or no overlap in ranges. The songs are reported to be quite different (Knox *et al* 2008).

Very rare vagrant from Siberia in autumn and winter.

Status: Eight records.

1905 Nottinghamshire: Gunthorpe shot 13th October
1959 Cleveland: Hartlepool 1st-winter male 12th December-24th February 1960, trapped 10th January
1961 Shetland: Fair Isle 1st-winter female 18th-21st October, trapped 18th
1968 Shetland: Whalsay 24th September
1975 Shetland: Firth, Mainland 6th-13th November
1979 West Midlands: Majors Green 17th-19th February, again 27th-28th February and again 18th, 19th and 23rd March

1983 Cornwall: Coombe Valley, Bude male 13th November
1987 Pembrokeshire: Skomer 3rd-5th December

Discussion: Despite the paucity of British records there are a sizeable number of European sightings, the majority in central Europe and Italy. As with other rare thrushes this species is rare in Scandinavia where there are six from Finland (23rd October 1980, 17th May 1983, 15th-17th November 2004, 27th December 2005-15th January 2006, 30th December 2006-7th January 2007, 5th-11th January 2008), seven from Norway (20th October 1889, 2nd November 1889, 26th October 1895, 20th November 1908, 17th March 1959, 15th April 1992, 17th October 2001) and two from Denmark (14th October 1888, 9th September 1968). There is a single record from the Faeroes (8th December 1947) to an unspecified taxa.

In Germany there are about 10 records (only two in the 20th century: 22nd-23rd April 1987, 29th-31st December 1996), two from Poland (one specimen from 19th century, 21st October 1966), two from the Netherlands (20th November 1899, 20th February 1955) and five from Belgium (15th October 1853, 5th November 1905, November 1906, 11th November 1956, 3rd-24th January 2009). There are three records from Austria (two from the 19th century, 6th January 2005). The species was formerly on the Hungarian list due to a male c1820 from the northern Carpathian mountains, but the location no longer lies within the boundaries of Hungary (Tamás Zalai *pers comm*).

In France there has been a total of 15 Dusky/Naumann's Thrush, 13 of them in the 20th century. There are four records of Dusky Thrush (December 1856, November 1971, 23rd November 1983, 31st January 1994), four Naumann's plus three intermediate individuals (18th or 19th November 1978, 16th November 1997, 27th October 1999) and four records where attribution to one of the two species was not assessed (5th December 1910, 21st November 1957, 24th October 1964, 20th-26th January 1979).

Elsewhere in the Mediterranean there are c29 birds from Italy, 20 of which have been since 1950 (all in the period November-February including an exceptional 12 in 1981), three from Cyprus (10th-11th November, 25th April 1958, 1993 and 28th December 1994) and another three from Croatia (two in the 19th century and a male in December 1906). Farther east there is one from Kuwait (16th January-13th February 1987) and two from the UAE (12th-24th March 1998, 27th February-3rd March 2005). One from Israel (November 1982) was not allocated to taxa. One from the Azores (16th January 2008) has not yet been assessed by the PRC.

Being the more northerly breeding of the pair, Dusky has a longer migration than Naumann's and is consequently the more frequent of the two in Europe. Its rarity in Britain is difficult to explain in light of the comparatively generous number of European records. Could it be that it slips through undetected amongst the masses of Redwings that pour into our lands during the autumn? Although less prone to substantial movements than many Siberian vagrants, it is clearly capable of reaching western Europe with some regularity. Despite the fact there are eight records, the next Dusky Thrush is highly desired.

Naumann's Thrush *Turdus naumanni*

Temminck, 1820. Breeds farther south than Dusky Thrush, from central Siberia east to Amurland and Sakhalin and south to northern China. Winters from extreme eastern Russia through north-east China south to Korea and Taiwan.

Monotypic. For a discussion of taxonomy see Dusky Thrush.

Very rare vagrant from Siberia.

Status: Two records, both in January from Greater London.

1990 London: Woodford Green male 19th January-9th March
1997 London: South Woodford 1st-winter 6th-11th January

Discussion: This species occurs in the forest zone of central and eastern Siberia. It breeds in the extreme east of the Urals/western Siberia region on the Yenisey. The range limits are poorly known and it is reported to have bred in the southern Yamal Peninsula in July 1975 (Ryabit-sev 2008).

Naumann's undertakes shorter movements than Dusky and is not surprisingly the rarer of the two in Europe. In Scandinavia there are four from Finland (27th April 1988, 19th-26th November 1994, 17th-18th April 1999, 6th-19th December 2005) and five from Norway (25th November 1996, 15th June 1997, 10th October 1997, 5th April 1998, 29th November-11th December 2008). Seven have been recorded from Poland (two in the 19th century, one in 1908, 16th January 1955, 22nd January 1965, 10th April 1967, 14th January 1970), four from Germany (the last around 1906), two from Austria (one 19th century, 8th April 1984) and singles from Hungary (winter 1820) and the Czech Republic (7th February 1999).

There is one record from Belgium (26th October 1951) and four from France that have been attributed to Naumann's (September 1845, September 1901, 13th-17th February 1985, 7th January 1996), but see Dusky Thrush for indeterminate birds. There are five records from Italy (2nd November 1901, 21st March 1904, 10th November 1977, 5th February 1978, 13th December 2005). One trapped and photographed in Spain (Ebro Delta, 23rd-24th November 2005) no longer forms part of the Spanish List as the record has been withdrawn by one of the observers.

The fact that both British Naumann's occurred within a short distance is perhaps unnerving, for they form two of only a small number of European records since the 1980s, but they could simply be one of those curious quirks of fate that birding seems to accumulate over time.

Black-throated Thrush *Turdus atrogularis*

Jarocki, 1819. Breeds in central and northern Urals east across southwest Siberia and east Kazakh-stan to northwest China. Winters from Iraq to northern India, east through Himalayan foothills to Bhutan.

Monotypic. The decision to treat Black and Red-throated Thrushes as separate species by the *BOU* is a recent one (Knox *et al* 2008), though formerly Stepanyan (1978, 1990) and Collar (2005) treated them as distinct species. The two are morphologically distinct regarding plumage, though birds showing characters of both forms occur where the ranges overlap (from the upper Lena River to the Russian Altai and Sayan mountain ranges), but it has not been proven that such individuals are true hybrids (Clement and Hathway 2000). The two taxa have quite different songs (Arkhipov *et al* 2003), albeit based only on small samples from parts of the range. Treatment as separate species is based on their markedly different appearance, vocalisations and habitat preferences.

Voelker *et al* (2007) felt that a formal split was premature. They examined two male *ruficollis*, one from well within the Russian range of atrogularis and the other from just west of the westernmost edge of the range defined for *ruficollis*. The males exhibited zero differentiation for the genes sequenced and both Voelker *et al* (2007) and Collar (2005) suggest that further study is warranted. Red-throated Thrush is a bird of sparse mountain forest and scrub; Black-throated favours lowland forest and dry woodlands in subalpine steppes.

Rare, but increasingly regular vagrant from Siberia with a preponderance of autumn records on Northern Isles and east coast and regular inland wintering records.

Status: There have been 64 British records to the end of 2007; none in Ireland.

Historical review: The first was a male shot near Lewes, East Sussex, on 23rd December 1868. The next was an immature male from Perth and Kinross in 1879. The wait for the third took until a male took up residence on Fair Isle from 8th December 1957 to 22nd January 1958.

Another lengthy gap ensued before the next, an adult female at Toab, Mainland Shetland, from 5th-7th October 1974. It was followed swiftly by the first modern birds on the mainland, both from Norfolk: a 1st-winter female at Holkham from 21st-24th October 1975 and a very well watched male at Coltishall, from 21st February-3rd April 1976. These two birds readily installed this attractive thrush into the consciousness of observers and began what was to become an almost annual rate of detection. After just three previous records, twice as many were found between 1974 and 1979, including two one-day birds in 1978.

This rate continued during the 1980s, with 10 accepted in five years. These included a record four in 1987. One well inland in urban Sheffield, South Yorkshire, early in the year was

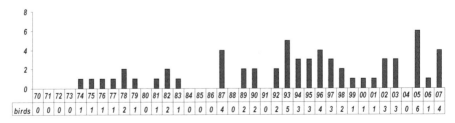

Figure 46: Annual numbers of Black-throated Thrush, 1970-2007.

followed by three late autumn records at different ends of the country (one on Scilly and two in Shetland).

Arrivals increased significantly during the 1990s, a decade during which only 1991 failed to produce any birds. A new record five were found in 1993, including a spring bird in Kent and four October birds comprising two in Shetland and singles from Scilly and Norfolk. In 1994 a 1st-winter male on Fair Isle on 16th October was joined by a second bird on the following the day, the first and only multiple sighting in Britain.

Four in 1996 was unusual in that they were found away from the offshore islands and all occurred in the first part of the year. Males found in January in both Worcestershire and Cambridgeshire put in extended and well appreciated stays through to mid-February and mid-March; the first long staying mainland birds for 20 years. In the same year one was present for two days in Bristol, Avon, in February and another, a 1st-winter female, stayed in Norfolk from mid-March to early April. After a long absence of obliging mainland records no fewer than three could have been collected at leisure within a matter of weeks.

Five between 1997 and 1998 included two more inland males, in Derbyshire in January and February 1997 and Berkshire from late December 1998 through to early March 1999. The decade had amassed a whopping 26 birds, including a number of crowd-pullers that frequented suburban housing estates.

Between 2000 and 2007 there were a further 19 birds. These included a record six in 2005, which comprised three typical October records from Shetland and inland birds located in December from Northumberland, Somerset and Glamorgan that remained in 2006. The accessible and obliging bird in Swansea was present to mid-March and represented the first for Wales.

The annual average in the 1970s was 0.7 per annum. In the 1980s it increased slightly to 1.0 per annum. From 1990-2007, this statistic has increased markedly to 2.5 per annum. Whether this reflects increased observer awareness or a real change in status is unclear, but the enhanced number of opportunities to see this delightful thrush is to be welcomed.

Where: Shetland is the place to connect with this attractive thrush, accounting for 27 birds (44 per cent), 11 of them from Fair Isle. The only other areas to yield more than singletons are Scilly, Norfolk and Yorkshire, each with five birds; there have been three in Orkney and two in Northumberland. Additional singles have occurred in 13 more English counties and three areas of Scotland. The sole Welsh record was from Glamorgan.

The 62 birds since 1957 show a clear pattern of occurrence. Arrivals are found on the Northern Isles and other offshore islands during the autumn period. More sedentary birds, often associating with Redwings *T. iliacus*, are found at inland sites during the winter months. These birds presumably made landfall in Western Europe during the previous autumn. There is a small spring presence along the east coast of birds exiting Britain or western Europe.

When: The two earliest autumn records have both been in September on the Northern Isles (1st-winter female on Foula, Shetland, on 23rd September 2000 and a 1st-winter on North Ronaldsay from 26th-28th September 1990). October is the peak month for occurrence, accounting for 47 per cent (29 birds) of all arrivals, peaking between 15th and 21st (12 birds). All records

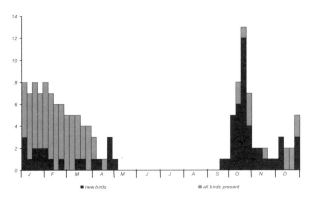

Figure 47: Timing of British Black-throated Thrush records, 1957-2007.

have been from islands and coastal hot-spots: 20 on Shetland; five from Scilly; two in Norfolk; and one each from Orkney and Cornwall.

Just five have been found in November. Two of these are from Shetland early in the month, after which the emphasis shifts to inland sites. The earliest inland sighting was one grounded for a short period with other thrushes at the famous visible migration watch point of Redmires Reservoir, South Yorkshire, on 13th November 1995. The remaining November records are one-day sightings late in the month from Greater Manchester and Staffordshire.

Fourteen mid-winter records span the period 24th December to 8th February, with one additional bird in late February the last find of the winter period (the male at Coltishall, which remained to early April). These birds are well distributed and the vast majority of them inland. Birds at this time of year have been found in suburban/urban settings and often put in protracted stays, to the delight of visiting birders.

A smattering of spring records presumably reflects the withdrawal of birds from Britain or Western Europe, and all but one of the eight have been along the east coast. Two have been in March (Norfolk and East Yorkshire) and five in April (two on Shetland, singles on Orkney and Norfolk and one in Shropshire). The latest spring bird was a female at Lydd, Kent on 2nd May 1993. All spring records have been since 1993, perhaps reflecting increased numbers of birds reaching and presumably wintering in Britain and Western Europe.

Discussion: Elsewhere in Europe there are large numbers of records, presumably reflecting the closer proximity of the western part of the range where some 50,000-55,000 pairs breed in the Ural foothills and in the Komi Republic (Hagemeijer and Blair 1997). Unlike Eyebrowed, Dusky/Naumann's and the *Zoothera* thrushes, there is clearly a push into northwest Europe. There are 30 records from Sweden (over half of which have been since 1994 and around a third of them in spring), 35 from Finland (a quarter of them in spring) and 27 from Norway. There have been nine from Denmark (one October, others November- February including four in December; these include a returning female which was first present from 3rd February-4th April 2007 then again from 8th November 2007-18th April 2008 and yet again from 15th November 2008). Two have reached Iceland (13th-22nd November 2005, 12th October 2008).

In Germany there have been more than 40 birds, including a number of small groups such as 14 at the end of October 1806, 12-14 on 1st February 1902 and three males in mixed flock of thrushes on 27th March 1958. There are nine records from Austria (singles April and September, rest October to January). In the Netherlands there were six records between 1981-2002 (mostly winter and spring) and four from Belgium (three October and one March). Elsewhere there are three records from Latvia, 13 records from Poland (eight in the 19th century, four between 1966-1982 and one in 2003; all recent records between October and February) and singles from Estonia (February) and the Czech Republic (19th century).

France has 12 records (four in 19th century and eight between 1982-1999, six from October-February, plus one on 29th March 1982 and a singing male in southern France on 16th May 1996). The 1999 record involved a female found dead in Ardèche (southern France), on 20th February 1999 that had been trapped and ringed in Italy on 14th November 1998 as a Mistle Thrush *T. Viscivorus*. An intermediate *atrogularis/ruficollis* was also present on Ouessant on 1st November 1989.

To the southwest birds have strayed to Spain (13th October 2002). Twenty-eight have occurred in Italy (14 since 1950 and all between 29th September-30th January, with singles in September and October, four in November, two each in December and January and three undated). There are also singles from Romania, Bulgaria (January 1964), Greece (4th March 1956) and Turkey (15th February 2006). It is a frequent enough visitor to the Middle East. There are 11 accepted records for Israel in the decade to 1988 (mainly November-December). It is a rare winter visitor to Kuwait, where the highest daily count was 24 on 18th January 1985. There were at least 15 records from UAE between 2004-early 2008, including a presumed hybrid *atrogularis* x *ruficollis* in January 2003. Farther south there are two records from Egypt (male collected in autumn 1833, male on 3rd January 1982).

The spread of records across Europe is impressive, indicating a sizeable dispersal northwest and southwest of the breeding range. It has been proposed that Black-throated Thrushes in Europe result from random dispersals from their breeding areas (Gilroy and Lees 2003). Inexperienced birds consort with Fieldfare *T. pilaris* and Redwing flocks heading westwards from central Siberia. Ringing recoveries indicate that Fieldfare and Redwings from Russia and Siberia winter in the same western European areas as Fennoscandian birds. These Fieldfares, along with Siberian Redwings, are among the greatest east-west migrants on earth, with round-trip migrations often exceeding 12,000 km (Milright 1994). Birds from these regions could be responsible for carrying Black-throated Thrushes into Western Europe. However, those heading southwest and into the Middle East clearly do so without these carriers to take them there. This species appears to have the greatest vagrancy potential of any of the eastern thrushes.

The increase in British records is presumably a function of the increased numbers of observers routinely working the Northern Isles in autumn. It also seems likely that increased numbers of birds are wintering in Western Europe, a hypothesis supported by the increase in spring records. The inland records may also be associated with this increase, but could also reflect increased awareness by observers during the winter months. The periodic inland occurrences of the past decade have no doubt stimulated observers to get out and scrutinise local thrush flocks that bit more closely.

Red-throated Thrush Turdus ruficollis

Pallas, 1776. Breeds to the south and east of atrogularis in south-central Siberia. Winters in east Himalayas and southern fringe of the Tibetan plateau from Nepal to southwest China, and north to northeast China.

Monotypic. For a discussion of taxonomy see Black-throated Thrush.

Very rare vagrant from Siberia.

Status: One record.
1994 Essex: The Naze 1st-winter male 29th September-7th October

Discussion: This is a great rarity throughout Europe, which is not surprising considering the more easterly breeding and wintering ranges compared to *atrogularis*. This differing distribution is reflected in the paucity of records.

In Scandinavia there is one from Sweden (30th April 2001). In northern Europe there are about four records from Germany (last 1836) and five (possibly six) records from Poland (one or two specimens were in bird collections in the 19th century; immature male 4th March 1979 and males on 1st April 1984, 5th May 1985 and 9th February 1986). In southern Europe there are two from France (Camargue on 3rd April 1969, with another possibly present at the same time, and a male in central western France on 2nd January 1990). There is one accepted record from Italy (1st-winter male on 21st September 1999). An old report from Madeira (Schmitz 1903) was not accepted by the PRC. Three widely quoted records from Norway are also not accepted (8th September 1980, 31st October 1981, 16th October 1987).

No doubt those who travelled to see the Essex bird of 1994 remain happy that they chose to do so in light of the recent split by the *BOU* (Knox *et al* 2008). Others have been claimed over the years. This taxon was long overdue as an addition to the *British list*, and the timing and location provided impeccable credentials for its acceptance.

American Robin Turdus migratorius

Linnaeus, 1766. Breeds throughout North America from the tree line of Alaska and northern Canada south to southern Mexico. Winters from southern Canada to southern USA and Central America south to Guatemala.

Polytypic, seven subspecies. Records are attributable to nominate *migratorius* from the north and northeast of the range, which is one of the darker races, though exceeded in depth of pigmentation by *nigrideus* Aldrich and Nutt, 1939 of eastern Canada in which black of the head of the male extends to mantle (*BWP*).

Very rare vagrant from North America, the majority in autumn and winter.

Status: There have been 34 records, 24 in Britain and 10 in Ireland, to the end of 2007.

Historical review: The first three records all came from Ireland, the first of which was shot at Springmount, Shankill, Co. Dublin, on 4th May 1891. This was followed by singles shot in Co. Sligo on 7th December 1892 and Co. Leitrim in December 1894.

Several other records from Britain for the period were not considered acceptable, the record arbiters of the time grappling with the difficulty of record assessment for this highly attractive thrush.

A difficult case of a highly migratory bird which is also a cagebird and has been the subject of several attempts at introduction in Britain. Lord Northcliffe's attempt to introduce them near Guildford, Surrey, about 1910, was presumably responsible for the bird which appeared in Richmond Park in May 1912 and began to build a nest.
Alexander, W.B and Fitter, R. S. R. *British Birds* 48: 10.

The next record to survive the stigma of the release experiments did not occur until over half a century later and represented the first British record, a 1st-winter, on Lundy from 25th October-8th November 1952. Notably this individual occurred 10 days after the discovery of a White's Thrush *Zoothera dauma* on the island. Both departed the island on the same date. Two more Irish birds followed in 1954 (Co. Wexford) and 1955 (Co. Kerry).

The 1960s were boom years for this gaudy thrush with no fewer than nine recorded. These included two in 1963, on the Isles of Scilly at the same time, just prior to 20th December. In 1966 three were recorded: in January in Dorset; mid-February to early March in Surrey; and a one-day bird in Dumfries and Galloway in May. The last of the decade was on Foula, Shetland, for a week in mid-November 1967.

The 1970s yielded just three, which arrived in consecutive years. An eight-year gap elapsed before the next took up an inaccessible residence for a month on St. Kilda from mid-January 1975 and was of but passing interest for a generation of new listers. The following year in 1976 one on St. Agnes for just under two weeks in late October was adored by an appreciative audience, unlike the third of the decade, a one-day bird in Co. Cork in January 1977.

The 1980s fared much better, accomodating eight individuals. Two in 1981 were both one-day birds. A record three in 1982 included one found dying in Co. Down and short-stayers on offshore islands in November. Singles followed in 1983 and 1984. None of these were accessible for the masses, being one-day birds or on offshore islands. The exception was an Irish record in 1983 in Co. Offaly, which was present from 8th June to the end of July, but news did not break until after it had departed.

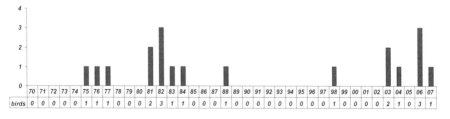

Figure 48: Annual numbers of American Robins, 1970-2007.

The situation changed in 1988 when two occurred. The first, a three-day bird on Scilly, was one of the highlights of October. More widely appreciated was a 1st-winter male at Inverbervie, Northeast Scotland, from 24th to 29th December. This bird attracted large numbers of visitors during its festive stay. The mercurial nature of this species then delivered a 10-year gap until the next, the only one of the 1990s, on Scilly from 26th-28th October 1998.

One on Bardsey from 11th to 12th November 2003 (possibly since 9th November) was the forerunner of a modern surge of records that has banished any notion of true rarity for a fresh generation of observers. This individual, the first for Wales, was gratefully received by a small number of birders. The appearance of this bird and the two subsequent records that followed were linked to a massive easterly displacement of this species and other late migrants from the North American midwest to the eastern seaboard of the USA during the second week of November 2003. This displacement delivered two much more accessible and obliging individuals: a 1st-winter female at Godrevy, Cornwall, from 14th December 2003 to 2nd February 2004 and 1st-winter female was found in Grimsby, Lincolnshire, on 1st January 2004, where it remained until killed by a Eurasian Sparrowhawk *Accipiter nisus* on 8th March.

Three more in 2006 included one for several days in late March in London that it transpired had been present since January. This was followed by an early May sighting on two dates, from two locations in Highland. The final sighting of this record showing was a well watched long stayer on Scilly for 18 days in October. Yet another obliging bird in West Yorkshire from early January to mid-February 2007 brought the total number of birds between 2000 and 2007 to seven.

Where: Perhaps not surprisingly it is the southwest and Ireland that have accounted for the majority of records. Scilly has hosted five and Co. Kerry and Devon three each. In Scotland there have been seven records from six areas, though just two have reached Shetland. Surprisingly, the only other county with multiple records is Surrey, with two. Singles have occurred in seven Irish counties, five English counties and on one Welsh island.

When: Unusually the distribution of arrivals presents three apparent peaks. As would be expected a number have been in autumn. Five in October includes three from Scilly. The earliest autumn arrival was a 1st-winter at Haslemere, Surrey, on 12th October 1984. Six more in November are followed by the peak period for finding birds from early December to late January, when no

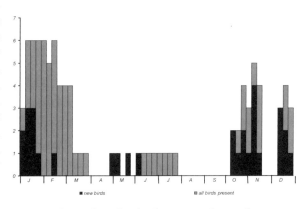

Figure 49: Timing of British and Irish American Robin records, 1950-2007.

fewer than 48 per cent (16 birds) were found. A single February record is the last of the winter period. There are five spring and early summer birds between 4th May and 8th June: three from Scotland and two from Ireland.

Discussion: Elsewhere in Europe there have been five in Iceland (14th March 1958, 28th October 1969, 13th October 2001, 6th October 2003, 18th-30th October 2008), two from Sweden (24th April 1988, 10th April 1994) and singles from Norway (3rd October 1983), Denmark (16th November 1994) and Finland (31st January 1999).

There have been six from Germany (five between 1851 and 1913; the only recent one on 30th November 2000 in Saxony-Anhalt) and two from Belgium (January-7th February 1965, 29th April 1994). In addition there are three 19th century records from Austria and one from the Czech Republic (one shot between 1857 and 1874). A record formerly accepted from Poland has recently been rejected after reconsideration. The only record from southern Europe was from Spain (10th-15th December 1999).

For some records there is the possibility of ship-assistance. In late April 1961 one was observed on board the RMS *Queen Elizabeth* 1,120 km out from New York. It was last recorded when the boat was off the Isles of Scilly on 30th April (Durand, 1961). A further category E record involved the fresh corpse of a 1st-winter male found on board ship in Felixstowe Docks, Suffolk, on 2nd November 1994 and now at Ipswich Museum.

In North America, southwards movement begins in mid-September, although many tend to remain on the breeding grounds. Later movement of these lingering birds occurs when food supplies diminish or severe weather prompts dispersal. The trend for late arrivals probably reflects southbound birds encountering Atlantic depressions in the early winter period leading to occasional multiple arrivals, such as those in 2003. The small peak of spring birds presumably reflects northbound birds that made landfall in Europe the previous autumn or winter period, though some could involve a spring overshoot.

For a long time American Robin was one of the most eagerly sought Nearctic vagrants. A recent glut of long-staying and accessible birds has dramatically changed that perception for the time being at least. Although this is one of the most abundant birds in eastern North America, there is no reason to believe that future occurrences will be anything other than erratic in their arrival. Observers may as well enjoy the current spree of records.

Zitting Cisticola Cisticola juncidis

(Rafinesque, 1810). Resident throughout Mediterranean basin and north along Atlantic seaboard of France. Recent northwards expansion curtailed by severe winters. Elsewhere, other races breed in Indian subcontinent, south China and southern Japan to southeast Asia and northern Australia, and in sub-Saharan Africa.

Ray Scally

Polytypic, up to 18 subspecies (for example, Baker 1997). The racial attribution of occurrences is undetermined. Geographical variation is clinal, with those in the Western Palearctic including: *juncidis* (Rafinesque, 1810) of Mediterranean France, Corsica and Sardinia east to Crete and western Turkey, also Egypt; *cisticola* (Temminck, 1820), western France, Iberia, Balearic Islands and north-west Africa; and *neurotica* (Meinertzhagen, 1920), Cyprus, Levant, Iraq and western Iran (*BWP*). The most probable to arrive in Britain and Ireland would appear to be *cisticola*, but *juncidis* has been suspected in Austria, Germany and Switzerland.

Very rare vagrant, probably from the near Continent.

Status: Eight records, six in Britain and two in Ireland.

1962 Co. Cork: Cape Clear 23rd April
1976 Norfolk: Cley 24th August; presumed same, Holme 29th August-5th September
1977 Dorset: Lodmoor in song 24th-28th June
1985 Co. Cork: Cape Clear 18th April
2000 Dorset: Portland 15th-16th May
2000 Dorset: Hengistbury Head in song 20th-30th May
2006 Kent: Bockhill, St. Margaret's 25th August
2008 Kent: Swalecliffe 13th September*

Discussion: Not surprisingly there is a south coast bias to the records, with one in Norfolk the only bird away from the English Channel coastline and the Celtic Sea. Four have been in spring between mid-April and late May, one in June and three between August and September (the last pending acceptance). The two Dorset birds in 2000 could easily have been taken to be the same bird. However, the individual in song at Hengistbury Head was in possession of a full tail, whereas that several days earlier at Portland was lacking one.

It is perhaps surprising that this nondescript species has not had more of an impact on the *British List* following a rapid range expansion in the 1960s and 1970s. During this time it spread north from Spain up the Atlantic coast of France and along the Channel to Belgium and The Netherlands. At one time it seemed that colonisation of Britain was a formality (Ferguson-Lees and Sharrock 1977). In contrast the expansion of the eastern Mediterranean population northwards has been much slower, with the first Bulgarian breeding record in 1984 (Hagemeijer and Blair 1997).

For the time being at least, this species seems likely to remain a very rare visitor to our shores, though once again numbers have increased across the Channel. It was rendered almost extinct in France after the severe winter of 1986-87, following a huge decrease in the winter of 1984-85. In the Netherlands the first was noted in 1972 and breeding first proven in 1974. Extinction there followed the cold winters of the 1980s, but breeding recommenced in 1990 and regular sightings began from 1993 onwards. There were at least 20 singing birds in summer 2007 and a record year noted in 2008, some of which were in the extreme north (Arnoud van den Berg *pers comm*). In Belgium the first was in 1975 and breeding occurred in 1977; 27 singing males were in the Zeebrugge area in May 2004.

On the Channel Islands the first for Jersey was recorded as recently as 1996 and breeding occurred in 2005, though it was suspected on Alderney in 2004. Recent mild winters will doubtless aid its consolidation in these areas and it may well be that a northwards surge in the future will lead to the (belated) fulfillment of the prophecy made in the 1970s. Its reluctance to cross expanses of water or stray too far north is illustrated by the fact that there is just one record from Sweden (male 11th-16th August 2000) and three from Denmark (singing birds between July 2001 and August 2002).

The species is only conspicuous when singing and it is not surprising that extralimital records from northern Europe, including some of those in Britain, are from the late summer period. Males, including juveniles, seek out new territories following post-fledging dispersal and are at their most conspicuous during their monotonous display flights. Thus, these late summer periods might offer the best opportunity for observer rewards in suitable areas of habitat within distance of the Channel coastline.

Pallas's Grasshopper Warbler *Locustella certhiola*

(Pallas, 1811). The race rubescens breeds across central and eastern Siberia north to 64° N from Irtysh River east to Yakutia and Sea of Okhotsk. Three other races breed to the south, from northeast Kazakhstan through Mongolia to Ussuriland and northern and Nnortheast China. Winters from northeast India to southern China, and south throughout southeast Asia.

James Gilroy

Polytypic, four subspecies. Records are attributable to *rubescens* Blyth, 1845. Racial variation is largely clinal. The darkest race, *rubescens* of northern Siberia, has rufous-brown upperparts and deeper rufous rump than other races, little grey on hind-neck, streaks on upperparts rather heavy and sharp, and underparts (especially under tail-coverts) more rufous-cinnamon (*BWP*).

Molecular analysis (Drovetski *et al* 2004) based on the mitochondrial ND2 gene concluded that *Locustella* may need revision as it would appears to comprise two clades. Pallas's Grasshopper Warbler belonged to a clade that included Gray's Grasshopper Warbler *L. fasciolata*, Middendorf's Grasshopper Warbler *L. ochotensis* and other eastern Palearctic species. The second clade included Savi's Warbler *L. luscinioides*, River Warbler *L. fluviatilis*, Lanceolated Warbler *L. lanceolata* and Common Grasshopper Warbler *L. naevia*, plus two Asian *Bradypterus*, Chinese Bush-warbler *Bradypterus tacsanowskius* and Chestnut-backed Bush-warbler *B. castaneus*. Common Grasshopper Warbler was more closely related to the Asian *Bradypterus* than to other *Locustella*.

Very rare vagrant from Siberia, typically from late September into early October, with most records coming from Shetland.

Status: This highly sought-after Sibe skulker has accumulated a total of 39 records to the end of 2007; 37 in Britain and two from Ireland.

Historical review: The first was found dead at Rockabill Lighthouse, Co. Dublin, on 28th September 1908. The first British bird made landfall on Fair Isle from 8th-9th October 1949, starting a long-standing association between the island and this species. Another was trapped there on 2nd October 1956.

Twenty years elapsed before the next, which also constituted the first mainland bird. It surprised a single observer along the West Bank at Cley on 13th September 1976 and was followed seven days later by another trapped on Fair Isle on 20th September that remained to 24th.

There was less of a wait for the next in 1981, which was predictably located on Fair Isle. This individual heralded the end of this skulker's reign as a mega, as the decade went on to produce seven birds. Singles graced Shetland in 1983 and 1986. The second English bird was on the Farne Islands on 26th October 1985. An unprecedented arrival in 1988 involved three betraying their presence on Fair Isle over a one-week spell between 5th and 12th October, accelerating the total number of records into double figures.

The 1990s continued the hastening of occurrences with no fewer than 13 birds. Predictably, ten records came from Shetland, but three were trapped away from that archipelago. Welcome strays to North Ronaldsay, the second for Ireland on Cape Clear and the first and only record for the south coast, at Portland, offered glimmers of hope for those extracting from mist-nets elsewhere. The Cape Clear bird was identified in the field before being trapped to confirm the identity. The decade culminated in a flourish of these inveterate skulkers. Four Shetland finds in late September and early October 1997 were a record showing, but they were rapidly eclipsed by a superb five on Shetland between 30th September-8th October 1998, no fewer than four of them on Fair Isle.

The next were not until 2001, but two late September records from this batch were even more welcome deviations from the norm. The first, on Blakeney Point during 22nd-24th September, was not only away from Shetland but was also found on the mainland without a mist net. It went on to become the most well watched example of this species ever, due to location and a three-day showing. It was tentatively suggested that this bird may have belonged to the

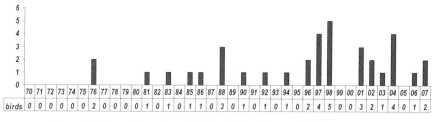

Figure 50: Annual numbers of Pallas's Grasshopper Warblers, 1970-2007.

variable (and poorly differentiated) southern Siberian form *sparsimstriata* rather than the northern form *rubescens* (Stoddart and Joyner 2005). On 29th September another in Northumberland was less obliging, but was welcome confirmation that it is possible to find the species in the field away from the scant vegetation of Fair Isle. All 10 since have been from Shetland, four of them in 2004 when three made landfall on Foula on 1st and 2nd October. Since 1996, there have been just three blank years. Records away from Shetland remain exceptional. The median arrival date for finds on Fair Isle and Shetland as a whole is 2nd October.

The weather conditions during the third week of September 2001 provided Norfolk birders with a rare opportunity to find vagrants usually associated with the Northern Isles. A combination of slow moving low pressure systems over northern Germany and an anticyclone over Scandinavia produced easterlies through the Baltic region and strong northerly winds down the North Sea.

With these possibilities in mind, Andy Stoddart and I decided to work Blakeney Point whilst the conditions remained favourable. On our first visit on 18th September, we saw a good selection of common migrants, plus Ortolan Bunting, Hoopoe, and brief views of a small Locustella which was surely a Lanceolated Warbler. The following two days produced a Bluethroat, four Barred Warblers, a Red-breasted Flycatcher and several Grasshopper Warblers, while two other sightings of possible Lanceolated Warblers on the north Norfolk coast pointed to an unusual influx of Locustella warblers…

On 22nd September, whilst working the suaeda bushes before Blakeney Point's famous 'Halfway House' landmark, I flushed a roughly Grasshopper Warbler-sized Locustella which appeared to have a pale tipped tail. On a second flight view, I noticed that the rump was rufous and contrasted with the darker upper-parts and, particularly, with the very dark, almost black, tail. AS joined me and we flushed the bird twice more before losing it in an area of suaeda cut off by the incoming tide.

Realising that it must be a Pallas's Grasshopper Warbler, we mobile-telephoned for reinforcements and waited for the tide to fall. Other birders quickly began to arrive and, with greater numbers, the bird was soon relocated.

Obtaining good views in the thick cover was impossible, however. The skulking nature of the species only allowed brief flight views and tantalising glimpses of a mouse-like bird running through the bushes. Nevertheless, a pale supercilium, yellow wash to the underparts, warm-toned rump contrasting with an almost black tail, and clear off-white tips to the spread tail on alighting were all noted. It clearly was a Pallas's Grasshopper Warbler.

The bird remained for the following two days and was eventually seen well on the ground, and the diagnostic features (including the white spots on the tips of the inner webs of the tertials) were observed unequivocally.

Steve Joyner. Birding World 14: 382-384.

Where: No fewer than 31 'PG Tips' have occurred in Shetland, where Fair Isle alone, with 18 birds, is responsible for just under half of all national records. All of those trapped on Shetland have been aged as 1st-winters.

Of the eight non-Shetland records, it is surprising that two have reached Ireland in the light of the apparent reluctance of other rare *Locustella*s to betray their presence there and on Scilly; in addition to the first record in Co. Dublin in 1908 another was seen and trapped at Cape Clear on 8th October 1990. Elsewhere, Norfolk (Cley on 13th September 1976 and Blakeney Point from 22nd-24th September 2001) and Northumberland (Farne Islands on 26th October 1985 and Newbiggin-by-the-Sea on 29th September 2001) have had more than one. Singletons have visited Dorset (Portland on 13th September 1996) and Orkney (North Ronaldsay 23rd-25th September 1992).

Outside of the review period the first for Yorkshire was trapped at Spurn on 14th September 2008.

When: Arrivals are quite tightly packed into late September and early October, with 34 arriving between 17th September and 14th October. Those seeking guidance on when to head northwards in search of their own should note that no fewer than 12 have arrived between 28th September and 2nd October. Stays are typically

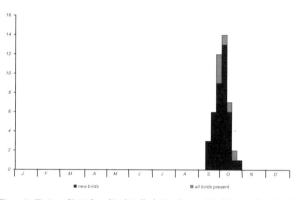

Figure 51: Timing of British and Irish Pallas's Grasshopper Warbler records, 1908-2007.

brief, but this probably says more about the elusive nature of individuals than their chosen length of stopover.

Given the species' affinity with Shetland, it is surprising that the two earliest records both made landfall elsewhere: one at Cley on 13th September 1976 and another trapped at Portland on the same date in 1996. Just two have occurred later than 14th October, one on Fair Isle on 19th October 2001 and the last, on the Farne Islands, on 26th October 1985.

Discussion: European records are relatively few, but in line with British records there has been a marked increase in the last decade. Two August records, from Germany and France, pre-date the earliest British records. There are 12 from Norway (three 1986-1988, nine 1992-2003; all 15th September-9th October), six from the Netherlands (5th October 1991, 21st September 2002, 27th September 2002, 12th September 2003, 30th September 2004, 20th September 2005), five from France (31st August 1987, 10th-13th October 1998, 2nd October 2003, 19th-21st October 2002, 19th September 2008), three from Belgium (28th September 1989, 10th September 1997, 25th September 1999), plus singles from Germany (Helgoland on 13th August 1856), Latvia (15th September 1971) and Poland (12th September 1989). The only other record for the Western Palearctic was from Israel (trapped at Eilat on 25th February 1983). The dearth of records from eastern Scandinavia and Denmark suggests that the vector into Europe does not have as much of a northwesterly orientation as Lanceolated Warbler *L. lanceolata*.

There is some evidence of a westward range extension, which may be associated with the increasing number of records in northwest Europe. East of the Urals, it was reported as common in the Vakh valley and has occurred at Surgut in the Ob' valley (Gordeev 1977). Between 1968 and 1971 it was first noted at Khanty-Mansiysk in the Irtysh valley and also on the Ob' valley where singing birds have been noted as far downstream as the village of Oktyabr'skoe (Gordeev 1977). More recently, in 2003, a singing male was found in the Ural river valley, 3 km upstream of the settlement of Ural in Kizil'skiy district (52º37'N, 59º00'E), on the evening of 15th June and another was seen in the valley of the Yuryuzan' river (55º17'N, 58º09'E) on 16th June. The latter was in the southwest Urals, north-east of Ufa, in what is now called Bashkortostan (Zakharov 2003; Valyuev and Valyuev 2003). These represent the westernmost records of singing males and are perhaps indicative of breeding (Roselaar *et al* 2006).

This species, like several other eastern vagrants, is experiencing an apparent upsurge in fortunes. There were just 15 records prior to 1996, but 24 since. It is one of the eagerly anticipated specialities for visitors to Fair Isle, but the island is no longer the epicenter of attention for seekers of this species. Locations elsewhere on Shetland, notably Foula, are starting to deliver finds, implying that many more birds are out there. It seems likely that much of the increase can be accounted for by increased observer effort on the islands.

The secretive nature of the species and the hint of occurrences farther south in Britain, as well as the Low Countries, suggest that more must make landfall along the east coast or filter farther to the west, where comparatively lush habitats mitigate against their discovery.

Lanceolated Warbler *Locustella lanceolata*

(Temminck, 1840). Singing males regular in eastern Finland. To the east, breeds discontinuously from central Urals east across much of Siberia to Kamchatka, Kuril islands, Hokkaido and northeast China. Winters in Indian subcontinent from Nepal east through northeast India to southeast Asia and the Philippines.

Monotypic. Drovetski *et al* (2004) concluded that Lanceolated Warbler is part of a clade within *Locustella* that includes Common Grasshopper *L. naevia*, River *L. fluviatilis*, and Savi's Warblers *L. luscinioides*, plus two species of Asian *Bradypterus*.

Rare annual migrant in September and early October, mostly to Fair Isle. Exceptionally rare and presumably overlooked elsewhere.

Status: A total of 116 were accepted to the end of 2007, all from Britain. Seven were between 1908-1938 and 109 in 1957-2007. Surprisingly, perhaps, none have yet had their landfall detected in Ireland.

Historical review: The first was shot by William Eagle Clarke on Fair Isle on 9th September 1908, though even he did not guess at the affinity that Lanceys would develop with this magical island over the following century. The next followed a year later, this time an exceptionally late male at North Cotes, Lincolnshire, on 18th November 1909. The third was caught at Pentland Skerries, Orkney, on 26th October 1910. Four were documented from Fair Isle between 1925 and 1938, yielding a cumulative total of seven pre-*BBRC* records.

When walking along the sea-bank at North Cotes, Lincolnshire, on November 18th, 1909, I shot an example of the Lanceolated Warbler (Locustella lanceolata). I first observed the bird in the long grass on the side of one of the marsh-drains, out of which it ran on to the short grass of the adjoining field. I watched it for a short time as it ran about the ground like a mouse, and I noticed that it kept its tail depressed, and not erected over the back, as is usually the case with the Grasshopper-Warbler (Locustella naevia) when running over open ground. At one time it flew up to a barbed-wire post close by, up which it climbed with the facility of a Tree-Creeper. It soon flew back to the ground, and I shot it just as it reached the long grass again. Unfortunately the bird was much shattered by the shot, and I had great difficulty in making a skin of it.

G. H. Caton Haigh. British Birds 78: 200

The first of the 'modern era' was trapped on 21st September 1957 on Fair Isle (where else?), followed by two more there in 1960 and one in 1961.

Formerly an extreme rarity, this popular species has been almost annual since 1972. There were 17 records in the 1970s, 13 of them on Fair Isle and two from Out Skerries. Four in 1975 included three on Fair Isle on 11th October and a new bird there on the 14th. On 14th October 1978, a 1st-winter was found inside a toilet bowl on a ship in the Forties Oilfield, 145 km east of Aberdeen, Northeast Scotland; it was released at Aberdeen. The most staggering record of the decade was one pulled from a mist-net at Damerham, Hampshire, on 23rd September 1979. In a matter of just a few years, the the status of this sought-after Sibe had changed forever.

The last clean sheet on the record books was as long ago as 1983. The 1980s as a whole fared comparably with the decade before, producing 19 birds, 15 of them from Fair Isle, two from elsewhere on Shetland and singles from Isle of May and Northumberland. The record annual total was matched in 1984, with three birds on Shetland from 18th-23rd September, two of them on Fair Isle followed by a very late bird at Tynemouth in mid-November. A then-record five were found in 1987, three of them on Fair Isle between 17th and 20th September.

If the 1980s consolidated the species' status as a rare visitor to Shetland, mostly Fair Isle, then it was the 1990s that changed this perception dramatically. The decade produced 36 birds,

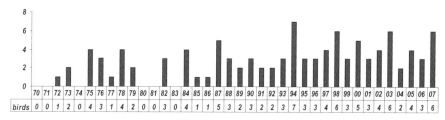

Figure 52: Annual numbers of Lanceolated Warblers, 1970-2007.

18 on Fair Isle and eight from elsewhere on Shetland. More noticeable was the delivery of 10 trend-breaking birds away from Shetland. These included no fewer than three from Bardsey and seven on the east coast between Northumberland and Suffolk, including two each in Norfolk and Yorkshire. An unequalled record seven in 1994 had arrival dates spread over a month from mid-September onwards; only four came from Shetland. One at Filey occurred the day after a Pechora Pipit *Anthus gustavi*, fleetingly moving Fair Isle a tad farther south. By way of redressing the balance, an influx of six in 1998 was restricted to Shetland with five arriving on Fair Isle between 26th September and 6th October.

Following a substantial overnight fall of Robins, Song Thrushes and Goldcrests, DR and KS enjoyed a busy first net-round at Sheringham, Norfolk, early on 29 September 1993. The very last bird of the round, extracted by DR, was a Locustella warbler caught in the bottom panel of a mist-net running through a narrow belt of rough grass fringing the seaward edge of a copse 400m from the sea. KS's "Anything good?" received the reply "Got a Gropper here – seems a bit small though ….." before the bird was quickly bagged for processing back at base.

Both DR and KS returned to the ringing laboratory to find an excited SV bringing news of a Hoopoe, well settled on the nearby cliff-top. Dividing the workload, the "Gropper' was handed to SV for thorough examination whilst DR and KS set about ringing commoner migrants. As SV removed it from the bag, we all had a quick look at it. It certainly looked small – and interesting. All fell strangely silent. And the longer the silence prevailed, the greater the anticipation grew. Occasional glances at SV revealed detailed sketches of individual tertials appearing in the notebook and a host of precise measurements being carefully recorded. Despite the unimaginable outcome, things looked extremely promising and, as DR released the last Robin, SV confidently announced his conclusion: "It's a definite Lancey!"

Struggling to maintain composure, we all took a deep breath and stared at the almost unbelievable. One of British birders' most sought after vagrants was right here – and we were nowhere near Fair Isle!

As we were familiar with the species' exceptionally skulking habits, it was obvious that, were we to release it in the copse, it would never be seen again. We all agreed that there was a large area of suitable habitat for it on the nearby cliff-top and therefore we decided to release it there where, hopefully, there was at least an outside chance that others would see it. It was ringed, photographed and released, and the news put onto Birdline.

Upon release, it flew strongly for about 40m before dropping into an extensive stand of Marram grass – and we were convinced it had gone for ever. Miraculously, however, as soon as the first birders arrived, it reappeared right beside the cliff-top footpath, near to where it was to remain faithful for the rest of the day. It was extremely skulking at times but regularly gave good views. A large number of birders arrived from far and wide and everyone saw it, with the more patient observers enjoying superb views. Unfortunately it could not be found the following day despite careful searching, so it had probably moved on overnight.

Dave Riley, Kevin Shepherd and Steve Votier. *Birding World 6: 396-397*

The period 2000 to 2007 continued in a similar upbeat tempo, with 33 skulkers detected. Unlike the previous decade, when it was beginning to appear as though east coast birders might be in with a shout of finding one, all but one of these came from the Northern Isles. Shetland boasted 30 birds, 22 of them on Fair Isle; two came from Orkney. The only deviation was the long over-due first for Scilly on Annet from 22nd-23rd September 2002. The magnetism of Fair Isle was further exemplified by five more creeping about the island between 4th-16th September 2000 and five more in 2007 condensed into the period 27th September-2nd October, including three present on the first date.

Where: Shetland accounts for over 80 per cent of British records and Fair Isle weighs in with 67 per cent (77 birds) of the archive. Away from Shetland there have been just 19. Just five Scottish records have surfaced away from Shetland: three on Orkney (Pentland Skerries on 26th October 1910, North Ronaldsay on 8th September 2003 and Sanday on 29th September 2003), one in Fife (Isle of May on 2nd October 1987) and one at sea in Sea area Forties in 1978.

Northeast England musters four records: two in Northumberland (Prior's Park, Tynemouth on 13th November 1984 and the Farne Islands from 16th to 17th September 1985) and two in Yorkshire (Filey on 10th October 1994 and Spurn on 21st September 1996). Lincolnshire boasts records in 1909 (North Cotes) and 1996 (Rimac on 22nd September). The well-watched Norfolk coastline contributes records in 1993 (Sheringham on 29th September) and 1994 (Mundesley on 21st September) and Suffolk one in 1997 (trapped at Landguard on 26th September).

Considering its rarity elsewhere it is perhaps surprising that Bardsey has produced three birds (18th October 1990, 8th October 1994 and 27th September 1997) in contrast to the well-watched Isles of Scilly which yielded one only as recently as 2002. Thin pickings away from Shetland make the one extracted from an inland mist net at Damerham, Hampshire, on 23rd September 1979 even more extraordinary.

When: Around half of the British records have been located during the two-week period between 17th and 30th September, making this endearing *Locustella* one of the earlier quality autumn rarities. The earliest records both occurred in 2000, when one was on Out Skerries on 1st September and another on Fair Isle on 4th September.

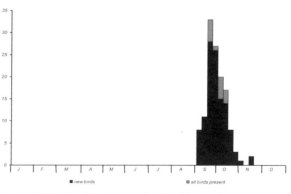

Figure 53: Timing of British Lanceolated Warbler records, 1908-2007.

Records decline sharply after mid October, with just nine in the second half of the month. Three November records include one on Fair Isle on 1st in 1960, and the records at Tynemouth

(Northumberland) on 13th November 1984 and (the latest ever) at North Cotes (Lincolnshire) on 18th November 1909.

Despite the increasing number of arrivals reaching these shores, there has not been a trend towards earlier arrival dates (see Table 19). This is surprising given a westward range expansion in Europe. Taking records from Fair Isle since 1970, the median arrival date over four time periods varies between 22nd and 28th September, with recent records intriguingly about five days later than those in the 1970s and 1980s.

Table 19. Median arrival dates of Lanceolated Warblers on Fair Isle since 1970.

Period	1970-1979 (n=13)	1980-1989 (n=15)	1990-1999 (n=18)	2000-2007 (n=22)
Median date	22nd September	23rd September	28th September	27th September

Discussion: In Russia it is considered to be local in the taiga zone of the Arkhangel'sk Region but is perhaps extending its range northwards (Red'kin 1998). It is rare west of the Urals, but is common or very common in many parts of western Siberia. This is a secretive species and nests are difficult to find, so the range limits are poorly known (Ryabitsev 2008). Of 97 birds recorded in Finland up to the end of 2007, 93 involved singing males and just four were in autumn (Lindblom 2008). Elsewhere in northern Europe this species remains a rarity. For example, just 10 have been in Sweden where all but one have been since 1987 (three summer males found between 22nd June-6th July and seven in autumn: four 28th September-6th October, rest 13th-19th October), 10 from Norway (29th September 1980, c15th September 1982, 6th October 1991, 3rd October 1994, 27th-30th September 1997, 19th-22nd September 2002, 15th October 2003, singing male 1st-16th June 2007, singing male 13th-30th June 2007 and a trapped bird from 1st October 2008 that is yet to be submitted), six from Denmark (2nd October 1932, 5th October 1935, 2nd October 1943, 7th October 1991, 2nd October 2004, 17th October 2008). One alighted a vessel c70 nautical miles north of Bear Island in the Arctic Ocean (15th September 1982). There is one record from Iceland (9th October 1983).

Farther south it is very rare, in keeping with the few records from southern Britain. There are four from Germany (13th October 1909, 25th September 1920, 13th October 1979, 13th October 1993), three from the Netherlands (11th December 1912, mid September 1958, 20th September 2002), five from Belgium (10th-12th September 1988, 5th October 1991, 14th October 1994, 7th October 1996, 1st October 2000) and four from France (one singing 15th-16th August 1986, found dead on 11th September 1986, 28th October 1990, 26th October 2005). There is also an exceptional record from southern Montenegro on 12th November 1907.

Evidence suggests that the autumn passage of all populations is via northeast and east China. Western populations undertake an ESE heading via north-east and eastern China while Japanese and north-east Russian populations head southwest. The mechanism that brings the species here with such regularity is presumably one of random post-breeding dispersal, though full reverse migration of 1st-winter birds has also been implied. All of those trapped on Shetland have been 1st-winter birds. Autumn migration is prolonged. Departure dates from western parts of the breeding distribution occur earlier than those farther east; the north-east Altai begin to be vacated as early as mid- to late July (*BWP*). It seems logical to associate the peak of British sightings in the early part of the autumn with birds originating from closer populations. The relatively few late records could come from populations farther east.

The present breeding distribution of Lanceolated Warbler may be less extensive than it was in the 1890s or even the 1960s. It has been suggested that it may undergo long-term fluctuations, as it is absent from locations and habitats in the southern Urals where it occurred in the 19th century, and it was regularly found during the breeding season in much of east European Russia up to the 1960s (Hagemeijer and Blair 1997). During the latter part of the 20th century, a northwesterly range expansion took place, as indicated by increased numbers of singing males in eastern Fenno-Scandia, with 20 singing males present in Finland in 2000, but numbers have since declined. In total 63 singing males were found there between 2000-2006. It is considered to be the latest arriving of all spring migrants. Most activity by singing males occurs during the darkest hours of the night (Lindblom 2008).

This range extension, together with the increased rarity-hunting effort in the Shetland Isles, has presumably contributed some way to the recent change in status here. It is now an expected annual rarity. With rare bird hunting probably at its highest ever levels of coverage, it is perhaps surprising that there have not yet been any proven spring occurrences in Britain; one on Fair Isle on 4th May 1953 is no longer considered acceptable. The first summer records for Norway, two males, both occurred in July 2007. Maybe a singing summer male is not so far away.

The seeming rarity of this species away from the open Fair Isle landscape is presumably due to its skulking behaviour. Even in scant vegetation, it remains a devilishly difficult species to obtain good views of in the field. Riddiford and Harvey (1992) acknowledged the Houdini-like nature of this species in an identification article: *"Fair Isle vagrants habitually burrow, mouse-like, into tussocks of grass and other dense vegetation until lost to view, only to emerge at some other point a few seconds, or even minutes, later."*

The recent bird on Scilly in 2002 was the first for this intensely watched archipelago. One can only wonder how many more of these and other *Locustella*s go undetected in the dense cover of the British mainland. As with the first a century ago, those realistically seeking this enigmatic species as a vagrant must travel north to Fair Isle, where tales abound of confiding birds scuttling around and over the feet of their bemused observers (see for example, *British Birds* 84: plate 269). Such anecdotes elicit envy from those of us watching well-vegetated coastal patches farther south.

River Warbler Locustella fluviatilis

(Wolf, 1810). Patchy and local distribution across central and eastern Europe, but is spreading into northwest Europe: from Germany to central Finland and east through central Russia to western Siberia. Southern limit extends to Croatia and Ukraine. Migrates through Middle East and north-east Africa to winter in East Africa.

Monotypic. Closely related to Savi's Warbler *L. luscinioides*, and within a clade that includes Lanceolated Warbler *L. lanceolata* and Common Grasshopper Warbler *L. naevia* (Drovetski *et al* 2004).

Very rare vagrant from eastern Europe mostly involving singing summer males in northern Europe and autumn vagrants on Shetland. Exceptionally rare during the last decade.

Status: Presently the rarest of the *Locustella*s, there was a total of 33 records by the end of 2007, all in Britain.

Historical review: The first occurred as recently as 1961 with one on Fair Isle from 24th to 25th September. Two more followed in 1969, both autumn birds and occurring within a day of each other: Fair Isle on 16th September and a moribund bird on Bardsey the following day.

This short-lived flurry was not repeated until 12 years later, when one of three birds in 1981 propelled this rare warbler onto the lists of a modern generation of birders. One on Fair Isle from 23rd-24th May was found dead the following day. An August bird from Spurn was extracted from a mist-net on 24th and never seen again. Another bird put on a public performance: a male in song at Roydon Common, Norfolk, from 29th May-6th June delighted many hundreds of observers and was the harbinger of a number of obliging and accessible individuals on the mainland during the 1980s and 1990s.

Between 1981 and 1984, no fewer than seven birds were recorded and in 1989 two more followed, bringing the total for the decade to an exceptional nine. Two in 1982 came within a few days of each other in late September on Fair Isle and involved an adult and a 1st-year bird. Both 1984 records were in summer, including a one-day early June bird on Fair Isle and an inland bird in song at Pettistree, Suffolk, from 13th July-3rd August. Two in 1989 were similar occurrences, the first a one-day late May bird at Flamborough Head and the second an inland bird in song at Boughton Fen, Norfolk, from 8th-21st July.

The 1990s boasted an impressive 17 birds, momentarily relegating this once great rarity to a year tick for the listing fraternity. There were records in every year bar the extremes of the decade. A singing male briefly at Wicken Fen, Cambridgeshire, in 1992 was the first for three years. Unlike its predecessors at inland sites, this male remained only for two early June days. Three in 1992 were equally brief in appearance, all on Shetland with one in late May and singles in late September and early October. Another mid-July male performed well at Clatto Reservoir, Fife, from 16th-25th July 1994 and was duly admired by the masses. A deluge of seven in 1995 represented the pinnacle of fortunes for the species on our shores. These included a spring bird on Fair Isle, but much more widely received was a male at Scotsman's Flash, Greater Manchester, for a month from 11th June. Another male sang at Wicken Fen for four days in early July; the

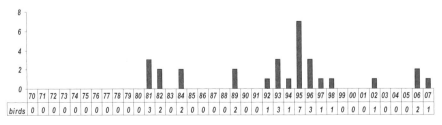

Figure 54: Annual numbers of River Warblers, 1970-2007.

returning male from 1992, perhaps? Quite exceptional though was a fall of four birds between 14th and 15th September on the Northern Isles, including simultaneous arrivals of birds on Foula and Lerwick on 14th and birds trapped at Sumburgh and North Ronaldsay on 15th. Three in 1996 included a male in song near Bellingham, Northumberland, for two weeks in late June and a male at Doxey Marshes, Staffordshire, from 20th June-22nd July. Singles in 1997 (Buckinghamshire) and 1998 (Lincolnshire) saw out the decade and there was an air of expectation that the presence of the species would appear each year.

Then there was a four-year gap until the first of the Millennium in 2002, an autumn bird from Fair Isle in late September. Just three have occurred since, all from Shetland (one-day June birds from Fair Isle in 2006 and 2007 and an early October bird from Foula in 2006). After a glut of obliging mainland records throughout the 1980s and 1990s it is remarkable to consider that it is now nearly 12 years since the last long-staying bird on the mainland.

Following the review period, two were found in 2008. The first, for a day a Beachy Head on 30th May frustrated observers, but a male at Evie, Orkney, from 9th-18th June was a welcome long-stayer for the islands.

Where: As with its congeners the barren Shetland landscape has made River Warblers easier to detect and accommodated the majority of sightings with 17 records (12 on Fair Isle). Two each have come from Northumberland, East Yorkshire, Norfolk and Cambridgeshire. Singles come from Orkney, Fife, Lincolnshire, Suffolk, Buckinghamshire, Staffordshire, Greater Manchester and Caernarfonshire. All bar three of the 14 autumn passage birds have been found on Shetland (six on Fair Isle), the exceptions all from other bird observatories: Bardsey (moribund), Spurn and North Ronaldsay (both trapped). The odds for self-find along the east coast appear slim.

A number of inland birds have proved deservedly popular, these include singing males: Roydon Common, Norfolk, from 29th May-6th June 1981; Pettstree, near Wickham Market, Suffolk, from 13th July-3rd August 1984; Boughton Fen, Norfolk from 8th-21st July 1989; Wicken Fen, Cambridgeshire, from 10th-11th June 1992; Clatto Reservoir, Fife from 16th-25th July 1994; Scotsman's Flash, Wigan, Greater Manchester from 11th June-12th July 1995; Wicken Fen from 7th-10th July 1995; near Bellingham, Northumberland, 16th-30th June 1996; Doxey Marshes, Staffs from 20th June-22nd July 1996; and Linford GPs, Bucks, from 15th-16th June 1997.

When: The bulk of records have occurred in the spring and summer: 14 between the earliest, on Fair Isle, from 23rd to 24th May 1981 through to the end of June. These include six on Shetland between 23rd May-6th June and eight dispersed singing males. The spread of the latter birds reflects the opportunism of the species in that many sang well

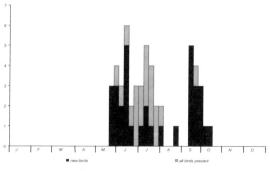

Figure 55: Timing of British River Warbler records, 1961-2007.

away from migrant hotspots in inland locations; many more may well have been missed during the glory years when singing males appeared prevalent. There is a further cluster of five singing males between 7th July and 2nd August.

A 1st-year bird at Spurn on 24th August 1981 is the earliest of the autumn records by some stretch. The remaining birds are accounted for by 11 between 14th-26th September and two more in early October, the latest of which was one on 9th October 1993 on Out Skerries (to 10th). Of the 13 latest records, no fewer than 11 have been from Shetland, one on Orkney and the only Welsh record, a moribund bird from Bardsey, on 17th September 1969. Not surprisingly considering the species' skulking habits, 10 of these were located through trapping.

Discussion: The Western Palearctic breeding range is smaller and lies to the east of Common Grasshopper *L. naevia* and Savi's Warblers *L. luscinioides*, with 75 per cent of the range in central and eastern Europe. Early records suggest a gradual range expansion in central Europe in the early 20th century, followed by a retreat (Hagemeijer and Blair 1997). Since the 1950s, the range has again extended west and north.

Breeding in Finland was first confirmed in 1974 and the tiny population there doubled in the late 1980s. In Sweden the second breeding occurrence took place in 1988, since when around 200 males have been recorded in most years; at the turn of the century the increasing breeding population was put at 40-60 pairs. This is a difficult species to survey as paired males cease singing while unpaired birds continue to advertise their location. In Estonia, it was a scarce breeding bird until the 1950s, but by the end of the 20th century the population was perhaps as large as 10,000 pairs (BirdLife International 2004). This range expansion and increase in numbers accords well with the rise in British records over the period.

A recent decline in records is less obvious to explain as core populations across Europe were considered stable in the last decade of the 20th Century (BirdLife International 2004). In the Netherlands there were 45 between 1924-2006 and singing males were annual between 2003 and 2008; breeding has been suspected, but even the strong rumours in 1997 remained unsubstantiated (van den Berg and Bosman 1999; Arnoud van den Berg *pers comm*).

Over the years the oscillating fortune of this *Locustella* in Britain has prompted much speculation regarding the possibility of colonisation. Since the turn of the century there have been just a handful of records and a long-staying mainland bird is once again long overdue, though one on Orkney in summer 2008 was well appreciated. This modern trend towards renewed rarity is quite the opposite of its Siberian congeners; speculation over future breeding appears to have been premature (see *British Birds* 92: 594).

Spring and summer birds presumably represent overshoots, the short-stays of birds earlier in the summer perhaps reflecting migrants still keen to find mates; those in mid-summer may represent failed breeders from elsewhere. Some of those in the mid-1990s may well have involved returning birds, as suspected in the Netherlands at that time (van den Berg and Bosman 1999).

Not surprisingly all autumn records have come from sparse Shetland and a trio of bird observatory captures elsewhere. Silent birds are unlikely to betray their presence by other means amid the well vegetated mainland and it seems probable that many pass clandestinely on their way along the east coast.

Savi's Warbler *Locustella luscinioides*

(Savi, 1824). Breeds discontinuously in Europe from Iberia to Netherlands; range contracting to southeast, although expanding northeast into Baltic countries. To the east occurs through temperate Russia south through Ukraine to Black Sea coasts. European birds winter in West Africa from Senegal to northern Nigeria. Central Asian race, fusca, breeds from Caspian Sea east across Kazakhstan to northwest China and western Mongolia, wintering in northeast Africa.

Polytypic, three subspecies. Records are attributable to nominate *luscinioides* (Savi, 1824), but there is one pending record of the eastern race *fusca* (Severtzov, 1872), which has less brown on upperparts and paler underparts than *sarmatica* Kazakov, 1973 of European Russia and nominate *luscinioides* (*BWP*). Savi's Warbler forms part of a clade with River Warbler, Lanceolated Warbler and Common Grasshopper Warbler *L. naevia* (Drovetski *et al* 2004).

Formerly a scarce passage migrant, now an increasingly rare migrant, especially so away from core areas in southern England, where it is a rare breeding species.

Status: Between 1950 and 2007 there were *c*638 birds recorded from Britain and a further 13 from Ireland, where the last was in 1996. As a result of its increasing rarity, Savi's Warbler was readmitted to the list of species considered by *BBRC* from 1999 onwards.

Historical review: This popular *Locustella* was formerly quite widely if thinly spread in the Norfolk Broads and fens of East Anglia. Extensive drainage contributed to its extinction as a breeding bird in 1856. The last recorded survivor of the original breeding population was shot at Surlingham, Norfolk, on 7th June of that year.

Following its extinction as a breeding bird in the mid-19th century, the only subsequent accepted records of this species in Britain up to 1950 were one shot and another seen on Fair Isle on 14th May 1908. There were also three insufficiently documented observations towards the end of the 19th century and a Sussex record in 1916, which was later rejected with the other Hastings Rarities (Nicholson and Ferguson-Lees 1962)

Since the 1940s the range of Savi's Warbler has expanded north and west in Europe, resulting in the recolonisation of England during the 1960s. A singing male at Wicken Fen, Cambridgeshire, from 2nd June to mid-August 1954 was the first to summer in England for almost a century. Savi's was found breeding in 1960 at Stodmarsh, Kent, where three singing males were present and two pairs bred, though it was thought that breeding at this locality may have been taking place since 1951 (Pitt, 1967). An additional record in 1960 referred to a singing male at Chew Valley Lake, Somerset, on 24th and 30th July.

The colonisation of Stodmarsh took place rapidly, with singing males in the following years, but no evidence of breeding again until 1964 when there were four singing males and two pairs bred. In 1965 no fewer than 12 singing males were recorded and at least one pair bred. In 1966 there were eight or nine singing males and two pairs bred. Elsewhere, a singing male at Minsmere, from 20th April-30th May 1964 and a singing male at Coate Water, Wiltshire, from 6th to at least 20th May 1965 were signs that the number of birds reaching England was increasing. By 1967 Savi's were recorded in summer in various other counties. One at Pyewipe Marsh, near

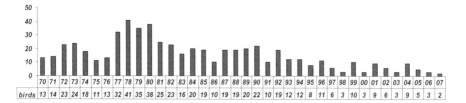

| birds | 13 | 14 | 23 | 24 | 18 | 11 | 13 | 32 | 41 | 35 | 38 | 25 | 23 | 16 | 20 | 19 | 10 | 19 | 19 | 20 | 22 | 10 | 19 | 12 | 12 | 8 | 11 | 6 | 3 | 10 | 3 | 9 | 6 | 3 | 9 | 5 | 3 | 2 |

Figure 56: Annual numbers of Savi's Warblers, 1970-2007.

Lincoln, from 9th-11th May 1967 was the most northerly mainland record at the time. It was soon succeeded by another Lincolnshire record in 1969 and one in North Yorkshire. Breeding was thought to have occurred at a second locality in Kent from 1969.

Outside of Kent, only two sites held birds with any regularity and both were in Suffolk. The first record for Minsmere was in 1964 and the first record for Walberswick was in 1968, where breeding was first proven in 1970. The first proven breeding record for Minsmere occurred the following year (Axell and Jobson 1972). Juveniles trapped at Walberswick during the 1970s confirmed that breeding took place with some regularity.

During the 1970s 224 birds were accepted to the national archive, the majority from the Suffolk reedbeds and newly re-colonised breeding areas, such as the Norfolk Broads. Fortunes boomed during the late 1970s, when up to 30 singing males were recorded. A further 209 were documented during the 1980s, a slight decline, though the possible maximum number of pairs remained in the teens and peaked at 20 in 1987.

Two ringing returns occurred during this period. An adult male ringed at Westbere, Kent, in June 1986 was re-trapped there in June 1987 and June 1988 and one ringed at Brandon Marsh, Warwickshire, in May 1989 was re-caught at Tring, Hertfordshire, two months later (Wernham *et al* 2002).

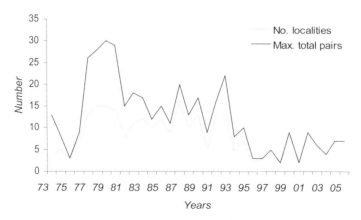

Figure 57: Breeding activity for Savi's Warbler reported by the Rare Breeding Birds Panel.

245

Reasonable numbers were still being reported in the early 1990s, but overall the decade produced a total of only 113 birds; since then numbers have declined markedly, plummeting from a peak of 22 possible pairs in 1992 to just 10 possible pairs in 1994, the last year in which a double-figure number was reported by the Rare Breeding Birds Panel. The trend has continued downwards; just 40 birds were recorded in 2000-2007.

The Norfolk Broads continues to attract birds on a regular basis, though even here records are no longer annual. Kent hosts some birds and a nesting attempt was made in 2003. In neighbouring Sussex a pair raised two broods in 2000 in the last proof of success. The wetlands of Somerset have emerged as a potentially new area in recent times. Breeding has been confirmed on just one occasion since 1994.

A Savi's Warbler trapped on Fair Isle on 30th September 2003 had plumage that suggested it belonged to the eastern race *L. luscinioides fusca* Eastern Savi's Warbler. There is overlap in measurements between *fusca* and the nominate form and differences may be clinal, so ascertaining racial identity with certainty is difficult. The record remains under consideration.

Where: Away from the reedbeds of Norfolk, Suffolk and Kent, this remains an exceptionally rare species even in the south of the country. In East Anglia, Cambridgeshire has recorded just 12 birds (despite being the county that hosted the first record of the recolonisation period) and Essex three. From the southwest, there are a number of sightings from Dorset (15 birds), Somerset and Bristol (12 birds), Devon (11 birds) and one each from Cornwall and Scilly. Sussex (18) leads the way for the southeast, Hampshire (11 birds) is also into double figures, but two from Wiltshire and one from Berkshire are the only other records away from breeding areas.

In the northeast, there are 20 records from Yorkshire, five in Lincolnshire and one from Northumberland. In the northwest records are even rarer, with just five from Lancashire and north Merseyside and one from Cheshire. In the Midlands, records have come from Warwickshire (eight), Worcestershire and Hertfordshire (four each), Nottinghamshire (three), Leicestershire (two) and Northamptonshire, Gloucestershire and Staffordshire.

In Scotland, 10 birds have been recorded: six from Fair Isle (two on 14th May 1908, 24th June 1981, 7th June 1986, 4th-6th May 1993 and 24th-31st May); three from elsewhere on Shetland (Whalsay, 29th-30th May 1995, Foula, 28th-29th May 2002, Unst, 28th May-3rd June 2006); and one from Perth and Kinross (10th-16th May 2005).

In Ireland there are seven records of nine birds; surprisingly these have possibly included summering birds. The first was at Shannon Airport Lagoon, Co. Clare, from 17th-23rd June 1980. The next was near Youghal, Co. Cork, from 17th-23rd June 1985 with another present there on 19th June. One was at Ballycotton, Co. Cork, from 3rd-15th May 1988. One was at Pollardstown Fen, Co. Kildare, from 13th-17th June 1989. In Dublin one frequented Castleknock from mid-May to late August 1990 and two were seen together on one occasion in August. There was one at Lough Atedaun, Co. Clare, on 9th May 1992 and the last Irish record was a male in song at Tacumshin, Co. Wexford, from 18th-22nd July 1996.

Wales has just five records, the first of which was on Skokholm on 31st October 1968. The subsequent records comprise a male at Dowrog, Pembrokeshire, on 18th June 1983, singles at

Oxwich Marsh, Gower, from 13th to 20th May 1987, one trapped at Llangorse Lake, Powys, on 4th July 1994 and a male at Malltraeth, Anglesey, from 8th to 11th June 1999.

Discussion: The state of the English breeding population reflects fluctuations elsewhere at the periphery of its range. Recent declines have been observed along the western and southern range limits, though conversely increases have occurred in east Europe and the Baltic states. Elsewhere in northern Europe, breeding took place for the first time in Latvia in 1972 (now 400-800 pairs) and Estonia in 1977 (now 150-300 pairs). In southern Finland singing males appeared during the 1980s, with a pair breeding there in 1984 and about 75 singing males reported in 1986. The Finnish population is now down to a handful of pairs. A similar number is in Sweden. In southern Norway singing males have appeared, but failed to colonise, with females seemingly reluctant to catch up with pioneering males.

During the period it was lost as a breeding species in Britain, the Netherlands retained a healthy population and it seems likely that this was the source of the birds that recolonised southeast England. However, the population in the Netherlands declined by 50-75 per cent from 1965-75 to 1993, with thinly occupied areas losing their entire breeding populations as also happened in Britain. Reasons for the declines are thought to be linked to wintering conditions in the Sahelian belt (Hagemeijer and Blair 1997, BirdLife International 2004).

Return from the African wintering quarters is undertaken via two routes. Those arriving from the southwest via north Africa and Iberia appear earlier than those returning farther east. Passage from the southeast of birds heading for east Europe and Scandinavia continues in mid-May and early June. Birds arriving from a more southeasterly orientation presumably account for later records on Shetland and the east coast.

Paddyfield Warbler Acrocephalus agricola

Ian Wallace

(Jerdon, 1845). In Europe restricted to Black Sea coasts from Bulgaria and Danube delta east to Ukraine. Breeds widely across steppes of southern Russia and southwest Siberia, Kazakhstan, Mongolia and north-west China, south to Uzbekistan and northern Pakistan. Winters throughout Indian subcontinent.

Treated as either polytypic or monotypic. Three subspecies *agricola*, *septimus* and *capistrata* have been recognised (for example, Baker 1997), but the first two are genetically near-identical and all are morphologically non-diagnosable, hence Paddyfield

Warbler is treated as monotypic by the *BOU* (Dudley *et al* 2006). Paddyfield Warbler forms a clade that is well separated from the rest of the reed warbler group, and it is probably a sister species to Manchurian Reed Warbler *A. tangorum* and Blunt-winged Warbler *A. concinens* (Leisler *et al* 1997).

Status: There have been a total of 71 records to the end of 2007, 67 from Britain and four in Ireland.

Historical review: The first was on Fair Isle from 26th September-1st October 1925 and the second was trapped there on 16th September 1953. Sixteen years elapsed before the next, and the first to be seen by many, at Hartlepool from 18th-21st September 1969. Autumn birds in 1974 comprised a long-stayer on St. Mary's from 30th September-15th October and one trapped at Low Hauxley, Northumberland, on 12th October.

A fascinating record from this period includes the bird that was until very recently considered to be the only extant record of Moustached Warbler *A. melanopogon* for Britain. This bird was trapped at Wendover, Buckinghamshire, on 31st July 1965 and was considered to be wearing a little-known plumage of that species (Bradshaw 2000). The Wendover record was subsequently re-circulated around *BOURC* and, after much debate, it was decided unanimously that the bird was not acceptable as a Moustached Warbler as the available evidence made it impossible to exclude an adult Paddyfield Warbler. Thus the last possible Moustached Warbler was removed from the *British List*. Although all the in-hand biometrics and the description supported the identification as Paddyfield Warbler better than Moustached Warbler, the *BOURC* decided that a certain identification of the bird as the former species was not justified (Melling 2006).

As the rapid increase in birdwatching activity through the 1970s had done little to change the status of this subtle warbler, there was little expectation that two one-day birds in 1981 were the start of anything different, other than one of them was an inland bird (trapped at Tring Reservoirs, Hertfordshire, on the late date of 9th November). The decade went on to amass 13 records, with birds occurring in all but one of the years between 1981 and 1988. The first for Ireland was an adult trapped at North Slob, Co. Wexford, on the extremely late date of 3rd December 1982; it died the following day. Most sightings during the decade were brief with the exception of one on Fair Isle for just over a week from late September 1986.

The first of the 1990s was in 1991, since when records have been annual. No fewer than 29 were recorded during the decade, including an inland bird trapped at Thatcham, Berkshire, on

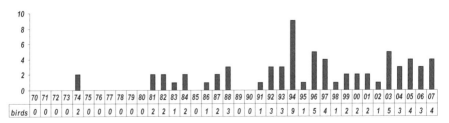

Figure 58: Annual numbers of Paddyfield Warblers, 1970-2007.

7th September 1997. The pinnacle of this change in status was an exceptional nine in 1994, eight of which occurred in localised falls along the east coast. Two were found in Cleveland between the 17th-18th September and then four on Shetland from 22nd-30th September followed by two on Orkney from 22nd-23rd October. Records during the decade were markedly aligned to the axis from the Northern Isles along the east coast. No fewer than 13 came from Shetland and four from Orkney, with seven along the coast from Cleveland to Essex. There were also two in the southeast extracted from mist nets, two from Ireland and a single in Cornwall. A further interesting record from the period involved a 1st-winter that came aboard a fishing boat 409 km southwest of Ireland on 14th September 1993. It was exhausted and caught by hand, but died on 16th; this record lies outside the national recording area and is not included in national totals.

The occurrence pattern has remained similar, with 24 recorded between 2000 and 2007. These included the first for Wales, trapped inland at Llangorse Lake, Powys, on 11th September 2004, and birds trapped at Icklesham, East Sussex on 28th September 2003 and 7th October 2007. During this period there has been a slight shift in emphasis to the southwest, where five have been seen on Scilly and two in Cornwall, along with singles from Hampshire, Powys and Co. Cork. In the southeast, two each have been recorded in Kent and East Sussex. Surprisingly, in contrast with the previous decade, there were just two east coast records (East Yorkshire and Lothian). In the north, Shetland accumulated seven records and one bird reached west to the Outer Hebrides. The reasons for the change are unclear, though many of the records from the west and southwest tend to occur later, having presumably made landfall farther north, and so may reflect an increasing tendency for individuals to head southwest.

Since 1981, the 66 birds have arrived at an average of 2.4 per annum, this statistic increasing to nearly three per annum when only the last 10 years is considered.

Where: As for many species, Shetland leads the way with 25 records, no fewer than 17 from Fair Isle. Elsewhere in Scotland there are four from Orkney and singles in Fife, Lothian and the Outer Hebrides. The southwest is the next best area with 11 records: eight from Scilly and three from Cornwall, seven of which have been since 2001. The northeast has acquired nine records, including a seemingly disproportionate five from Cleveland and two each in Northumberland and Yorkshire. Eight have reached the southeast, with three trapped at Icklesham, East Sussex, two from Kent and singles in Hampshire, Berkshire and Hertfordshire. Five East Anglian birds comprise two each in Norfolk and Essex and one in Suffolk. The sole record from the northwest is from Cumbria and the Welsh record is from Powys. In Ireland birds have reached Co. Cork twice and once each to Co. Wexford and Co. Donegal.

When: Spring records are scarce, though nine have been found between the earliest on 30th May (Fair Isle in 1984) and 13th June (seven on Shetland and singles in Fife and Suffolk). This pattern is consistent with overshoots from southeastern Europe. One trapped at Holm, Orkney, on 18th July 1994 is less easily explained, but it may have involved a summering bird that had made landfall on the islands earlier in the spring.

In autumn a 1st-winter trapped at Kilnsea, East Yorkshire, on 13th August 2006 is almost a month earlier than the next earliest at Thatcham, Berkshire, on 7th September 1997, a record that is also exceptional in location. Could the Yorkshire bird have originated farther east than most other records? Birds around the Black Sea begin to migrate

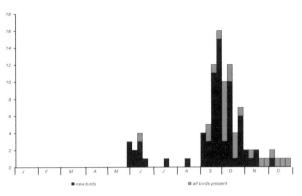

Figure 59: Timing of British and Irish Paddyfield Warbler records, 1925-2007.

in mid August. Migration from Siberia occurs much earlier, with adults departing in late July and young later (*BWP*). A further 33 records are crammed into the rest of September. The September records have the typical distribution of many westbound vagrants with Shetland (15 records) and along the east coast (three of which were in Cleveland) dominating proceedings. Three were found on Scilly and others in Cumbria, Co. Donegal and the Outer Hebrides.

The 13 in the first two weeks of October have more of a south coast bias, five between East Sussex and Scilly, two in Co. Cork, just three on Shetland and singles in Lothian, Northumberland and Essex. Eight in the rest of October include three on Orkney, two each on Scilly and Cleveland and one in Essex. Four in November include two in Cornwall (Cot Valley 15th November 2001 and Marazion from 16th November to at least 28th December 1996), singles on Scilly (Gugh, 1st to 6th November 2002) and an inland bird in Hertfordshire (Tring Reservoirs on 9th November 1981). An adult trapped at North Slob (Co. Wexford) on 3rd December 1982 died the following day, but represents the latest detection for the species (though one can only guess where and when it made original landfall). The pattern of records suggest initial arrivals along the east coast and Northern Isles followed by spill down to the southwest.

Discussion: The increase in records is associated with several factors. Increased observer coverage will have presumably played its part, especially on Shetland. The European range has spread northwest since about 1975 and the European breeding population increased between 1970 and 1990. Sizeable populations in Romania and Russia were stable or increased during 1990-2000 (BirdLife International 2004). Associated with the recent range expansion, breeding was first noted in Latvia in 1987 and Finland in 1991 (*BWP*; Nankinov 2000). In Russia breeding in the Kirov Region northeast of Moscow and over 1,000 km north of the main breeding range was first noted from 1995 (Sotnikov 1996). By the late 1990s, the northern limit of the breeding range ran from the east bank of the Volga at 50°45'N east to the border with Kazakhstan at 51°20'N (Yakushev *et al* 1998).

More pertinent to British observers is the successful breeding on Vlieland, the Netherlands, in 2007. Three (adult and two juveniles) were trapped and ringed on 21st August and another juvenile was trapped on 28th August 2007. The three juveniles must have fledged shortly before they were ringed and it was assumed that breeding took place at the site (Ovaa *et al* 2008). Spring records for Britain, bar one at Landguard, have been on offshore islands. There is no hint yet of birds in suitable breeding habitat, but this could be a breeding species of the future for Britain.

Despite the relatively small sample there have been some interesting movement data from birds ringed or controlled in Britain. A 1st-winter trapped on Fair Isle on 19th September 1996 had been ringed in Lithuania on 8th September 1996 as a Blyth's Reed Warbler *A. dumetorum*. In 2003, a 1st-winter trapped on Foula, Shetland, on 29th September remained on the island to 8th October. The bird moved 45 km in an east-northeast direction and was relocated at Kergord, Shetland, from 11th to 12th October.

The number of inland records is impressive, with birds trapped in Essex, Berkshire, Hertfordshire and Powys. Also of note are three trapped in the reed beds at Icklesham in East Sussex. This represents a high proportion of records intercepted by mist nets at inland sites. Many others must pass un-noticed in reed beds across southern Britain.

Blyth's Reed Warbler *Acrocephalus dumetorum*

Blyth, 1849. Breeds in southern Finland, Estonia, Latvia and European Russia to 64° N. To the east, found across central Siberia to Lake Baikal and upper Lena River, south through Kazakhstan and Tajikistan to northern Pakistan. Winters Indian subcontinent south to Sri Lanka and east to northwest Burma.

Ray Scally

Monotypic. The taxonomic position of Blyth's Reed Warbler is unclear. Both Leisler *et al* (1997) and Helbig and Seibold (1999) found it lay outside the main Reed Warbler clade – Marsh Warbler *A. palustris*, Eurasian Reed Warbler *A. scirpaceus* (including Caspian *A. s. fuscus*), African Reed Warbler *A. baeticatus* and Mangrove Reed Warbler *A. avicenniae* – despite occasional hybrids with Marsh Warbler *A. palustris*. Blyth's Reed Warbler is closely related to the recently rediscovered Large-billed Reed Warbler *A. orinus* (Bensch and Pearson 2002; Round *et al* 2007).

Rare, but increasingly regular vagrant from east Europe and west Asia. More spring birds encountered in recent years, but most occur in late September and early October.

Status: There were 89 British records to the end of 2007. The first for Ireland, in 2006 (revealed by photographs, but not yet accepted formally), was followed by two more in 2007. This is yet another rarity that is undergoing a rapid change in status; formerly revered, now expected.

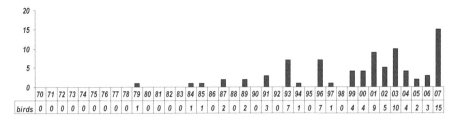

birds	70	71	72	73	74	75	76	77	78	79	80	81	82	83	84	85	86	87	88	89	90	91	92	93	94	95	96	97	98	99	00	01	02	03	04	05	06	07
	0	0	0	0	0	0	0	0	0	1	0	0	0	0	1	1	0	2	0	2	0	3	0	7	1	0	7	1	0	4	4	9	5	10	4	2	3	15

Figure 60: Annual numbers of Blyth's Reed Warblers, 1970-2007.

Historical review: The first was on Fair Isle from 29th to 30th September 1910, when it was shot. In 1912 no fewer than seven birds were obtained, an amazing number for such a rare warbler at the time. These comprised birds shot at Spurn on 20th September and Holy Island on 25th September, plus sightings on five dates from Fair Isle between 23rd September and 1st October (four were shot). The final record of this extraordinary influx was one found dead on the Dudgeon light vessel, 30 km off Wells, Norfolk, on the night of 20th-21st October. The next, and final bird prior to the present recording period, was shot on Fair Isle on 24th September 1928.

There followed a 51-year wait until the first of modern times, when one was trapped on Holm, Orkney, on 5th and 13th October 1979. A spring male at Spurn on 28th May 1984 was the first to enjoy mass appeal, providing an unexpected bonus for birders heading to the site for a late May fall of migrants, including a Thrush Nightingale *Luscinia luscinia*. The sense of disbelief at the appearance of such a rarity was widespread through the crowd that day, but this bird was to set the trend for subsequent years: this subtle *Acro* has failed to make the national archive in only six years since.

At 0700 hrs. on 28th May 1984 B.R.Spence heard an unfamiliar song from a thick tangle of Sea Buckthorn on the east side of the Point area. For the next 30 minutes, during which he was joined by N.A.Bell and M.G.Neal, the bird continued to sing. Eventually, it gave brief views in flight and, shortly afterwards, equally brief views at rest, when all that could be noted was that it appeared to have olive-grey upperparts, a poor superciliary stripe, a stoutish bill and a steep forehead. All three observers, tentatively suggested that the bird might be Olivaceous Warbler, though N.A.B., who had heard the bird in Israel, thought that the song did not ring true for Olivaceous. The song was rather slow, mimetic and with some quite melodic 'pwee-pwee' notes and more scratchy Acrocephalus/Hippolais "material". It seemed the only way to identify it was to catch it. A mist-net was set up and, by 0800 hrs, the bird was in the hand.

The problem of the bird's identity was not solved immediately it was caught. The tail shape was checked and seemed to be square, even though it was wet. Back at the Warren, this fact did not help the initial inquiries which were on Hippolais warblers in the B.T.O Identification Guide no 1. It looked as if it should be Olivaceous Warbler but the wing formula was not right. Upcher's Warbler was too big. Could it be Booted Warbler? Then, while Svensson's Identification Guide to European Passerines (third edition) was being checked, again for Hippolais, a footnote was seen to refer to the possibility of difficulty in separating the SE race of Olivaceous Warbler, H. pallida elaeica from Blyth's Reed

Warbler. Was the bird actually an Acrocephalus and not a Hippolais? The tail, now dried out was found to be rounded though not obviously so as it was slightly abraded. The third primary was checked and found to be slightly notched on the inner web. This notch and the one on the second primary were measured; these together with the very short first primary clinched the bird's identity as Blyth's Reed Warbler

When the bird had been "processed", photographed and exhibited to the surprisingly large number of visitors who had "materialised" during the deliberations in the Ringing Laboratory, the question was where to release the bird. Ideally, a migrant should be released where caught but, in this case, this was out of the question if the waiting watchers were to have any chance of seeing the bird in the field. It was decided to release the bird into the hedge and bushes at the end of the 'canal zone' where the Thrush Nightingale caught on 23rd was still claiming plenty of attention, where there seemed more chance of the Blyth's Reed being seen and where it might, hopefully start to sing again. But, this was not to be. Unfortunately, as soon as the bird was released towards the west side of the hedge, most of the watchers present rushed right up to the hedge. Not surprisingly, the bird was lost at once. Followed by a line of people, it presumably went straight through the hedge and away. It was re-located about two hours later near Warren Cottage from where it was "pushed", by the ever-increasing numbers of visitors, as far as 'big hedge' where it apparently remained for the rest of the day. It is not known whether anyone had more than the briefest views in flight or rest.

Barry Spence. Spurn Birding website (www.spurnbirdobservatory.co.uk)

This ground-breaking bird cleared the way for others. One was trapped in Shetland in the following year. Two were trapped in both 1987 and 1989, one of them the second spring record, on Portland on 12th June 1989. Seven had been added in just 10 years, and its rarity status was starting to erode rapidly.

The 1990s continued in a similar vein, eventually adding 23 birds. Seven in 1993 equalled the record set more than 80 years earlier: one in spring and six in late September and late October, including the simultaneous arrival of birds in Suffolk and the Outer Hebrides; all were trapped with the exception of one found dead. One ringed on Fair Isle on 22nd October was relocated at Sumburgh, Shetland, from 23rd-31st October, when it was killed by a cat. Seven in 1996 was another good showing, and included two on Fair Isle in early June and four along the east coast between 21st-25th September and a mid-October find.

Notable in this flurry of records were two birds that finally laid to rest the notion that this species had to trapped to be identified (Golley and Millington 1996; Thomas 1996). Well watched birds at Filey, North Yorkshire, and Warham Greens, Norfolk, were documented well enough to prove that field identification – previously considered nigh on impossible – was achievable given good views. These two birds became the first to be accepted without trapping or death.

There has been no let up in the pace of records since 2000, with 53 recorded from 2000-2007. The last blank year was as long ago as 1998. Three spring birds in 2000 included the first long-staying singing male, at Nigg Ferry, Highland, from 7th-22nd June. It was followed later in the same year by the first record away from the east coast when one was trapped inland at Woolston

Eyes, Cheshire. A new record influx took place in 2003 with 10 birds recorded: two in spring and eight autumn birds in Scotland (two mid-September and six between 26th September and 5th October). The new record was surpassed by 15 birds in 2007 (three in spring and the rest 29th September-13th October). Ireland got in on the act with the second and third records: Mizen Head, Co. Cork, from 10th (possibly 8th) to 15th October and The Mullet, Co. Mayo from 10th to 14th October. The first record has since been identified from photographs taken on Cape Clear on 20th October 2006.

Where: Most records have been on Shetland (38 birds including 21 on Fair Isle) and Orkney (eight records). Most eastern counties have accommodated the species at some point, but it remains a quality find on the mainland. Northumberland and Yorkshire lead the way for the mainland with eight apiece. One or two have been found in each of the counties between Lincolnshire and Kent. On the west coast three have reached the Outer Hebrides and four have filtered south to Portland. Singles have reached Wales (Bardsey in 2001) and Scilly (in 2002). Records along the south coast are few in number and tend to appear much later than northern records, suggesting that southern birds have trickled down after landfall.

Inland birds are rare with just three such records: trapped at Woolston Eyes, Cheshire, on 26th August 2000; Canary Wharf, London 6th-28th October 2001; and a male in song at Fisher's Green, Essex on 16th June 2003. The Canary Wharf individual is surely one of the most urban of rarity records.

When: Spring records are scarce but increasing, most presumably hailing from the expanding Scandinavian population. A total of 18 such birds (seven on Shetland) are spread between 11th May (Portland in 2004) and 16th June. Spring records away from Shetland are rare, but include two each from Spurn and Portland plus singletons on Orkney and in Highland, Norfolk, Essex and Kent.

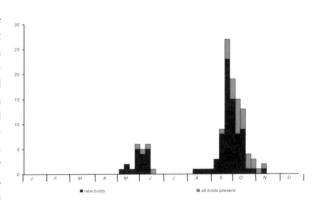

Figure 61: Timing of British and Irish Blyth's Reed Warbler records, 1910-2007.

As would be expected, this species appears most often in autumn. The earliest was trapped at Noss, Shetland on 14th August 1985 and is one of just three August records (all trapped). The peak in sightings falls between 24th September and 6th October (38 birds), followed by a further small peak between 15th and 21st October (nine birds). Just three have occurred after this date, two in late October. The latest was trapped at Portland on 12th November 2001 (the second at the site that autumn after one from 16th October-3rd November).

254

Discussion: In many ways, this species epitomises the huge advances that have been made in field identification for many tricky species since the mid-1980s. It was once impossible to identify without the aid of a mist net, but field identifications are now routine. Modern rarity-finders are tuned in to listening for the distinctive *Sylvia*-like call and looking for subtle differences in plumage and wing structure. Occasional birds still defy consensus, such as one on Unst, Shetland, in autumn 1997 (for example, Bradshaw 2001).

The increase in spring records over recent years reflects an increasing number of overshoots from the expansion of the breeding range north and west into the Baltic States and Finland. Breeding was first proven in Estonia in 1938 (now 2,000-4,000 pairs), Latvia in 1944 (now 3,000-6,000 pairs) and Finland in 1947 (now 5,000-8000 pairs). A very small population is established in Sweden (breeding first occurred in the 1970s). In 2007 a record 15 singing males were present in Norway, all but two of them in the southeast (Hagemeijer and Blair 1997; BirdLife International 2004). In the Netherlands, a successful pairing with a Marsh Warbler *A. palustris* occurred in 1998.

Most British records occur later than the main departure from the western part of the breeding range (vacated in July and August). It is likely that many autumn birds are from farther east. Records tie in with migration dates through south-central Siberia and southwestern central Asia.

Those at Spurn on that overcast day in 1984 would have not realised that the bird before them was the forerunner of a change in fortunes for what was, up to that point, an extreme rarity. With its suite of identification features now well appreciated by rarity finders, plus a burgeoning European breeding population, there seems every reason to expect this challenging *Acro* will become even more frequent in the coming years.

Great Reed Warbler *Acrocephalus arundinaceus*

(Linnaeus, 1758) Breeds discontinuously through continental Europe from Iberia to Greece, north to southern Sweden and Finland and east across southern Russia, Turkey and Caucasus to western Siberia. A. a. zarudnyi breeds in central Asia from Volga to northwest China. Winters central and southern Africa.

Polytypic, two subspecies. Records here are attributable to nominate *arundinaceus* (Linnaeus, 1758).

Regular overshoot from southern Europe (some may be from Scandinavia). Spring records are the norm. Pioneering males often put in extended summer stays.

Status: There have been a total of 230 birds: 227 from Britain and three from Ireland; since 1950, 219 in Britain and two in Ireland.

Historical review: The first British record, shot near Swalwell, Co. Durham, on 28th May 1847, was followed by five more from 1853-1900. The next was the first for Ireland, found dead at Cosheen, Castletownshend, Co. Cork, on 16th May 1920. Two more occurred prior to the

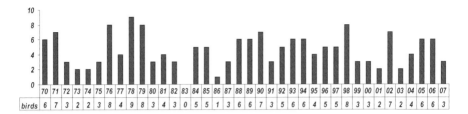

Figure 62: Annual numbers of Great Reed Warblers, 1970-2007.

modern recording era, from Northamptonshire in June 1943 and a one-day bird in Hertford-shire in April 1946.

Seven followed in the 1950s, including two males in song at Rye, East Sussex, in 1951 (one from 24th May to 28th July and another on 10th June). All records bar one during the decade were found between Dorset and Suffolk, the exception was one on Shetland.

The 1960s added 40 more to the total, records coming in each year except 1968. A bird at North Warren, Suffolk, from 9th June to 2nd July 1961 was joined by a second bird on 12th. The five recorded in each of 1960, 1963 and 1965 were eclipsed by a record nine in 1969. This arrival comprised seven spring birds between early May and mid-June. The only birds to linger were June visitors to Dorset and Norfolk and one-day birds in August and October in the southwest.

The 1970s kept up with the pace, producing 52 birds, eight each in 1976 and 1979 and a record-equaling nine in 1978. These comprised mostly short-stay birds with the exception of a male at King's Lynn, Norfolk, from 23rd May-10th June 1978.

Pickings in the 1980s were comparatively lean: just 36 birds and with 1983 the only blank year in the last 40. The yearly totals for the decade included six birds in each of 1988 and 1989. There was a simultaneous arrival of inland birds in East and North Yorkshire on 19th May 1984. Three 1988 birds arrived on Fair Isle between 27th May and 3rd June

The 1990s equalled the 52 birds of the 1970s. Notable among these was a male at Elmley, Kent which not only sang from 27th May-3rd July 1993 but also returned to the same patch of reeds from 12th-25th June 1994. It was seen carrying nest material during its residence. Another (or the same) was there from 30th May-16th June 1996. Consecutive appearances at the same site are not unusual in this species. Breeding has never been proven, but there remains the intrigu-ing possibility that some of the long-staying males have paired up with a secretive female.

The period 2000 to 2007 has added a further 33 birds, slightly down on the previous decade, but still indicative of healthy numbers of overshoots to our shores. The annual average since 1960 is roughly 4.5 birds per annum, and since 1990 this is marginally higher at 4.7.

Where: Most records are in southeast England, primarily the coastal counties between Nor-folk and Sussex (with just over one third of all records). Kent leads the way with a relatively whopping 35 birds. The only other English counties with double-figure aggregates are Sussex (16 birds), Norfolk and Suffolk (13 each) and Devon (10).

The region with the second-highest tally is Shetland (29 birds). It has been suggested that some of the spring records in the Northern Isles are associated with easterly drift conditions and presumably involve birds heading for the relatively small Scandinavian breeding populations or larger populations in Russia. However, the often simultaneous arrival of birds on Shetland with those elsewhere suggests that spring records on the islands are equally likely to be overshoots from the declining European populations.

There have been records from many inland counties, though few have been recorded in the northwest. The paucity of Irish records is perhaps surprising.

When: The earliest record was at Slapton, Devon, on 13th April 1981, two weeks earlier than the next, at Marsworth Reservoir, Tring, Hertfordshire, on 27th April 1946. Following the review period an exceptionally early bird was on St. Mary's, Scilly from 23rd March-4th April 2008, three weeks earlier than any previous sighting.

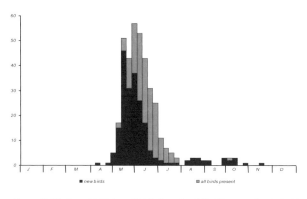

Figure 63: Timing of British and Irish Great Reed Warbler records, 1950-2007.

A large proportion of records feature singing males. Arrivals gather momentum through mid May and peak in the last week of May and early June. Males at this time may put in extended stays. After mid-June, records tail off, but a few birds have been found in each week through July.

Autumn birds are scarce or overlooked, with finders no longer guided by the male's grating song. Most records at this season involve birds extracted from mist nets during August and the first week of September.

Later autumn records are rare, with just eight between 24th September and 12th October, all one-day birds except a two-day bird at Dungeness (4th-5th October 1998) and a lingerer at Hengistbury Head from 24th September-2nd October 1989. Just three have occurred later: one trapped at Dunwich, Suffolk, on 14th October 2006; one trapped at Big Waters Nature Reserve, Northumberland, on 24th October 1990; and an exceptionally late individual trapped at Thurlestone, Devon, on 15th November 1972. All late records have been coastal and five have been between Shetland and Kent. Could these later birds originate from the eastern parts of the range?

Discussion: Despite increased observer coverage, there has been no noticeable change in trend since the 1960s. This is surprising given a decline in breeding numbers in many parts of Europe (BirdLife International 2004). With its loud and familiar song, a vocal male in spring and summer must be one of the easiest rarities to find. Silent birds are another matter, and many must must pass by undetected in our reed beds.

Thick-billed Warbler *Acrocephalus aedon*

Ian Wallace

(Pallas, 1776). Breeds in southern Siberia from Ob basin and northern Mongolia to Ussuriland and northeast China. Winters from Nepal east through northeast India to Indochina and central Thailand.

Polytypic, two subspecies. The racial attribution of occurrences is undetermined.

Very rare vagrant from Asia, typically in early autumn.

Status: Shetland has accounted for all four British records of this unusual *Acro,* one in May and three between 14th September and 6th October.

1955 Shetland: Fair Isle trapped 6th October
1971 Shetland: Whalsay 1st-winter trapped 23rd September, released Lerwick, 24th September, found dead 25th September
2001 Shetland: Out Skerries 1st-winter trapped 14th September
2003 Shetland: Fair Isle adult 16th-17th May, trapped 16th

Spring records of major Siberian rarities remain exceptionally rare, but such occurrences are increasing slowly.

> *Spring 2003 migration had been very slow, but all that was to change very early on 16th May.*
>
> *I was woken by a knock on the bedroom door at 5.30am: "Deryk, I need you in the ringing room". It was Alan Bull. "Okay!" I replied sleepily, sitting up immediately. "Don't wake the children!" came a sudden shout from my side. "Doesn't that boy ever sleep?" I moaned, looking at my watch. However, I knew it must be important – Alan would not wake me this early for anything trivial – he values his health too much!*
>
> *I left as quietly as I could and trudged down to the ringing room with conflicting thoughts running through my head: either he had caught a very good bird or (more likely) some catastrophe had happened – perhaps the ringing room ceiling had caved in and the guests in Room 12 were lying on the floor?*
>
> *I slowly opened the door and was pleased to see that all was intact, but was surprised to see Glen Tyler (ex Fair Isle Bird Observatory Assistant Warden and now island resident) standing there with AB. "Hi Glen. What's going on?" I asked, still puzzled but sure that bad news was imminent. I then spied the bird bag hanging on the rail and sensed the excitement of the two (presumed) insomniacs.*
>
> *"It must be a good bird", I thought, but they would not let on what it was! As they grinned at me, I took the bag down and put my hand in. The bird felt chunky, with large*

legs and feet. "Is it a crake?" I asked "No! No! We're not that bad" came the reply. I pulled it out in the ringer's grip. It was a huge warbler! "What is it?" asked Glen. It was the middle of May, I had heard that there had been one at Dungeness the day before and I was still half-asleep. "Brilliant! It's a Great Reed Warbler!" I declared. "Think again!" said Bully "Think rarer!" piped Glen. I held it up and began to wake up. I noted the olivegrey colour, the long graduated tail and the rose-coloured starling-like head. My heart rate quadrupled. "No! No! It can't be! I began to fumble for Svensson (1992) as the other two leapt around the room. Glen calmed down enough to help me, as I was so delirious I could not find the correct page. "I've measured it and it all fits," said Bully gleefully. I had to check for myself: 2nd P – 7/8; emarginated on 3rd, 4th and 5th wing point 4th primary (just); long first primary; wing length 82mm. Yes! Incredible! It is one – a spring Thickbilled Warbler!

Apparently, Glen had risen very early to watch the lunar eclipse that was due at 4.00am. After the show (he had taken some stunning photographs) he could not get back to sleep, so had decided to go for a wander to check some flowers near the Plantation. He noticed a large warbler on the fence by the Vaadal reservoir. Not knowing exactly what it was, he flushed it into the trap. He brought it to the bird observatory and, not wishing to risk waking my children, had woken Alan Bull with the news "I think I've got a Great Reed Warbler". The two deliberated over the bird for a few minutes, measuring the re-measuring it, before deciding that it was, in fact, a Thick-billed Warbler and they had better wake me! I finished processing the bird, taking a few more measurements (tail length, bill length/depth and 1st primary length) as Alan woke everyone else in the observatory. A bleary-eyed, disbelieving crowd soon assembled, a quick description was written, photographs were taken and the bird was released into the new plantation outside the observatory. Glen received a big hug and a pat on the back and I gleefully went to telephone the news out. The enormity of it all began to sink in – the frantic phone calls about to ensue and the planes and boats which were likely to follow.

The bird showed on and off, sometimes incredibly well, for the rest of the day, but it became more skulking and elusive late in the afternoon. Two planeloads and one boatload, some 25 twitchers in all – made it in to the island and all eventually saw the bird.

By mid afternoon, I had not had a chance to get out all day and by 4.00pm I was itching to get out, so I decided to go for a pre-dinner wander round the southwest of the island. Arriving at Shirva, a pipit jumped up onto the fence – a smart Red-throated Pipit! This became an added bonus (as if it were needed!) for all the birders present, but they also had a very confiding and stunning male Red-spotted Bluethroat and a fine male Red-backed Shrike thrown in. The weather was also glorious all day; calm, warm and sunny.

That night, the telephone continued to ring as the weather began to close in. The next morning dawned with blanket fog and a strong easterly wind. The MB Good Shepherd IV could not sail and the prospects for planes were not good. There was no sign of the bird around the observatory. It began to rain and the telephone continued to ring as anxious birders awaited news on a) the bird and b) the weather prospects for getting across – neither of which were very happy! The observatory staff headed out on the regular

migrant census nevertheless and, at 11am, Assistant Warden Rebecca Nason telephoned my mobile to say that she had relocated the Thick-billed Warbler towards the southern end of the island, in the Meadow Burn. It was "looking very bedraggled, but active". I immediately telephoned the observatory and asked them to put the joyous news out! All was in vain, however, as the bird could not be found again later and, although the two scheduled Loganair flights managed to get in (both of which were already full with a Shetlands Wildlife Tours group), unfortunately no others managed to.
Deryk Shaw. Birding World 16: 206-208

Discussion: There are just four other Western Palearctic records: Egypt, seen at St. Katerine Monastery, Sinai, 20th November 1991; Finland, 11th October 1994; and two from Norway (6th October 2004, 3rd October 2005). Next-nearest are two records from Oman (11th-13th April 2001, 23rd December 2007).

Although all British records have been brief, it is encouraging that, following a 30 year absence, two arrived at the turn of the century. These, and the recent Norwegian records, suggest that the next might be imminent. As with many mega Sibes, the trend would appear to be for more records. Our birds perhaps signal a distant change in range or population for this large skulking warbler.

At first glance, the dates in autumn would appear to be early for a rarity of such magnitude. However, the breeding grounds in the northwest of the range are vacated by August and arrival in the westerly wintering areas occurs from September onwards. The spring bird in 2003 most probably involved a bird that wintered somewhere in Europe or western Africa, rather than an overshoot.

Eastern Olivaceous Warbler *Hippolais pallida*

(Hemprich and Ehrenberg, 1833). Breeds throughout Balkans from Croatia to Greece and Turkey, southern Caucasus, southern Kazakhstan, Uzbekistan, Iraq, Iran and northern Afghanistan. Migrates through Middle East to winter in East Africa.

Polytypic, four subspecies. All records have involved the race *elaeica* (Lindermayer, 1843) breeding from the Balkans east to Iran and Kazakhstan. The *BOU* recently endorsed the view that Olivaceous Warbler should be split into two species on the basis of morphology, plumage characters, vocals, behaviour and DNA analyses: Eastern Olivaceous Warbler, *H. pallida* (polytypic, sspp. *reiseri* Hilgert, 1908, *laeneni* Niethammer, 1955, *pallida* (Hemprich and Ehrenberg, 1833), *elaeica* (Lindermayer, 1843)) and Western Olivaceous Warbler *H. opaca* (monotypic) (Knox *et al* 2002). The DNA sequences of *elaeica* and *opaca* differ as much as that between Icterine Warbler *H. icterina* and Melodious Warbler *H. polyglotta* (Parkin *et al* 2004).

The four races of Eastern Olivaceous Warbler might yet be better treated as two species (Svensson 2001). Molecular data have not yet been published for *pallida*, *reiseri* or *laeneni*, and their phylogenetic affinities with *elaeica* remain uncertain; *elaeica* appears to be slightly different from African *pallida*, *reiseri* and *laeneni* (Parkin *et al* 2004). The grey-brown race *elaeica* breeds from the Balkans to southwest Asia and winters in East Africa.

The three remaining races are found locally in North Africa. They are paler than *elaeica* and form a very similar but distinct group. *H. p. reiseri* is found in the Algerian Sahara, southern Morocco, Mauritania and Libya and may overlap in range with Western Olivaceous Warbler. The exact wintering areas of *reiseri* are inadequately known because of its similarity to *laeneni* and *pallida*, but may lie between Senegal, Mali, Niger and northern Nigeria, with a suggestion of differential migration of first-year and adult birds. *H. p. laeneni* occurs in the northern parts of Niger, Chad and Nigeria to western Sudan. Nominate *pallida* breeds in Egypt and winters in southern Sudan and Ethiopia (see Salewski *et al* 2005 and Salewski and Herremans 2006 for detailed discussion of the migration phenology of *H. opaca* and *H. p. reiseri* in the western Sahara). Western Olivaceous Warbler and *reiseri* have sympatric or parapatric ranges, but hybrids or intermediates between the two are lacking in museum collections and they are treated as separate species (Svensson 2001).

Phylogenetic analyses of mitochondrial DNA sequences indicate that the Olivaceous Warblers, Booted Warbler and Sykes's Warbler are most closely related to species traditionally included in Acrocephalus rather than *Hippolais* (Leisler *et al* 1997; Sangster 1997). As a result the Dutch CSNA has placed these four species within the genus *Acrocephalus* (Sangster *et al* 1999). That treatment has not yet received support here. A review by the *BOURC* retained these species within *Hippolais*, but acknowledged that the genus fell into two groups (Knox *et al* 2002).

Status: There are 16 records (including two pending from 2008), 13 in Britain and three in Ireland.

1967 **Fife:** Isle of May trapped, 24th-26th September, when killed by a Great Grey Shrike
1967 **Kent:** Sandwich Bay trapped 27th September
1977 **Co. Cork:** Dursey Island trapped 16th September
1984 **Scilly:** St. Mary's 16th-26th October
1985 **Scilly:** St. Mary's 17th-27th October
1995 **Shetland:** Fair Isle 5th-13th June, trapped 5th
1995 **Suffolk:** Benacre 12th-13th August
1998 **Scilly:** St. Agnes 24th September-8th October
1999 **Dorset:** Portland trapped, 4th-5th July
1999 **Co. Cork:** Cape Clear trapped, 18th September-9th October
2000 **Northeast Scotland:** Collieston 1st-winter 13th-21st September, trapped 15th
2002 **Shetland:** Sandwick/Hoswick, Mainland adult 18th to at least 28th August, trapped 18th
2003 **Dorset:** Portland 1st-winter trapped 31st August
2006 **Co. Cork:** Cape Clear 24th September-1st October
2008 **Dorset:** Portland 17th May, trapped*
2008 **Shetland:** Foula 23rd-26th September*

Discussion: A review of records by the *BBRC* in 1999 found a number of older records, including two Western Olivaceous Warbler *H. opaca*, to be unacceptable. A similar review by the *IRBC* in 2003 also found an older record to be unacceptable. All acceptable records were considered to belong to Eastern Olivaceous Warbler H. *pallida*.

Elsewhere in northern Europe records are rare. From Scandinavia there are three from Finland (4th-5th October 1983, 1st-17th June 1996, 3rd-4th July 2007), three from Sweden (26th August-2nd September 1993, 6th July 1997, 9th-10th October 2004) and singles from Denmark (1st-2nd June 2003), Norway (trapped 12th September 2004) and Iceland (15th-22nd September 2008).

In northern Europe there are three from Germany (two from Helgoland: 20th September 1883, 1st October 1936; and one caught near Berlin on 29th October 2000 which was released on 30th and identification confirmed by DNA analysis), three from Austria (9th May 1998, 9th September 2000, 29th May 2005) and one from Belgium (trapped 22nd September 2005). There are three records from France (6th September 1961, 22nd May 2001, 24th July 2001), plus two further Eastern or Western (13th April 1990, 14th May 2000). There is also a record of Eastern or Western from Estonia (30th June 2004).

Interestingly, there is a record of Western Olivaceous Warbler *H. opaca* from Sweden (trapped 25th September 1993) and three from France (21st May 1960, 1st May 2004, 19th May 2005). These records indicate that there is still a chance for this species, which lacks the classic tail-dip of *elaeica*, to reach Britain and Ireland.

Unusually for a rare warbler it exhibits an erratic spread of arrival dates in Britain and Ireland. There are singles in May, June and July, three in August, eight between 13th-27th September and two in mid-October. The spring records probably involve overshoots (*elaeica* do not arrive back on southeast European breeding grounds until May). Departure for wintering areas takes place from mid July to late September. British and Irish records during the autumn would appear to involve migrants on a northwesterly orientation. Conveniently, several have lingered for extended stays; the three birds on Scilly, plus the Benacre individual, have all attracted crowds, as did the long-staying bird in Northeast Scotland.

Hippolais identification used to be rather more straightforward, but nowadays an expanded number of trickier possibilities need consideration during the ID process. The early individual at Portland from 4th-5th July 1999 was noteworthy. The August birds at Benacre in 1995 and Portland in 2003 offer further guidance to would-be finders that the late summer period may be a good time to seek out an encounter with this pallid warbler.

Booted Warbler Hippolais caligata

(M.H.C Lichtenstein, 1823). Breeds central Russia and western Siberia to Yenisey valley, central and northern Kazakhstan to western Mongolia and western Xinjiang province, China. Range expanding west to southern Finland. Winters northern and peninsular India south to Karnataka.

Monotypic. Previously treated as conspecific with Sykes's Warbler *H. rama*. A *BOU* review of molecular, morphological and vocal

evidence indicated that these taxa should be treated as separate species (Knox *et al* 2002). The Dutch presently consider Booted Warbler to be within *Acrocephalus*; see Eastern Olivaceous Warbler above.

Rare, but regular, migrant from northeast Europe or Asia, mostly in the early autumn.

Status: There have been 115 records to the end of 2006, 111 in Britain and four in Ireland.

Historical review: The first was a female shot on Fair Isle on 3rd September 1936. Thirty years later the first of the modern recording period, another on Fair Isle from 28th August-17th September 1966, was quickly followed by one on St. Agnes on 23rd October 1966. The last of the 1960s was trapped on Fair Isle on 8th September 1968. It was accepted as a Booted Warbler, but is now considered to have been either Booted or Sykes's Warbler *H. rama*.

The next came from the Isle of May in 1975, followed by Shetland records in 1976 and 1977. The first for the mainland was trapped at Kilnsea, East Yorkshire, on 9th September 1978. Since this run of finds there have been just four years with a blank return.

In the 1980s, 20 birds (including a then record four in both 1981 and 1987) created a rapid change in status. The southwest accounted for the bulk of records, with six on Scilly and three in Dorset. Three graced Kent and four singletons made landfall between Norfolk and North Yorkshire. There were just two in Shetland and one in Orkney.

As with many other eastern species, the 1990s produced a relative glut of records. No fewer than 50 were found during the decade. Off the back of an exceptional spring influx of eastern drift migrants in 1992 came the first spring records with three birds over a 10-day period in mid-June from Cleveland, East Yorkshire and Cumbria. Another prolonged easterly airflow in autumn 1993 revised the record books again: no fewer than 13 arrived, with 11 between 7th and 27th September. Records were spread along the east coast between East Sussex and Northeast Scotland. One on Skokholm from 25th-28th September was the first for Wales. A Booted Warbler ringed at Spurn on 16th-17th September 1993 was recovered in Belgium on 5th October. A further spring bird was in song at Beachy Head on 5th June 1994. Seven in 1994 and eight in 1998 were creditable annual counts.

The period 2000 to 2007 continued in a similar theme with 38 birds noted, including the first for Ireland, on Tory Island, Co. Donegal, on 27th September 2003 part of another good

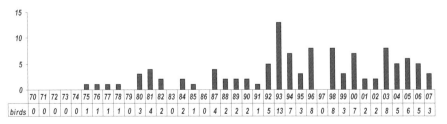

Figure 64: Annual numbers of Booted Warblers, 1970-2007.

arrival. The occurrence of three more in Ireland in recent years is interesting given the paucity of records in western Britain.

Once a great rarity, there have been roughly 4.6 Booted Warblers per annum since 1987. Its occurrence is now an expected opening act of each early autumn period.

Where: Despite increased numbers, this remains primarily a Shetland (31 birds) and east coast specialty. Yorkshire boasts 15 birds, though sightings are curiously rare elsewhere in the northeast with four from Cleveland, three from Northumberland and two each from Co. Durham and Lincolnshire. The Spurn area accounts for nine birds, second only to Fair Isle's ten.

In the southwest, the Isles of Scilly has amassed 14 birds, Dorset five and Devon one. The southeast has fared well, with eight from Kent and four from Sussex. In contrast, there have been only seven in Norfolk and one from Essex.

The sole record from the northwest was from Cumbria. Surprisingly, two have penetrated west to Argyll and one to the Outer Hebrides. There have been four offshore Welsh records (two from the islands of Pembrokeshire and two from Bardsey) and four from Ireland.

When: Five spring records included three during an exceptional continental airflow in 1992, including males at Hartlepool Headland from 7th to 8th June, Spurn from 10th to 22nd June and South Walney on 17th June. These were followed by a further male at Beachy Head on 5th June 1994. A lengthy gap has lapsed before the next sing-ing male at Blakeney Point on 2nd June 2007, the earliest spring record ever.

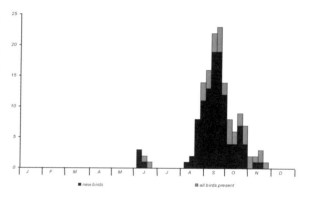

Figure 65: Timing of British and Irish Booted Warbler records, 1936-2007.

The earliest autumn record was at Hartlepool Headland on 10th August 2004. Fifteen further sightings have occurred between 15th-31st August Most records fall in September, the middle two weeks (10th to 23rd) accounting for the lions share. October records are much rarer, with just 19 (10 of them on Scilly). Two exceptionally late records involve singles at Collieston (Northeast Scotland) from 11th to 13th November 1994 and the latest ever, at St Margaret's, Kent, on 12th November 1996.

The increase in autumn records over the past two decades is matched by a shift to an earlier arrival pattern (see *Table 20*).

Table 20: Average arrival dates of autumn Booted Warblers.

Period	Pre 1990 (n=27)	1990-1999 (n=46)	2000-2007 (n=37)
Average Date	24th September	20th September	11th September

Discussion: Prior to 1975, Booted Warbler was an extreme rarity. The dramatic increase since then coincides with the westward expansion of the breeding range across European Russia and into southeastern Finland. It is widely distributed in the Urals/WesternSiberia region, especially in the south. It is rare in the southern taiga subzone and further north; unevenly distributed. There has been an increase in numbers of Booted Warblers on the edge of the range in (Russian) Kareliya and the Leningrad Region, where breeding first occurred in 1997 and was followed by the discovery of two nests in 2000 (Shirokov & Malashichev 2001). Marked fluctuations in breeding density occur (Ryabitsev 2008).

The first record for Finland was a singing male in June 1981. The first territorial male was found in June 1986. Breeding was proven for the first time in 2000, when two breeding pairs in a colony of five males produced 11 fledglings. Breeding is not confirmed every year, but several territorial males are found annually. Most are in eastern Finland and some are probably breeding (Lindblom 2008). Booted Warbler has also become established as a rare breeding species in Estonia from 2002 (Anon 2008).

The increase in records here also correlates with increased numbers of observers and a greater appreciation of what were formerly considered quite subtle identification criteria.

Sykes's Warbler *Hippolais rama*

(Sykes, 1832). Breeds central and southern Kazakhstan to western Xinjiang province, northwest China, southwards locally to Persian Gulf states, Iran, Afghanistan and northern Pakistan. Winters north and west India, occasionally south to Sri Lanka.

Monotypic. Previously treated as conspecific with Booted Warbler *H. caligata*. A *BOU* review of molecular, morphological and vocal evidence indicated that these taxa should be treated as separate species (Knox *et al* 2002). The Dutch presently consider Sykes's Warbler to be within *Acrocephalus*; see Eastern Olivaceous Warbler (above).

Very rare vagrant from central and southwest Asia.

Status: There have been 11 records; one from 2008 is yet to be assessed by *BBRC*.

1959 Shetland: Fair Isle adult 29th-31st August, trapped 29th
1977 Shetland: Fair Isle 1st-winter 20th-27th August, trapped 20th
1990 Co. Cork: Cape Clear 17th October
1993 Shetland: Seafield, Lerwick, Mainland adult 22nd October-9th November, trapped 24th
2000 Dorset: Portland trapped, 1st July

2002 **Norfolk:** Sheringham 23rd August
2002 **Orkney:** North Ronaldsay trapped 26th August
2002 **East Sussex:** Beachy Head 31st August
2003 **Orkney:** North Ronaldsay 29th September-1st October, trapped 29th
2003 **Shetland:** Baltasound, Unst 4th-8th October, trapped 5th
2008 **Shetland:** Sumburgh 25th September*

Discussion: The field identification of Hippolais warblers has been regularly refined since the early 1960s (Wallace 1964). Booted and Sykes's Warbler were widely treated as conspecific, with the exceptions of Stepanyan (1978; 1983) and then later by Sibley and Monroe (1993) who treated them as separate species. Consequently interest in *rama* intensified and the separation of this recent entry into birders' consciousness has been aided through careful study of breeding birds and several identification papers (for example, Svensson, 2001; Svensson and Millington 2002; Svensson 2003). These provided a detailed overview of the differences between *rama* and *caligata* and led to their treatment as separate species (for example, Knox *et al* 2002; Parkin *et al* 2004).

The taxon was first established on the *British List* on the basis of a record from Shetland in 1993 (*BOURC* 1999). After its recognition as a new species, a review of all British records of trapped Booted Warblers where the biometrics suggested that *rama* might be a possibility was undertaken. The review resulted in a Fair Isle bird of 1959 being honoured as the first British record.

Five of the British records have occurred in August, a month in which just 17 Booted Warblers have occurred. Such odds clearly make any early Booted-type warbler worthy of extra scrutiny. The early peak coincides with period when birds rapidly depart their breeding grounds. Two of the others have occurred during the third week of October, again during a time that Booted Warblers are thin on the ground. The early July 2000 bird in Dorset recalls the Eastern Olivaceous Warbler *H. pallida* there on a similar date a year earlier. That coincidence teases self-finders with the possibility that some ultra rare warblers occur at times when few would traditionally be out bush bashing.

The identification of field-only records has been at the forefront of recent bird identification advances. It is likely that more Sykes's Warblers will be accepted without the aid of a mist net as criteria become better refined and applied. Although much is left to learn regarding the identification of this subtle species, well documented birds lacking the support of biometric data have been accepted (Norfolk and East Sussex in 2002). Another, from Shetland in 2008*, was well photographed. But, even some trapped birds present biometrics that may not prove specific and some may still evade an easy identification.

It remains a rare bird elsewhere in Europe. There are two accepted records from Norway (11th September 1983 and 20th September 1997, plus one from 22nd September 2008 that is not yet submitted). There is one from the Netherlands (11th October 1986), two from Sweden (3rd September 1995, 19th August 2002) and singles from Finland (9th October 1997), Iceland (14th September 2002) and Germany (30th September 2003).

The glut of nine records in Europe between 2000 and 2003 led to confident predictions of a short shelf life as an extreme rarity for the new species. The subsequent dearth of records was a surprise, but is in keeping with the pattern of most central Asian rarities: periodic pulses of glut followed by famine; surely more will follow sooner rather than later. And when they occur, birders will be better prepared to identify them.

Olive-tree Warbler *Hippolais olivetorum*

(Strickland, 1837). European breeding range restricted to the eastern Mediterranean, from coastal Croatia south through the Balkan Peninsula to Greece and along Black Sea coast in Bulgaria and southern Romania. To east, breeds west and south Turkey, northwest Syria and northern Israel. Migrates south to winter in east Africa from southern Tanzania to northern South Africa.

Monotypic.

Status: One record.
2006 Shetland: Boddam 1st-winter 16th August

Discussion: There have been two previous claims, both of which were considered to be inadequately documented for an addition to the *British List*: Scilly in September 1972 and St. Kilda in August 1999. The species was long ago mentioned by Wallace (1980) as a potential vagrant to our shores.

This is the first vagrant Olive-tree Warbler to be accepted to the north and west of the breeding range. Initially identified as an Icterine Warbler *H. icterina*, it was re-identified as an Eastern Olivaceous Warbler *H. pallida* later in the day. Presumably it would have been accepted as the latter, but for the fact that a small number of photographs were sent to identification consultants. Their responses allowed the bird to be re-identified as an Olive-tree Warbler (Harrop *et al* 2008).

The breeding grounds are vacated from mid-July onwards. Southerly passage through Israel and Syria takes place from mid-July to mid-August. The arrival of the 1st-winter on Shetland correlates with the peak migration period through the Middle East. Analysis of the weather patterns preceding the sighting revealed conditions conducive to vagrancy across Europe (Harrop *et al* 2008).

The arrival of the Shetland bird accords well with those of other vagrants that breed only in southeastern Europe and the eastern Mediterranean and wander to Britain and northwest Europe. Some have also been recorded in the early autumn period, for example Eastern Olivaceous Warbler H. *pallida*, Eastern Bonelli's Warbler *Phylloscopus orientalis*, White-throated Robin *Irania gutturalis* and Rüppell's Warbler *Sylvia rueppelli*. These co-vagrancies not only allow the thought of another jumbo Olive-tree but also other mouthwatering species from southwest Asia.

Lesser Whitethroat *Sylvia curruca*

Linnaeus, 1758. Breeds extensively across the Palearctic from Britain and France to eastern Siberia, eastern Mongolia and the Tian Shan in China. Winters in Africa south of the Sahara from the upper Niger east to Sudan and Eritrea, also Egypt and Arabia.

Polytypic. Most authors (for example, *BWP*; Shirihai *et al* 2001) recognise *curruca* (Linnaeus, 1758), *halimodendri* Sushkin, 1904, *margelanica* Stolzmann, 1897, *minula* Hume, 1873 and *althaea* Hume, 1878. These authors treat *blythi* Ticehurst and Whistler, 1933 as a synonym of *curruca*. Shirihai *et al* (2001) treat *halimodendri* as a subspecies of *curruca* and demonstrate that *margelanica* Margelanic Lesser Whitethroat, *minula* Desert Lesser Whitethroat, *althaea* Mountain Lesser Whitethroat form a separate clade from *curruca/halimodendri* Lesser Whitethroat. Of these *margelanica*, *minula* and *althaea* are as mutually divergent as other taxa of *Sylvia* that are treated as species. Martens and Steil (1997) reported bioacoustic differences between *curruca*, *minula* and *althaea* that merit treatment as allospecies. Shirihai *et al* (2001) conclude that further field studies are required.

The precise attributions of vagrant taxa are unknown, though at least two have been claimed in the autumns of recent years.

Status: The difficulty of identification has left the various claims of differing taxa on the *British List* in a state of limbo for the time being. Despite increased coverage in the literature in recent years, there seems little consistency about which races are recognised or into which subspecific groups they may fit.

S. c. blythi, regarded as an invalid taxon by Shirihai *et al* (2001), is currently listed as a scarce migrant by *BOURC*. Despite several candidates having been trapped, there are no positively identified records of any other vagrant races of Lesser Whitethroat. However, around 20 reports of birds of other eastern races stretching back over 20 years are currently awaiting assessment or review; it is possible that some eastern races may turn out to be scarce but regular visitors to Britain (Kehoe 2006).

Distinctive looking and sounding individuals occur, for example birds on Fair Isle in June 1999 (Holt and Turner 1999) and South Gare in November 2000 (Money 2000), and are considered by the *BOURC* probably to represent one or more vagrant races, perhaps *minula*. Molecular evidence exists for some past claims, but it needs to be considered in relation to a wider taxonomic framework not yet in place. Any proclamation on such records set against present knowledge would be premature and speculative.

Orphean Warbler *Sylvia hortensis*

(JF Gmelin, 1789). Western Orphean Warbler Sylvia hortensis hortensis breeds North Africa from Morocco to northwest Libya, north through Iberian Peninsula to southern France, southern Switzerland and Italy. Winters sub-Saharan Africa from southern Mauretania and northern Senegal to Chad. Eastern Orphean Warbler Sylvia hortensis crassirostris breeds Slovenia and Croatia south through Balkans to Greece and east through Turkey to at least Armenia, also northeast Libya and Israel. The form balchanica breeds in southern Transcaspia, Turkmenistan and Iran and jerdoni

in Persian Baluchistan east through to Pakistan and Afghanistan and north to Tien Shan Mountains in southeast Kazakhstan. Apparently jerdoni winters in the Indian subcontinent; others range from Sudan east through Eritrea and Ethiopia and the coasts of the Arabian peninsula.

Polytypic, four subspecies. Only nominate *hortensis* (JF Gmelin, 1789) has so far been accepted as a vagrant here, plus four additional records of undetermined subspecies. Shirihai *et al* (2001) recommended splitting Orphean Warbler into two species on the basis of DNA differences, upperpart tone, undertail pattern, bill biometrics, extent of hood in males, whiteness of underparts and differences in song. Under this proposal nominate *hortensis* is Western Orphean Warbler *S. hortensis*, a monotypic species. The remaining subspecies *crassirostris* (Cretzschmar, 1826), *balchanica* (Zarudny and Bilkevitch, 1918) and *jerdoni* (Blyth, 1847) form the polytypic Eastern Orphean Warbler *S. crassirostris*. This split was adopted by the Association of European Rarities Committees (*AERC*) in the 15th draft of the *AERC* Checklist of bird taxa occurring in the Western Palearctic region, but is not yet adopted by the *BOURC*.

Very rare vagrant from southern Europe.

Status: There are five accepted records. Only one has been positively identified, as Western Orphean Warbler *hortensis*, a further four accepted Orphean Warbler records are, as yet, unassigned to a race/species.

Western Orphean Warbler
1955 Dorset: Portland trapped 20th September

Four birds so far unassigned:
1967 Cornwall: Porthgwarra trapped 22nd October
1981 Scilly: St. Mary's 1st-winter male 16th-22nd October
1982 Northeast Scotland: Seaton Park, Aberdeen immature trapped 10th October
1991 Cornwall: Saltash male 20th-22nd May

Discussion: The original description of the Portland bird of 1955 contained a single tail feather that fell out during the processing of the bird. Fortuitously, this had been stuck to a piece of paper as part of the record submission. The DNA was recently analysed and the results showed a strong match for *hortensis*. Before the DNA results were known, the bird was expected to be of the eastern form *crassirostris*, largely because it had a very large bill. Its final identification signifies the caution required with assigning individuals to race/species. All unassigned birds are undergoing review by *BBRC* and the results are imminent.

Records of either form in northern Europe are rare during modern times. There have been two recent Eastern Orphean Warblers from Norway (3rd-8th October 2004, 12th August-9th September 2006).

There are five modern Western Orphean records for Germany (Helgoland on 1st September 1964, two males and one female possibly breeding in Baden-Württemberg from

19th July-9th August 2003, Helgoland on 9th-10th June 2004, 29th May 2006). There is a recent record, of male Western Orphean from Switzerland on 16th June 2006 (the first since 1994). Additional Western Orphean Warblers include the first for the Netherlands (29th October-5th November 2003) and one on Ouessant, France (4th October 2004).

Both Western and Eastern Orphean Warblers depart from their breeding grounds early in the autumn. These are vacated from the first half of July onwards. Return migration is similarly early, with birds passing through northwest Africa in late February and reaching Spain in late March (Western Orphean Warbler) and peak passage through Israel in late March and early April (Eastern Orphean Warbler). Quite why extralimital records are so rare is difficult to explain. Western undergoes a relatively short migration, which presumably makes it less prone to navigational error; Eastern is a longer-distance migrant but is nowhere common as a breeding bird in southeast Europe where there are around 70,000 pairs, compared with 100,000 pairs of Western in Spain (BirdLife International 2004).

Birders who did not happen to be on or travel to Scilly in October 1981 would be grateful for a modern bird of either form/species upon which to feast their eyes. This remains a surprisingly rare visitor to our shores and elsewhere in northern Europe.

Asian Desert Warbler *Sylvia nana*

(Hemprich and Ehrenberg, 1833). Breeds northern Iran and north Caspian Sea through deserts of Kazakhstan and Mongolia to northwest China. Winters in deserts and arid regions from northeast Africa, through Arabian Peninsula to Pakistan and northwest India.

Monotypic. Recently split from African Desert Warbler *S. deserti*, a local resident of deserts in southern Morocco and Algeria, on basis of differences in plumage and song by Shirihai *et al* (2001). In December 2003, *BOURC* announced its recommendation that Desert Warbler *Sylvia nana*, comprising the subspecies *nana* and *deserti*, be split into two monotypic species: Asian Desert Warbler S. *nana* and African Desert Warbler S. *deserti* (*BOU* 2004). In light of this change all previous records were reviewed and, as expected, all British records were found to be attributable to Asian Desert Warbler.

Very rare vagrant from central and western Asia, mostly in late autumn but two spring birds have occurred recently.

Status: Eleven British records: two May, six October, two November and one in December.

1970 **Dorset:** Portland 16th December-2nd January 1971, trapped 16th December.
1975 **East Yorkshire:** Kilnsea trapped, 20th-24th October
1975 **Essex:** Frinton-on Sea 20th-21st November
1979 **Cheshire and Wirral:** Meols, Wirral 28th October-22nd November
1988 **Isle of Wight:** Bembridge Pools 30th October
1991 **East Yorkshire:** Flamborough Head 1st-winter 13th October-5th November, trapped 13th October
1991 **Isle of Wight:** Bembridge 27th October-9th November, trapped 27th October

1991 Kent: Seasalter 3rd-5th November
1992 Devon: Mount Gould, Plymouth 19th-26th October
1993 Norfolk: Blakeney Point male in song 27th May-1st June
2000 East Yorkshire: Sammy's Point, Easington 7th-11th May

Many of the records have been obliging long-stayers. The mini-influx of three in 1991 was well received, especially so the obliging individual at Flamborough Head. The first nine were all found in the late autumn period; the last two were spring birds. The male in Norfolk was often in full song and observed nest building.

> On 27th May 1993, the temptation of a Common Rosefinch and two Icterine Warblers on Blakeney Point, Norfolk, proved irresistible, so I set off on an evening trudge in pursuit of these two potential year-ticks. I reached the area of dunes known as 'the Hood' at about 6.15pm, but I was undecided whether or not to press on to the point for the Rosefinch, or to look for the Icterine which earlier had been in the suaeda bushes around the Hood. In the event I decided to skirt the dune on my way to the point, in the hope that I might then chance upon the warbler as I passed.
>
> Actually, my mind had wandered, and I happened to be musing over the recent Black Lark sighting in Sweden, when a small, pale bird appeared beside the path and immediately arrested my attention. It flicked across in front of me, and I knew instantly that it was going to be something good. In the split second that it took me to get my binoculars onto the bird, it dropped out of view, but displayed an orangey tail as it did so – it had to be a Desert Warbler!
>
> Shortly afterwards, the bird was hopping confidently about in the low scrub nearby, apparently unaware of my frantic efforts to attract the attention of other birders. A few were fortunately further along the point, but it had gone 7.00pm by the time the first birders arrived, in various states of exhaustion, from Cley and even Norwich.
>
> Fortunately for the great majority, however, the bird remained until at least the month's end, holding territory, singing and even nest-building.
>
> **Mark Golley.** *Birding World 6: 182*

Discussion: Asian Desert Warbler is a long-distance migrant that leaves its breeding areas on a southwesterly heading. The earliest populations to depart are those of central Asia, which begin to vacate breeding areas in August and September to arrive on the wintering grounds in India and Pakistan from mid to late September and in the Arabian Gulf from mid August. Western populations migrate later, leaving Kazakhstan and Iran in September and November, with some present into December. It seems probable that most British records have involved birds from the western populations. The two spring birds are unlikely to be overshoots and most probably spent the winter in Europe or North Africa.

Elsewhere in northern Europe, all desert warbler records have also been Asian Desert Warblers. Scandinavia has a combined total of 26 records: 13 in Sweden (two in May, eight in October and three in November), 10 in Finland (one in May, seven in October and two in November) and three from Denmark (11th-12th November 1989, 13th November 1994, 10th

May 1998). There are two each for the Netherlands (30th October-3rd November 1988, 8th-9th October 1994) and Germany (21st June-7th July 1981, 24th-27th May 2002) and one from Estonia (20th-24th October 2005).

Most north European records share the late autumn arrival patterns of British records and rarer spring sightings. Six spring birds comprise two from Sweden (singing male on 20th May 1982, 29th May 2000), singles from Finland (22nd May 1992) and Denmark (10th May 1998) and two from Germany (a singing male on the Baltic Sea coast built two nests between 21st June-7th July 1981; 24th-27th May 2002).

African Desert Warbler is more sedentary and an unlikely vagrant, but its occurrence is a possibility. It has been recorded in Italy, Malta and the Atlantic islands. A movement from such a supposedly sedentary species to northwest Europe would be exceptional, but several other species from the same region have also occurred here, most notably Moussier's Redstart *Phoenicurus moussieri* and White-crowned Black Wheatear *Oenanthe leucopyga*. Tristram's Warbler *Sylvia deserticola* has reached Gibraltar. Separation of the desert warblers is relatively straightforward and, as any desert warbler is likely to be well scrutinised and photographed, an African bird would most likely be noticed.

Spectacled Warbler *Sylvia conspicillata*

Temminck, 1820. Breeding distribution centred on the Mediterranean, with two distinct breeding populations. The nominate race breeds from Iberia and North Africa east to Italy and Libya. A further population occurs in Cyprus, Levant, Israel, Jordan and possibly some southern and western parts of Turkey. Western populations short-range migrants. Southern limit in winter poorly known, but appears to be Senegal, southwestern Mauritania, southeastern Algeria, and northern Niger. Subspecies orbitalis resident on Madeira, Canary Islands and Cape Verde Islands.

Polytypic, two subspecies. The racial attribution of occurrences is undetermined, but is likely to have been nominate *conspicillata,* rather than *orbitalis* (Wahlberg, 1845).

Very rare spring overshoot from southern Europe. One autumn record.

Status: Five records, four between late April and early June and one in October.

1992 North Yorkshire: Filey male in song 24th-29th May, trapped but not ringed 24th May
1997 Suffolk: Landguard male 26th April-2nd May, trapped 26th April
1999 Devon: Roborough Down, near Clearbrook 1st-summer male in song 3rd-6th June, possibly since mid May
2000 Scilly: Tresco 1st-winter 15th-21st October
2008 Suffolk: Westleton Heath 1st-summer male 10th May*

Discussion: There has been a recent slight northward range expansion in Spain, France and Italy (Hagemeijer and Blair 1997) which may be linked to the recent cluster of British records. In northern Europe there are three accepted records from Germany (Helgoland collected on

10th September 1965, 6th June 2001 and a male in Niedersachsen from 20th-23rd May 2008), one from the Netherlands (2nd November 1984) and a male from Belgium (14th June to 23rd July 1999 paired with a Common Whitethroat *S. communis* and raised two young). In France it is very rare outside the limited breeding range, though extralimitals include one on Ouessant, Brittany, on 16th October 1984 and a singing male in Bay of Mont-Saint-Michel, Ille-et-Vilaine on 20th May 1989. In Switzerland, where it is a rare vagrant, a male was seen feeding young in July 1989 in the south of the country. Breeding again took place in 2005 and 2008.

Spectacled Warbler has an on-off-on association with the *British List*. The first accepted record from Spurn from 21st-31st October 1968 was re-assessed years later from in-the-hand photographs and found to have been a 1st-winter female Subalpine Warbler *S. cantillans*. Subsequent records had also been accepted: a male at Porthgwarra on 17th October 1969 and a male on Fair Isle from 4th-5th June 1979. The rejection of the Spurn bird meant possible promotion of a later record to the first for Britain, but following a review neither met the watertight standard demanded of a first and they too were rejected (for a discussion of these records see Lansdown *et al* 1991; *BOU* 1991).

After three false starts in 24 years this diminutive *Sylvia* finally made it on to the *British list* in 1992, courtesy of the fine male at Filey during that year's classic spring influx of Mediterranean overshoots. Infamously this bird was trapped but not ringed in the excitement of the event. Three more surprises came in rapid succession over the next eight years, but for their finders the clouds of identification confusion had finally been banished. The surge of records seems to have lost momentum, although a well-documented spring bird in Suffolk 2008 might be the start of another flurry.

Marmora's Warbler *Sylvia sarda*

Temminck, 1820. Restricted to Corsica, Sardinia and small islands off the west coast of Italy, and northern Tunisia. Winters in northern Algeria, Tunisia and Libya.

Polytypic, two subspecies. The racial attribution of occurrences is being looked at by *BBRC*, but is unlikely to have been other than nominate *sarda* Temminck, 1820. Nominate *sarda* of the central Mediterranean is reported to be genetically well differentiated from *balearica*, which is endemic to the Balearic Islands (Blondel *et al* 1996; Shirihai *et al* 2001) and there are differences in morphology and vocalisations.

The proposed post-split name for *sarda* is Marmora's Warbler and *balearica* is Balearic Warbler; *balearica* is believed to be sedentary. Most *sarda* also remain within their breeding range all year, but some winter in North Africa from Algeria to Libya and even as far south as the edge of the Sahara.

Vagrants of this race/species have reached mainland Spain, Gibraltar, mainland France (for example, 11 during 5th April-3rd June 1997), mainland Italy and even Greece (eight records).

Very rare late spring overshoot from the Mediterranean.

Status: There have been five British records, all males.

1982 **South Yorkshire:** Midhope Moor male in song 15th May-24th July
1992 **East Yorkshire:** Spurn male in song 8th-9th June, trapped 8th
1993 **Borders:** St Abb's Head male in song 23rd-27th May
2001 **Norfolk:** Scolt Head male in song 12th, 18th May
2001 **Suffolk:** Sizewell male in song 29th May

Discussion: Aside from the British records the only others for northern Europe are from Belgium (3rd-12th May 1997) and Denmark (one trapped on 12th June 2005). There have been 24 records on mainland France, all from the Mediterranean coast, especially Alpes-Maritimes (eight records) and Bouches-du-Rhône (five records).

The male on an upland moor in the Peak District in South Yorkshire in 1982 was widely considered, at the time, to be a freak birding one-off, so the arrival of the second was almost as startling as the first. Two in 2001 was quite exceptional. More can perhaps be expected. One wonders how many others have passed through undetected due to the skulking nature of this small *Sylvia*. It may not be coincidence that the first three records were far away from areas that host Dartford Warbler *S. undata*, where any long-tailed *Sylvia* would be examined closely. Not surprisingly all have been attention seeking spring males in song.

With four further east coast records and a better understanding of the vagrancy potential of this species, its status as a rare vagrant no longer draws gasps of disbelief, but the birds have performed extraordinary feats of migration all the same.

Rüppell's Warbler *Sylvia rueppelli*

Temminck, 1823. The small breeding range is restricted to the northeast Mediterranean, including southern Greece, Crete, western and southern Turkey and northwestern Syria. Winters largely in northeast Africa, mainly in the central Sahel zone of Chad and Sudan.

Monotypic.

Very rare vagrant from southeast Europe.

Status: There have been five British records.

1977 **Shetland:** Dunrossness, Mainland adult male in song 13th August-17th September, trapped 15th August
1979 **Devon:** Lundy male 1st-10th June, trapped 4th
1990 **Shetland:** Whalsay male 3rd-19th October, trapped 3rd
1992 **Norfolk:** Holme 1st-winter 31st August-4th September
1995 **Caernarfonshire:** Aberdaron male 21st June

The 1992 bird was faithful to one small area of the reserve and proved exceptionally popular during its stay.

At 7.30am on Monday 31st August 1992, with the wind blowing from the southwest, I began looking around the scrub known as 'The Forestry' at Holme Dunes Norfolk Naturalists' Trust Reserve in northwest Norfolk. As on the previous few days, there was minimal bird activity but, expecting to bump into only the usual Lesser Whitethroat and Reed Warbler, I checked the most sheltered spot in the Sea Buckthorn at around 8.00am. After I had been waiting at the southern end of a secluded ride through the bushes for about ten minutes, a strange grey warbler with pale underparts popped into view in a nearby dead elder.

The view was very brief, but the general appearance and the strikingly pale orangey legs indicated that it was a rare Sylvia warbler. However, I needed another look at it before I could attempt to put a name to it and, frustratingly, it disappeared. Fortunately, however, after about three-quarters of an hour, it re-appeared close-by in a bramble bush.

It was pale bluish-grey above and off-white below, with the wing feathers (especially the tertials) distinctly edged with white, somewhat reminiscent of a diminutive Barred Warbler, and the eye glinted a distinctive wine red. Feeling sure that I had seen enough to identify the bird, I rushed back to reserve headquarters to alert the warden, Bill Boyd, and Neil Lawton.

A little later, at the reserve centre, Bill and I consulted both The Shell guide to the Birds of Britain and Ireland (Ferguson-Lees et al 1983) and The Macmillan field guide to bird identification (Harris et al 1989). The pictures in the former suggested that the bird was a Subalpine Warbler, and the text in the latter seemed to support this. We were mindful of the cautionary statement in Harris et al that "with their rather featureless appearance, 'colourless' 1st-winter female Subalpines, have, in the past, often been misidentified as rarer Sylvia", and settled for it being a Subalpine.

We promptly phoned the news to Birdline, and visitors soon began to arrive. The bird was seen by about a hundred people during the next hour or so, but then it disappeared again for a while. By early afternoon, a small crowed hoping to see a Subalpine Warbler had gathered at the end of the ride, but when the bird eventually hopped into view, Richard Millington, who was then standing next to Bill Boyd in the jostling 'rear gallery', immediately exclaimed "it's a Rüppells'!" – thereby injecting a certain amount of panic into the situation, for the warden at least!

Obviously further clarification was needed and Sardinian Warbler needed to be fully eliminated, so I set off to fetch Lars Jonsson's Birds of the Mediterranean and Alps from my flat, while Richard called his wife, Hazel, on his mobile phone to discuss the text of a Mystery Photograph article in a recent Dutch Birding which had dealt with female Rüppell's Warbler. An on-site debate amongst several experienced birders raised a number of conflicting opinions on the exact colour of the wing-feather fringes and their relevance, but reality finally set in when I opened the pages of the Jonsson guide; the illustration of female Rüppell's Warbler fitted almost perfectly.

With the identification firmly settled and Birdline broadcasting the glad tidings, plans were made to accommodate the many birders that we knew would soon be wanting to see the bird. Fortunately, from the reserve's point of view, the bird was content to spend most of its time in the only sheltered spot in the centre of the Sea Buckthorn clump, at the north end of a ringing ride, and was fairly confiding. Up to about 30 people at a time were able to view the bird and, apart from the first evening (when about 150 birders were beaten by failing light), just about all visitors were able to enjoy good views of this major rarity during its five-day stay.

All the birders were very patient with the queuing system that we had arranged, even though a few had the misfortune of having to wait up to four hours for their turn (on the Tuesday). The Rüppell's Warbler finally departed during the cold, clear night of 4th September.

Gary Hibberd. Birding World 5: 336-337

Discussion: Vagrants elsewhere in northern Europe include two from Finland (7th-8th June 1962, 30th May 1985) and singles from the Faroe Islands (mid May until mid June 1974) and Denmark (adult female 7th-8th May 1993). Farther south there are two males from France (south-eastern France on 20th March 1970, the Camargue on 20th May 1996) and c40 from Italy (mostly March-April through southern Italy where it is a scarce migrant). Shirihai *et al* (2001) lists extralimital records for Romania, Ukraine, Iran and Kuwait.

The first three British records were all on offshore islands, but conveniently put in extended stays. The first Shetland bird was found to be in heavy moult when it was trapped. Since its moult had been in progress for several weeks, it seems likely that it had arrived on the islands at some point in the spring and avoided detection until its fortunate observer stumbled across it. The Norfolk bird of 1992 was easily the most accessible and crowd pulling of the records.

Rüppell's Warbler is a medium-distance migrant. A loop migration route takes it south to southwest across the eastern Mediterranean in autumn, but in spring it takes a more easterly, mostly overland, route similar to that of Eastern Subalpine Warbler *S. c. albistriata*. Some must take a more direct route, as illustrated by the records from Italy. The British records accord well with spring overshoots from southeast Europe.

Subalpine Warbler *Sylvia cantillans*
(Pallas, 1764).

Polytypic, three subspecies recognised by Vaurie (1959) and Svensson (1992) and four by Shirihai *et al* (2001). All three authors accept: nominate *cantillans* (Pallas, 1764) of southern Europe from Iberia east to Italy, including Sicily and Sardinia; *inornata* Tschusi, 1906 of northwest Africa; and *albistriata* (C L Brehm, 1855) from southeast Europe, east from Slovenia and western Turkey. Shirihai *et al* (2001) add *moltonii* Orlando, 1937 of the west Mediterranean islands for which the recently resumed name for this taxon is *S.subalpina* (Temminck,1820; see Baccetti *et al* 2007), though in keeping with recent publications its former name is used here. Shirihai *et al* (2001) concluded that differentiation between Western Subalpine Warbler nominate *cantillans/inornata*, Moltoni's Subalpine Warbler *moltonii* and Eastern Subalpine Warbler *albistriata* is more advanced than between most subspecies of other Sylvia, but that more study is required to assess whether the complex should be treated as a superspecies.

Eastern *albistriata* differs in plumage from other forms and *moltonii* differs in song and contact call and possibly moult. Individuals of *moltonii* and southern *cantillans* sampled in sites of sympatry in central Italy exhibited perfect concordance between phenotypic and genetic identifications, indicating the taxa are not interbreeding and that *moltonii* should be ranked as a distinct species *Sylvia moltonii* (*S. subalpina*; see Baccetti *et al* 2007 for nomenclatural proposed changes), a split endorsed by the Dutch CSNA. Eastern *albistriata* is phylogenetically related to southern *cantillans* and it was proposed that it should provisionally be retained as a subspecies of *S. cantillans* (Brambilla *et al* 2008).

Nominate *cantillans* Western Subalpine Warbler was removed from the list of species considered by the *BBRC* at the end of 2005, by which time it had occurred 540 times. Vagrant *albistriata* and potential records of *moltonii* are considered by the *BBRC*. The first British record of *cantillans* was obtained from St. Kilda on 14th June 1894. The first Irish bird was an immature male found dead on the lighthouse balcony at the Hook Tower, Co. Wexford, on 17th September 1933.

Eastern Subalpine Warbler *Sylvia cantillans albistriata*

Breeds southeast Europe from Slovenia and Croatia though Albania and Greece to western Turkey. Migrates through the Middle East to winter along the southern edge of the Sahara south to Sudan.

Spring vagrant of unknown status (only males identified with confidence at present).

Status: There are 37 records so far considered to involve this form that are either accepted by the *BBRC* or treated as such by *Birds of Scotland*. Just 21 are so far published by *BBRC*. Of these, 36 have been in Britain and one has recently been accepted for Ireland by the *IRBC* but has not yet been formally published. A review of the racial attributions of British records by the *BBRC* is underway.

Historical review: The days of a Subalpine Warbler being as just that are long gone. Attention focused on the racial attribution of vagrants with greater intensity following the announcement by Shirihai *et al* (2001) that the four forms of the Subalpine Warbler complex may be candidates for splitting.

The status of this form is poorly understood at present. A male on Fair Isle from 20th-27th May 1951 is presently regarded as the first British record, followed by another male there from 23rd-24th April 1964. There is one Irish record so far, a male at The Mullet, Co. Mayo, on 3rd May 2007.

Accepted records have involved pulses of arrivals, with three each in 1988 and 1989, six in 1993 and four in 2004 and 2007. Given the greater interest in ascribing individuals to form, many more can be expected. With them may come a fuller understanding of its true status, rather than the current bias towards recent males.

Where: Twenty of the 37 records have been in Scotland, with nine on Shetland (eight from Fair Isle), two from the Outer Hebrides and one from Orkney (Forrester *et al* 2007). There are nine records from England – two from Dorset and singles in Cornwall, East Sussex, Kent, London, East Yorkshire and Cleveland – and five from Wales – four Bardsey and one Ramsey Island – two from the Isle of Man and one from Co. Mayo.

When: The earliest record was from Penless, near Rame Head, Cornwall, on 16th April 2007, followed by one at Lonsdale Road Reservoir, London, on 21st April 2003. The bulk of records have occurred in the first three weeks of May, with a suggestion that those on the Scottish Islands occur marginally later than those in England and Wales, Isle of Man and Ireland.

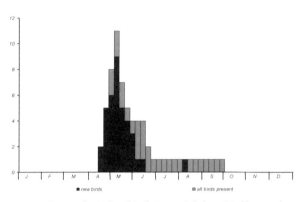

Figure 66: Timing of British and Irish Eastern Subalpine Warbler records, 1951-2007.

There is also a suggestion that birds arrive later than Western Subalpine Warblers (Williamson 1974; Vinicombe and Cottridge 1996), implying that many of the other Subalpine Warblers occurring in the late spring period may also be Eastern Subalpine Warblers (Forrester *et al* 2007). The median arrival date on the Scottish Islands is 10th May compared with 4th May away from the Scottish Islands, offering some support for this hypothesis. But, the sample size is small and such conjecture may be premature until the full suite of records incorporating all age classes is understood. Presently males and 1st-summer males make up the bulk of records, though three trapped females have also been attributed to this race (Bardsey in 1984 and 1985 and one from Dungeness in 2004).

The only autumn find was a male at Sumburgh, Shetland, from 11th August-30th September 1971.

Discussion: Eastern Subalpine Warbler performs a long-distance loop migration (anticlockwise), requiring a northwestly correction during the final approach to the breeding grounds. The heading makes them prime candidates for overshooting to northwest Europe, as illustrated by the records so far assessed. In both seasons the migratory route extends east of the known breeding range and spring passage extends farther east than autumn.

It would be futile to draw too many conclusions about Eastern Subalpine Warbler occurrences. The numbers involved are small and comprise only the more obvious birds that have been accepted at a national level. The *BBRC* is attempting to establish criteria for the separation of ages other than adult males. Once that process is complete, records of other age/sex groups will be assessed. Autumn birds may prove problematical. A fuller interpretation of the records following the *BBRC* review may be more illuminating regarding where and when to look for this bird.

Moltoni's Subalpine Warbler *Sylvia cantillans moltonii*
Breeds on the Balearic Islands, Corsica, Sardinia and possibly continental Italy. Precise wintering area unknown, but birds with moult patterns typical of the race have been ringed in Nigeria.

Potential vagrant of unknown status. One well-documented record pending acceptance.

Status: As with the previous race, a review of the racial attributions of Subalpine Warbler records by *BBRC* is underway. One record is pending acceptance.

Historical review: A bird at Portland in April 1975 was widely suspected of belonging to this race (Gantlett, 2001) but assessment of the call proved that the bird was a *cantillans* (Kehoe 2006). The diagnostic rattling call was heard from a male at Burnham Overy Dunes, Norfolk, from 30th September-5th October 2007 (Golley 2007), but this record has not yet been submitted.

A male at Skaw, Unst from 1st-10th June 2009 has been well documented and sound-recorded and seems likely to represent an unequivocal first for Britain. During its stay it was observed nest-building.

Discussion: Presently the only records in northwest Europe are single males in the Netherlands (23rd-26th May 1987) and Belgium (20th-21st May 2001).

Moltoni's has a different moult strategy to the other races. Thus, any Subalpine Warbler in spring with fresh primaries and a characteristic rattling call is likely to be a *moltonii*; other differences include the brownish-pink of the underparts reaching the sides of the belly on spring males.

Many observers are now paying closer scrutiny to any Subalpine Warbler. Although, as yet, there is no formal acceptance of *moltonii*, an incipient species, on the *British List*, it is surely only a matter of time before one does make the grade. Given the distribution of Moltoni's, it seems likely that it may turn out to be a regular vagrant to our shores.

Sardinian Warbler *Sylvia melanocephala*
(J.F. Gmelin, 1789). Largely resident or dispersive throughout Mediterranean basin, from north-west Africa and Iberia to southern France, northern Italy and east to western Turkey and Israel. Some winter in North Africa from Sahara south to Mauritania and southern Libya.

Polytypic, four subspecies. Records attributable to nominate *melanocephala*.

Rare vagrant from southern Europe, typically in spring, but with a broad spread.

Status: There have been 76 records (74 in Britain and two in Ireland), to the end of 2007.

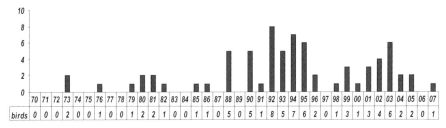

birds	70	71	72	73	74	75	76	77	78	79	80	81	82	83	84	85	86	87	88	89	90	91	92	93	94	95	96	97	98	99	00	01	02	03	04	05	06	07
	0	0	0	2	0	0	1	0	0	1	2	2	1	0	0	1	1	0	5	0	5	1	8	5	7	6	2	0	1	3	1	3	4	6	2	2	0	1

Figure 67: Annual numbers of Sardinian Warblers, 1970-2007.

Historical review: The first British record made landfall as recently as 1955 when a male was trapped on Lundy on 10th May. The next was a male on Fair Isle from 26th-27th May 1967. The year after a male was trapped on Skokholm on 28th October. Just three offshore records ensured that this was a highly sought-after Sylvia at the beginning of the 1970s.

Four were seen in the 1970s. Two were in April 1973, including the first female, which was trapped at Dungeness on 17th April, and a male at Waxham from 28th-29th April. Single males in 1976 (Beachy Head) and 1979 (Gibraltar Point) put in extended stays of two and nearly three months. The decade closed on a cumulative total of seven, though the two widely appreciated individuals had eroded this skulking Sylvia's rarity value.

Two autumn birds in 1980 were the first of 12 during that decade, four of them from the Isles of Scilly. An unprecedented five in 1988 included birds in Orkney and from Scilly to Kent between 24th April and 22nd June. Little did observers realise that this multiple arrival would set the trend for subsequent years.

In the 1990s, the species failed to put in an appearance only in 1997. Eight in 1992 and seven in 1994 led the way, as the decade collected no fewer than 38 birds. A wider spread of sightings included birds away from the favoured south and east coast. In the 1992 glut one reached Lancashire (Formby Point from 28th-31st May). Even more impressive was a female trapped in Surbiton, Greater London, on 2nd June. The following year Ireland bagged its first, as two April males visited Co. Cork in the space of four days (Cape Clear from 10th-12th and Knockadoon Head from 14th-21st April). In 1994 the second for Wales was present on Bardsey from 2nd-7th June. In a short time this Mediterranean warbler had metamorphosed from extreme rarity to an expected arrival, with four per annum during the 1990s.

Since 2000, numbers appear to have stalled, with just 19 between 2000-2007 and the first gap in sightings for eight years in 2006. An interesting series of records occurred in Norfolk in 2003, when a male (or males) was seen on the north Norfolk coast at Holme/Old Hunstanton from 16th-24th March, with possibly the same at Beeston Bump from 29th March-6th April and at Winterton on 4th June. A male had been seen at Old Hunstanton from 27th September-15th October 2002; could this bird have returned to Norfolk in 2003? Also in 2003, Skegness, Lincolnshire, hosted a male from 2nd October through to 11th January 2004 and then a female from 10th November through to at least 4th January 2004; an unprecedented wintering attempt and the first multiple sighting.

Where: The south coast between Scilly and Dorset and the east coast between Norfolk and Yorkshire account for the majority of records. In the southwest 22 birds comprise eight from Scilly, six in Cornwall, four in Dorset, three in Devon and one in Somerset. In the southeast there have been 12 birds, six in Kent, five in East Sussex and one in London. In East Anglia there have been 15 birds, 13 from Norfolk and two in Suffolk. Farther north there have been four in Lincolnshire and two each in East and North Yorkshire. Seven of Scotland's 13 birds were on Shetland, three Orkney, two Fife and one the Outer Hebrides. There are two records apiece for Wales and Ireland, and singles from Lancashire and the Isle of Man.

When: This visitor from the Mediterranean is largely a resident species and exhibits quite a curious pattern of occurrences. Only three winter months lack a new arrival. Most (74 per cent) have been in spring and summer.

Six March records illustrate the propensity for early spring arrivals. The earliest was a male at Stratton, Cornwall, from 8th-22nd March 1990, 10 days earlier than a male on The Lizard, Cornwall, from 18th-26th March that year. There are other late March records from Scilly and Norfolk (two). The main detection period extends from early April to early June, with new birds found to July. With such a notoriously skulking species, it is difficult to be sure that a first date of occurrence actually represents an arrival.

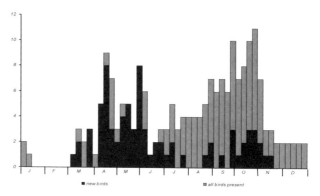

Figure 68: Timing of British and Irish Sardinian Warbler records, 1955-2007.

A much smaller cluster of autumn discoveries (26 per cent) is evident from late August through to mid November. A feature of the species is the frequent extension of its stays, which are among the most protracted of any passerine rarity. Two have wintered together at Skegness from October and November 2003 into January 2004. Wintering, or extended stays, have also been recorded in northern France and the Netherlands.

Discussion: Elsewhere in northern Europe there are five records from Sweden (all since 1980: two May, singles June, August and October), three from Finland (all since 1986: two May, one June), two from Norway (26th July 26th 1981, 3rd and 20th October 1982) and six from Denmark (all since 1978: three May, two June and one October).

There are eight records from the Netherlands (all since 1980: two April, three May, plus singles October, November and December; one wintered 14th December 1980-22nd February 1981 and another was present 12th November-29th December 1995), there have been five from Belgium (two in May, singles in June, October and November). In Germany there were eight between 1969 and 1999. The first for Poland was on 16th April 2001 and the second was in May 2005. There have been two on the Channel Islands (22nd May 1976, 17th October 1987).

There has been dramatic increase in sightings of this species, one of the commonest warblers in much of the coastal Mediterranean. It is presumably linked to the northwards expansion of the breeding range since 1970 in Italy, southern Bulgaria, parts of northwest Spain and southern France. Its distribution began to extend as long ago as the end of the 19th century (Hagemeijer and Blair 1997).

Western populations are resident or short-distance migrants, but the extent of the winter range implies that some must move considerable distances. Eastern populations are wholly migratory. Migratory movements in southwest Europe take place in February and March. These birds are presumably responsible for the early British records.

Greenish Warbler *Phylloscopus trochiloides*

(Sundevall, 1837). The European and west Siberian race viridanus expanded west during the 20th century to eastern Poland, Baltic countries and southern Finland, with sporadic breeding in Germany, Czech Republic, Sweden and Norway. To the east, breeds Russia and western Siberia to Yenisey River, south through northwest Mongolia to northern Afghanistan and northwest Himalayas. Winters Indian subcontinent. Other races in Himalayas to southwest China winter Indian subcontinent to Indochina.

Polytypic, five subspecies. Records attributable to two races. Western Greenish Warbler *viridanus* Blyth, 1843 was removed from the list considered by the *BBRC* at the end of 2005 after more than 448 records since 1950. Ireland recorded 22 birds during the same period. The first British record was shot at North Cotes, Lincolnshire, on 5th September 1896. The first Irish bird was shot on Great Saltee Island, Co Wexford, on the 25th August 1952. There are four British records of Two-barred Greenish Warbler *plumbeitarsus* Swinhoe, 1861.

The Greenish Warbler complex is a popular quarry for birders and a considerable puzzle for taxonomists. Morphological variation and distribution within Asia led to it being referred to as a ring species that encircles the warbler-free high-altitude desert of the Tibetan Plateau (Collinson 2001; Collinson *et al* 2003; Irwin *et al* 2005). A gap in the distribution occurs between *obscuratus* in the southeast and *plumbeitarsus* in the northeast. West Siberian *viridanus* and east Siberian *plumbeitarsus* coexist without interbreeding in central Siberia

Differences in plumage, structure and song suggest that the two forms can be considered separate species. Variations in plumage, morphology and song form a cline between the other subspecies in the ring (Irwin 2000; Irwin 2001). Treatment as separate species is presently considered unsatisfactory, despite the fact that molecular studies suggest that populations at the two extremes are effectively reproductively isolated (Irwin *et al* 2001; Irwin *et al* 2005).

Although *viridanus* and *plumbeitarsus* behave as distinct species in their zone of overlap, the conundrum for taxonomists is that the five subspecies *viridanus*, *ludlowi*, *trochiloides*, *obscuratus* and *plumbeitarsus* form a cline within which it is not possible to define species boundaries (Collinson *et al* 2003).

Two-barred Greenish Warbler *Phylloscopus trochiloides plumbeitarsus*

Breeds from the Yenisey valley east to northern Mongolia and Ussuriland, south to northeast China. Details of the Russian breeding distribution have not been clarified. Winters southern China to northern Indochina and central Thailand.

Very rare vagrant from Siberia, all from late September to late October.

Status: Four records.

1987 Scilly: Gugh 1st-winter 22nd-27th October
1996 Norfolk: Holkham Meals 15th-16th October
2003 Scilly: Bryher 27th-28th September
2006 North Yorkshire: Filey 16th-18th October

A cracking weekend at Filey produced a Richard's Pipit, Short-toed Lark and Isabelline Wheatear in one field! Those of us returning to work on Monday 16th October 2006 went with no little trepidation. Almost inevitably Phil Cunningham texted news of a cracking Radde's Warbler in the Top Scrub; not only that, but it was showing well. After a brief curse I brought my meeting to a close, travelled over to Scunthorpe for a second meeting; wrapped that up early and dashed back to the coast. Arriving at 1500hrs, ten minutes at the Radde's proved fruitless, but then rumours circulated of an Arctic Warbler 400m away in Arndale. Upon arrival, only two other birders were present, Phil and David Gilroy; the latter having had two views of a wing-barred phyllosc and put out the news as an Arctic; however, he hadn't seen the bird for thirty minutes.

Within a minute the warbler started to flick about on the edge of the trees. The first feeling was one of surprise; the date really pointed to this being an Arctic, but surely this wasn't one. It looked comparatively small, appeared a clean, bright green above and whitish below; more importantly, it showed a whopping greater covert bar. Moving forward, closer views revealed pale edges to the median coverts forming a second wing-bar. Confident that this was going to be a Two-barred Greenish, I rang Birdguides to correct the erroneous Arctic Warbler message. With several present still convinced it was still an Arctic Warbler, news circulated as 'showing the features of Two-barred'.

For the next 45 minutes as much detail as possible was gleaned, but the bird remained mute. That evening was a complete nightmare. The only photograph showed a phyllosc with a great super and apparently a long bill. Surely this wasn't going to be an Arctic Warbler after all! Safe to say, I hardly got any sleep that night, dreading the bird giving a monosyllabic call next day.

Next day the warbler performed from dawn to a group of 25 birders; key features were cross-checked, but it wasn't for an hour that it finally found its voice, giving four calls in a two-minute period. Without any question those 'ch-ee-wii' notes, like a slurred version of a Greenish and perhaps without the piercing quality, were some of the sweetest sounds we'd ever heard!
Craig Thomas. *Birding World 19:435-436.*

Discussion: Debate regarding the taxonomic status of *plumbeitarsus* will doubtless continue, but it remains a quality find. The statistics would imply that it is an exceptionally rare vagrant, but given that, despite the difficulties of identification, three have occurred in the space of 10 years it is perhaps not as rare as its four discoveries suggest. Elsewhere in Europe there have been two each from Sweden (5th July 1991, 6th October 1999) and the Netherlands (17th September 1990, 2nd October 1996) and singles from Finland (2nd-3rd October 2002) and France (26th October 2004). An exceptional record photographed on Madeira (29th April 2008) awaits assessment by the CPR. The concentration of all European records into a 20-year timeframe suggests an artifact of increased knowledge and improved detection rates.

Three of the four British records have been in October, as have five of the other European records, showing an arrival pattern a month later than the peak period for *viridanus*. This is not surprising as *plumbeitarsus* shares the range and migration patterns of a number of far-eastern vagrants that occur periodically in October. Greenish Warbler is much rarer than Arctic Warbler in October, and late birds are likely to receive close scrutiny nowadays in the expectation of a Two-barred.

Any suspected Greenish, or indeed Arctic Warbler, in late autumn would repay a second look with Two-barred in mind. The first was initially identified as an adult Yellow-browed Warbler *P. inornatus*, with the age used to explain the odd colour and lack of tertial edgings (Bradshaw 2001). No doubt the recent refresher courses in identification will be put to good use in coming years, given the increasing prevalence of rare far-eastern vagrants reaching us.

With little evidence of extreme vagrancy, *viridanus* would appear to be a classic drift migrant and spring overshoot, in varying numbers, whereas its sibling taxa *plumbeitarsus* appears to be a classic far-eastern vagrant of the late autumn period.

Green Warbler *Phylloscopus nitidus*
Blyth, 1842. Breeds northern Turkey, Caucasus and west-central Asia to northeast Iran. Winters India and southeast Asia.

Monotypic. As with Two-barred Greenish Warbler the treatment of allopatric *nitidus* and Greenish Warbler *P. trochiloides* has oscillated between separate species (for example Vaurie 1959; Voous 1977; *BWP*; *HBW*) and conspecific (Baker 1997; Sangster *et al* 2002). Geographically isolated outside the Greenish Warbler ring, the mtDNA of *nitidus* differs from *P. t. viridanus*, albeit based on a small sample (Helbig *et al* 1995, Irwin *et al* 2001a). Diagnostic plumage characteristics include vivid yellow supercilium, face, throat and upper breast compared with any Greenish Warbler. The song differs, with a dry trilling element in nearly all phrases that is not evident in *viridanus* (for example, Albrecht 1984). The level of difference between *nitidus* and other Greenish Warblers is similar to or greater than that between other closely related *Phylloscopus* species.

The *BOU* previously lumped Green and Greenish, but now consider *nitidus* as a monotypic species (Knox *et al* 2008).

Very rare vagrant from Southeast Europe.

Status: One record.
1983 Scilly: Bryher 1st-winter 26th September-4th October

Discussion: Elsewhere in northern Europe there remain just three accepted single records from Germany (shot on Helgoland 11th October 1867), Faeroe Islands (trapped on 8th June 1997 and identification confirmed with a blood sample) and Sweden (trapped on 29th May 2003). The Swedish bird was initially identified as an unusually yellow Greenish Warbler. Although caught, ringed and subsequently suspected of being a Green Warbler, the bird was not formally identified until a DNA analysis of the collected blood-sample was conducted (Irwin and Hellström 2007). In southern Europe, there are two records from Greece (18th September 1998, 27th September 2000). Farther east there are three from Israel (27th October 1987, 24th-29th August 2004, 1st May 2008).

Two widely referenced reports from Germany (8th June 1997, 1st June 1998) were not accepted (Peter Barthel *pers comm*), nor were two from Italy (Andrea Corso *pers comm*). One from France (Pas-de-Calais on 21st September 2003) is under review and will shortly be removed from the French list (Marc Duquet *pers comm*).

Migratory, moving mostly southeast to winter in southern India and southeast Asia. Autumn records presumably involve dispersing migrants and spring records overshoots en route to their southwest Asian breeding grounds.

Arctic Warbler *Phylloscopus borealis*

(JH Blasius, 1858). Breeds locally in northern Scandinavia, becoming widespread across northern Russia east to extreme northeast Siberia, south to Baikal region, Ussuriland and northeast China. Other races breed in Alaska, and Kamchatka through Kuril Islands to north Japan. Migrant through east China to winter widely in southeast Asia to Java, Philippines and Sulawesi. Arctic Warbler is the only Phylloscopus that breeds regularly in the New World.

Polytypic, three subspecies recognised by recent authors. The racial attribution of occurrences is undetermined, but probably involves nominate *borealis* (JH Blasius, 1858).

Rare migrant from northeast Europe eastwards, mostly late August through September.

Status: Numerically this is one of the most frequently occurring species still to be considered a rarity, with a cumulative total of 290 records by the end of 2007. Of these 283 were in Britain and seven in Ireland.

Historical review: The first British record was killed at Sule Skerry Lighthouse, Orkney, on 5th September 1902. Ten more British birds followed between 1908 and 1932, seven of them from Shetland, including a male obtained on Fair Isle, on 30th July 1928, and others in Norfolk, Lincolnshire and Northumberland. The occurrence pattern has changed little since.

The 1950s added a further 11 records. Seven were from Shetland, with singles in Norfolk, Northumberland and Northeast Scotland. At the time an influx of four from 1st-17th September 1959 was quite exceptional. These included three trapped on Fair Isle and the first away from Shetland or the east coast, trapped on Lundy on 6th September 1959.

During the 1960s a further 31 were added; a blank year in 1963 was the last time this species failed to put in an appearance. Among these were the firsts for Ireland (trapped Tory Island, Co. Donegal, on 1st September 1960) and Wales (trapped Bardsey on 13th September 1968). Just under half (14 birds) came from Shetland, 11 of them from Fair Isle. Two in Fife, five in Northumberland, three in East Yorkshire and two in Norfolk provided indications that Arctic Warbler could be found elsewhere. Singles in Avon, Scilly, Bardsey, Cape Clear and Co. Donegal showed that it could also stray west. The peak annual counts were five in each of the years 1964, 1967 and 1968.

Despite the substantial increase in observers during the 1970s, numbers only increased slightly. A new annual record number was established in 1970, with eight birds including six trapped on Fair Isle. Shetland continued to dominate reports, with 28 (60 per cent) of the decade's 46 birds making landfall there. Five others came from the east coast of Scotland, four from northeast England between Northumberland and Lincolnshire, and three in Norfolk. More intense watching on Scilly produced four. Cheshire and Co. Cork each added singles.

As for many Sibes, the pace quickened markedly during the 1980s. Sixty were recorded at an average of six per annum. Two in 1980 was a below par start to the decade, but this was quickly rectified by bumper counts of 17 birds in 1981 (still a record annual tally) and 11 in 1984. The Shetland stranglehold on records was loosened, but with 23 (38 per cent) birds found there it was more reflective of greater activity elsewhere than any fall-off in numbers making landfall on the archipelago. Elsewhere in Scotland, seven birds included four from Orkney. On the English east coast, 14 were located between Co. Durham and Kent. In the southwest, intense autumn coverage on Scilly added eight and a further five were found between Dorset and Cornwall. In Ireland two reached Co. Cork and one Co. Clare.

The increase in finds continued in the 1990s with 78 logged (just fewer than eight per annum). These included consecutive double-figure accumulations between 1993 and 1996 (14, 12, 12 and 11 birds). The decade saw Shetland increase its share with 37 (47 per cent) birds; 12 other Scottish records included nine from Orkney. A total of 21 were found between Northumberland and Kent, including five of in Yorkshire and four in Norfolk. In contrast with the previous decade, the southwest fared badly with just seven birds, including just three each from Cornwall and Scilly. Whether this reflected reduced observer attention in these areas or fewer birds is difficult to ascertain. The most notable record of the decade, and one which is yet to be repeated, was the

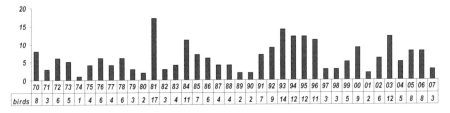

Figure 69: Annual numbers of Arctic Warblers, 1970-2007.

first inland bird: at Blithfield Reservoir, Staffordshire, from 8th to 11th September 1993. In light of the species' rarity away from Shetland and the east coast, this remains a remarkable find.

Since 2000 numbers have declined slightly, the 53 birds arriving at an average of 6.6 per annum over the eight autumns from 2000-2007, with 12 in 2003 the only double-figure count of the decade. Once again, 26 birds (48 per cent) came from Shetland, with the other Scottish records involving six from Orkney and one from the Outer Hebrides. Just eight were found along the east coast, four of them in East Yorkshire. In the southwest Scilly produced eight birds and there were two in Cornwall and one in Dorset. Two reached Ireland.

Since 1970, birds have arrived at just over six per annum. The recent fall in numbers despite more birders seeking out rarities on the Northern Isles could reflect a genuine decline. This is perhaps best illustrated by the fact that there have been just three autumn records on Fair Isle since 2000 compared with 11-14 birds in the four prior decades.

Where: The template of arrival for this long-distance migrant sketched out in the early 20th century was to hold true for the following 100 years. Arctic Warbler is still associated with Shetland (where it is more frequent than Greenish Warbler) and to a lesser extent the east coast (where Greenish is much commoner) and the southwest (where Arctic outnumbers Greenish). Along the south coast, it remains exceptionally rare, as it does in the west and northwest. This presumably reflects the more northerly breeding range of Arctic and the subsequent filtering through of birds to the southwest following landfall on the north and east coast. Greenish has a more southerly European breeding range and quickly reorientates following short-distance drift across the North Sea.

Shetland has accounted for 123 birds (48 per cent) – 72 from Fair Isle – with Orkney adding 21 and the Outer Hebrides just one. Six have occurred in Fife, three in Caithness and two each in Northeast Scotland and Lothian. The total number of birds seen in Scotland stands at 177.

The northeast of England is the region to have fared next best, with 44 birds, though even in some well-watched coastal counties it is still quite scarce with, for example, just 15 in Yorkshire, 14 in Northumberland, eight in Lincolnshire, four from Cleveland and three in Co. Durham.

Farther south it becomes even more rare. In East Anglia, the exceptionally well-watched Norfolk coastline accounts for 13 birds and there have been three in Suffolk and one in Essex. The only county in the southeast to have recorded the species is Kent, where three have been found.

The paucity of records along the south coast – with just two on Portland and one from Start Point, Devon – is intriguing, especially as Arctic Warblers reach the far southwest with some regularity. In this context an early bird at Sand Point, Somerset, on 17th August 1965 is an impressive find. There are three records from Lundy and seven have reached Cornwall. The saturation coverage of Scilly has produced 24 birds, making it second only to Shetland in number of birds.

The sole Welsh record was on Bardsey as long ago as 13th September 1968. One trapped at Meols, Cheshire, on 3rd October 1971 was notable, though not quite so impressive as the Staffordshire record.

When: In complete contrast with Greenish Warbler, spring overshoots are exceptionally rare, being late in date and all bar one restricted to Shetland. Three have occurred in June: Foula on 21st June 1996 and Fair Isle from 22nd-23rd June 2005 and 27th June 1995. Seven have arrived in July, all but one of which have been from

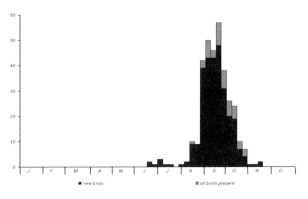

Figure 70: Timing of British and Irish Arctic Warbler records, 1902-2007.

Shetland. Three in the first week of July (two on Shetland and one at Titchwell, Norfolk) are presumably spring overshoots.

Three more July records on Shetland in 2002 were an intriguing phenomenon, with one ringed on Foula on 10th July followed by others on Fair Isle on 19th and 30th July. The remaining July record involved one on Fair Isle on 30th July 1928.

Arctic is predominantly an autumn visitant to our shores, usually occurring later than the peak period for Greenish Warbler, although autumn records start on 30th July. Just seven have been seen in the first half of August (one on Fair Isle on 7th, the rest from 12th onwards). Sightings increase during the second part of the month, peak in September and decline quite rapidly by mid October. Only 14 have been found in the latter half of October, seven of them in the southwest and Ireland.

Just two November birds have been located. One was at Baltasound, Unst, Shetland, on 10th November 2007. The latest ever was in 1984 and comprised two records of the same bird in Co. Durham, with one at Whitburn from 12th to 14th November and presumably the same bird seen at Seaham on 17th November.

An analysis of autumn records from Fair Isle since 1950 (Table 21) reveals that numbers remained broadly constant per decade until the turn of the millennium, since when there have been just three autumn records. The median arrival date on the island over the period has varied little around 12th-16th September. In contrast, the 24 records from the Isles of Scilly have a median detection date of 1st October. Intriguingly a third of records from the archipelago have been since 2000 at a time when Fair Isle has recorded its lowest numbers for over 40 years. The later date of Scilly sightings presumably reflects birds that originally made landfall farther north and filtered down to the southwest.

Table 21: Median arrival dates of autumn* Arctic Warblers on Fair Isle since 1960.

Period	1960-1969 (n=11)	1970-1979 (n=13)	1980-1989 (n=13)	1990-1999 (n=15)	2000-2007 (n=3)
Median Date	12th September	16th September	12th September	13th September	14th September

*defined as 1st August onwards

Discussion: Although most birders think of Greenish and Arctic as a pair, they have quite different temporal arrival patterns and differing geographical spreads. These reflect the different distributions and migration strategies of each species and the more northerly breeding distribution of Arctic. Arctic is a rare spring vagrant here, unlike the regular spring records of Greenish, which breeds much farther west.

Arctic Warbler has the greatest migratory journey of any Fennoscandian passerine. Departure from their winter quarters is late, mostly from April to early May. Breeding birds arrive in Finland in mid June, though the migration is prolonged during warm springs (Hagemeijer and Blair 1997). The relatively few spring records from Britain accord well with them being late spring overshoots.

Autumn migration commences early, with departure from late July and early August onwards, some two weeks or so earlier than Greenish. Despite the few extremely early autumn Arctic arrivals, Greenish is on average much the earlier of the two to arrive here. Greenish is prone to large-scale westward displacement as evidenced by several east coast falls of the species in recent years (c40 in both 1995 and 2005). Arctic, on the other hand, occurs later, with late September and early October arrivals not unusual.

The majority of British Arctic Warblers presumably originate farther east than those breeding in Fenno-Scandia to northwest Russia, where autumn departures take place mainly in August. Eastern birds depart later and correlate with the peak arrivals here. The majority of birds to be aged were 1st-winters. The preponderance of records from the Northern Isles relative to elsewhere could suggest that many are migrants on a reverse orientation. Those farther south, with much later detection dates, are likely to be birds that have filtered down following an earlier landfall farther north.

Julian R. Hough

Hume's Leaf Wabler *Phylloscopus humei*
(WE Brooks, 1878). Breeds in Altai mountains to western Mongolia, south through Tien Shan and Pamirs to northeast Afghanistan, northwest Himalayas and mountains in northwest China. Winters in southern Afghanistan to northern India, east to West Bengal. Another race, mandellii, breeds in central China from Shanxi to southern Yunnan, west to lower slopes of Tibetan plateau.

Rare but increasingly frequent late autumn migrant, including occasional wintering. It has a tendency to arrive later than Yellow-browed Warbler and is more usually associated with arrivals of Pallas's Leaf Warblers.

Polytypic, two subspecies. Records are attributable to the nominate *humei* (WE Brooks, 1878). Analysis of mitochondrial DNA sequences (Irwin *et al* 2001) revealed a substantial divergence between Yellow-browed Warbler *P.inornatus* and Hume's Leaf Warbler (*P.h. humei* and *P.h. mandellii* (Brooks, 1879)) as well as marked differences in song and call. Playback studies showed that *humei* and *mandellii* reacted strongly to each other's songs, but neither reacted to that of *P.inornatus*. The BOURC, in its 23rd report (1997) announced the treatment of *P.inornatus* (monotypic) and *P.humei* (including *mandellii*) as separate species, though the systematic position and status of *mandellii* is unclear, with Irwin (2001) proposing species status for this taxon; *mandellii* from China would appear to be an unlikely vagrant to the Western Palearctic.

Status: With a reputation for being a bit of a birders' bird, it is only recently that many observers have become better acquainted with the identification subtleties for this solemn *Phyllosc*. Even so, many are often logged first as the far commoner Yellow-browed Warbler before re-identification takes place. A total of 96 have now been identified (94 in Britain and two in Ireland).

Historical review: British birdwatchers first became aware of the (then) Hume's Yellow-browed Warbler in the mid 1950s (Alexander 1955). Treated as a race of Yellow-browed Warbler, there was little clarity of the identification features and its true status was obscured amid a number of unresolved claims in the 1970s and 80s. It was nudged briefly back into the limelight for some in the 1970s (Wallace 1973). The identification clouds truly started to dissipate in the late 1980s and early 1990s following a number of identification papers for birders eager to assimilate the information necessary to claim their own.

In its 24th report (1998), the BOURC added the species to the *British List* on the strength of a sight record at Beachy Head during 14th-17th November 1966. As a result of the reconsideration of other old records during the assessment of the first, more were added, the second at Cley from 2nd December 1967-7th January 1968, another trapped at Low Hauxley, Northumberland, on 7th November 1970 and four from the 1980s (Cumbria in 1982, East Yorkshire in 1985 and Northumberland and Norfolk in 1989).

With a greater awareness of the (then) race and the potential for a future split, birders started to pay more attention from the 1990s onwards and this led to 28 more records. Four in 1991 included a two-week January bird in Devon, initiating the discovery of wintering potential that was to become even more apparent from subsequent records. Two in 1993 included the first for Wales at Strumble Head on 20th November. Seven in 1994 came off the back of a good easterly autumn and were typically late, all in November, with the exception of an exceptional inland find at Westport Lake, Staffordshire, on 20th December. Observers had clearly got their eye-in on this tricky sprite.

A superb easterly airflow in early November 2000 produced a fall of birds, eight of them between East Yorkshire and Fife from 7th-11th, which was associated with more than 40 Pallas's Warbler *P. proregulus*. The initial arrival was followed by a west coast bird on Anglesey just over a week later and two south coast birds in early December. The total of 11 represented an exceptional arrival, many of which had been well watched and heard. Three each in 2001 and 2002 was a return to normal. Few were prepared for the events of 2003, which ended up producing an astounding 30 birds. Unlike the two previous influxes, this autumn influx occurred early, from mid-October onwards, and was allied with a record autumn for vagrant *Phylloscs*.

The arrival was shared with over 300 British Pallas's Warblers and more than 850 Yellow-browed Warblers. For many observers, the combination of these scarcities at patches along the east coast made them the commonest migrants on offer in the third week of October.

No fewer than 16 Hume's Leaf Warblers were logged between Suffolk and Northumberland, plus one in Cornwall, from 15th-27th October 2003. In early November one in East Sussex on 6th was the forerunner of another push of arrivals with six more located between 11th-30th November, including the first three records for Shetland. Three more were found elsewhere in December, all wintering birds from western locations: Caernarfonshire and the first for Ireland at Knockadoon Head, Co. Cork, from 18th December, followed rapidly by the second in Co. Wexford; both Irish birds remained into early 2004. Another was found at Fairlop Waters, Essex/London, from 11th January-25th April 2004, when it was in song. What was thought to be the same individual was relocated in song at Brent Reservoir, London, on 1st May. Between 2000-2007 no fewer than 61 birds had been logged, and the status of this subtle Phylloscopus continues to change apace.

Where: This is very much an east coast specialty. There have been 17 in Norfolk, 12 in East Yorkshire and 10 in Northumberland. Elsewhere in the northeast four have come from North Yorkshire, three Cleveland, two Lincolnshire and one in Co. Durham. Elsewhere in East Anglia two have been seen in Suffolk and one in Essex, and in the southeast five in East Sussex and three from Kent.

The southwest has attracted few records. These include five apiece in Dorset and Cornwall, three in Devon, two from Scilly and one on the Isle of Wight. This is a surprisingly rare bird away from Portland (four birds) on the south coast. It is also rare in Scotland, where there have been just 12 records (three each from Shetland, Northeast Scotland and Fife) and singletons from Angus and Dundee, Lothian and Orkney.

The sole record from an inland county was from Staffordshire and the only record for the northwest was from Cumbria. Four Welsh birds included two from Caernarfon and singles from Pembrokeshire and Anglesey. Two have reached Ireland, both associated with the 2003 influx.

This species appears quite happy to winter along the North Sea coastline, and several birds have done so successfully. Thus far it appears to be the only scarce Phylloscopus to entertain the notion of regular wintering in this geographical area (one from Northeast Scotland, two in Northumberland, two in Yorkshire, four in Norfolk and one from Essex; other records of short-stay birds in late winter in these areas may have involved wintering birds). Wintering birds are rare

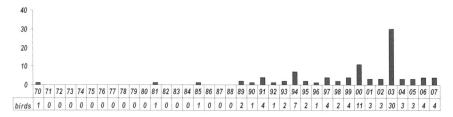

Figure 71: Annual numbers of Hume's Leaf Warblers, 1970-2007.

along the south coast (one in East Sussex) and there have been relatively few in the southwest (two Devon, one each in Cornwall and a short-staying bird in Dorset). The dearth of records from this area is surprising given the number of wintering Yellow-browed Warblers there. Farther west, both Irish birds attempted to winter. One has overwintered in the Midlands.

When: The earliest record was at Holkham Meals, Norfolk, from 6th-11th October 2007, almost a week earlier than the next at Auchmithie, Angus and Dundee, on 13th October 1991 and Skinningrove, Cleveland, on 15th October 2003. These constitute the only birds seen prior to mid-October.

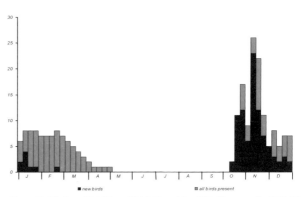

Figure 72: Timing of British and Irish Hume's Leaf Warbler records, 1966-2007.

Late October and more especially November is the peak time for arrivals. This fact is perhaps exaggerated by several influx years, but one that still holds true even in years when few are seen. Around 20 per cent of all birds have been located in December and January, a number of which have wintered through to March and April; one remained to May. In contrast none have yet been detected on a spring exit out of the country, even at an east coast location, unlike a number of other Sibes and increasing numbers of Yellow-browed Warblers and occasionally Pallas's Warblers.

Discussion: Elsewhere in northern Europe there have been over 20 from Norway, 42 in Sweden (more than half since 2000), 41 in Finland (34 since 1991; one spring record, others 7th October-27th November), over 20 from Denmark and over 32 in the Netherlands since 1980 and up to and including 2007. In Germany the first record was in 1990, but it is now nearly annual. Farther south there have been seven in Italy and there are single records from Turkey (28th December 1994), Hungary (11th November 2007 to 1st January 2008) and Spain (6th-8th December 2008). In the Middle East it is an annual winter visitor to Israel in small numbers.

The occurrence pattern here for Hume's differs markedly from Yellow-browed Warbler, being generally a much later arrival more closely correlated with Dusky *P. fuscatus* and Pallas's Warblers *P. proregulus*. The reasons for this association are unclear. Their breeding ranges overlap only very slightly, in the Altai Mountains, but it is presumably because all depart from their breeding grounds later than Yellow-browed. Hume's Leaf Warblers begin to vacate breeding grounds from August onwards, but the main movement out of western Siberia takes place from September to mid-October. These correlate well with the later autumn arrivals here which follow the peak late September to mid-October arrival period for Yellow-browed Warbler.

Quite why Hume's undertakes such vast migrations into Western Europe is unknown. It shares both Dusky and Yellow-browed Warblers' propensity for wintering here.

This species is now firmly established as a regular vagrant, albeit of fluctuating occurrence between years. Any late apparent Yellow-browed is worth a closer look. Despite increased observer competence with the species a number are still initially misidentified; it seems likely that many will have been disregarded as the commoner species in the past.

Western Bonelli's Warbler *Phylloscopus bonelli*

(Vieillot, 1819). Breeding range centered on southwest Europe from Iberia to northern France, southern Germany, Italy, Austria, and locally in mountains of North Africa. Winters along southern edge of Sahara, from Senegal and southern Mauritania to northern Cameroon.

Monotypic. Formerly treated with Eastern Bonelli's Warbler *P. orientalis* as a single species, Bonelli's Warbler. In British ornithology this is the first species for which DNA data contributed to a split. Helbig *et al* (1995) studied the molecular divergence between western (*bonelli*) and eastern (*orientalis*) populations of Bonelli's Warbler and found the degree of differentiation to be as great as that between these and Wood Warbler *P. sibilatrix*. Clear vocal differences between the two species (Geroudet 1973; Helb *et al* 1982) are supported by slight, but consistent, differences in morphology (for example, Page 1999). The proposal to treat them as two species was accepted widely (for example, Baker 1997; Beaman and Madge 1998, Snow and Perrins 1998) and was embraced by the *BOURC* in its 23rd Report (*BOURC* 1997). The species were named Western Bonelli's Warbler and Eastern Bonelli's Warbler; Balkan Warbler has been suggested for the latter (Beaman and Madge 1998), but has not received widespread use.

After the split it became necessary to review all previously published British records to establish which species, in light of new knowledge, featured within them. The tests were essentially based on the wing formula for trapped birds or the call note for field observations. Records failing to meet these criteria were adjudged indeterminate (either *bonelli* or *orientalis*), though most of these were felt to probably relate to Western Bonelli's Warbler (Rogers *et al* 1998).

Rare vagrant from southwest Europe or northwest Africa, mostly in autumn.

Status: A total of 96 Western Bonelli's Warblers have been identified in Britain and Ireland – 85 in Britain and 11 in Ireland – between 1948-2007. One Irish record from 2005 is yet to be accepted.

A further 80 birds have been recorded that have not been assigned to species, 74 of these indeterminate birds have been recorded in Britain and six in Ireland.

Historical review: The first British record was a female trapped on Skokholm on 31st August 1948. It was subsequently killed for identification purposes, the last such collection of a rarity at a Bird Observatory.

At 19.30 B.S.T. on August 31st, 1948, a warbler, was caught in the Garden trap on Skok-holm Island, Pembrokeshire. After a long examination that evening and again on the following morning it was decided to kill the bird. The specimen was sent to Mr. R. Wag-staffe, Director of the Yorkshire Museum, who identified it as a female Bonelli's War-bler (Phylloscopus bonelli).

P. J. Conder and Joan Keighley. British Birds 42:215-216.

The 1950s produced five records of Bonelli's Warbler, three of which are still acceptable as Western. One was trapped at Portland, Dorset, on 29th August 1955 and in 1959 two reached Bardsey: one trapped and retrapped five times between 18th August and 5th September and a second bird trapped on 10th September. Birds now regarded as indeterminate include one trapped on Lundy on 1st September 1954 and one at Marazion Marsh, Cornwall, on 14th September 1958.

Of the 1960s records, 11 were subsequently found to be acceptable as Western Bonelli's War-blers and four were no longer attributable to species. Two in 1961 included the first Irish record trapped on Cape Clear and present from 2nd to 3rd September 1961. Three in 1962 included two autumn birds on Bardsey plus an indeterminate bird in Co. Wexford. The decade also pro-duced the first inland bird, a male in song at Delamere Forest, Cheshire, from 19th May to 9th June 1963.

Seventeen from the 1970s have stood the test of time, all coastal birds. A further 23 indeterminate birds occurred, making up a creditable 40 for the decade. Eleven of these occurred in 1976 (five were Western Bonelli's). Most were on the south coast, 25 between Scilly (five Western and four indeterminate birds) and Kent. An indeterminate bird from Reading, Berk-shire, on 23rd August 1975 was a quite exceptional inland record.

There followed 43 more in the 1980s, 20 of them Western Bonelli's Warbler's and 23 indeter-minate birds. Once again the majority were located along the south coast, but northwest records came from Cumbria and the Isle of Man. Southern Ireland weighed in with six (just one not acceptable to species level): three in Co. Cork, including an indeterminate bird, plus singles in Co. Waterford and two in Co. Wexford.

An interesting bird from this period included the 'Penllergaer Phylloscopus' at Penllergaer Forest, Glamorgan, from 17th May-12th June 1985. Initially the song of this bird suggested Bonelli's Warbler, as did the grey-and-white plumage, but the bold head pattern, long wings and occasional upright posture was more in keeping with Wood Warbler *P. sibilatrix*. The bird was eventually trapped and on wing-formula found not to be a Bonelli's Warbler. This male was

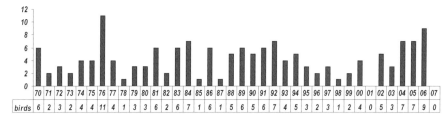

Figure 73: Annual numbers of Western Bonelli's Warblers and unassigned bonelli's warbler sp., 1970-2007.

associating with a breeding female Wood Warbler and it seems likely that it was an aberrant Wood Warbler with no yellow pigment (Moon and Herbert 1989).

The 1990s accumulated a further 38 birds: 13 Western Bonelli's Warblers and 25 indeterminate birds. The spread of records centered along the south coast, but a number were found on the east coast between Suffolk and Northumberland, including four in Yorkshire.

Greater observer awareness has resulted in most records during the period 2000-2007 being allocated to species level, 31 of them Western Bonelli's Warbler. However, despite increased knowledge some still defy identification and four such indeterminate birds have also been recorded over the period.

The number of birds recorded has remained relatively stable since the 1960s. Since 1970 the average number of Western Bonelli's or indeterminate birds recorded per annum has been just over four.

Where: The arrival pattern tends to be typical of Mediterranean overshoots. Spring records have a southerly bias; autumn birds occur on the east and south coastlines.

The southwest is the premier region for occurrences, accounting for no fewer than 61 birds (31 Western Bonelli's/30 indeterminate). The majority have been on Scilly with 37 (15/22), followed by 13 in Cornwall (9/4), eight in Dorset (5/3) and three from Devon (2/1).

Although it is a rare species away from southern Britain, Scotland has weighed in with 27 birds (14 Western Bonelli's/13 indeterminate), 16 of which are from Shetland (7/9), seven from Orkney (4/3) and two from Argyll (1/1). There have also been Western Bonelli's Warblers in the Outer Hebrides and Borders.

The southeast has yielded 22 birds (10 Western Bonelli's/12 indeterminate): 11 from Kent (6/5); eight in East Sussex (4/4); and two indeterminate birds from the Isle of Wight plus one in Berkshire.

Seventeen have been recorded in Ireland (11 Western Bonelli's/6 indeterminate), 10 from Co. Cork (6/4) and four from Co. Wexford (3/1), with two Western Bonelli's Warblers from Co. Waterford and an indeterminate bird from Co. Kerry.

In East Anglia 17 have been recorded (10 Western Bonelli's/7 indeterminate), Norfolk accounting for 11 (7/4), Suffolk five (3/2) and an indeterminate bird in Essex. The northeast has produced 17 birds (10 Western Bonelli's/6 indeterminate), 11 in Yorkshire (6/5) and six in Northumberland (4/2).

Wales has accumulated 11 (7 Western Bonelli's/4 indeterminate), seven from Caernarfonshire (5/2) and two from Pembrokeshire (1/1) plus a Western Bonelli's Warbler in Radnorshire and an indeterminate bird in Glamorgan.

Four in the northwest (3 Western Bonelli's/1 indeterminate), comprise two from the Isle of Man (1/1) and single Western Bonelli's Warblers from Cheshire and Cumbria. A noteworthy record from the Midlands involved an indeterminate bird in Leicestershire.

The majority have been coastal sightings and birds away from the coastal counties are notable. A male Western Bonelli's' Warbler was in song at Delamere Forest, Cheshire, from 19th May to 9th June 1963 and another sang at Gwastedyn Hill near Rhayader, Radnorshire, from 17th to 18th May 2006, both quite exceptional records. Indeterminate birds well inland include finds at

Small Mead Farm Gravel-pit, Reading, 23rd August 1975 and Hambleton Wood, Rutland Water, Leicestershire and Rutland, on 2nd May 2006.

When: Despite a large breeding population in Western Europe wintering south of the Sahara, spring overshoots remain surprisingly rare. Just 16 per cent of all Western Bonelli's Warblers and indeterminate records have occurred during spring and the majority of sightings have been in the autumn (84 per cent).

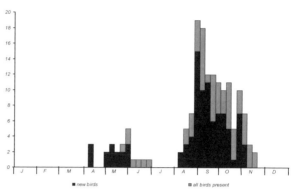

Figure 74: Timing of British and Irish Western Bonelli's Warbler records, 1948-2007.

Four spring records of Western Bonelli's Warbler have occurred in April, three on the same date, the 9th: Beachy Head in 1972; Lundy in 1976; and Holme, Norfolk, in 1988; and one on 30th April 2000 on St. Agnes, which remained to 5th May. A further indeterminate Bonelli's Warbler was at St Margaret's, Kent, on 28th April 1987. Another 16 have occurred through to early June, the latest both identified as Western: a 1st-summer trapped at Spurn on 2nd June 1999 and one at Sandwich Bay, on 3rd-29th June 1967.

Autumn records for all Bonelli's Warblers run from early August (the earliest Western Bonelli's at Holme, Norfolk, 7th to 13th August 1970 and earliest indeterminate bird at Alfriston, East Sussex, 7th August 1973). Peak numbers arrived from late August through September and into early October.

Two Western Bonelli's Warblers on Orkney in October have remained to 15th November (one at Graemeshall, Holm, Mainland, from 26th October

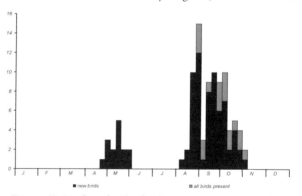

Figure 75: Timing of British and Irish indeterminate bonelli's warbler sp. records, 1948-2007.

1986 and another at Herston, South Ronaldsay, from 29th October 2004). Two early November records represent the latest arrivals: an indeterminate bird on St. Mary's from 2nd to 4th November 1980 and a Western trapped at Hauxley, Northumberland, on 4th November 1967.

Discussion: Western Bonelli's Warbler increased slowly in number and range during the 20th century, reaching northeast France and then Belgium (late 1960s) and the Netherlands (mid-1970s). The core population is in Spain, which accounts for over 65 per cent of the European breeding population. Between 1990 and 2000, the sizeable French population suddenly halved. Declines were reported in a number of countries, though trend data was not available for the Spanish population (Hagemeijer and Blair 1997; BirdLife International 2004). The decline in records in Britain and Ireland over the same period doubtless mirrored this downturn in fortunes.

Despite the fact that Western Bonelli's Warbler has a large European population, put at over 1.4 million pairs (BirdLife International 2004), and a West African wintering range, records in Britain and Ireland are relatively rare. Spring birds presumably represent overshoots; the concentration of records in the southwest and along the south coast would appear to support such an assumption. In autumn the main migration route takes birds on a south to southwesterly heading, with the peak southerly movement occurring in August. Vagrants at this season presumably involve inexperienced young birds orientating away from the expected southerly migration route to Africa.

Eastern Bonelli's Warbler *Phylloscopus orientalis*

(C.L. Brehm, 1855). Breeding confined to eastern Mediterranean from Balkans to southern Bulgaria and Greece east to southern Turkey. Winters in northeast Africa from Sudan to Ethiopia.

Monotypic. Taxonomy discussed under Western Bonelli's Warbler.

Status: Four British records, plus one pending.

1987 **Scilly:** St. Mary's 30th September; presumed same, 1st-winter, 8th-10th October
1995 **Northumberland:** Whitley Bay 20th-29th September
1998 **Shetland:** Sumburgh, Mainland 27th-28th August, trapped both days; same, Grutness 29th August-5th September
2004 **Devon:** Lundy 26th April
2009 **Dorset:** Portland 1st May*

Discussion: At the time of the announcement of the separation of Western and Eastern Bonelli's Warblers, three claims of *P. orientalis* were under consideration by the *BBRC* and the *BOURC*. Of these, only the 1995 Northumberland bird was accepted. One of the rejected birds was the 1987 Scilly bird. *BOURC* were asked to review this bird again in light of new correspondence (which involved a review of contemporary field notes, which proved to include unambiguous renditions of *orientalis* calls) and was found to be acceptable as the first for Britain (Wilson and Fentman 1999). The morphological differences between the two are subtle. Eastern is greyer on the head and upperparts, but this feature is variable; *orientalis* also has more contrast between the darker tertial centres, tertial fringes with whitish or silver-white edgings and a more obvious supercilium. The best feature for separation is the call, a sparrow-like hard 'tyip' in Eastern compared to the disyllabic 'hu-it' of Western (Page 1999).

The population of Eastern Bonelli's Warbler is estimated at only 15,000-40,000 breeding pairs, just one per cent that of the Western Bonelli's Warbler population (Hagemeijer and Blair 1997). The winter quarters are poorly known, but spring migration starts in late February, earlier than *bonelli*. The paucity of British records of *orientalis* relative to *bonelli* can be attributed to the differing magnitude of the populations and the fact that the migration routes of *orientalis* are largely north-south, making spring overshoots less likely.

Elsewhere in northern Europe there have been four from the Netherlands (15th-16th May 1983, 17th May to late June 1983, 13th-16th July 1986, 30th April-1st May 1993), three from Norway (21st-22nd September 2002, 8th-9th May 2004, 22nd-24th September 2007; the latter still pending but likely to be accepted), plus singles from Sweden (15th May 1992), two from Finland (2nd June-1st July 1997, 9th October 2004) and one from France (16th September 2000).

Interestingly, three of the four British records have been in the autumn, whilst in northern Europe spring and early summer records outnumber autumn birds 2:1. This presumably reflects the propensity for spring overshoots to reach Scandinavia, though it is interesting that all four records from the Netherlands have been in spring and summer, outside of the autumn period. In Britain the occurrence pattern parallels that of Western Bonelli's which is far more frequent in autumn.

Surprisingly, perhaps, despite a greater awareness of what to look (and listen) for, this remains an exceptionally rare vagrant to our shores.

Iberian Chiffchaff *Phylloscopus ibericus*

Ticehurst, 1937. Breeds locally in French Pyrenees and in most of Iberia. North African range restricted to northwest Morocco and northern Algeria to northwest Tunisia. Non-breeding range imperfectly known, but majority probably migrate to tropical West Africa. Recorded widely in northwest Africa in non-breeding season, mostly Tunisia, but exact numbers uncertain due to confusion with wintering Common Chiffchaff P. collybita. Many of those breeding in Iberia and southwest France are thought to move short distances to lower levels, usually in valleys and along the Mediterranean coast.

Monotypic. Iberian Chiffchaff was split from Common Chiffchaff *P. collybita* by the *BOURC* in January 1998, based on differences in structure and plumage, vocalisations and mitochondrial DNA sequences (Helbig *et al* 1995, Clement *et al* 1998). Iberian Chiffchaff was formerly known under the name *brehmii*, but it was later established that the type specimen of the taxon was not an Iberian Chiffchaff (Svensson 2001).

Status: Nearly all 15 British records of this subtle chiffchaff were sound recorded and/or trapped. The rise in records since 1999 is doubtless due to increased awareness, though this remains a tricky species to identify. Several mixed singers in recent years have masqueraded as Iberian only to fall short on their final assessments.

1972 **Greater London:** Brent Reservoir 3rd June
1992 **Scilly:** St. Mary's 14th April-21st May
1999 **Dorset:** Verne Common, Portland 25th April-8th July, trapped 9th May
1999 **Devon:** Start Point 6th-14th May
2000 **Cornwall:** Dunmere Woods, near Bodmin 13th-31st May
2001 **Kent:** Dungeness 14th-17th April
2001 **Oxfordshire:** Great Tew 27th April to at least 15th May
2003 **Devon:** Kingswear 19th May to at least 17th June, possibly since 6th May
2004 **Northumberland:** Woodhorn 18th-19th April
2004 **Cornwall:** The Lizard 30th April-3rd May
2004 **East Yorkshire:** Easington trapped, 17th May
2006 **Devon:** Challacombe Common, Dartmoor 1st May-6th June
2006 **Lothian:** Pitcox 5th May; presumed same, Pressmennan Lake, 6th–13th May
2007 **Norfolk:** Colney Lane, Norwich 21st April-7th June
2007 **Devon:** Beer Head 28th April

Discussion: Salomon *et al* (2003a, 2003b) proposed two subspecies of Iberian Chiffchaff, a treatment that was questioned by Elias (2004). The southern form, nominate *ibericus*, breeds "in an area composed of relatively wet Mediterranean habitats, mainly cork-oak formations sparsely spreading from central Portugal (Coimbra region) to southern Andalusia"; the northern form *biscayensis* breeds "between the extreme north of Portugal and Galiciato the French basque country across the Cantabrian Cordillera (northern and southern slopes), the Spanish Basque provinces and Navarra". This area is characterised by a wet Atlantic climate, moist and mild throughout the year. Salomon *et al*'s analyses indicated biometric differences with *biscayensis* having significantly longer wings, longer 10th primaries and, relative to body size, shorter tarsi (both sexes) and shorter and wider bill (males only); a cryptic split for the future?

Given the identification conundrums, documentation of an out-of-range Iberian Chiffchaff should include sound-recordings of a variety of song phrases and the distinctive call. Morphological differences to note include a relatively long tail, wing and bill, as well as bright plumage coloration and well-marked supercilium, but none are diagnostic on their own. In the hand, biometric differences can be found in the wing formula (Rogers *et al* 2004; Slaterus 2007).

The typical song is characteristically made up of three phrases comprising short rising and falling notes concluded by a simple loud rattle. The call also differs from Common Chiffchaff *P. collybita*, having a 'sad' downward inflection. Without proof of these differences, preferably sound recorded, rarities committees are unlikely to accept claims (Rogers *et al* 2004).

In recent years there have been a small number of Iberian Chiffchaff claims that have not presented acceptable characters of this species (for example, Skelmersdale, Lancashire, and Dibbinsdale, Cheshire, in 2004, plus others in the Netherlands). Quite what these birds are is presently uncertain. It has been suggested that Common Chiffchaffs performing odd song phrases may involve first-year birds that have yet to crystallise their song (Slaterus 2007). So-called mixed singers have been reported from a small contact zone where Common and Iberian Chiffchaffs in the Pyrenees are sympatric. These birds have structures pertaining to both taxa

(Salomon 1987: Salomon *et al* 2003b) and may be either of hybrid origin or they could be Iberian Chiffchaffs that have picked-up Common Chiffchaff song (for example, Salomon 1989; Salomon and Hemim 1992; Collinson and Melling 2008).

Elsewhere in northern Europe the number of records is growing steadily in line with observer awareness. In Scandinavia records are still rare with, for example, five in Denmark (two in late April, two in May and one from late June-July), four in Sweden (two in April and two from May onwards). The first for the Netherlands was in 1967 and a total of 18 birds have been accepted there, with most birds found between 27th April-10th May. There are also 10 for Germany, three from Belgium and singles for Poland (June) and the Czech Republic (July).

Based on northern European records, Iberian Chiffchaff would appear to be an annual over-shoot. The two British records in 1999 allowed many observers to learn the distinctive song of this species. Doubtless more will be found in coming years.

Asian Brown Flycatcher *Muscicapa dauurica*

Pallas, 1811. Breeds across northern Asia from western Siberia to Japan, also locally in southern Asia. The nominate race, in the northern part of the range, is a long-distance migrant that winters across India and southeast Asia to the Philippines.

Polytypic, six races. Nominate *dauurica* Pallas, 1811 from central Asia to Japan, and also India, has occurred in the Western Palearctic.

Very rare vagrant from Siberia.

Status: Two records, both awaiting acceptance by the relevant bodies.

2007 **East Yorkshire:** Flamborough Head adult 3rd-4th October*
2008 **Shetland:** Fair Isle 1st-winter, 24th-25th September*

Thankfully for those who saw the first of these, it put on a two-day show in the famous Old Fall Plantation at Flamborough Head, an easily accessible mainland location for a change.

October 3rd dawned overcast on Flamborough Head with a light ESE and promise of rain. After a check of the garden Phil Cunningham and I set out for a morning's birding. News of two freshly arrived Yellow-browed Warblers nearby filled us with optimism.

After seeing out a light rain shower the first few migrants appeared; Redwings, a male Ring Ouzel and a couple of Brambling. We arrived at Old Fall an hour later and, as is often the case, the most birds we had seen all morning were in the wood. The rain had stopped and despite being wet we were in prime position. Phil searched the wood whilst I checked the fields.

After birding this small area we met on the north side of the plantation. Phil and I both had brief views of an unusual small flycatcher. We waited for it to appear again with Kevin Barnard.

When the flycatcher appeared in front of us we were expecting to clinch its identity straight away. None of us were quite expecting it to look like a Brown Flycatcher! Looking back I remember how cool we all pretended to be, and how we were almost in denial of the obvious! Could it be an Eastern Spotted Fly? Were we missing features which would make it a Red-breasted? It was as if we wanted to believe these possibilities rather than the evidence of our own eyes. I decided we needed to rule out all the Asian/Siberian flycatchers, and as we were so close to home I strode quickly up Old Fall path in search of the truth.

Having no photos of the bird I rang Mike Pearson. Andrew Lassey was with him and after hearing my description wasted no time in finishing his tea and getting down to the plantation. Staring at my bookshelf for help took too much time so I was straight on to Google Images. Photos of Brown Flycatcher screamed at me … I was at Phil's house like a shot where he was coolly making tea! On to his computer and bang! Phil's cool exterior gave way to "Sh**, sh**, sh** …!" We abandoned the tea and raced to the plantation.

Down at the wood it took a nervous 30 minutes to relocate the flycatcher. Andrew got some quick shots and we discussed and ruled out species such as Sooty and Grey-streaked. It was time to contact Simon Waines, the farmer of Head Farm. Fortunately Simon is accommodating in such situations, and was happy to have the news put out providing we managed the crowd.

Up to 800 birders gained views of the bird up to dark on the 4th. A car park was opened in a nearby stubble field. £1,000 was collected for use of the field, local charities and non-profit organisations. Everyone associated with the twitch had only good words for the behaviour of visiting birders, including Simon, who over the past two years has committed a lot of land to Higher Level Stewardship.

Richard Baines. *Flamborough Bird Observatory Report for 2007*

Discussion: The breeding grounds are vacated from mid-August to early October, with the peak departure in late September. Thus the credentials for the Yorkshire bird would appear to be impeccable in terms of date and location, and there was a fine supporting cast of Sibes from Shetland southwards. The only slight taint might be the bird's age, given that insectivorous vagrants from so far east are almost all 1st-winters. The Fair Isle bird lends it further weight.

An acceptable British example of this familiar species for birders visiting eastern Asia has been long overdue. There are three other accepted European records: Denmark (24th-25th September 1959), Sweden (27th-30th September 1986) and Greece (4th September 1993). In the Middle East, there was a published record from Iran (21st September 2001) that was considered most probably of this species (Sehhatisabet 2006) and one from Oman (1st-winter 25th-26th October 2006).

Historically there were other claims of Brown Flycatcher. These included birds on Holy Island on 9th September 1956 and in Co. Wexford on 6th September 1957, both of which were not accepted because key field characters were missing from the descriptions. A more recent claim involved one trapped and ringed on Copeland Island, Co. Down, on 24th October 1971. No sketches or photographs were taken and the bird's bill structure and weight fell outside the range for Brown Flycatcher. There was also a report of some darker feathering in the crown. The record was rejected.

An even more recent record of a 1st-summer on Fair Isle from 1st-2nd July 1992 was widely expected at the time to represent the first confirmed record for Britain (Harvey 1992). After much deliberation, the BOURC decided controversially to place the record in Category D, because of the escape potential and an unusual date that did not conform to the expected autumn arrival pattern for insectivorous Siberian vagrants (Parkin and Shaw 1994). Siberian insectivores in summer are incredibly rare; such species usually have a prior track record of autumn showings.

The increase in eastern vagrants over recent years suggests that further records of this long-distance migrant are to be expected, though few would have predicted the arrival of birds in consecutive years.

Taiga Flycatcher *Ficedula albicilla*

Pallas, 1811. Breeds Siberia from Ural mountains east to Sakhalin and Kamchatka, south to northern Altai, Baikal region, northern Mongolia and Amurland. Winters from northwest India east to southern China and south to Malay Peninsula.

Monotypic. Formerly treated as conspecific with Red-breasted Flycatcher *F. parva*, Taiga Flycatcher was elevated to full species status by the BOURC in 2004 on the basis of differences in song, plumage pattern, moult sequence and divergent mtDNA (Sangster *et al* 2004; Svensson *et al* 2005). At the time of both British records it was considered a subspecies, but these records highlighted the need for an assessment of its taxonomic status. The subsequent split had been suggested several years previously following the first European record (Cederroth *et al* 1999).

Very rare vagrant from Siberia.

Status: Two records
2003 East Yorkshire: Flamborough Head 1st-summer male, trapped 26th-29th April
2003 Shetland: Sandgarth, Mainland 1st-winter trapped 12th-15th October

Discussion: Since the first record for Europe in Sweden (26th October 1998) it has been more eagerly sought. In France, there are three accepted birds, all 1st-winters (Sein Island, Finistère, on 14th October 2000 and 28th September 2002 and The Camargue, 3rd November 2002). A cited record from Denmark (Lassey 2005) is not accepted (1st-winter trapped and photographed 16th September 2002).

Although long expected as a vagrant to our shores, it still came as a bit of a surprise that the first was a cracking male, albeit a 1st-summer individual. It was followed closely by a less striking 1st-winter on Shetland. The Flamborough bird was appreciated by large numbers of observers over its extended four day stay. Presumably this bird had overwintered somewhere in Western Europe, even Britain, or possibly in western Africa. This bird exemplifies the relatively recent phenomenon of increasing number of Sibes on the east coast in spring.

Given that its breeding range extends to the eastern boundary of Europe, it is perhaps surprising that it took until 2003 for Taiga Flycatcher to make the *British List*, and equally unexpected that others have not followed in their wake. A lack of observer awareness has probably masked its true status over the years and its former treatment as a race of Red-breasted Flycatcher stifled interest in the past. Given greater attentiveness, more are surely to be expected.

Collared Flycatcher Ficedula albicollis

(Temminck, 1815). Small numbers breed west to eastern France and southern Germany, but most occur from central and eastern Europe to temperate regions of European Russia west of the Urals. Isolated populations breed on the Baltic islands of Gotland and Öland, and southern Italy. Winters in east and central Africa from Tanzania to Zimbabwe.

Monotypic.

Very rare vagrant from central or eastern Europe, nearly all in spring.

Status: This striking species was recorded just 28 times in Britain during the period (and two obliging males in spring 2009 were well received). It has never been recorded in Ireland.

Historical review: The first was an adult male shot on Whalsay, Shetland, on 11th May 1947. The next was on Bardsey on 10th May 1957. Three males were logged in the 1960s, from Orkney on 30th May 1963, found dead in Cumbria on 2nd June 1964 and in Norfolk from 4th-6th May 1969.

The status of this eye-catching flycatcher changed little during the 1970s, when four were recorded. Three of these came from Shetland, one of which was the first female (Out Skerries, 25th May 1976). Two in 1979, from Shetland and Essex, provided the first ever multiple arrival.

For a period in the mid-1980s, multiple sightings become the norm with six males found in just three years. Two in 1984 included the first long stayer, a male at Foreness, Kent, from 24th May-9th June. During spring 1985 an easterly airflow delivered three east coast birds between 12th and 21st May. These included cracking males that were well watched in Lowestoft, Suffolk, and Filey, North Yorkshire, and well scrutinised, but less easily identifiable, a 1st-summer male at Holkham Meals, Norfolk, from 12th-13th May. This individual was greatly debated, with some believing that it was a possible Collared x Pied Flycatcher *F. hypoleuca* hybrid. Following

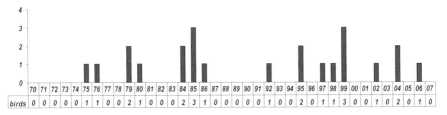

Figure 76: Annual numbers of Collared Flycatchers, 1970-2007.

the discovery of a new series of photographs, the bird was belatedly accepted in 2002 (Rogers *et al* 2003). One trapped on Fair Isle on 8th October 1986 was the first, and so far only, 1st-winter to be found acceptable.

Eight in the 1990s, three of which were in 1999, was a match for events of the preceding decade. All bar three were on Scottish islands, but two on the east coast of Scotland were widely appreciated (1st-summer male at Ethie Mains, Angus and Dundee, from 31st May-1st June 1997 and a male at Cove, Angus and Dundee, from 30th April-1st May 1999). A noteworthy record on North Ronaldsay on 31st May 1999 was the second female to be accepted.

A further four between 2002 and 2006 were all from Shetland, with the exception of a one-day male at Church Norton, West Sussex, on the late date of 20th June 2002.

Relatively few of the 28 records have been accessible to the masses. Since the spring arrival of 1985 just two have put in extended stays on the mainland.

Where: This species is very much a specialty of the Northern Isles and the east coast. In Scotland, 11 birds have made landfall on Shetland, three on Orkney and two on the east coast. The only other Scottish record is from the Outer Hebrides.

In England, five have come from East Anglia, three of them between Holme and Cley on the north Norfolk

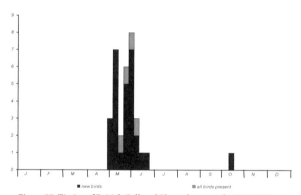

Figure 77: Timing of British Collared Flycatcher records, 1947-2007.

coastline and singles in Essex and Suffolk. There are single records from Kent and West Sussex. The only other east coast find was from North Yorkshire. Records in the west are exceptionally rare: one on Scilly, the sole Welsh bird from Bardsey and a male found dead near Ravenglass, Cumbria, on 2nd June 1964.

When: The 27 spring sightings fall in the range 30th April (Angus and Dundee) to 20th June (West Sussex). There were seven records between 9th and 13th May and 12 between 21st May and 2nd June. In 2009 a male at Portland (Dorset) was the earliest ever (28th April-2nd May); it was followed by another male in Fife in May.

The sole autumn bird was a 1st-winter trapped on Fair Isle on 8th October 1986. A bird trapped at St Abb's Head, Borders, on 9th September 1995 was considered to have been possibly a hybrid Collared x Pied Flycatcher (Leyshon and Kerr, 1996).

Discussion: All but two of the spring birds have been males. The two females came late to the Northern Isles in the last week of May. It seems unlikely that a dazzling male could escape detection for long, but they often prefer the tops of taller trees and can be surprisingly elusive. Thus, one on the mainland might be easy to overlook; a male in the more barren landscape of the Scottish islands would be far more obvious. The paucity of females in the records implies that they have been overlooked. Likewise, the difficulty of identifying autumn birds probably means that their true status at this season is under-represented and may never be fully realised.

Despite sizeable breeding populations in eastern and central Europe northwards to Sweden and a wintering range south of the Sahara, Collared Flycatcher remains an exceptionally rare vagrant here. It has been suggested that their rarity may be explained by their migration orientation (Cottridge and Vinicombe, 1996). In autumn Collared Flycatchers head on a narrow south-southeasterly, moving from their central European breeding areas through the eastern Mediterranean and the Levant to their wintering grounds. Such a route creates a narrow vagrancy shadow (Cottridge and Vinicombe, 1996), but this does not satisfactorily explain their rarity here. They are rare as autumn vagrants anywhere within the vagrancy shadow and it seems more likely that the difficulty of identification is masking their true occurrence patterns in autumn.

The spring migration is on a broader front, with the heaviest passage through the eastern Mediterranean. It seems that only occasionally do spring overshoots get deflected westwards by prolonged periods of suitable weather systems and make landfall here.

Long-tailed Tit Aegithalos caudatus

Linnaeus, 1758. Breeds from Ireland across the Palearctic to Japan, and south to Iran and central China.

Polytypic, up to 19 subspecies. These typically comprise four subspecies grouping (*europaeus, glaucogularis, caudatus* and *alpinus*) of which the Northern Long-tailed Tit, nominate *caudatus* (Linnaeus, 1758), is the only white-headed form. Both *caudatus* and *europaeus* (Hermann, 1804) of western and central Europe to the south of nominate *caudatus*, have been recorded as vagrants in addition to the endemic *rosaceus* Mathews, 1938, of Britain and Ireland.

Northern Long-tailed Tit Aegithalos caudatus caudatus

Breeds from Fennoscandia and northeast Europe across northern Asia to Mongolia, Siberia and Japan. Occasionally dispersive.

Very rare vagrant from Northern Europe.

Status: The first British record of a Northern Long-tailed Tit was from Tynemouth, Northumberland, in November 1852. Although its occurrence has not been closely monitored, it seems genuinely rare (with perhaps fewer than 30 British records). In the modern recording period from 1950, however, it has been prone to occasional small invasions associated with continental irruptions.

Criteria for identifying *caudatus* were published by Kehoe (2006). Key features include head pattern (although some pure *caudatus* do show some faint grey streaking on the head sides behind the eye), tertial pattern (the precise extent of white on each feather), and the colour of the underparts. Continental *europaeus* is also on the *British List*, but is extremely difficult to identify.

British records appear to be concentrated in southeast England. Notable records from Norfolk include 10 at Cley on 2nd October 1961 and a flock of 11 to 13 at Blakeney on 8th October 1966. Farther north records are rare (for example, just one published from Yorkshire prior to two birds in the Spurn area from October 2005 through to April 2006). The subspecies was removed from the Scottish List following review (Forrester *et al* 2003). Two reports of white-headed birds from Northeast Scotland in 2004 (Collieston on 17th October and Aberdeen University on 9th December) would represent the first records for Scotland if substantiated as pure *caudatus*.

The first birds of this form to be appreciated by the masses occurred between January and March 2004 when up to six birds graced Westleton Heath, Suffolk, and four more were at Lewes, East Sussex. During the same period at least 11 were seen in the Netherlands from December 2003 into early 2004.

For those eager to assign an apparent *caudatus* too quickly a cautionary tale is provided by a white-headed male which frequented a garden in Southend-on-Sea, Essex, from March 2006. He paired with a *rosaceus* Long-tailed Tit and was seen feeding young. The bird was photographed and although initially suspected of being a Northern Long-tailed Tit, the pictures showed the ghost of head markings indicative of *rosaceus*. Kehoe (2006) noted that some pure *caudatus* show some faint grey streaking on the head sides behind the eye. Careful attention to this and other features is required to secure a claim.

These attractive northern birds are certainly newsworthy nowadays and more than capable of enticing a crowd.

Crested Tit Lophophanes cristatus cristatus and L. c. mitratus
Linnaeus, 1758. Breeds from Scandinavia south to Spain and Greece and east to the Urals; isolated and sedentary population in central Scotland is an endemic race scoticus.

Polytypic, seven subspecies. Two continental taxa are widespread. Nominate *cristatus* Linnaeus, 1758, is found from Scandinavia east across Russia to the central Urals and south to Romania and northern Ukraine, and *mitratus* C L Brehm, 1831, occurs across most of west central Europe. Endemic *scoticus* (Prazák, 1897) is found only in Scotland. This species was formerly placed in *Parus*, but the distinctness of *Lophophanes* is well supported (Gill *et al* 2005) and now recognised by *AOU* and *BOU* as a distinct genus.

In addition to *scoticus,* both continental taxa are on the *British List*. Most of the (few) records of Crested Tit away from core breeding areas of *scoticus* probably refer to one of the continental races.

Very rare vagrant from central and northern Europe.

Status: There have been 20 records from England, only two of which were positively identified as of the continental races. The most recent record, from Northumberland, most closely resembled *scoticus* when comparing photographs with museum skins, but it could not be identified with certainty (Tim Melling *pers comm*).

Historical review: A bird of the race *mitratus* was shot at Yarmouth on the Isle of Wight before 1844 and one of the nominate race *cristatus* was obtained near Whitby, North Yorkshire, in March 1872 (Brown and Grice 2005). No other precise racial attributions are on record. Witherby (1912) mentions a record of one shot near Stanpit Marsh, Christchurch, Hampshire, in 1846.

Elsewhere in England Witherby (1912), Witherby *et al* (1938) and Brown and Grice (2005) list 11 year-dated records of the species between 1829 and 1899. Six appeared in East Anglia and five in Dorset/Hampshire, Devon and Cornwall. Undated records came from North Yorkshire, Wiltshire, Berkshire and Surrey (two). Although nearly all the 19th century records are located away from the traditional centres of bird keeping, with all but one in southeast and southern counties, it should be remembered that continental birds were occasionally imported and exhibited at cage bird shows in the years before 1910 (Robson 1911).

During the early part of the 20th century, there were another six birds: Westonbirt, Gloucestershire in December 1930; Bridford, Devon in 1936; Torquay, Devon, on 25th and 27th January 1945; Godstone, Surrey, on 10th April 1945; St. Mary's, Scilly, on 15th September 1947; and Dawlish, Devon, on 28th December 1947. Curiously, unlike the 19th century records, these occurrences include none from East Anglia and five from the southwest.

There have been just four records since the formation of the *BBRC*.

1964 Lancashire: Lytham 24th October
1968 London: Kensington Gardens 25th April
1971 Scilly: St. Mary's 11th November
1984 Northumberland: Hauxley 1st-year, trapped, 24th August-29th September.

Discussion: The quest to see this species is always a key part of observers' visits to the coniferous woodlands of the Scottish Highlands, where the endemic population numbers between 1,000-2,000 pairs and has a winter community of 5,600-7,900 individuals (Summers *et al* 1999). These birds are sedentary, adults being resident within territories throughout the year and juveniles undergoing short dispersals from their natal area (up to 7km). Just three Scottish birds have been reported away from their usual range during recent times: Sauchi Craig, Stirling, Forth, in February 1981; Loch Tulla, Argyll, in November 1991; and Tobermory, Mull, Argyll, from 19th October-9th November 2002 (Forrester *et al* 2007). There are two records for the Channel Islands, both from Jersey, on 27th October 1985 and 16th February 2005.

Continental vagrants away from the breeding areas are exceptionally rare and puzzling finds. With just four during the latter part of the 20th century unlikely to be tainted by the prior possibility of escapes, they are far rarer than many species that we routinely consider to be vagrants.

Willow Tit *Poecile montanus*
Conrad, 1827. Found from Britain in the west across the Palearctic to Anadyr on the Pacific in the east. Its southern limits are Greece in Europe and Japan in the east; generally found between 50° and 63°N.

Polytypic, 11 subspecies. *P. m. kleinschmidti* Hellmayr, 1900 is endemic to Britain, but it was not until 1900 that *kleinschmidti* was first described, from specimens collected near Finchley in London in 1897 (Harrop and Quinn 1996). There are two currently accepted records of the north European *borealis* De Selys-Longchamps, 1843; several other claims are under review. Central European *rhenanus* Kleinschmidt, 1900 has been suspected in Britain at least once. It would be difficult to confirm *rhenanus* without a ringing recovery (*BWP*; Harrop and Quinn 1996; Kehoe 2006). The separation of *borealis* from other eastern races would be equally difficult.

Harrap and Quinn (1996) divided Willow Tit into three subspecies groups: *salicarius* of northern Eurasia comprising six subspecies; *kamtschatkensis* of the Pacific rim in eastern Asia comprising four subspecies; and *montanus* of the mountains of central Europe with one subspecies. Songar Tit *P. songarus* of central Asia was considered a separate species.

Northern Willow Tit Poecile montanus borealis and P. m. rhenanus

P. m. borealis occurs from Fennoscandinavia and Denmark east to the Baltic States and European Russia, and south to the northern foothills of the Carpathian Mountains in Ukraine and to Penza in southeast European Russia. Northern populations are partially migratory. P. m. rhenanus is found in central Europe east to westernmost Germany and northern Switzerland.

Vagrant of unknown status.

Status: There are two accepted records of one or other of these two races; both are under review.

1907 Gloucestershire: Tetbury shot in March
1975 South Yorkshire: Thorne Moors 8th February

The first record looks unusual, but there is nothing suspicious other than the westerly location. The shooter donated it without fee, as he had done with other specimens (Tim Melling *pers comm*). The second was a sight record.

Discussion: Little attention has been paid to the formal reporting and identification of vagrants of continental Willow Tit taxa. In consequence, their status and the veracity of the accepted records are uncertain. Treatment in county and national avifaunas reflects this.

Limbert (1984) mentions two other previously published records of possible *borealis*, from Welwyn, Hertfordshire, in January 1908 and on Fair Isle on 3rd November 1935. The latter was seen in the company of three immigrant Blue Tits *P. caeruleus*, but is no longer considered to be acceptable as an unequivocal nominate *borealis* (Pennington *et al* 2004). Forrester *et al* (2007) state that *borealis* has been reported elsewhere in Scotland on a few occasions. None have been substantiated adequately.

There are three additional claims from Suffolk (Minsmere from 15th-16th September 1974, Landguard on 25th September 1983 and Worlingworth on 10th November 1990). None of these has been submitted to *BBRC*, but the racial identity of the Minsmere individual was considered to be certain and the appearance of the 1990 bird was considered to support the racial attribution (Limbert 1984; Piotrowski 2003). Two birds reported to show characteristics of *borealis* have appeared in Norfolk, at Sandringham on 23rd November 1980 and East Ruston on 21st November 1996 (Taylor *et al* 1999). In addition to the Thorne bird, there have been five additional birds reported from Yorkshire between 1978-1985, including one reportedly showing characteristics of *rhenanus*, though none have been assessed by the *BBRC* (Wilson and Slack 1996).

It seems likely that the review of records will find most if not all records wanting. New guidance for observers on what to look for might stimulate a fresh wave of observer activity whilst the steep decline of Willow Tit across much of their British range may prompt observers to investigate sightings with more intcrest. Meanwhile a well documented bird is awaited.

Red-breasted Nuthatch *Sitta canadensis*

Linnaeus, 1766. Breeds across much of central and southern Canada, south-east Alaska and northeastern and western USA. Partial migrant with northernmost populations migrating south; remaining populations resident most years but subject to irruptive movements typically every two to four years when the cone crop fails.

Monotypic.

Very rare vagrant from North America.

Status: One record, the second for the Western Palearctic.
1989 Norfolk: Holkham 1st-year male 13th October-6th May 1990

Julian R. Hough

Not surprisingly this bird proved exceptionally popular.

> *…people descended on Holkham Meals in their hundreds to see the bird. Indeed, it developed into one of the largest gatherings of birdwatchers so far seen in the UK, a situation not helped by the narrowness of the paths, the elusive nature of the bird and the desperation of a few observers literally fighting for a better vantage point. Some people visited eight or more times without seeing it at all, and many needed a second or third attempt before catching a glimpse. For others, however, it performed remarkably well. By listening for its distinctive call, or by following the tit flocks, or by simply waiting in a favoured spot for the Nuthatch to appear, many people were able to see and photograph it during its stay of several months.*
> **Rob Hume and David Parkin**. *British Birds 88:151.*

Discussion: One was reported on board the *Aquitania* a day out from Halifax, Nova Scotia, in October 1943. Another was on board the *Queen Mary* 350km out from New York on 1st October 1963 along with a White-breasted Nuthatch *S.carolinensis* (Durand 1972).

Extralimital records have also occurred on Bermuda (three between 1975 and 1978; one March and two October). The first and only other Western Palearctic record is of one found dead in Iceland on 21st May 1970.

Migration begins in late July or August and may continue until November, with most irruptive birds arriving on wintering grounds between the end of September and early November. The Atlantic coast appears to be used heavily in invasion years and many thousands of birds move south through Quebec and New Brunswick. Birds migrate during night and day and have been recorded flying over the Gulf of St Lawrence at night (Ghalambor and Martin 1999).

The appearance of this bird was initially met with incredulity and then panic. The ensuing spectacle constituted one of the most popular twitches of all time, with an estimated 2,000 birders paying homage over its extended stay. Clearly there is potential for this species to occur here again as it is not averse to sea crossings and ship-assistance is possible. The lucky finder of the next one will not be met with quite so much disbelief.

Wallcreeper Tichodroma muraria
(Linnaeus, 1766). Breeds in mountainous regions of southern Europe from the Pyrenees and Alps discontinuously east to China. In winter dispersive and prone to altitudinal movements.

Polytypic, two subspecies. Racial attribution undetermined, but likely to have been *muraria* (Linnaeus, 1766) from Europe east to western Iran.

Very rare vagrant from southern Europe.

Status: This highly sought-after species has amassed 10 records: six prior to modern times; and four since 1969.

1792 **Norfolk:** Stratton Strawless shot 30th October
1872 **Lancashire and North Merseyside:** Sabden shot 8th May
c. 1886 **East Sussex:** Winchelsea shot in late spring
1901 **Somerset:** Mells seen in September
1920 **Dorset:** Chilfrome seen on 24th April
1938 **East Sussex:** Rottingdean seen early June
1969 **Dorset:** Worth Matravers male 19th November-18th April 1970
1976 **Somerset:** Chelm's Combe Quarry male early November to 6th April 1977
1977 **East Sussex:** Ecclesbourne Glen 6th-10th April
1977 **Somerset:** Chelm's Combe Quarry, Shipham Quarry and Cheddar Gorge male early November to 9th April 1978; presumed same as 1976-77
1985 **Isle of Wight:** St Catherine's Point male 16th May

Those among the large numbers of observers who saw either the wintering male in Dorset or the returning male in Somerset have on their lists one of the most spectacular and desperately desired birds on Earth.

> ....the first for over 30 years, was at Winspit, near Worth Matravers, on the Dorset coast. The first authentic sighting of this individual was on 19th November 1969, though there is some evidence that it had been present since at least 9th September. It stayed throughout the winter, working its way each day back and forth through the abandoned stone quarries that extend along the cliffs for about a mile east from Winspit. Hundreds of birdwatchers travelled to see it, local inhabitants and holiday-makers borrowed binoculars, and it even put in an appearance on television, but at times it could be remarkably elusive and no one seems to have met with much success in photographing the bird. By the end of March it had already acquired the black throat of the male's breeding plumage, and it was last seen on 18th April.
>
> **H. Löbrl.** British Birds 63: 163.

Discussion: Recent records of a returning male at Liège, Belgium, between 1986-1990 and a wintering female in Amsterdam, Netherlands, from 13th November 1989-11th April 1990 and 27th November 1990-5th April 1991 raised the possibility of another crossing the English Channel. Early in 2008, hopes soared once again when a bird wintered just across the Channel at Pas-de-Calais.

With global warming, this is one species that might become less dispersive due to ameliorating conditions in their high mountain homes and, therefore, less likely to reach our shores. When the next one arrives, let us hope it is as accommodating as some of its predecessors.

Short-toed Treecreeper Certhia brachydactyla

C.L. Brehm, 1820. Widespread resident of continental Europe from southern Spain north to Denmark and east to Poland, western Ukraine and Greece. Elsewhere, resident in mountains of North Africa, western Turkey and western Caucasus.

Polytypic, six subspecies. The racial attribution of occurrences is undetermined but probably involves *megarhyncha* C L Brehm, 1831, which is found as close as the Netherlands, northern and western France and the Channel Islands.

Very rare vagrant from continental Europe.

Status: There have been 25 birds, all in England and most in Kent.

Historical review: The first was trapped at Dungeness on 27th September 1969, where it remained until 30th September when it was again trapped. The record was not formally admitted to the *British List* until 1975 (Scott 1976). The observers of that bird could not have realised that in subsequent years 'Dunge' would become the epicenter of the relatively few accepted occurrences of this species on this side of the English Channel.

Two in 1970 were both away from Kent and provided county firsts for Yorkshire (trapped at Hornsea Mere on 26th October) and Dorset (Portland from 23rd November through to 21st March 1971; previously rejected but re-evaluated 30 years on and found to be acceptable). One from Epping Forest, Essex, on 26th May 1975 was the only other to avoid being extracted from a mist net and was the one success from a careful four-year hunt for the species at the site (Ian Wallace *pers comm*). The whole decade produced eight records, including another two together at Dungeness from 7th-10th October 1978, both trapped on 7th. Also from that decade was an intriguing comment in *The Atlas of Breeding Birds in Britain and Ireland* that: "claims were made that pairs had bred in Dorset 1971-72, but these were subsequently withdrawn." (Sharrock 1976).

After the raised hopes and expectancies of the 1970s, the 1980s were disappointing. Just three were logged, all of which blundered into mist-nets in Kent at St Margaret's, Dungeness and Sandwich Bay.

Four trapped at four different Kent sites in 1990 opened the decade's account with something of a bang. An intriguing record of a female at Reculver on 29th July was only the second ever in a breeding season. Only four more followed during the decade, all immigrants to Dungeness, to bring the total for the 1990s to eight birds.

Five more were found between 2001 and 2005, all from Dungeness with the exception of one at Bradwell-on-Sea, Essex. This individual was present from 6th-10th April 2005; the first to be recorded outside Kent since 1979, it provided a glimmer of hope for obsessive scrutinisers of treecreepers elsewhere.

Where: Based upon accepted records, there would appear to be only one county where observers can expect to encounter this species: 20 of 25 birds were in Kent; and 13 from Dungeness alone. Elsewhere, there have been two each in Essex (May 1975 and April 2005) and Dorset (November 1970 to March 1971 and May 1979). One trapped at Hornsea Mere, East Yorkshire, (26th October 1970) remains the only one found away from southern and southeastern localities.

When: There is a slight peak in spring (late March to early April). Autumn records present a much more defined peak (late September to end of October). Puzzlingly, birds have also been found in January and July.

Given such a cryptic species it is probable that the recorded occurrence pattern does not reflect the true status of the species. Others must surely make landfall and be overlooked, especially

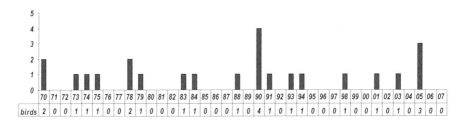

Figure 78: Annual numbers of Short-toed Treecreepers, 1970-2007.

in areas of deciduous woodland where Eurasian Treecreeper *C. familiaris* is present. Familiarity with the birds' distinctive song and Coal Tit-like call hold the key to identifications by alert observers. The propensity for extended stays might be indicative of a species that may colonise.

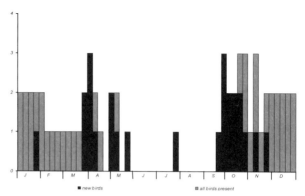

■ new birds   ■ all birds present

Figure 79: Timing of British Short-toed Treecreeper records, 1969-2007.

Discussion: Across most European countries, populations remain stable. The species is slowly expanding north in Denmark (where breeding first occurred in 1946). Range expansion has also been noted in Poland. Significantly, the population in the Netherlands has increased by 41 per cent and the nearest community, in France, has grown by 19 per cent. Short-toed Treecreeper is normally considered to be a resident, with few birds moving more than 10 km from their ringing site, but it has been suggested that it is a partial migrant in recently occupied breeding areas along the northern edge of its breeding range (Bauer and Kaier 1991; Hagemeijer and Blair 1997).

The first major identification paper for this tricky species was published over 30 years ago (Mead and Wallace, 1976). Surprisingly for such ground-breaking material, it did not provoke a subsequent surge of accepted records. The British records are probably associated with the expansion of both range and population on the near continent. There has likely been a real increase in the numbers arriving here, rather than records being the product of improved detection skills, especially as most have been identified at just one observatory. Eurasian Treecreeper is absent from the extensive patch of coastal scrub at Dungeness, so any treecreeper there invites attention. Once again, it is a scenario that invites the question of how many are overlooked elsewhere.

Eurasian Penduline Tit Remiz pendulinus

(Linnaeus, 1758). Wide but local distribution in central and eastern Europe from Denmark, Germany and Italy northeast to central Sweden and Estonia. Absent from much of northwest Europe. Locally numerous in Spain. To the east breeds from southern Russia to Volga River. Largely resident or dispersive in Europe. Other races occur in central Asia and from southern Siberia to northeast China and winter in Pakistan, southern China and southern Japan.

Polytypic, 12 subspecies. Records attributable to nominate *pendulinus* (Linnaeus, 1758) of western and southern Europe east to Tambov and Rostov (central European USSR), northern foothills of western Greater Caucasus and western Asia Minor.

Formerly very rare vagrant from central and eastern Europe, now increasingly regular but still very rare away from the south coast and East Anglia.

Status: A total of 210 birds of this delightful species had been recorded by the end of 2007, all in England and Wales.

Historical review: The first appeared as recently as 1966, when one was found at Spurn on 22nd October. After 10 minutes it disappeared. On the 28th, presumably the same bird was relocated at nearby Kilnsea, where it was watched for about four hours. Attempts were made to catch it in a mist net, but on two occasions it bounced out and on a third it found a small hole and escaped (Raines and Bell 1967); a clear case of ornithological sod's law!

The next did not occur until 1977, when one was on St. Agnes, Scilly, on 25th October. The third British record, at Stodmarsh, Kent, on 18th May 1980, was followed by three in 1981 (including the first for Wales, a male, on Bardsey from 9th-13th May), one in 1982 and four in 1983 (one of which was relocated in March 1984). Since the late 1980s, the floodgates have opened. Although one in Cheshire and Wirral in 1986 was the first for a couple of years, five in 1987 were rapidly eclipsed by 10 in 1988 from October onwards and then 14 in 1989, bringing the cumulative total for the decade to 39 birds.

The 1990s saw the onslaught continue with double-figure totals noted in six years, including a new record of 19 in 1994. Notable among these was a male from an undisclosed locality in Kent from 21st April to 1st May 1990, during which time he built one nest and three-quarters finished a second (Spencer *et al* 1993). A party of four (possibly five) at Land's End, Cornwall, on 12th October 1993 was the first to exceed more than two birds together. It was matched the following year by four at Titchfield Haven, Hampshire on 4th November 1994. A feature of these incursions was the simultaneous arrival of birds at dispersed sites. The decade amassed a whopping total of 105 birds.

Since 2000 there has been a slight reduction in numbers recorded, but 64 have been documented to the end of 2007 including 19 in 2004 which equalled the record set a decade earlier. In 2005, six were recorded together at Rainham Marshes, London, on 19th December: a new record count of birds at a single site.

Over a very short time this species has gone from extreme rarity to a scarcity of regular occurrence. Just 12 had been recorded by the end of 1986, but since then birds have arrived at an average of 9.4 per annum, a figure that rose to 10.5 per annum in 1990-1999. Given their choice

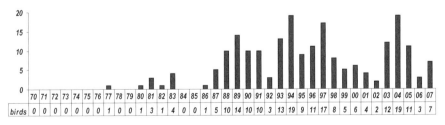

Figure 80: Annual numbers of Penduline Tits, 1970-2007.

of habitat many more may have visited undetected in reed beds. Breeding has been suspected, but not yet proven. Surely such an event is not far away?

Where: Despite the increasing regularity of sightings in East Anglia and southern England, Penduline Tit remains a rare vagrant elsewhere. Birds away from the core areas are notoriously short-stayers, a frustrating trait for county listers.

The southeast has produced the greatest concentration of records (97 birds or 46 per cent of all records), with Kent alone yielding 49 birds (23 per cent). Elsewhere in the region, there have been 21 (10 per cent) in East Sussex, 11 in London, nine from Hampshire, four from West Sussex and singletons on the Isle of Wight and inland one-day birds in Bedfordshire (September) and Berkshire (October).

The southwest is the next most popular region with 48 birds (23 per cent). The Isles of Scilly has provided landfall for 18 birds (9 per cent) and other double figures have come from Somerset (11 birds) and Dorset (10 birds). Seven have reached Cornwall and two Devon. East Anglia has a total of 41 birds (20 per cent): 20 (10 per cent) in Suffolk, 16 (8 per cent) in Norfolk, three in Cambridgeshire and two in Essex.

Despite the northeast producing the first-ever record, there have been just 16 birds (7 per cent): five each in Lincolnshire and East Yorkshire; four in Cleveland (two of which were also seen in Co. Durham); and one inland in North Yorkshire. A juvenile found offshore at sea (54°01'N 02°01'E) in sea area Dogger died in care. The Midlands have been penetrated by one-day late-October singles in Nottinghamshire and Northamptonshire. Northwest records are exceptionally rare, with just two one-day birds in Cheshire (September) and Lancashire (November). The total for Wales the total is also low, but four birds have reached either end of the country in remarkable westward pushes. These included a one-day October bird on Anglesey, a five-day May bird on Bardsey and two together from November to March at Kenfig, Glamorgan.

When: Records have occurred across all months, but as this is a species that can remain hidden in large areas of habitat for extended periods, a discovery date may not necessarily represent an arrival date. A few have been found between January and April, and a number of long-staying wintering birds have remained to early spring.

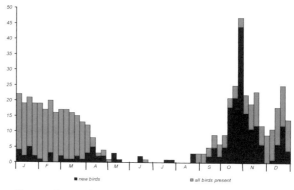

Figure 81: Timing of British Penduline Tit records, 1966-2007.

A small cluster of records in spring might provide evidence of birds exiting the country, as several have been at coastal sites, but could just as easily represent birds becoming more conspicuous, as several relate to males.

Just two new arrivals have occurred in June (Lincolnshire and Essex; a bird in Suffolk in June was presumed to be the same individual as one the previous April at the same site) and July (Cleveland and Suffolk). Three juveniles on Scilly in 1994 make up the sole August sighting. Two more have been found in the first half of September, and six more in the second half of the month. Most records (58 per cent) fall between mid-October and early November, mostly in the last week of October. Many autumn birds relate to juveniles and are presumably associated with post-breeding dispersal from the continent.

Away from its favoured regions, these bandit-faced reed mace dwellers remain a quality find sought eagerly by dedicated patch watchers.

> April 10th 2001 was yet another in a long run of glorious spring days with a biting cold northerly wind and the threat of rain! Having quickly lost the teenage thrill of "trespassing" on closed Nature Reserves I decided to opt for a pep-up of any breeding bird activity on my census site at Water's Edge, Lincolnshire. The chances of finding any sort of migrants seemed about on par with getting some common sense from the Ministry of aggravated farce and fiction but at least I would get some fresh air.
>
> At about 15:00 hrs having had my daily fix of the drake Ferruginous Duck I moved to the south-eastern part of the site where tall hawthorns provide shelter from the north wind and a nice pool of reed mace and willows offers the ever present chance of a Penduline Tit. For twenty years, since the Blacktoft marathon, I had been scanning the three small areas of reedmace that exist in my 120ha local patch of clay pits. Autumn and winter scanning had turned up numerous Blue Tits, Reed Buntings, Wrens and the occasional Long-tailed Tit scattering the bulrush fluff skywards but never a sign of the bandit masked tit.
>
> As I took stock of the breeding bird activity in the shelter of the hawthorns I noticed one of the local pair of Long-tailed Tits sitting motionless on the outer limb of a small hawthorn on the edge of the pit. For such a hyperactive species its lack of movement seemed somewhat strange but I guessed its nest was nearby and concentrated on locating it. After a couple of minutes of watching the static Long-tail I picked up a strange song that failed to ring any bells? Surely I had not missed the song of Long-tailed Tit for 33 years? Then I thought I heard a high pitched sreeee call descending at the end! It was repeated and a small bird dropped from behind the Long-tailed Tit into the base of the adjacent reed mace. Still not really believing the possibility at hand it was only when a cracking male Penduline Tit climbed up the reed mace stem and settled to feed on the fluffy head that it really sunk in – I had found a long-hoped for bird not only in Lincs but on my very local patch.
>
> **Graham Catley.** BBRC submission.

Discussion: The range in Europe has expanded steadily northwards in recent decades and its western border extended 300 km during 1930-1965. Breeding first occurred in Denmark and Sweden in 1964. Following a short lull, the border of the range moved rapidly 250 km west and 200 km north between 1975 to 1985. In the Netherlands the main colonisation took place from 1981 onwards (there are now 140 to 210 pairs). Breeding was first confirmed in Belgium in 1993 (now 13-20 pairs), the same year that the species bred in Norway for the first time. Following a marked population increase in Sweden at the end of the 1980s, recent estimates show a decline there to between 50 to 100 pairs at the turn of the century. A similar decline has been noted in Denmark, where 25 to 50 pairs remain after a 50 per cent to 79 per cent decline (Burton 1995; Hagemeijer and Blair 1997; BirdLife International 2004). Despite these declines, habitat change has assisted the overall expansion with wetland eutrophication providing an increase of late summer food sources in reed beds such as the mealy plum aphid *Hyalopterus pruni* (Valera *et al* 1990).

This inherently dispersive and irruptive species' rapid expansion of range and population is reflected in its hasty change of status in Britain. Many records refer to juveniles, which lead the winter irruptions and move farther afield than their parents. As the species is loyal to previous winter quarters, colonisation of new breeding areas, mostly by birds under a year old after a spring migration, completes the expansion process (Valera *et al* 1993). Breeding has not yet been proven but has been suspected in Britain. Given the extensive areas of impenetrable Penduline Tit habitat, and the propensity for birds to go missing for extended periods, breeding may have taken place undetected several times.

Penduline Tit exhibits an unusual approach to parenthood. Unmated males start building their structurally superb nests and sing to attract a female. Over the course of a breeding season more than half of all clutches are cared for by the female and up to one fifth cared for by the male. The rest, roughly one third, are abandoned by both parents before incubation commences (Szentirmai *et al* 2007). Solitary birds in an reed bed during the breeding season may represent a breeding attempt.

The origin of some birds in Britain has been traced by ringing recoveries. Remarkably there have been three confirmed exchanges between Sweden and Sussex. The first was an adult female ringed in East Sussex on 15th October 1988 and retrapped in a Swedish colony in May 1989. The second was a juvenile ringed in Sweden on 7th July 1997, trapped in East Sussex on 25th October of the same year and then present in Kent from 26th-28th October. The third was a colour-ringed bird from Sweden trapped at Filsham, East Sussex, on 9th April 2004. In Devon an adult male at Slapton Ley from 22nd December 2003 to 23rd February 2004 and then at Paignton, Devon, from 21st to 22nd March 2004 was trapped on 24th January 2004 and found to have been ringed as an adult at Baarlo, Limburg, Netherlands, on 12th July 2003, 697 km to the east.

Additionally, a bird ringed at Orfordness, Suffolk, on 4th November 2003 was trapped at Bouches-du-Rhone, France, 963 km SSE almost exactly a year later on 2nd November 2004. The five ringed birds constitute an amazingly high recovery rate for any bird, let alone a rarity that inhabits large tracts of inaccessible habitat. The movements of these birds suggest that a proportion of our vagrants are arriving from recently colonised parts of Europe. Together they may illustrate the current dispersal strategy of Penduline Tit, whose breeding numbers on the Continent fluctuate annually.

Brown Shrike *Lanius cristatus*

Linnaeus, 1758. Four races breed from Ob river basin east to Sea of Okhotsk and Kamchatka, northern Japan and much of eastern China. Nominate cristatus winters Indian subcontinent and southeast Asia; other races winter southeast Asia, Philippines, northern Borneo and Indonesia.

Polytypic, usually comprising four subspecies (Svensson, 1992). Worfolk (2000) considered three distinct subspecies, each of which could be considered as diagnosable species: *L. cristatus*, *L. lucionensis* and L. *superciliosus*. Nominate *cristatus* Linnaeus, 1758, breeds in Mongolia and northeast Asia and is a long-distance migrant to Pakistan and India east to Thailand. It is the most likely to occur in Europe. Another form, *lucionensis* Linnaeus, 1766, of Korea and parts of north, east and southeast China, grades into *confusus* in southern Ussuriland, southern Manchuria, and North Korea. The third, *superciliosus* Latham, 1801, occurs in southern Sakhalin, Kunashir (Kuril islands) and Japan The disputed form *confusus* Stegmann, 1929, occurs south and east from lower Amur and Zeya basins, northern and central Manchuria, and eastern Mongolia, grading into nominate *cristatus* in south-east Transbaykalia and middle Amur and Zeya basins. Its validity requires further research (*BWP*; Lefranc and Worfolk 1997; Worfolk 2000).

Brown, Isabelline *L. isabellinus* and Red-backed *L. collurio* Shrikes have variously been treated as one, two, three or even four species (Lefranc and Worfolk 1997). They are largely allopatric, but with areas of sympatry and hybridisation in central Asia. Very few hybrids involving Brown Shrike have been found (Lefranc and Worfolk 1997; Svensson 2004).

The racial attribution of Brown Shrike occurrences here is undetermined, but *cristatus* would seem most likely. The *BOU* is presently investigating the racial attribution of the British birds. The trapped bird from Fair Isle looks closest to *cristatus*, and the Sumburgh bird of 1985 and the 2008 Flamborough individual also appear to be of that form.

Very rare autumn vagrant from central or eastern Asia.

Status: Seven records, two of which in 2008 have yet to be assessed by *BBRC*.

1985 Shetland: Sumburgh, Mainland adult 30th September-2nd October
1999 Co. Kerry: Ballyferriter adult female 22nd November-10th December
2000 Shetland: Fair Isle 1st-winter female trapped 21st October
2001 Scilly: Bryher 1st-winter 24th-28th September
2004 Shetland: Skaw, Whalsay adult male trapped 19th-24th September
2008 East Yorkshire: Flamborough Head adult 24th-25th September*
2008 Outer Hebrides: Vallay Strand, North Uist 1st-winter 18th and 24th November*

Due to its prolonged stay the Co. Kerry individual of 1999 proved very popular. This bird could not be assigned with certainty to either of the likely vagrant subspecies as it showed characteristics of both *cristatus*, the most probable vagrant on migratory and geographical grounds, and *lucionensis*. It was considered to be a possible intergrade between the two from the area of breeding range overlap (Milne *et al* 2002).

The Scilly bird of 2001 was the first major rarity to receive a retrospective identification based on pictures placed on the internet; its picture originally appeared on Surfbirds (*www.surfbirds.com*) labelled as a Red-backed Shrike.

Monday 24th September 2001 dawned bright and sunny on Bryher, Isles of Scilly, and I set off birding unusually early for me whilst on Scilly – at 8.00am. Within just a couple of minutes, I found a shrike in a large weedy field adjacent to the road that runs down from the town towards hell Bay Hotel.

It was on the far side of the field, at a distance of about 150 yards, and I did not have my telescope with me. Through binoculars, the bird looked quite concolourous and not particularly barred. I put this down to the distance of the observation and decided to walk along the road towards the town to get better views. On reaching the area where the bird was, it flew past me, up the hill towards the Vine café, perched briefly, completely silhouetted, and then flew directly away and out of sight. I telephoned RBA at this point and told them I had found a Red-backed Shrike on Bryher. There was no reason to think anything to the contrary and I carried on with my circuit of the island.

As far as I know the shrike was not seen again that day until mid afternoon, when Pete Simpson came back into our rented chalet and said it was showing well just 50 yards up the road. So, with telescopes and Nikon Coolpix in hand, off we went. The bird was in the gorse and bracken, and would occasionally perch up and allow us to get some photographs.

That evening, whilst looking at the images on the digital camera, I noticed that the outermost tail feather looked much shorter than the others, and the primary-projection looked short and rounded. These features, together with the lack of barring on the upperparts and the almost black 'bandit' mask made me wonder about its true identity. Checking the field guides that we had with us, and looking first at Red-backed Shrike, it was obvious that the outer tail should show some white. But could the bird be in moult?

I then turned my attention to Brown Shrike. There were several features our bird showed that fitted well, but there was nothing conclusive. Svensson and Grant (1999) helpfully says 'short primary – projection distinguishes it (Brown) from Red-backed', but at what ratio? I was a little concerned and telephoned my brother at home and asked him to have a look through his guides and try to find a clinching feature mentioning, in particular, the tail pattern or primary-projection. He discovered much the same as us, not having either Lefranc and Worfolk (1997) or Worfolk (2000). There really was not much to encourage me to stick my neck out on such a mega-rarity. I asked him not to contact anyone else, but to let me know if he found out anything more.

The following morning, I bumped into one of the other birders who was staying on Bryher and asked him if he had any previous experience of Brown Shrike, saying that I had my concerns over the identity of the bird. He had seen Brown Shrike in Asia before and felt this bird did not look right for Brown and, in particular, that the bill was not heavy enough. He said if he saw the bird again he would scrutinise it further to be sure. Despite much searching, the only time the bird was seen again that day was by myself later in the evening near the Fraggle Rock pub. Again I just had binoculars at the time so I could not determine any more features than we had already seen.

The next day, we again searched for the bird. At about 11.30am, I noticed a heavy rain front moving in from the west and so, as I was near the pub, I decided to seek refuge there. I was joined by three of the other group of birders who were staying on the island and, as we were discussing the shrike, I noticed a bird land on the large thistle outside and was surprised to see that it was the shrike.

The bird was wet and now looked far more muddy brown. The birder with previous experience started to consider Brown a little more. None of us had telescopes, so a telephone call was made to the remaining member of the group to join us as soon as possible. He was with us with us within twenty minutes, but the bird had completely disappeared; it was not to be seen again that day or the next.

On Friday morning (28th September), I bumped into a local birder leading one of his tour groups and, whilst having a chat, mentioned to him to keep his eyes open for the shrike, as it had a short outer tail feather and short primary-projection, and its upperparts appeared concolourous with a distinct lack of barring. He and his group later saw a 'Red-backed Shrike' near Fraggle Rock. It was not seen next day or subsequently despite continued searching.

As the bird had been seen by a number of birders, none of whom had even hinted at the bird being anything other than Red-backed, I did not want to stick my neck out on such a rarity (especially given the reaction of some individuals a couple of years previously when I made a genuine mistake over a 'Bonelli's Warbler' on St Martin's). So I decided my best course of action would be to post a photograph of the bird on the Surfbirds website as a Red-backed (the photograph with the outer-tail feather exposed seemed most likely to provoke reaction) and see what happened. The photograph appeared on the web early on Monday morning, 1st October, but it was not until the following Friday morning that Surfbirds first labelled it as a probable Brown Shrike. Finally, following comments from several birding luminaries, later that day it was announced as England's first Brown Shrike.

Marcus Lawson. *Birding World 14: 428-431*

Discussion: All but one of the British and Irish records have occurred in just under 10 years, correlating with an increase in European sightings over the same period. The first European record from Sweden on 3rd October 1984 was originally accepted as an Isbelline Shrike and subsequently reidentified. The next European record was from Denmark (15th October 1988). Since the Irish record of 1999, a further five have been recorded elsewhere in Europe: France (24th October 2000, 3rd-5th November 2004), Germany (15th-16th October 2001), Italy (29th November 2002 to February 2003) and Norway (2nd-3rd October 2005).

Given the run of recent sightings, there is now an increased appreciation of the identification criteria, especially for the less obvious 1st-winters. It seems likely that this sometimes subtle species has overlooked in the past. Before finders get too confident, a couple of cautionary experiences came up in autumn 2008. A claimed Brown Shrike on Scilly was eventually reidentified as a Red-backed Shrike and the North Uist individual was initially identified as an Isabelline Shrike. Debate over these two birds has usefully reminded us that naming an extralimital bird remains a challenging proposition requiring an assessment of a number of criteria. In all plumages, Brown

Shrike has a shorter primary projection than Red-backed (typically with only five primary tips visible on the closed wing), a thicker and stubbier bill and a longer, narrower and more graduated tail (Worfolk 2000; Moores 2004; van der Laan 2008).

Curiously, four of the records have involved adults, an age bias shared with other shrikes that occur here in autumn (see the records of both Isabelline and Lesser Grey Shrike). Observer alertness, combined with the upsurge in far eastern vagrants identified in northwest Europe, ensure that it is a fairly safe bet that more Brown Shrikes will be recorded in due course. The first mainland bird at Flamborough was a popular draw; those who missed out might not have to wait long for another.

Isabelline Shrike
Lanius isabellinus

Julian R. Hough

Hemprich and Ehrenberg, 1833. Nominate isabellinus Daurian Shrike breeds from the Russian Altai into northern China and Mongolia and winters from southern Arabia to eastern and central Africa; phoenicuroides Turkestan Shrike breeds from Iran and Transcaspia to Pakistan and just into northwest China, wintering in southern Arabia and east Africa. The population arenarius includes tsaidamensis Chinese Shrike and breeds in northwest China. It winters in a markedly different area than the other two subspecies, reaching Iran, Pakistan and northwest India.

Polytypic, four subspecies. The racial attribution of occurrences is currently under review, but would seem to involve both *isabellinus* Taczanowski 1874 and *phoenicuroides* Schalow 1875. This species has had, still has and may always have a complex taxonomic history. In central Asia it hybridises readily with Red-backed *L. collurio* (Harris and Franklin 2000). It has formerly been treated as conspecific with Red-backed Shrike (for example, Vaurie, 1959) or with both Red-backed and Brown Shrike *L. cristatus* (for example, Dement'ev and Gladkov, 1954). Indeed, it was only relatively recently that Isabelline Shrike, Brown Shrike and Red-backed Shrike were considered to be separate species (Voous 1977 and 1979; BOU 1980).

As if its taxonomy was not complicated enough already, Pearson (2000) suspected that the type (individual specimen first used to describe the species) of Isabelline Shrike *L. isabellinus isabellinus* (collected in western Arabia and described by Hemprich and Ehrenberg 1833) was of the form *speculigerus*. The rules of nomenclature dictate that for polytypic species, the type should be of the nominate subspecies (in this case *isabellinus*). Further investigation confirmed that the type specimen was of the form previously known as *speculigerus* Taczanowski

1874 which was renamed as *L. i. isabellinus*. The form previously known as *isabellinus* was renamed *L. i. arenarius* Blyth 1846. Only the attributions to *phoenicuroides* and *tsaidamensis* remained unchanged.

Status: A total of 80 Isabelline Shrikes have occurred to the end of 2007; 77 in Britain and three in Ireland. Around 26 per cent of these have involved adult or adult-like birds.

Historical review: Isabelline Shrike was formerly considered a race of Red-backed Shrike (Voous 1977 and 1979; *BOU* 1980). As such, the first birds were not warranted the tick attention afforded to full species by observers. The first was an adult male on the Isle of May on 26th September 1950. The second was at Portland on 10th September 1959. It was followed a year later by the third, an adult male on Fair Isle, from 12th-13th May. Records have been almost annual since 1975, which correlates with a greater interest and awareness of an impending split followed by attention growing further following elevation to full species level.

A 15-year gap ensued before the next two in 1975. One of these was a long-staying male in song at Sidlesham, West Sussex, from 1st March-20th April, which enticed observers from near and far. An unexpected trio in 1978 in the second part of October included another long-staying bird, at Winspit, Dorset, from 14th to at least 24th October. A single in 1979 brought to six the number of birds seen during the decade. An adult male from Hemingford Grey, Cambridgeshire, on 8th October 1978 was accepted, but subsequently withdrawn by the observer who cautiously considered that he could not eliminate the possibility of a hybrid with Red-backed Shrike *L. collurio* (Rogers *et al* 2003)

As with many other species, it was the 1980s that transformed the status of this by then popular shrike. There were six between 1980 and 1986, including the first for Wales at Holyhead, Anglesey, on 25th October 1985, but still little hint of an imminent change in status. Three in 1987 equalled the 1978 record total. The easterly airflow in autumn of 1988 was good for a number of Asian species and brought seven Isabelline Shrikes to six coastal counties from Dorset to Shetland between 12th October and 4th November. Three more in 1989 closed the decade's haul of 19 birds.

The 1990s continued the momentum set during the latter years of the previous decade and yielded 23 birds in total. A blank year came during this period, but was as long ago as 1992.

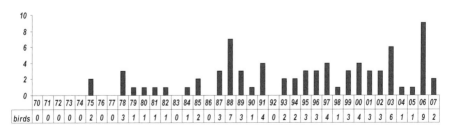

Figure 82: Annual numbers of Isabelline Shrikes, 1970-2007.

No annual return matched that of 1988, but four birds were recorded in 1991 and 1997 and three in 1995, 1996 and 1999.

From 2000 to 2007 no fewer than 27 were been recorded. These included the first for Ireland, a 1st-winter at North Slob, Co. Wexford, from 20th November-9th December 2000. Six in 2003 came close to equalling the record. Nine in 2006 – a one-day spring bird and eight autumn birds including three between 20th September-4th October and five between 15th-27th October – surpassed it by two.

Since 1987, the annual average has been 3.1 birds.

Where: Most, as would be expected, have been along the east coast, with Shetland, East Yorkshire, Norfolk, Kent and Scilly accounting for the majority of sightings.

The most likely regions are at opposite ends of Britain. The Scottish total of 19 comprises nine from Shetland, four from Orkney, two each in Fife and Borders, and singletons on the Outer Hebrides and Angus and Dundee.

The same regional total has been recorded from the southwest. Scilly weighs in with six, Devon and Dorset four each, Cornwall and Somerset two each and, inland, Wiltshire one.

A total of 15 have occurred in the northeast, including six from East Yorkshire, four from Lincolnshire, three from Northumberland and two in Co. Durham. Surprisingly, only nine have been uncovered in East Anglia: six from Norfolk and singles in Essex, Suffolk and Cambridgeshire. The southeast has produced a mere eight: six from Kent and singles from Surrey and West Sussex.

In the west, six have reached Wales: three in Caernarfonshire, two on Anglesey and one in Pembrokeshire. The sole northwest record is an inland bird from Lancashire. The three Irish records comprise the first from Co. Wexford and two from Co. Cork.

Given the species' rarity away from the east and south coasts it is perhaps surprising that several have been located inland. These include a male found long dead at Richmond, Greater London/Surrey, on 21st March 1994, a 1st-winter at Stocks Reservoir, Lancashire, from 5th-11th November 1996, an adult female on the Nene Washes, Cambridgeshire, from 8th-9th September 2000 and a 1st-winter at Cotswold Water Park, Wiltshire, on 28th October 2001.

When: Two March records have presumably featured birds that had wintered in Britain or elsewhere in Western Europe. Four found between 2nd May and 3rd June presumably relate to birds exiting Western Europe. Three of these came from the east coast (Shetland, Durham and Norfolk) and there was a stray in

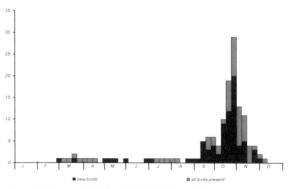

Figure 83: Timing of British and Irish Isabelline Shrike records, 1950-2007.

Somerset. Two midsummer records, a well appreciated female at Cemlyn Bay, Anglesey, from 2nd July-8th August 1998 and a first-summer male at Porthgwarra, Cornwall, on 26th June 2002, are less easy to explain.

The peak period for arrivals is the late autumn, particularly the period 8th to 28th October (51 per cent of all records). Overall, autumn records indicate a protracted spread of arrivals, with the earliest an adult female on Fair Isle, from 23rd-24th August 1994. The only other sighting in that month was a male of uncertain age from Lundy on 28th August 2001. The latest involved a 1st-winter on North Ronaldsay on 30th November 1997.

Racial attribution: The taxonomy of the Isabelline Shrike complex was revisited by Pearson (2000), who outlined the morphological differences between the races in adult plumage. Increasing interest from the birdwatching community in racial identification in recent years has led to attempts to assign individuals to subspecies. These, and a review of all records by the *BBRC*, have enabled some birds to be assigned with some degree of confidence. Presently this is considered to be possible only with adult birds. Unequivocal criteria for birds in 1st-winter plumage have yet to emerge. Svensson (2004) found at least 75 per cent of adult females and a higher proportion of immatures in Kazakhstan impossible to allocate to subspecies.

The current consensus is that two of the forms of Isabelline Shrike have occurred as vagrants: nominate *isabellinus* (Daurian Shrike) and *phoenicuroides* (Turkestan Shrike) have been identified in their more obvious adult plumages. Some authorities treat these two taxa as separate species. The decision of the Dutch committee for systematics (CSNA) to split the Isabelline Shrike complex into three species was published in 1998 (Sangster *et al* 1998). This view is not one presently endorsed by the *BOURC*. Chinese Shrike *arenarius* has not yet been recorded and, as it travels the shortest distance to its wintering areas, would appear an unlikely vagrant. However, birds thought to show some characteristics of *arenarius* have been reported in Europe from Italy (a juvenile on 19th October 1998 was considered probably *arenarius/isabellinus* and an adult female showing characters of *arenarius* from 4th-9th January 2000) and, farther east, from both Iraq and Syria (van der Laan 2008).

A disputed form or morph *'karelini'* is also thought to have occurred in Britain. These birds are found commonly at Lake Alakol and Lake Zaysan (Altai) and vary in appearance. They resemble *phoenicuroides*, but are paler and have a grey crown. Presently it is uncertain whether *'karelini'* is a variant of *phoenicuroides*, an ancient hybrid population with Red-backed Shrike or a distinct taxon. The biometrics are not supportive of the hybrid theory. It seems more probable that *'karelini'* is a form of *phoenicuroides* that has adapted to the clay and salt deserts that predominate to the north and west of Lake Balkhash (Andrew Lassey *pers comm*).

British birds considered possibly to relate to *'karelini'* include an adult male on Fair Isle from 12th-13th May 1960 (trapped on 13th). Based on the assumption that *'karelini'* is a form of *phoenicuroides*, this bird would represent the first British record of the *phoenicuroides* group. However, if further research concludes that *'karelini'* is a hybrid population with *collurio* or a distinct taxon, then a new unequivocal first will need to be determined. In 1982 a male in Lincolnshire at Anderby Creek on 7th-8th November, and at nearby Gibraltar Point on 15th, was also thought to be *'karelini'* but the possibility of *collurio* genes in this individual confuses the matter and reinforces the caution required when attempting to match any individual to a given form. More recently a 1st-winter, trapped and present from 29th September-5th October 2007 at Buckton, East Yorkshire, was also considered to resemble *'karelini'* (Andrew Lassey *pers comm*).

There are also two birds showing characters fitting *'karelini'* from the Netherlands (1st-6th October 2000 and 13th-27th August 2002). The latter looked almost exactly the same as the bird in 2000, leading to speculation that it could have been a returning individual, but it was accepted as a separate bird (van der Laan 2008).

On the basis of an ongoing review by the *BBRC*, plus three additional records from the *IRBC*, the following records have been assigned to *isabellinus* Daurian Shrike.

1991 Orkney: North Ronaldsay adult male 28th October-2nd November
1994 Greater London/Surrey: Richmond male found dead 21st March
2000 Cambridgeshire: Nene Washes female from 8th-9th September
2000 Co. Wexford: North Slob 1st-winter, trapped, 20th November-9th December
2002 Shetland: Fetlar adult male 14th-17th September
2006 Co. Cork: Old Head of Kinsale 1st-winter 17th-20th October
2007 Co. Cork: Mizen Head adult 19th-20th October

Forrester *et al* (2007) list an adult male on Fair Isle from 9th-12th October 1981 as belonging to *isabellinus*, but this potential first record has not yet been assessed by *BOURC*. Other records previously published by *BBRC* as belonging to this form (Fraser *et al* 2006; Fraser *et al* 2007) featured adult females at Portland on 15th–23rd September 1985 and Dunglass, Borders, on 13th September 1989. These records are being reassessed in the light of emerging identification criteria.

The following records have been assigned to *phoenicuroides* Turkestan Shrike.

1960 Shetland: Fair Isle adult male, trapped, 12th-13th May, thought to be *'karelini'*
1994 Shetland: Fair Isle adult female 23rd-24th August
1995 Norfolk: Snettisham adult male, 2nd May
1998 Anglesey: Cemlyn Bay female 2nd July-8th August
2002 Cornwall: Porthgwarra 1st-summer male 26th June
2003 Somerset: Porlock Marsh male 3rd June
2006 Co. Durham: Whitburn male 14th May
2007 East Yorkshire: Buckton 1st-winter, trapped, 29th September-5th October, thought to be *'karelini'*

Presently, for adults at least, these two forms remain equally rare. Based on the small sample of assignable individuals, Turkestan Shrike appears to exhibit a propensity for spring and summer arrivals; with one exception, Daurian Shrike comes in autumn. However, since the majority

Table 22: Records of adult or adult-like birds from Europe presently attributed to phoenicuroides or isabellinus*

	May	June	July	August	September	October
phoenicuroides	4	1		3	3	1
isabellinus	3			1	3	3

* table excludes records from Britain and Ireland.

comprises currently unidentifiable first years in autumn, the full occurrence pattern for these forms may be obscured. It will be interesting to see if the situation changes as the subspecific identification criteria evolve and are applied to suitable new arrivals in the coming years.

Discussion: Elsewhere, records are distributed throughout Europe. There are 27 from Scandinavia (11 from Sweden, 10 from Norway, five from Finland and one from Denmark). From the Baltic there is one from Estonia, three from Latvia and five from Poland. Farther south there are eight from Germany, 10 from the Netherlands, three from Belgium and 12 from France. In central Europe, three have reached Austria and there are single records from Switzerland and Romania.

A good number of records from the Mediterranean are testament to the broad arrival pattern of the species: with five from Spain, nine from Italy and three from Greece. Additionally, there have been 17 records from Turkey (one January, two April, five May, two July, four August, two September and one in November; of these five are attributed to *phoenicuroides* and four to *isabellinus*), plus four more reported in autumn 2008 (three October and one December) (Kirwan *et al* 2008; Kirwan 2009).

European records to 2006 were reviewed by van der Laan (2008). The attributions for adult and adult-like birds in this paper, plus those subsequently made by various rarities committees, are summarised in Table 22. Various committees attributed 12 individuals to *phoenicuroides* and 10 to *isabellinus*, with one bird considered to show characters of *arenarius*. It should be noted that different committees across Europe treat 'isabelline shrikes' differently. A number of committee's have also attributed 1st-year birds, but when criteria are revised such attributions may no longer be valid.

Daurian Shrike, being a longer distance migrant, may end up being more frequent than Turkestan Shrike, especially in autumn when its greater migratory potential could make displacement to northwest Europe more likely. The southwesterly movement of both taxa to their wintering areas should make them rare here and in Europe, but this is clearly not the case, as with all the classic steppe/desert vagrants that perform 90° errors with alacrity.

Clearly much is still to be learnt regarding the identification and arrival pattern of these birds in Europe. Many attempted racial attributions may, for immature birds at least, be a little premature.

Long-tailed Shrike Lanius schach

Linnaeus, 1758. Race L. s. erythronotus breeds Afghanistan, southeast Kazakhstan, southern Uzbekistan, Kyrgyzstan, Tajikistan and northeast Iran to northern India. Northern populations migrate to Indian subcontinent, straying to the Middle East and Arabia. Other races occur in southern China, southeast Asia and highlands of Papua New Guinea.

Polytypic, nine subspecies. Several of these are island, or near-island, endemics within southeast Asia. The race recorded is undetermined, but likely to be *erythronotus* (Vigors, 1831), which has previously been treated as a separate species (Rand and Fleming 1957) based on a lack of interbreeding with *tricolor* (Hodgson, 1837) of north-central India and Nepal to Yunnan in southwest China; this treatment questioned by Biswas (1962).

Very rare vagrant from Asia.

Status: One record.

2000   Outer Hebrides: Howbeg and Howmore, South Uist 1st-winter 3rd-4th November, probably since 27th October

Discussion: The breeding range of the species is vast and centred on the Indian subcontinent, southern China and southeast Asia. The distribution of western breeding *erythronotus* is quite unlike that of any other vagrant on the *British List*, with only a narrow arm extending a short way northwest out of Afghanistan. Most of the migratory population of *erythronotus* breeds in Turkmenistan, Uzbekistan, Kazakhstan and Afghanistan and winters to the southeast in the Indian subcontinent; those communities of *erythronotus* farther south and across much of northern India are resident (Lefranc and Worfolk 1997). Thus, although its global range is quite extensive, the migratory potential of this shrike stems from a much smaller area than those inhabited by the other Asian shrikes such as Brown *L. cristatus* and Isabelline *L. isabellinus*.

As befits a less-likely vagrant to Western Europe, it has presented just two additional records: a 1st-summer male in Sweden (11th June 1999) and 1st-winter male in Denmark (15th-17th October 2007).

Elsewhere, to the southwest of the breeding range birds have occurred in the United Arab Emirates (five records), Oman (four) and Kuwait (two). Farther west, there are singles from Israel (male at Sede Boqer from November 1982 to February 1983), Turkey (1st-winter male trapped and collected Biricek on 24th September 1987) and Jordan (11th-13th April 2004). A former record from Hungary on 11th April 1979 is now considered to have been a hybrid Woodchat Shrike *L. senator* x Lesser Grey Shrike *L. minor*.

The South Uist individual was considered to be probably of the partly migratory race *erythronotus*. However, the largely resident *caniceps* of the Indian subcontinent could not be ruled out due to the lilac tinge to the neck and white above the eye. The latter form would be an unlikely candidate for natural vagrancy; despite the apparent plumage anomalies *erythronotus* seems much more likely. It may be that within *erythronotus* there are greater plumage variations than is presently appreciated. During migration through Central Asia, *erythronotus* mixes with Isabelline Shrike *L. isabellinus* of the races *phoenicuroides* and *isabellinus* migrating to wintering areas in Arabia and East Africa. Presumably these Isabelline Shrikes are the carriers of the Long-tailed Shrikes that reach the Middle East. Could they also have delivered the individuals to northwest Europe? The Scottish bird arrived during an exceptional period for other eastern vagrants, with Brown *L. cristatus*, Southern (Steppe) Grey *L. meridionalis pallidirostris* and Isabelline Shrikes *L. isabellinus* also found in Britain, Ireland and elsewhere in Europe.

The Scottish record was an unlikely addition to the *British List*. But, as it was both preceded and followed by other examples in northern Europe, the odds of a second might not be impossibly long.

Lesser Grey Shrike *Lanius minor*

JF Gmelin, 1788. European range centred east of Balkans to eastern Poland, with small numbers west through the northern Mediterranean to southern France and northeast Spain. To the east, breeds locally from Black Sea coasts across southern Russia and Kazakhstan to extreme northwest China and southwest Siberia. Migrates through East Africa to winter in southern Africa from Namibia to southern Mozambique and northern South Africa.

Monotypic.

Rare migrant from Europe, declining in frequency.

Status: There have been 182 birds recorded to the end of 2007; 176 in Britain and six in Ireland. Since 1950 there have been 155 British and six Irish records. Around 34 per cent of autumn birds were adults.

Historical review: The first record was shot near Heron Court, Dorset, in September 1842. Five more were obtained during the 19th century and a further 15 were recorded between 1900 and 1944, eight of them from Shetland between 1927 and 1944. The split in records involved nine in spring and 11 in autumn, plus a seemingly unlikely record of a bird picked up dead in Bedfordshire on 25th January 1907. Although presently accepted this bird must surely have referred to a misidentified Great Grey Shrike *L. excubitor*.

During the 1950s a total of 17 were recorded, followed by 35 in the 1960s, including the first for Wales at South Stack, Anglesey, on 26th May 1961. Numbers increased to a peak of 40 in the 1970s, including an Irish first on Great Saltee Island, Co. Wexford from 30th-31st May 1978. The 1980s mustered just 24 birds and the 1990s a comparable 27, which given the increased numbers

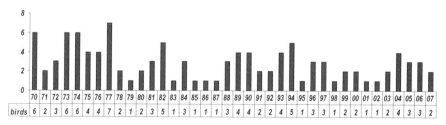

	70	71	72	73	74	75	76	77	78	79	80	81	82	83	84	85	86	87	88	89	90	91	92	93	94	95	96	97	98	99	00	01	02	03	04	05	06	07
birds	6	2	3	6	6	4	4	7	2	1	2	3	5	1	3	1	1	1	3	4	4	2	2	4	5	1	3	3	1	2	2	1	1	2	4	3	3	2

Figure 84: Annual numbers of Lesser Grey Shrikes, 1970-2007.

Table 23: Distribution of Lesser Grey Shrikes by region and season 1951-2007.

Region	Spring	Autumn*
Scottish east coast (Shetland to Borders)	23 (45%)	28 (55%)
English east coast (Essex to Northumberland)	30 (59%)	21 (41%)
South coast (Scilly to Kent)	20 (57%)	15 (43%)
West coast (northwest, Wales, Ireland and west Scotland) and Midlands	14 (58%)	10 (42%)
Total number of records	**87**	**74**

* Autumn defined here as 1st August onwards.

of observers suggests a decrease in arrivals over the period. Since 2000 there have been a further 18, an average of 2.5 per annum since 1980.

The best year on record was 1977 with seven, five of which arrived between 1st and 8th June. 1970, 1973 and 1974 each produced six records, as did 1961. The best annual total since the mid-1970s was five each in 1982 and 1994.

Where: Nearly two-thirds of the sightings have been in the Northern Isles and along the east coast (see Table 23). The Shetland harvest (41 birds in total and 20 per cent of sightings since 1950) might be expected for a species with an eastern European breeding range and prone to overshooting to northwest Europe and across the North Sea. Just two other counties boast double-figure totals: Norfolk has fared best with 19 birds (15 of them since 1956) and there have been 10 in East Yorkshire since 1959. In contrast to those elsewhere, the Scottish records exhibit a slight autumn bias, perhaps indicative of migrants from an easterly origin.

Kent has produced seven records (six between 1956 and 1995). Along the rest of the south coast this species has been distinctly scarce: four in Dorset (three since 1965) and Devon (three since 1961) and just two in Cornwall. The Isles of Scilly, a magnet for all things rare, boasts six records, but interestingly five of these were prior to 1956 and just one since (in 1976).

A number of inland counties have accommodated the species over the years. Farther west nine have reached Wales since 1961. Only one, however, has reached northwest England (Lancashire) and just five have penetrated to the west coast of Scotland, including one on the Outer Hebrides.

Curiously all six Irish records occurred in poor years in Britain, three of which produced just one British record and in one year none at all. The sole Outer Hebrides bird and one of the two Cornish individuals were also loners for the year. Did these birds originate from the near-relict westerly populations of the species?

When: Spring records peak during 21st May-10th June, this period accounting for 28 per cent of all records. The earliest sighting was a bird that landed on a yacht 18 km southeast of Start Point, Devon, on 6th May 1952. The next-earliest records were all on 12th May (Spurn in 1959; Kirton Marsh, Lincolnshire, in 1990; and Fair Isle in 1993).

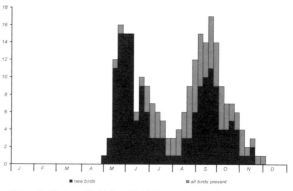

■ new birds   ■ all birds present

Figure 85: Timing of British and Irish Lesser Grey Shrike records, 1950-2007.

Midsummer records are not unusual. Autumn records increase from mid-August onwards, with the peak occurring in September, a month that accounts for 25 per cent of all records. Fewer are seen during October (11 per cent of records). There are four November arrivals. The latest of these were a female at Jarrow Slake, Co. Durham, from 17th to 28th November 1984 and a 1st-winter female found dying on 18th November 1973 at Haslemere, Surrey. The identification of the Co. Durham bird was hotly debated and was only resolved when it was trapped on the 23rd. How quickly times change! Such a long-staying female would nowadays present few identification difficulties.

One 1st-winter male performed a remarkable act of continuing disorientation. First seen and ringed at Monks' House, Seahouses, Northumberland, during its stay from 13th-28th September 1952, it flew north to be found alive down a chimney at Aberdeen, Northeast Scotland, on 15th October 1952 and later died.

Following an abrupt increase in spring occurrences in the 1960s and 1970s there has been a shift to autumn records (see Table 24). The increasing scarcity of spring overshoots is presumably linked to the eastward contraction of the breeding range.

Discussion: Lesser Grey Shrike is a loop migrant, moving in spring farther east than in autumn and performing one of the longest migratory movements among passerines of around 10,000 km to wintering areas in South Africa. In autumn westward breeding birds head south or southeast over Greece and the Aegean Sea and enter Africa on narrow front, principally through

Table 24: Distribution of Lesser Grey Shrikes by decade and season.

	1950s (n=17)	1960s (n=35)	1970s (n=40)	1980s (n=24)	1990s (n=27)	2000 (n=18)
Spring	35%	60%	65%	58%	44%	44%
Autumn*	65%	40%	35%	42%	56%	56%

* Autumn defined here as 1st August onwards.

Egypt. In spring birds head north on a more easterly bearing, first following the African Rift Valley and then crossing into Arabia, largely avoiding the autumn route taken through Egypt and Sudan (Lefranc and Worfolk 1997).

Spring records presumably feature overshoots returning to western parts of the breeding range. Autumn movement begins in west and central Europe from late July and peaks in mid to late August, a period in which there have been surprisingly few British records. Autumn arrivals are now dominant (56 per cent of records since 1990) despite little change in the actual numbers recorded. Although the eastern populations may supply the later autumn arrivals, these communities also leave their breeding areas early and the engine powering this shrike's vagrancy remains unclear.

This attractive shrike is one of the few rarities that in real terms has become much rarer in the last couple of decades in contrast to the general trend or artifact for increasing numbers of southern and eastern vagrants found here. The increasing rarity correlates with the long-term decline in breeding populations across Europe, which showed a decline of more than 10 per cent during 1990-2000 (BirdLife International 2004). The decline has been particularly evident in the west of its range, where it is now extinct or nearly so as a breeding species in a number of countries in which it was formerly quite abundant. It disappeared as a breeding species in central and northeast France by 1979. The remaining French population declined from 56 pairs in 1999 to 21 pairs in 2003 (Rufray and Rousseau, 2004). Climate fluctuations (particularly cooler and wetter summers), pollution and insecticide use on the breeding grounds are thought to be the main causes of poor breeding success and decline. And the main wintering areas lie in the Kalahari, where there has been a chronic drought since the 1970s that may be applying an additional survival test upon a species already in trouble in its breeding areas (Hagemeijer and Blair 1997; Herremans 1998; Giralt and Valera 2007).

Southern Grey Shrike *Lanius meridionalis*

Temminck, 1820. Nominate meridionalis breeds in Iberia and southern France. Six races occur in Africa including algeriensis in northwest Africa others across Africa south to Chad (koenigi, elegans, leucopygos). There are four races in the Middle East and Arabia (theresae, aucheri, buryi and uncinatus) and two in Asia (lahtora in the Indian subcontinent and the more northerly and most widespread race pallidirostris). Most races are fairly sedentary with comparatively local dispersal, but pallidirostris is a long-distance migrant.

Polytypic, at least 11 races recognised (Lefranc and Worfolk 1997); Vaurie (1959) recognised 15. All surviving records have involved the race *pallidirostris* Cassin, 1852, but nominate *meridionalis* Temminck, 1820 has been claimed in northern Europe. One historic British claim fell with the Hastings rarities (Nicholson and Ferguson-Lees 1962).

Research into the taxonomy of the grey shrike complexes is ongoing. The form *pallidirostris*, colloquially referred to as Steppe Grey Shrike, was treated as a subspecies of Great Grey Shrike *L. excubitor*. In 1996, Great Grey Shrike and Southern Grey Shrike were divided into two polytypic species, given qualitative differences in plumage, breeding ecology and behaviour. Accordingly the latter species was re-admitted to the *British List* by *BOURC* in its 23rd Report as *pallidirostris* was assigned as a race of Southern Grey Shrike.

The Dutch CSNA treat *pallidirostris* as a monotypic species *L. pallidirostris* (Panov 1995; Panow 1996; Lefranc and Worfolk 1997; Harris and Franklin 2000). This split has attracted support from Hernández *et al* (2004) through the analysis of mitochondrial DNA. Should it be accorded full species status by the *BOURC*, there will be a requirement to agree a distinct English name. Steppe Grey Shrike may not be appropriate, since it its habitat preference is for Saxaul scrub. There is a growing tendency by authors to refer to *pallidirostris* as Saxaul Grey Shrike, a name first proposed by Svensson (2004).

Steppe Grey Shrike *L. m. pallidirostris*
Breeds in central Asia from lower Volga east to southern Mongolia and extreme northeast China, south to northern Iran and northern Pakistan. Winters Sudan, northern Ethiopia and Somalia through Arabian peninsula to western Iran.

Status: There have been 19 British records, one of which from 2008 is yet to be assessed.

1956 **Shetland:** Fair Isle 1st-winter trapped 21st September

1964 **Shetland:** Fair Isle 1st-winter trapped 18th October

1986 **Suffolk:** Landguard trapped 6th December

1989 **North Yorkshire:** Bishop Monkton probably 1st-winter, 31st October-1st November

1989 **Dorset:** Portland 1st-winter 1st November

1992 **Cornwall:** Cape Cornwall and Kenidjack Carn 21st-23rd April

1992 **Suffolk:** Easton Bavents and Southwold age uncertain, 4th-7th October

1993 **Wiltshire:** Swindon 1st-winter taken into care injured 23rd September, released 24th and present to 28th

1994 **Orkney:** North Ronaldsay 1st-winter male 14th September-16th October, trapped 5th October

1994 **Essex:** Great Wakering 1st-winter 26th-30th October

1994 **Cumbria:** South Walney 1st-winter 2nd November

1994 **Shetland:** Boddam, Mainland probably 1st-winter, 7th-10th November

1994 **Orkney:** Papa Westray 11th-26th November

1996 **Essex:** Holland-on-Sea 1st-winter 18th November-4th December

1997 **Northamptonshire:** Long Buckby 1st-winter 3rd-4th November

2000 **Orkney:** Windwick, South Ronaldsay 1st-winter 22nd September

2003 **Isle of Man:** Ballaghennie Ayres 1st-summer male 17th June-12th July

2004 **Kent:** Ash 1st-winter 6th-7th November

2008 **Lincolnshire:** Grainthorpe Haven 1st-winter 7th-26th November*

Discussion: Greater knowledge of this subspecies, its taxonomy and the imminent possibility of a split was simultaneous with a surge of 12 records between 1989 and 1997. These suggested at the time that birds had been overlooked previously. Despite that increased awareness, just four more followed between 2000 and 2008, but these have included two very welcome and obliging birds.

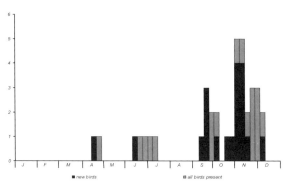

Figure 86: Timing of British Steppe Grey Shrikes, 1956-2008.

The 1st-summer male on the Isle of Man in the summer of 2003 was an oddly dated but extremely popular and educative attraction for a newer generation. This bird initially caused problems because of the misconception that all *pallidirostris* at any age should have pale lores and bill, this expectation being fuelled by the relatively poor literature for *pallidirostris* in which most field guides only illustrated 1st-year birds. These features on the Ballaghennie Ayres bird were largely black (the lores not wholly so). The confusion cleared when it was finally shown that pale lores and bill were typical only of 1st-year birds. Five years later, an extremely confiding 1st-year put on a superb show in Lincolnshire in November 2008. It was even photographed perching on observers' heads.

Elsewhere in Europe records remain quite rare with a total of 30 *pallidirostris* looking paltry against our 19. In Fenno-Scandia there are six records from Norway (5th September 1953, 16th October 1976, 29th October 1987, 5th-14th November 2005, 21st-22nd November 2005, 19th-28th October 2008), nine from Sweden (4th June 1985, 6th-18th October 1998, 31st October-1st November 1998 and unspecified date November 1998, 30th August-6th September 1995, 27th-28th April 2000, 19th September 2004, 29th-30th September 2004, 20th May 2007) and four from Finland (31st October 1981, 22nd-25th May 2005, 9th June 2005, 20th-22nd May 2006). Farther south there are six from Denmark (22nd-23rd October 1995, 24th November 1995, 19th-26th November 1997, 6th-14th October 2000, 11th October 2002, 26th October 2003). There are just singles from the Netherlands (4th-23rd September 1994), Austria (1st November 2002), and Greece (21st September 2008), plus two from Italy (Sicily in winter 2000/2001, 21st November 2008). An adult Southern Grey Shrike in Greece (9th December 2006) was considered to be a southern race, either *aucheri* or *elegans*.

An overshoot by nominate *meridionalis*, of the Iberian peninsula and southern France, and wintering to North Africa, remains possible. Birds of this race have been claimed in northern Europe and were formerly accepted, but none now remain acceptable. One from Helgoland, Germany, prior to 1900 failed as there is no skin and the description does not allow subspecific identification. Others from southern Norway (5th October 1984) and Poland (18th October 1997) were rejected after re-assessment. If, or more likely when, *meridionalis* occurs in the British Isles it will be a popular quarry for insurance listers in hope that the *BOU* will endorse the stance taken by CSNA.

Woodchat Shrike Lanius senator

Linnaeus, 1758. Breeds central and southern Europe, North Africa and western Asia. Winters Africa and Middle East.

Polytypic, three subspecies. Nominate *senator* Linnaeus, 1758, is a regular overshoot and *badius* Hartlaub, 1854 a rare vagrant. A record of *niloticus* (Bonaparte, 1853) is under review.

The nominate race of Woodchat Shrike was removed from the BBRC list at the end of 1990, by which time over 500 records had been amassed. The first British record of the nominate race was shot at Bradwell, Suffolk/Norfolk, in April 1829. The first Irish Woodchat was killed when it struck the Blackwater Bank Lightship off Co. Wexford on the night of the 16th of August 1893. An intriguing report involved a male present at Little Heath, Suffolk, from 8th-16th June 1980 that was paired with a female Red-backed Shrike, but they failed to rear young (Sharrock *et al* 1982).

The BBRC still considers claims of Woodchat Shrike that are not of the nominate race. A review of records was undertaken to see which Balearic Woodchat Shrike *badius* claims contained the necessary identification criteria for acceptance.

Identification of *badius* is centred on the absence, or virtual absence, of a white patch at the base of the primaries and a much squarer, deep-based bill, compared with the slimmer bill of nominate *senator*. The diagnosis of Eastern Woodchat Shrike *niloticus* differences at all ages depends on the combination of extensive white on the base of the tail, whiter underparts and a larger white patch on the primaries.

Balearic Woodchat Shrike L. s. badius

Restricted to islands in western Mediterranean, including Balearic islands, Corsica and Sardinia, east to Elba and Capraia. Winters in West Africa from Ivory Coast to northern Cameroon.

Very rare overshoot from the western Mediterranean, probably overlooked.

Status: There have been six records of this distinctive west Mediterranean race in Britain and two in Ireland. The Somerset bird proved to be a major draw for insurance listers.

1980 **Suffolk:** Sizewell 15th-18th June
1986 **Dorset:** Portland male 10th May
1995 **Kent:** Dungeness female 15th-21st July, trapped 18th
1995 **Norfolk:** Great Cressingham male 2nd-6th July
2002 **Co. Cork:** Mizen Head adult 3rd-5th June
2005 **Somerset:** Weston-super-Mare 1st-summer 11th-13th June
2008 **Somerset:** Minehead 1st-summer 30th June*
2009 **Co. Wexford:** Duncormick 1st-summer male from 12th-23rd May, possibly since 10th*

Discussion: Elsewhere in northern Europe, there are relatively few records, although given the increased observer awareness it is likely that this race will in future be distinguished more frequently. A juvenile male in Norway from 26th-29th September 1972 was trapped and collected. Other records come from Sweden (23rd-27th May 1998) and Finland (26th-27th May 2005). Two records from The Netherlands were belatedly accepted from photographs (5th June 1983 and 6th June 1993). There are three recent records from northern France (4th June 1995, 12th August 1999, 21st April 2004). This subspecies is not common in France, with only 400-800 pairs breeding in Corsica (Marc Duquet *pers comm*).

Eastern Woodchat Shrike L. s. niloticus

Breeds from Turkey east to Iran and winters eastern Sudan, Ethiopia, Eritrea, into northern Kenya, Uganda and Somalia with a small population in Yemen.

Very rare vagrant from southeast Europe, possibly overlooked.

Status: One record is under consideration by *BOURC*.
2003 Shetland: Maywick juvenile 15th-18th September*

Discussion: There are three records from Sweden (9th October 1991, 13th-21st October 1997 and 3rd-10th November 2007) out of 56 Woodchat Shrikes of any race, and one from Finland (13th May 1989), from where there have been just 15 Woodchat Shrikes of any race since 1975. There are also three accepted records of birds ringed in the Mediterranean islands as vagrants (female on Aire Island, off Menorca, 4th April 1992, male on Aire Island 18th April 2002 and Columbretes islands, Castelló, 21st-23rd April 1995)

As with birds of the form *badius* it seems likely that increased awareness will lead birders to identify others with confidence. Moult is a useful identification aid for birds of this race which, unlike their western relatives, undergo a complete moult prior to starting the autumn migration (*BWP*). As such, any autumn birds in non-juvenile plumage would be worthy of closer inspection. The incidence of this race in records from Scandinavia suggests that some are overlooked here.

Masked Shrike Lanius nubicus

Lichtenstein, 1823. Breeding confined to the eastern Mediterranean, and locally east to western Iran. Passage through the Middle East to wintering area in narrow band of northern sub-Saharan Africa from River Niger in Mali east to Sudan and Ethiopia, and Red Sea coastline of western Saudi Arabia and Yemen.

Monotypic.

Very rare vagrant from southeast Europe.

Status: This is another Hastings rarity that eventually returned to the *British List* through two recent British records.

2004 Fife: Kilrenny juvenile, trapped, 29th October-14th November
2006 Scilly: St. Mary's 1st-winter 1st November

Discussion: Masked Shrike has a small breeding distribution confined to the eastern Mediterranean, Asia Minor and locally within the Middle East. The stronghold is in Turkey (30,000–90,000 pairs), but this community declined by more than 10 per cent during 1990-2000 (BirdLife International 2004).

Masked Shrike is a relatively short-distance migrant that has occurred three other times in northern Europe: Finland (23rd October 1982), Sweden (1st October 1984) and Norway 2nd-15th October 2006). Judging by these few records October is the month for vagrancy to the north and west of the restricted breeding range. There are single records for Malta (October 1985) and Spain (adult male Mallorca 22nd-26th April 1991) plus two from France (Alpes-Maritimes, south-eastern France, 18th April 1961 and the Camargue on 9th April 2008).

The first record in Fife was hugely popular. It prompted little surprise among predictors of additions to the national list in light of the Scandinavian records that pre-dated it. Perhaps more surprising was the quick return of the species, but as the first performed on the mainland for over two weeks, the discovery of the Scilly bird received little response. Two in three years gives hope of more to come.

Spotted Nutcracker *Nucifraga caryocatactes*

(Linnaeus, 1758). Nominate Thick-billed caryocatactes breeds from southern Fennoscandia and eastern France across central Europe to the Urals and is largely sedentary. Slender-billed macrorhynchus breeds from the Urals through Siberia to Kamchatka and Sakhalin and is prone to irruptive movements. A further eight to 10 subspecies occur across the Palearctic.

Polytypic, 10-12 subspecies. Both nominate Thick-billed Spotted Nutcracker *caryocatactes* (Linnaeus, 1758) and Slender-billed (Siberian) Spotted Nutcracker *macrorhynchos* C L Brehm, 1823 have been recorded.

Thick-billed Spotted Nutcracker has been recorded on only two occasions: a male shot at Vale Royal, near Northwich, Cheshire and Wirral on an unspecified date in 1860 and one obtained at Chilgrove, West Sussex, on 21st December 1900. The Cheshire bird was examined and confirmed by T. A. Coward, so the identity is not in doubt. But it occurred in an era when high prices were paid for rare specimens, a culture that doubtless fuelled fraud (Tim Melling *pers comm*).

The largely resident European form would seem an unlikely vagrant. However, there are four specimens fitting the biometrics of Thick-billed from the Netherlands (7th October 1911, 18th February 1936, 12th August 1968, 14th September 1968). The records in 1911 and 1968 occurred during some of the largest ever invasions of Slender-billed (Siberian) Spotted Nutcracker. Remarkably, the bird in August 1968 was one of a group of seven birds killed by a car. The six others in the group were Slender-billed (Siberian) Spotted Nutcrackers (van den Berg and Bosman 1999). There is overlap between the two forms in all published characters (*BWP*), so could it be that some Slender-billeds fitting the biometrics of Thick-billed occur in Europe during invasions? In the case of the British records, 1860 was not a known invasion year, though the late autumn of 1900 yielded three Slender-billed (Siberian) Nutcrackers shot between late October and mid November.

As expected from its irruptive behaviour, nearly all records have involved Slender-billed (Siberian) Spotted Nutcracker *macrorhynchos*.

Very rare vagrant outside of an exceptional invasion in 1968/69.

Status: The total of 375 recorded since 1950 obscures the typically rare status of this fine species, as nearly all came to Britain during the autumn and winter of 1968-1969. Since 1985, there have been just eight accepted birds:

1985 Kent: Northward Hill 23rd October
1985 Cambridgeshire: Wandlebury Park 24th-26th October
1985 Suffolk: Westleton 2nd November-7th December, when found dead
1987 Cambridgeshire: Cherry Hinton, Cambridge 13th February
1991 Staffordshire: Cocknage Wood, Stoke-on-Trent 14th October-9th November
1991 Hampshire: Copythorne 22nd and 26th October
1996 Hampshire: Denny Wood 29th December
1998 Kent: Kingsdown 6th-7th September

Historical review: Spotted Nutcracker has one of the highest aggregate number of records of any rarity, yet it is a true mega for the modern birding community.

A total of 64 birds remain extant prior to the formation of the rarities committee. The first British individual was shot at Mostyn, Flintshire, on 5th October 1753. One other 18th century sighting and a further 31 from the 19th Century remain acceptable. The latter include three autumn individuals in 1868 (two in October from Scotland and one November sighting in Dorset) that were part of a small influx into Europe. Four appeared between late October and December 1900 and there was an arrival of six to East Anglia and the southeast between early October and mid-November 1911. Another 17 birds were documented up to 1948. None were recorded during another European irruption in 1933, although two appeared in winter 1936/1937 at Bonchurch, Isle of Wight. This is the only accepted record of more than one bird prior to 1950.

The 1950s produced 10 birds from late August to December, all one-day birds except one at Wadworth Wood, near Doncaster, from 6th-10th October 1953. One in Gwent was the sole British representative of another European irruption in 1954. Three at Eccup Reservoir, West Yorkshire, on 1st November 1955 was a particularly unusual occurrence for a non-influx year.

Between 1961 and 1967, there were only five, all one-day birds (one in February and all others between late August and early October). The events of autumn 1968 and early 1969 completely rewrote the record book for the species and are discussed in detail below.

The 1970s added 13 birds, 12 of them were one-day birds from late August onwards; there were three October sightings in 1978. The 1980s and 1990s each produced just four records.

The closest modern British birders have come to experiencing an influx were arrivals of three between 23rd October-2nd November 1985 and two in October 1991. These included two lingering crowd pleasers: Westleton, Suffolk, in 1985 and Stoke-on-Trent, Staffordshire, in 1991. The confiding nature of the species encouraged these two birds on to many birders' favourite twitch list. The Suffolk bird entertained thousands of visitors during its stay, frequently making visits to apple trees in a cottage garden. Crowds of birders watched at close quarters while the bird smashed apples to bits to get at the pips. The Staffordshire bird's antics included landing on telescopes and hopping through observers' legs. One photographed at Kingsdown, Kent, in 1998 also attracted a crowd, but that bird was mobile and did not linger.

The influx of 1968-69: The statistics for this sought-after species are highly skewed by the events of autumn 1968 and early winter 1969, when a massive invasion of at least 339 Slender-billed Spotted Nutcrackers powered into Britain. Most arrived on the east coast between Dover and The Wash. Although they constituted an unprecedented and unequalled influx, the British occurrences were very much on the western periphery of this most spectacular eruptive movement of the species ever recorded. The British influx was tiny compared with events elsewhere: in Sweden 4,400 were observed passing one watch point on 11th August; and there were 6,000 sightings in the Netherlands. In autumn 1968 one was caught and ringed in Portugal. One even reached North Africa. Farther afield, Spotted Nutcrackers appeared in the Gobi Desert in autumn 1968, indicating that the species had erupted eastwards as well as west. An overview of the influx was documented in detail by Hollyer (1970) and is summarised below.

From a British perspective, the 1968 influx began in Norfolk on 6th August, with the first two birds from Trunch and Ditchingham in the east of the county. The third singleton came from Denton, Kent, on the 7th. These three formed the vanguard of an invasion that produced a succession of birds through to the summer and autumn of 1969.

During the rest of August, Spotted Nutcrackers reached Britain in at least two waves. The first major influx between the 6th and 17th was restricted to Norfolk, the north-east Suffolk coast and east Kent. It produced 27 birds: 21 in Norfolk, four in Suffolk and two in Kent. The birds arrived while southeast England was under a vast area of high pressure centred over northern Europe.

A more marked influx commenced on 21st August. Between then and the 31st, a total of 66 birds were recorded, most in East Anglia. These included 10 in Norfolk, 27 in Suffolk and seven in Essex. Eleven were in Kent and 11 elsewhere, including three in Wales at the end of the month. This part of the arrival included the only bird to reach Scotland (Shetland from 21st-23rd). High pressure from the Continent continued to have its effect and, with much of Britain blanketed in fog, the morning of the 23rd saw a localized fall of Spotted Nutcrackers to the Suffolk coast, concentrated mainly in the area between the rivers Deben and Aide. From 25th August to 11th September, the average number of Spotted Nutcrackers seen in Britain each day was 22. Each arrival phase was associated with north-easterly or light variable winds.

In September, R. P. Bagnall-Oakeley realised the magnitude of the invasion in Norfolk and Suffolk and appealed for records on the BBC and the Anglia television programme *Look East*. This appeal brought to light many further observations made by members of the public. The total for September rose to no fewer than 138: 47 in Norfolk, 41 in Suffolk, eight in Essex, 10 in Kent and 32 elsewhere. Numbers in eastern England began to fall sharply from mid September onwards; by the 26th even the most resident individuals of earlier influxes had departed.

During October 67 more were recorded: 17 in Norfolk, 11 in Suffolk, two in Essex, three in Kent, and 34 elsewhere. Between 4th and 9th October Norfolk and Suffolk experienced another small arrival in coastal districts where there had been absences for several days. The considerable numbers of Spotted Nutcrackers in new inland areas presumably involved healthier birds filtering inland following an earlier coastal arrival.

After 29th October coastal records were comparatively few, but seven appeared at new localities following easterly gales on 14th-16th November. In that month a total of 33 was recorded: two in Suffolk, two in Essex, three in Kent, and 26 elsewhere. In December, 17 were recorded: two in Suffolk, one in Essex, four in Kent, and 10 elsewhere. By the end of the year it is probable that only about 10 Spotted Nutcrackers remained, and three or four of them disappeared in January.

During the first few months of 1969, two established residents in Kent were seen regularly, two appeared (or reappeared) in a conifer forest in North Yorkshire, three frequented the Wirral, Cheshire, and seven wanderers were recorded. In June one was seen in an entirely new locality in Kent. In July two appeared near the Suffolk coast. Five more loiterers were recorded in the summer and autumn of 1969. Birds at two new localities in autumn 1969 may have been outliers of further movements reported on the Continent at that time.

The main axis of the influx was East Anglia (Norfolk 104 birds, Suffolk 94 and Essex 16) and Kent (34 birds). Many subsequently penetrated inland counties and the impetus took birds farther west. Seven reached Devon and five Cornwall. One reached the Isles of Scilly. Singles also pushed into both southwest and northwest Wales. In northwest England a single straggled to Lancashire. A group of three birds reached North Yorkshire (the farthest north of the English birds). Just one reached Scotland.

The mass arrival was concentrated in areas adjacent to the shortest sea-crossing of the North Sea. Perhaps farther north many more perished at sea; the lack of records from northern Britain is surprising for a species irrupting through Scandinavia. The route of most British immigrants would appear to have been through the Low Countries.

All the Spotted Nutcrackers reaching Britain during the irruption were of the Slender-billed race *N. c. macrorhynchus* of northeast Russia and Siberia. The age composition based on those described comprised about 60 per cent adults. Favoured habitats included gardens, public parks, churchyards, conifer plantations and, in at least 14 cases, roadside verges. Most birds made only short stays and most arrivals presumably died during the influx.

The only bird to reach Scotland (on Shetland) was observed predating House Sparrows *Passer domesticus*, killing at least three and eating one entirely. Another in Norfolk was seen to decapitate a freshly dead House Sparrow, split open its skull as though it were a nut and eat the contents. One in Suffolk returned three times to remove dead House Sparrows from wire covering a thatched roof. Such behaviour is an indication of the desperate hunger of many of the Spotted Nutcrackers that survived the North Sea crossing only to find themselves in areas with no seed-bearing cones.

When: Outside of the invasion of 1968/69, Spotted Nutcracker has been a great rarity. Analysis of the 36 records since 1951 shows a strong autumn and early winter bias. Just two have occurred outside this period, both in mid February. The earliest of the autumn records was at Beachy Head on 22nd August 1970, one of three August records.

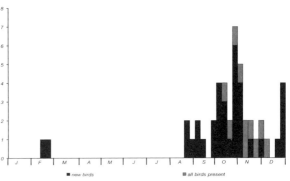

Figure 87: Timing of British Spotted Nutcracker records, 1951-1998 (excluding 1968/69).

A handful have been seen during September, though most occur in October, particularly so the last week of the month. Five December records include four in the last week of the year. As with the invasion, the occurrence pattern is concentrated on East Anglia and Kent, which account for 13 of the records, but as in 1968/69 several penetrated inland areas.

Discussion: Siberian *macrorhynchus* is dependant upon the seeds of the Siberian Stone Pine *Pinus sibirica* or, farther east, the Korean Stone Pine *P. koraiensis*. In years when these food sources are in short supply the birds vacate their regular range in huge numbers. Erupting Spotted Nutcrackers tend to show a strong westward orientation in their movements, bringing them to Western Europe (most other irruptive species from this area head in a southwesterly direction). This often leads to invasions being more obvious in Fennoscandia than farther south in Western Europe (Newton 2008).

Major irruptions in the 20th century occurred in 1911, 1933, 1954 and 1968; lesser movements were noted in 1900, 1913, 1917, 1947, 1971, 1977 and 1985. Late in 1991, 35 reached the Netherlands in late September and early October and others came to Belgium and northern France; one reached Helgoland and two pushed on west to Britain. In Finland the largest invasion since 1968 occurred between July and October 1995, with over 50,000 reported, but only small numbers penetrated Scandinavia and south to the Low Countries.

During the 20th century, Siberian *macrorhynchus* irrupted into Europe at an average interval of 7.7 years (range 2-16 years). Only during the 1968 irruption, which originated as far east as Lake Baikal (7,000 km away), did birds reach Britain in a sizeable number. Other irruptions have originated from different regions.

Studies of museum skins from irruption years established that the movements of 1864, 1911 and 1968 consisted solely of adults and those of 1885, 1913 and 1954 comprised 1st-year birds. Other irruptions have included a mix of age classes. During the invasion of 1968 many birds that reached Western Europe retraced their migration quickly during the same autumn, but probably did not return as far as their source. Recoveries of 14 birds ringed in Finland came from east and southeast of the ringing site, including birds up to 3, 300 km east. Birds trapped in Sweden were recovered in a subsequent year about 4,000 km to the east in Siberia (Newton 2008).

340

One can only speculate at the numbers that would be unearthed if a future invasion of similar proportions to the 1968 event were to occur.

Spanish Sparrow *Passer hispaniolensis*

Ray Scally

Scally '03.

(Temminck, 1820). Nominate hispaniolensis breeds from the Atlantic Islands east across North Africa, Iberia, Corsica, Sardinia and from the Balkans and Romania to Israel. Race transcaspicus is found from Iran eastwards to Kazakhstan and Afghanistan. Eastern populations are mostly migratory, wintering from Egypt to India, and those in the west of the range are largely sedentary.

Polytypic, two subspecies. Race undetermined, though *hispaniolensis* would seem the most likely visitor.

The Italian Sparrow *P. italiae* is found in Italy (except Sardinia) and adjoining regions of south-easternmost France (including Corsica), southern Switzerland, south-westernmost Austria and western Slovenia. Its taxonomic status remains uncertain and its treatment has changed repeatedly. It has been variously regarded as a stable hybrid between House Sparrow *P. domesticus* and Spanish Sparrow (Hagemeijer and Blair 1997) or as a subspecies of either House Sparrow (*BWP*) or Spanish Sparrow (Summers-Smith 1988). Töpfer (2006) suggests that it is a non-hybrid taxon and that Italian and Spanish Sparrow are conspecific. Following the rules of zoological nomenclature, if Italian were lumped with Spanish the taxa would be designated as *Passer italiae italiae* (Vieillot 1817) and *Passer italiae hispaniolensis* (Temminck 1820).

Very rare vagrant from southern Europe or farther east.

Status: Seven records, all involving striking males.

1966 Devon: Lundy male 9th-12th June
1972 Scilly: St. Mary's male 21st October
1977 Scilly: Bryher male 22nd-24th October
1993 Pembrokeshire: Martin's Haven male 18th May

1993 Orkney: North Ronaldsay male 11th-19th August, trapped 11th.
1996 Cumbria: Waterside male 13th July-13th December 1998
2000 Cornwall: Cawsand male 12th November

Discussion: The long-staying male in Cumbria ensured that for more than two years this rare sparrow was relegated to a year tick. The four British records in spring and summer accord well with the arrival pattern of birds elsewhere in northern Europe: two in Norway (13th May 1988, 21st July 1990); one in Finland (31st May 1996); and two in the Netherlands (4th-15th May 1997 and 13th May 2000). In the Netherlands an exceptional flock of 14 was found at Eemshaven, Groningen on 9th April 2009 (at least five males) with nine on 10th, six on 11th and finally four on 12th. Although not at an international harbour the flock is subject to acceptance as ship-assisted birds do not form part of the Dutch list, but another in the south of the country on 20th-21st and 26th April may hint at an overland influx.

The west-coast bias could implicate overshoots from the western part of the range and suggests a southern, rather than eastern, origin for sightings. Movements of the species are complex. Some southern populations are mostly sedentary and others partially migratory. Populations in northwest Africa are both migratory and nomadic. Even ocean crossings would appear to present little in the way of an obstacle. Colonisation of the Atlantic islands began in early 19th century and was completed during the second half of the 20th century with the first arrival on Madeira in May 1935 after persistent easterly winds. It is unclear whether colonisation of the Canary Islands and Cape Verde Islands was natural (Hagemeijer and Blair 1997).

Southeastern populations provide a possible source for arrivals as they show more regular migratory behaviour, moving north for successive breeding attempts in some areas. The breeding range in southeast Europe has expanded north in recent decades, correlating with the northern European sightings.

Given that House Sparrow flocks rarely receive much observer attention, and that immatures and females of the two species are near-identical, it seems likely that this attractive sparrow is under recorded as a vagrant.

Rock Sparrow *Petronia petronia*

(Linnaeus, 1766). Found across most of southern Europe, east through Turkey and Iran to the Himalayas; also northernmost Africa and the Atlantic Islands. Mainly sedentary, but undertakes regular short-distance migrations and dispersive movements.

Polytypic, seven subspecies. Race undetermined.

Very rare vagrant from southern Europe.

Status: One record.
1981 Norfolk: Cley 14th June

Discussion: This sighting represents the sole record in Britain and northern Europe in recent times. Unfortunately it showed to just a few fortunate observers before departing westwards towards Blakeney Point. It had been present for just 30 minutes in the company of a male Linnet *Carduelis cannabina* (Gantlett and Millington 1983).

Listed as accidental as far north as Belgium and the Netherlands by Wardlaw Ramsay (1923), the species had disappeared from southwest Germany and northeast France by the end of the 19th century (Hagemeijer and Blair 1997). A southward contraction across much of the range in the early part of 20th century brought with it a reduced chance of wanderings to northern Europe. With no evidence of such vagrancy elsewhere in Europe, there would appear to be little prospect of a reappearance in the near future of this unobtrusive occupant of sunny barren habitats.

Yellow-throated Vireo *Vireo flavifrons*

Vieillot, 1808. Breeds from central North America east through the Great Lakes and south to the Gulf coast. Some winter in southern USA, but winters primarily from southern Veracruz and eastern Oaxaca in Mexico through Central America to western Colombia and west and northwest Venezuela.

Monotypic.

Very rare vagrant from North America.

Status: One record.
1990 Cornwall: Kenidjack 20th-27th September

The Cornish bird was the first for the Western Palearctic.

I was birding in Kenidjack valley, north of St Just, Cornwall, during the morning of 20th September 1990. I had watched the area regularly during the previous seven years. Mentally, I tossed a coin and decided to check the bushes at the bottom of the valley just one more time. This paid off. The valley seemed very quiet, but, at the last bush, around 09.00 GMT, I noticed a movement at the back and focused on the wings of what I thought could be a Pied Flycatcher Ficedula hypoleuca. When it flitted around to the front of the bush, however, it showed a gleaming yellow throat. It proceeded to flycatch in the open. I was so excited and shaking that I had to sit down and study the bird with my telescope.

I quickly ruled out the American wood-warblers Dendroica already on the British List, as I was impressed by the remarkable yellow throat and face, unstreaked, bright green upperparts, broad white wing-bars and the heavyish blue-grey bill and legs. After making mental notes I rushed back to the house to consult the National Geographic Society Field Guide to the Birds of North America (1983). To my surprise, none of the wood-warblers seemed to fit. Only two, the Pine Warbler D. pinus and the Cerulean Warbler D. cerulea, seemed even remotely close. I tried unsuccessfully to phone several local people, but eventually succeeded in speaking to Richard Millington and described the bird to him. He telephoned some other people to try to confirm the sighting.

In the meantime, I went back for another look and, after 45 minutes' searching, I had brief, but good, views. I was reminded of Red-eyed Vireo Vireo olivaceus, by the jizz and the bluish bill and legs. Quickly turning my mind to vireos, I realised that it was, of course, a Yellow-throated Vireo V. flavifrons, a first for the Western Palearctic.

I left a friend, R. Ingham, at the site while I went back to the telephone. The first local birders arrived about midday, when the vireo was not showing. After an agonising two-hour wait, it reappeared and showed well to the few of us present that evening. Although quite elusive, the bird was seen by many hundreds of people during the weekend, was photographed, and stayed until 27th September.

Andrew Birch. British Birds 87: 362-365.

Discussion: There are three additional Western Palearctic sightings of Yellow-throated Vireo: one from Helgoland, Germany, (18th September 1998); and two on the Azores in autumn 2008 (Corvo 11th-18th October, São Miguel on 13th October).

The paucity of Yellow-throated Vireo records is perhaps surprising given that it is a long distance migrant from eastern North America to Mexico, the Caribbean and northern South America. It is, however, quite a scarce bird towards the north of its range. Autumn departure dates from breeding localities are not precisely known, but range from mid-August through to mid-September (Rodewald and James 1996), correlating well with both European records (though the two from the Azores were somewhat later).

Yellow-throated Vireo was not predicted by Robbins (1980), though both Philadelphia *V. philadelphicus* and White-eyed Vireo *V. griseus* were. White-eyed Vireo has occurred twice on Corvo, Azores, (22nd October-23rd November 2005, 24th October 2008). Two further vireo species were mentioned as candidates for vagrancy – Blue-headed Vireo *V. solilarius* and Warbling Vireo *V. gilvus* – neither of which has yet occurred in Europe; both of them should be remembered as possible strays.

Blue-headed Vireo is a short- to medium-distance migrant to wintering areas along the Atlantic and Gulf of Mexico coasts of southern USA and Central America. Although the short migration route makes it less likely than some of its relatives, movement along the Atlantic coast offers some hope of a stray (James 1998). Warbling Vireo is a medium- to long-distance migrant to western Mexico and northern Central America. Eastern populations follow an overland route along the Gulf Coast south and almost entirely west of Florida, rendering them less susceptible to weather systems suitable for transatlantic vagrancy. Even so, birds have occurred in western Cuba, Jamaica and Bermuda in the autumn (Gardali and Ballard 2000).

Philadelphia Vireo *Vireo philadelphicus*

(Cassin, 1851). Breeds from northeastern British Columbia across Canada to New Brunswick, also New England. Winters in southern Central America, primarily in southern Guatemala, Belize, Honduras, northern Nicaragua, Costa Rica and west and central Panama.

Monotypic.

Very rare vagrant from North America.

Status: Three records.

1985 Co. Cork: Galley Head 12th-17th October
1987 Scilly: Tresco 10th-13th October
2008 Co. Clare: Kilbaha 13th-14th October*

Discussion: Robbins (1980) attached a relatively low predictive value to Philadelphia Vireo, ranking it at the lower end of the likely cohort of potential vagrants and equally likely with Blue-winged Warbler *Vermivora pinus* and Prairie Warbler *Dendroica discolor*.

Philadelphia Vireo has the most northerly breeding range of all vireos and is a regular, if scarce fall vagrant in Bermuda (Amos 1991). Autumn migration extends farther east and south-east, but it remains a rare vagrant to the Western Palearctic where the only additional record was on the Azores (26th October 2005).

The Scilly bird was found during the islands' heyday for birds and birders. It involved the biggest ever mass movement of birders on Scilly, with up to 1,000 people ferried from St. Mary's to Tresco, most of them by breakfast time (Flood *et al* 2007).

Although numbers in the US have increased in recent decades (Sauer *et al* 2008), this species has a much smaller population size than Red-eyed Vireo (Rich *et al* 2004), which may explain its extreme rarity. As such, the autumn 2008 bird from Co. Clare was widely appreciated by a new generation of listers. It was part of an exceptional run of Nearctic vagrants.

Red-eyed Vireo *Vireo olivaceus*
(Linnaeus, 1766). Breeds throughout southern Canada and USA east of the Rocky Mountains. Migrates through eastern USA to winter in northern South America.

Polytypic. Enzyme data (Johnson *et al* 1988) from 20 species in four genera of family Vireonidae indicate that North American *olivaceus* is conspecific with South American *chivi* taxa and these are treated currently as two subspecies groups (Cimprich *et al* 2000). Two weakly differentiated subspecies are recognised in North America (*olivaceus* and *caniviridis*); up to nine South American subspecies are currently recognised. Records of birds in Britain and Ireland have not been assigned to subspecies, but *olivaceus* (Linnaeus, 1766) is most likely.

Regular autumn vagrant from North America, mostly to southwest Britain and Ireland.

Status: The most frequent Nearctic passerine vagrant with 153 records, 107 in Britain and 46 in Ireland, to the end of 2007.

Historical review: The first record was in Ireland, with one found dead at the Tuskar Rock lighthouse, Co. Wexford, on 4th October 1951. The next, and first for Britain, were two birds on St. Agnes from 4th-10th October 1962 (one seen again on 17th).

ST. AGNES, Part I: A quick check on the Lesser Golden Plover found it still eating earthworms and still on the church plough. Then quickly back to Periglis (the island's harbour cove) for high tide and the wader count. Almost as soon as we got there, a small dark wader popped up only a few yards away. After a few seconds, while his mind recalled his Canadian experience of eight years earlier, D.I.M.W. was extolling the virtues and beauty in miniature of an American Stint [now called Least Sandpiper]. Incredible as it sounds, its appearance was real and after a full hour and a quarter of close observation, all concerned were satisfied with the identification and once again delighted with St. Agnes. The stint was the third American wader of the week but unlike the plover and an emaciated White-rumped Sandpiper (two days before), it was very active, given to much calling and nervous flights. The last of the latter took it off to St. Mary's.

TRESCO: We had already arranged a trip to Tresco and set off in great glee, expecting to find a new wader for Britain. We landed at new Grimsby and were soon scouring the pools and southern beaches. Morning became noon, our feet tired and our spirits sank measurably. We listed 59 species but not one was uncommon, let alone rare. Even D.I.M.W. claimed nothing though he muttered about a duck rather bald on the pate [the alternate name for American Wigeon is Baldpate]. When we all tried for it, the duck got nervous and went off in an aerial circus. Not to have found any rare bird was utterly beyond reason and we all got soaked coming back. Inner thoughts of passing glories depressed.

ST. AGNES, Part II: Still with our fixation with waders, we were back at Periglis soon after 1400. There we added another 'new' wader but it was only a Curlew Sandpiper. Sort of cod roe, not caviar. Then in an offhand manner, the party split up to idle through the northern pieces (small flower fields). R.E.E. and I.J.F.L. had penetrated into the Parsonage Wood when D.I.M.W. and K.A.W noticed a 'large warbler' darting about in some tamarisk canopy on the edge. Clearly the clumping about of R.E.E. and I.J.F.L. was disturbing it, but D.I.M.W. managed only a strangled announcement of its presence before it plunged out of sight. What on earth was it? This question was answered almost too quickly because suddenly the bird came out on a tamarisk spray, showed a large blob-shaped bill and a striped head and then was gone, tearing across a piece and into deepest hedge. The island echoed to a shout of 'Vireo!' and R.E.E. and I.J.F.L. flew out of the Parsonage. D.I.M.W. recovered sufficient calm to add 'Red-eyed' to 'Vireo' and to assure the others that now they had an American passerine to contend with. Then something uncanny happened. The S.S. Rotterdam, passing close offshore, saluted the occasion with a glorious trump. We dashed in all directions trying to keep in contact with the delectable American. Its flights were rapid and lengthy but eventually the (more sensible) tactic of watching the Parsonage canopy paid off. We all got excellent views down to a few yards. Sharp-eared I.J.F.L. was able to note its progress through cover by listening for its Willow Tit-like call tchay. Not counting a wretched Irish corpse, this is a first for Britain. For this reason and the earlier presence of the exquisite Calidris minutilla [the Latin name *for American Stint], October 4th has whacked October 2nd into a cocked hat. Who needs Hippolais Warblers? (St. Agnes had produced all three found in Western Europe in the previous four days.) As a passing comment we feel that the Peterson drawing of Red-eyed Vireo does our bird no justice. From its erratic behaviour, it was not long in and we suspect that it had jumped ship."*

The log of the day went on to describe other migrants and argue that there had been a diurnal fall, surprising in view of the fine weather. On the next day, we found Britain's second Red-eyed Vireo alongside the original bird, and were totally beaten by what can only have been a Yellow Warbler (also from North America). This should have been an absolute first for Britain but its story is so full of frustration, I cannot bear to tell it.

Ian Wallace. Discover Birds

Four more occurred between 1966 and 1968 (two St. Agnes and others in Co. Cork) and the first for Wales, trapped on Skokholm on 14th October 1967.

Three during the late 1970s were low-key harbingers of an imminent change in status. These included the first mainland record at Aberdaron, Caernarfonshire, from 25th-26th September 1975 and two more, from Co. Cork and Scilly, in 1978.

The 1980s saw the onset of almost-annual autumn presences. Five in 1981 easily set a new annual record, but these were soon eclipsed by an exceptional 14 in 1985 that set a new scale for the maximum influx of an American landbird. This exceptional arrival delivered simultaneously the first and second records for the Netherlands, the second for France and one to Iceland. A series of depressions moved across the Atlantic and into Europe in early October. The southwest of Britain bore the brunt of arrivals from the 2nd October, with five found on Scilly, three in Cornwall and one on Lundy; three reached southern Ireland. The only records away from the southwest were the first for Scotland, at Wick, Caithness, and one on Bardsey.

Another multiple arrival in 1988 produced 12 birds, though with a much more dispersed arrival pattern. The southwest produced six, including two each on Scilly and Lundy. There were two in Co. Cork, plus the second Scottish record, on the Outer Hebrides. Remarkably three east coast records were found, with singles in Suffolk (almost simultaneously with several in the southwest), Northumberland and Caithness. The decade closed quietly with three in the southwest, but expectations had been transformed. The Atlantic depressions had cheered birders with a whopping 49 birds during the decade at an average of just under five per annum.

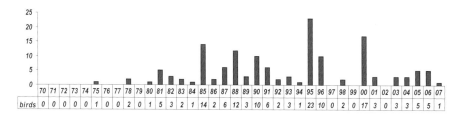

	70	71	72	73	74	75	76	77	78	79	80	81	82	83	84	85	86	87	88	89	90	91	92	93	94	95	96	97	98	99	00	01	02	03	04	05	06	07
birds	0	0	0	0	0	1	0	0	2	0	1	5	3	2	1	14	2	6	12	3	10	6	2	3	1	23	10	0	2	0	17	3	0	3	3	5	5	1

Figure 88: Annual numbers of Red-eyed Vireos, 1970-2007.

The 1990s opened in a similar fashion, with 10 in 1990. Five of the British birds included more east coast finds (East Yorkshire and Co. Durham) and more normally three in the southwest and five in Ireland. A blank British year in 1994 was a rarity in itself, but was soon corrected by an incessant delivery of Atlantic depressions in October 1995 that produced a record 23 birds (13 of in Britain and 10 in Ireland). The British records commenced with one on the M. V. *Scillonian III* off Land's End on 30th September. As usual, it was the southwest that hosted most birds. These included five on Scilly and three in Cornwall as well as singles in Dorset and Devon. One was in Pembrokeshire. More surprising was another simultaneous arrival in the middle of the influx of two in Suffolk; ship-assisted perhaps? In Ireland, four were found in Co. Cork, three in Co. Clare, two together in Co. Wexford and one in Co. Galway. One can only hazard a guess at the true magnitude of the arrival. Ten in 1996 was a creditable follow-up, but the arrivals were restricted to the southwest and southern Ireland. Two blank years out of the last three of the decade was unexpected, but the 1990s still closed with the cumulative discovery of no fewer than 57 birds.

A new millennium opened with the second-highest year total ever in 2000: 17 birds; 12 in Britain and five in Ireland. Once again, Scilly took the lion's share, with six birds, but four were found in Cornwall and one in Devon. In Ireland three were located in Co. Cork and singles logged in Co. Waterford and Co. Galway. There was a sole Scottish record, from the Outer Hebrides. Up to the end of 2007 a total of 37 had been recorded for the decade.

Since 1980 there has been an average of five per annum, a stark change in status given that there had been just 10 previous records. As with all Nearctic vagrants, weather conditions are key to whether arrivals take place and in how many stages they occur. While it would appear that this species has become even more prone to transatlantic vagrancy in recent years, some of the increase can be explained by greater coverage and and conducive weather systems. It is surely significant that counts on Breeding Bird Survey routes in Canada and northeastern USA have increased by about 50 per cent since the mid 1960s (Sauer *et al* 2008).

Following the review period the autumn of 2008 delivered good numbers of Nearctic vagrants to Britain, with Red-eyed Vireo once again the most plentiful species. These included three on the Isles of Scilly and in Co. Mayo, singles in Cornwall and Devon and more unusual records from the Isle of Wight and Tiree, Argyll. Two September birds were followed by six between 8th and 11th October and two more during the following week.

Where: As would be expected, the southwest and Ireland have accounted for the majority of records, harvesting 103 birds (68 per cent) in all. Local scores for the Isles of Scilly were 44 (29 per cent of records) and Cornwall 25 (16 per cent of records), while Devon had 10 (seven per cent) and Co. Cork 25 (16 per cent of records, including 14 on Cape Clear). Red-eyed Vireo occurs on the Isles of Scilly with a similar frequency to Radde's Warbler *Phylloscopus schwarzi* and around twice as frequently as Dusky Warbler *P. fuscatus* (Flood *et al* 2007).

Eight have reached Scotland: four on the Outer Hebrides, two from Caithness and singles from Argyll and Lothian. Unlike some Nearctic waders, there seems to be no suggestion of increasing numbers of the commonest American passerine farther north as a result of the northerly displacement of the Atlantic storm tracks. It remains a surprising absentee from the avifauna of the Shetland and Orkney Islands. Six have reached Wales: three from Caernarfonshire, two in Pembrokeshire and one on Anglesey. On the south coast three have reached Dorset and two Kent.

A surprising number have reached the east coast (14 records): two records from Caithness; singles from Lothian, Northumberland, Cleveland and East Yorkshire; two from Durham, two from Kent; and no less than four from Suffolk.

When: Both of the earliest records are from Cape Clear, where there were singles on 5th September 2004 and 14th (to 20th) September 1995. Six more have been recorded between 19th and 23rd September, including the earliest British record (found dead at Bardsey lighthouse on 19th September 1998) and the earliest from Scilly (St. Mary's from 20th-22nd September 1989).

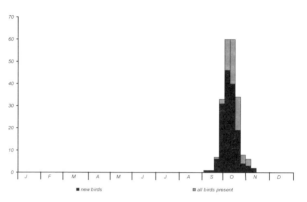

Figure 89: Timing of British and Irish Red-eyed Vireo records, 1951-2007.

The period between 24th September and 14th October has accounted for more than three-quarters of all records. The one-week period of 15th to 21st October has accounted for 12 per cent. There have been just eight new arrivals later in the year. Three have been found in the first week of November: two in the southwest (Loe Pool, Cornwall, from 3rd to 7th November 1998 and at Dawlish Warren, Devon, on 4th November 1996); and, the latest, in Scotland (Thurso, Caithness, on 8th November 1988).

The mean arrival date for British and Irish records is 7th October; for east coast birds it is 12th October. Possibly these individuals made landfall farther north and then made a southbound crossing of the North Sea. However, given the rarity of the species from northern areas, with none from Fenno-Scandia, ship assistance seems more likely for at least some.

Discussion: Red-eyed Vireo undertakes a long-distance complete migration between breeding grounds in USA and Canada and the Amazon basin of South America (Cimprich *et al* 2000).

The large number of birds elsewhere in the Western Palearctic have occurred across a deep latitudinal span from 65°N in Iceland south to 30°N in Morocco. Nineteen have been recorded from Iceland (16 since 1984 including five in both 1995 and 1997), with the earliest on 16th September 1951, 11 between 27th September and 8th October, six between 18th and 27th October 1995 and one found long dead on 3rd November.

The predominant southwest contingent of our birds is mirrored in France, with nine from Ouessant, Brittany (all 1983-2001; the earliest on 14th September 1998 and the rest between 8th and 21st October). Curiously, only one has been found in the Channel Islands (Guernsey 12th October 2006).

There is a surprisingly large total of eight in the Netherlands (13th October 1985, 19th October 1985, 24th September 1991, 2nd October 1991, 3rd-8th October 1996, 13th October 2001, 22nd October 2003, 9th October 2005). Others have reached Germany (4th October 1957), Belgium (13th October 1995) and even Poland (17th October 2000).

In the south there have been three in Spain (all along the Mediterranean coast; 19th October 1995, 25th October 1995, 30th October 2000) and singles from Morocco (one trapped in the reed bed of Oued Souss, near Agadir, October 1968) and Malta (ringed in October 1983). Ten have been found in the Azores (all since 2005; three between 11th and 13th October and seven between 21st and 28th October).

Citril Finch Carduelis citrinella

(Pallas, 1764). Breeds in the coniferous subalpine zone of the mountain systems of central and southwest Europe; northernmost breeding area is the Black Forest. Generally a short-distance and altitudinal migrant, wintering above 1,000m.

Monotypic. The Citril Finch was formerly considered conspecific with Corsican Finch *Carduelis corsicana*, of Corsica, Sardinia and the Italian islands of Elba, Capraia and Gorgona. However, they differ in morphology and vocalisation (*BWP*; Förschler and Kalko 2007) as well as mtDNA sequence (Sangster 2000) and are now treated as separate monotypic species (Sangster *et al* 2002).

Very rare vagrant (if accepted).

Status: One record awaiting adjudication by the relevant committees.
2008 Shetland: Fair Isle male 6th-11th June*

Discussion: One caught alive at Great Yarmouth Denes, Norfolk, on 29th January 1904 constituted the sole British record for 90 years, until the specimen was re-examined and re-identified as a male Yellow-crowned Canary *Serinus canicollis* (Knox 1994), presumably imported from southern Africa.

The arrival of the Fair Isle bird caught many observers on the hop given that the vagrancy potential of the species may have been rather underestimated. It prompted much speculation as to its origin. Trapped during its stay, it showed no obvious signs of captivity. Citril Finch is less of an unlikely vagrant than many may have thought, as it has been proven to migrate considerable distances. Birds from Garmisch-Partenkirchen in the German Alps have been recovered in the Cevennes and Mont Ventoux areas (about 600 km south-west of their breeding areas). A Swiss bird ringed in June was recovered near Barcelona, Spain, in October (a southwesterly movement of 625 km). One ringed at Col de la Golèze was recovered in central Italy (610 km to the southeast). Two Citril Finches captured in the Spanish exclave of Ceuta in North Africa

(31st March 1991 and 20th April 1991) (Navarrete *et al* 1991) confirm that the species is able to migrate longer distances and over sea (albeit a potentially a short crossing).

Recent records from northern Europe include three males trapped in Belgium (13th October 1993, 20th September 2000 and 8th April 2001) and two in Poland (12th July 1975 and two on 1st April 2001). In France there is a small number of records, the most northerly from Finistére (1971 and 10th February 1980) and the most recent from Île d'Yeu, Vendée (26th October 2006).

Most interesting was an adult female at a feeding site on the south coast of Finland from 17th May to 2nd July 1995. This bird arrived with Eurasian Siskins *C. spinus,* one of which was controlled at the same site just prior to the discovery of the Citril Finch and had been ringed just north of Rome, Italy, earlier in the year. This record was placed on category D by the Finnish rarities committee, as it was unclear when it had joined the Siskin flock, but it may have been as they passed north through the Alps.

The supporting evidence for the Fair Isle bird as a natural vagrant would appear to be compelling based on this analysis of movements within Europe. It is clearly not the case that there are very few records of Citril Finch vagrants at any great distance from known breeding areas (*contra* Knox 1994). However, arguments against the record may also be proposed. The bird's movement appears greater than those of other extralimital records, and only three of the others involved water crossings. The date of the record accords only with the bird recorded from Finland, and would not appear indicative of a near-European breeder. If the bird is accepted as wild, then it seems that many were blind to the potential for a genuine record in Britain and this bird certainly took most by complete surprise.

Arctic Redpoll Carduelis hornemanni

(Holboell 1843). Circumpolar distribution. European breeding range restricted to northern Scandinavia. Race C. h. exilipes breeds on tundra of Arctic Eurasia, Alaska and Canada to Hudson Bay. C. h. hornemanni breeds Ellesmere and Baffin Island to northern Greenland. Both races disperse south in winter, irregularly reaching northwest Europe.

Polytypic, two subspecies. Nominate *hornemanni* (Holboell 1843) Hornemann's Arctic Redpoll is a rare, but perhaps overlooked, vagrant from the Nearctic, the first record of which was from near Whitburn, Co. Durham in April 1855. The race *exilipes* (Coues 1862) Coues's Arctic Redpoll, from the northern Palearctic was removed from the list of species considered by the *BBRC* at the end of 2005. By that time more than 800 Arctic Redpolls had been recorded, most of them Coues's. Most notably, more than 400 were recorded in winter 1995/96, when huge numbers of the nominate race of Common Redpoll *C. f. flammea* (Mealy Redpoll) and the Eurasian Coues's Arctic Redpoll irrupted southwest. There are just three accepted Arctic Redpolls for Ireland. The first, at Dursey Island, Co. Cork, intermittently from 4th to 10th October 1999, was followed by others in 2000 and 2001; none of these three has been accepted to subspecies level. There have been just six birds recorded from Wales. The first, on Bardsey from 3rd-4th May 1987, was followed by four in early 1996 and one in 2002; all have been between February and early May.

A number of redpoll taxa have been described based largely on morphology and plumage. Presently these are placed in three species: Lesser Redpoll (*cabaret*); Common Redpoll (*flammea, rostrata, islandica*) variously named in English as Mealy, now Common (*flammea*), Greenland or Greater (*rostrata*) and Iceland (*islandica*), with the last two also referred to as Northwestern, due to recent indications of intergradations; and Arctic (*hornemanni, exilipes*).

Recent studies (Knox 1988; Herremans, 1990; Sangster *et al* 2002) tend to support the three-way split, but the variability of the redpoll group makes study of the taxonomy very much a work in progress. Based on enzyme and molecular studies there are no significant differences between *exilipes* and *hornemanni*. However, it has recently been suggested that since redpoll taxa have vocalisations that differ from each other and have diagnostic plumage characters, it is logical to apply the same taxonomic status for each of them (van den Berg *et al* 2007).

The true status of Hornemann's Arctic Redpoll is largely obscured by identification problems of its separation not only from Coues's but also from the highly variable Northwestern Common Redpolls that cross the North Atlantic regularly from Iceland, Greenland and potentially even the Nearctic. A proportion of the usually dark Icelandic birds are whitish with a white rump, so being near-identical to *hornemanni*. Only recently has observer attention focused closely on the sub-specific identification of Arctic Redpolls (for example, Pennington and Maher 2005). With increased attempts to assign these birds, backed by photographs and trapping, it may be that future analysts will report quite different patterns of occurrence and distribution for these two taxa.

Hornemann's Arctic Redpoll Carduelis hornemanni hornemanni
Breeds in the Arctic Canadian islands in Nunavut, including Ellesmere, Axel Heiberg Devon, Bylot, and Baffin Island, to northern Greenland south on east coast to Scorsby Sound. Disperses erratically south of breeding range in winter, irregularly reaching northwest Europe.

Status: The present *BBRC* statistics allocate 12 pre-1950 birds to Hornemann's Arctic Redpoll and at least 29 since then, nearly all in the past five years. Since identification issues cloud the taxa's true status, the totals surely represent underestimates of the numbers (for example, Forrester *et al* (2007) report 38 birds and 10 'probables' in Scotland up to 2004).

All records so far have been from Britain. There is not yet an accepted record of this subspecies from Ireland, although several recent claims from The Mullet, Co. Mayo await adjudication by the *IRBC*.

Some guidance on identification was offered in the *BBRC* report for 2003 (Rogers *et al* 2004).

> *Although identification to subspecies should be undertaken with caution, a combination of size (most easily judged against other redpolls) and plumage – notably a rich chamois-leather wash on the face, throat and breast and typically fairly uniformly frosty-grey upperparts (lacking the striking pale central panel and tawny-/brown-edged mantle so often shown by C. h. exilipes) – are useful pointers.*
>
> *Nonetheless, observers should beware of markedly cold-looking redpolls from the northwest, several of which have occurred in the Northern Isles in recent autumns and*

caused real problems. Critical observation of the rump and undertail-coverts (sometimes in the hand) was needed to confirm that these were indeed just Common Redpolls C. flammea. Exactly where these birds come from is something of a mystery as observers with recent field experience in Iceland suggest that birds matching this appearance simply do not occur there or, at best, are extremely rare.

Historical review: The first British record came from Whitburn, Co. Durham on 24th April 1855. A second English bird was seen at Kilnsea, East Yorkshire, on 25th February 1893. At least 10 more were obtained or seen between 1883 and 1935, including nine shot on Shetland, on Fair Isle and Unst, on dates from 18th September to 12th November. These included three on Fair Isle for several days from 18th September 1905 which remain the earliest ever autumn records. Two further birds were seen on the island during the autumn and another on Unst.

Inconsistent racial attributions followed, with some birds being attributed to *hornemanni* that are no longer counted in the national totals by the *BBRC*. Two in 1965 would appear to be the only extant records from the early part of the modern recording era: one on Foula from 13th-25th July and one trapped on Fair Isle and present from 4th-6th November

For the July record, the rarities report of 1965 stated:

So far as the one on Foula is concerned, it should be mentioned that the inhabitants of this island have the habit of catching small birds in autumn and releasing them in spring. Although enquiries failed to establish such an origin in this case, the possibility cannot quite be excluded in view of the most unusual date.

There was a 12-year gap until the next, one on Bryher from 19th October-5th November 1977. Famously, it shared a field with a Spanish Sparrow *Passer hispaniolensis*. However the *BBRC* reports for the time list several records from Kent and Norfolk that were attributed to *hornemanni* at the time but these racial attributions appear to be no longer upheld. One at Wells, Norfolk, from 28th September-10th October 1966 was filmed and Taylor *et al* (1999) report it as suspected of belonging to *hornemanni*, as was another in the county at Brancaster from 17th-18th April 1994. Neither were published as such by the *BBRC*.

A flurry of Scottish attributions followed in the 1980s: two on Fetlar, Shetland, on 13th October 1980; one on Whalsay, Shetland, from 28th April to at least 1st May 1987; one on Stronsay, Orkney, from 5th-7th November 1988; and 1st-winter trapped on Fair Isle, on 4th-18th October 1989. One at Carnoustie, Fife, from 27th December 1988 to at least 19th April 1989 was widely considered to belong to *hornemanni*, but has never been published as such by the *BBRC*.

Racial attributions on the official archive then stalled until 1997 with one on Unst, Shetland, on 10th October. It was followed in 1998 by one trapped on Out Skerries and present from 29th September to 2nd October and a mainland bird at Barrock, Caithness, from 6th-7th February 1999.

The next to be recorded came in an exceptional influx to Shetland in 2003 of at least 10. Two further birds at the same time were not assigned to race, but were felt most likely to also be *hornemanni*. All but one of these were seen between 22nd September and 17th October. These included two birds on 22nd September, one found dead at Lerwick and another on Fair Isle. These represent the earliest arrivals of the modern recording period. Another was on the Outer

Hebrides. The first record of this race for The Netherlands was also recorded during the influx.

In 2004, an unseasonable bird was present on Barra, Outer Hebrides, on 7th June. It was followed by four more between 29th September and 23rd October, two of them on the Outer Hebrides and two on Shetland.

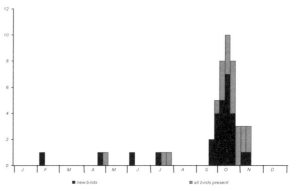

Figure 90: Timing of British Hornemann's Arctic Redpoll records, 1965-2007.

The good Nearctic autumn of 2008 delivered at least half-a-dozen birds of this race to Shetland, plus one to The Mullet, Co. Mayo, on 3rd October. Another was found on North Uist, Outer Hebrides, from 12th-13th October.

Discussion: Whether the recent flurry of records will define a clear pattern of occurrence for Hornemann's here, or whether they comprise one exceptional influx followed by more modest influxes is an unanswered question. Intriguingly, recent localities and dates reflect those of the very early specimens. It would appear to be mostly a bird of the Northern Isles and increasingly the Outer Hebrides. In these areas, it tends to arrive on earlier dates than those presented by Coues's, most of which tend occur in October and November.

Elsewhere in Northern Europe the status of this taxon is uncertain. The Norwegian status is unclear with few submissions, but it is probable that the taxon is regular in Norway in September and October; for example, 15-20 records submitted from Røst island in 2007 (Vegard Bunes *pers comm*). The status of the taxon in Iceland is unclear. There are three records from the Faeroe Islands.

The only other European records are one collected in Sweden (27th April 1934), a male trapped in Belgium (10th October 1937), two from France (one trapped at Pas-de-Calais on 1st February 1966 and one probably belonging to this race on Ouessant, Finistère, 19th-21st October 1986), a male trapped on Helgoland, Germany (20th-27th October 1991) and The Netherlands (11th-15th October 2003). In the Azores, where there have been 10 Common Redpolls, one was photographed on Corvo (20th-21st October 2005). It is perhaps surprising that just one has reached England in the last 30 years.

The possibility that Hornemann's has been overlooked in previous years seems likely, but based on present confirmed records it remains an irregular transatlantic vagrant. Since the massive influx of Coues's in the winter of 1995/96 and at a time of greater observer awareness, Hornemann's has been recorded almost twice as frequently as definite Coues's. Increasing coverage of the Outer Hebrides may reveal it to be more regular there, but it seems reluctant to filter down the east coast to England and beyond.

Two-barred Crossbill *Loxia leucoptera*

J.F. Gmelin 1789. Local resident within larch Larix forests of northern Eurasia from northern Russia to eastern Siberia, reaching Sea of Okhotsk and south to Baikal region. Irruptive dispersal leads to irregular breeding in Finland, and very occasionally in Sweden and Norway. Outside the breeding season, dispersal occasionally reaches northwest Europe. Other races (or species) breed across northern North America and on Hispaniola in the Caribbean.

Polytypic, three subspecies, variously treated as subspecies/species. Three forms of Two-barred Crossbill are currently recognised: Two-barred Crossbill *bifasciata* (C.L. Brehm 1827) in Eurasia, White-winged Crossbill *leucoptera* (Gmelin, 1789) in northern North America, and Hispaniolan Crossbill *megaplaga* Riley, 1916 in the mountains of Hispaniola. There are clear and consistent differences in measurements, structure, plumage and vocalisations that suggest the three forms would be best treated as separate species (for example, Benkman 1992; Elmberg 1993; *BWP*; *AOU* 1998). Hispaniolan Crossbill is considered a separate species to the White-winged Crossbill by the *AOU* (2003).

All British records have involved *bifasciata*. Nominate White-winged Crossbill has occurred through ship assistance and a presumed escape has been recorded in the Netherlands (Robb and van den Berg 2002; Harrop *et al* 2007).

Status: This popular finch has occurred 111 times since 1950, with around 95 from before then. The majority have been from Britain; four have occurred in Northern Ireland, but the species has yet to reach farther south in Ireland during its infrequent arrivals. Influxes usually occur following the coincidences of a good breeding season, followed by a failure in the Siberian Larch *Larix sibirica* cone crop.

Following the review period, an unprecedented influx during late summer and autumn 2008 presented over 60 birds (Harrop and Fray 2008).

Historical review: The first was a female shot at Grenville, near Belfast, Co. Antrim, on 11th January 1802 followed by a female shot near Brampton, Cumbria, on 1st November 1845. A number of other early records were too poorly documented to discount the rare morph of Common Crossbill *L. curvirostra* that shows marked wing bars. A recent review (Harrop *et al* 2007) of early records was undertaken by the *BOURC* to establish the first British record.

During the 19th Century there were several reports of flocks. These included nine shot at Castlesteads, Cumbria on 25th March 1846 and c20 at Ampthill, Bedfordshire on 3rd January 1890, four of which were obtained. The first documented major influx would appear to have taken place over winter 1889/1890 with eventually 34 birds documented following the first, an immature male shot at Easington, East Yorkshire, on 12th August. This invasion was also reported from the Netherlands where 32 birds were recorded in September 1889 and in Belgium 23 were collected in 1889/1890 (van den Berg and Bosman 1999). In Britain, six further singletons were reported between 1908 and 1912 and 21 between 1927 and 1948. These included the first for Wales, a male picked up dead near Llandrindod Wells, Radnorshire, in November 1912.

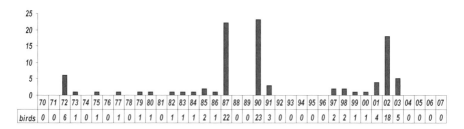

Figure 91: Annual numbers of Two-barred Crossbills, 1970-2007.

Just three were added in the 1950s (one in 1953 and two in 1959). Birds occurred in five years during the 1960s. The first widespread arrival of modern times was in 1966. It was a curious influx completely at odds with present expectations, in that all five males were found in England between 7th July and 22nd October and none recorded in Shetland. Two long-staying males included one at Holkham, Norfolk, from 21st September-25th November which came regularly with Common Crossbills to the famous (now dry) drinking pool. Another was present at Frensham, Surrey, from 22nd September-20th October. Three other singles came from Dorset, Hertfordshire and Cheshire and Wirral, the latter found dead.

The 1970s produced 10 males in five years, but the only multiple arrival involved six males in 1972. During this influx the first two made landfall on Shetland in July. They were followed by a late-July male in Northumberland and simultaneous finds in early August in North Yorkshire and on Lundy. A wintering bird was located in Shropshire in early November and relocated early in 1973. The decade ended with a hugely popular male at Cannock Chase, Staffordshire, from 16th December 1979-17th March 1980 that enabled many to enjoy their first experience of the species.

The early 1980s continued in a similar vein with singletons in four years, including a controversial male at Slufter's Inclosure in the New Forest, Hampshire, from 14th March-28th April 1984. This bird was eventually accepted despite doubts arising from the presence of a wing-barred Common Crossbill at the same site. Interestingly, the identification is no longer supported by one of the original finders, but the record remains accepted (Rogers *et al* 1999). In 1985 a male at Carron Valley Forest, Forth, on 13th October led to the discovery of a juvenile to 1st-winter female the following day, which remained through into mid-April 1986, the first collectable individual since the Cannock bird.

The events of 1987 changed perceptions of a new generation of observers on where and when to look for this species. A female killed by a cat on 9th August on Fetlar, Shetland, was the harbinger of at least 22 birds. All were in the Northern Isles. The only other adult was a female on Fair Isle on 28th August, the last of all; the rest were juveniles. During the influx at least 19 birds reached Shetland with an arrival between 10th and 18th August. On Foula, three were present on 11th August, two from 12th-16th August, six (possibly 10; six were accepted by *BBRC*) on 15th August and one to 18th August. Despite the large numbers on Shetland just three juveniles were located on Orkney (one North Ronaldsay and two at Finstown). A large invasion of the species into Norway was observed at the same time. The absence of records from farther south in

Britain was testament to the essentially westerly orientation of the irruption. Subsequent irruptions have come on a more southwesterly track, with birds making landfall on the mainland. Presumably the differences in the distribution of invasions here can be explained by birds from different breeding areas being affected by food shortages.

Two blank years followed, but in 1990 an exceptional invasion of Common Crossbills carried yet another welcome deluge of Two-barreds. These included 23 during the latter part of 1990 and three more early in 1991. This time, the influx was protracted and observed on a broad front. The first birds were once again detected on the Northern Isles (three adults between 25th and 31st July, a male and female on Fair Isle from 25th and a female at Hoy, Orkney, from 31st July). These were augmented by seven juveniles on the Northern Isles between 18th and 28th August. There were no further records until a juvenile was trapped at Kergord, Shetland, on 22nd September, the penultimate record of the influx from the islands.

The first on the mainland was a male at Sandringham Warren, Norfolk, from 30th September to 6th October and possibly 14th October. Twelve new birds were located by the end of the year, and a further three associated with the influx were found in spring 1991. Just one of these was on Shetland (in late October). Two were logged in Kent and two each in Norfolk and Northants. The spread of records extended to Northumberland, North Yorkshire and Derbyshire in the north and Devon and Somerset in the southwest. In the following spring birds were detected in Denbighshire, Kent and Dumfries and Galloway. Several long-stayers delighted the masses, notably a male at Harwood Forest, Northumberland, from 24th December 1990 through to 16th March 1991 and a female at Lynford Arboretum, Norfolk, from 24th November 1990 to 1st June 1991.

There was a gap of six years before the next birds were found in Britain. These included two in August 1997: a male on the Isle of May from 8th-11th August and a male trapped in Thetford Forest, Norfolk, on 21st. There were no further finds until a long-staying female was found near Parkend, Gloucestershire, from 15th February to 17th March 1998. The last of this modest influx was a 1st-summer female at Sandringham, Norfolk, from 23rd-24th May. Four birds was a poor showing compared to the influx of 181 reported in the Netherlands (almost three times the previous national total there). The records in the Netherlands during this influx were considered to arrive from a source between north-northwest and north-northeast, rather than the more usual easterly or northeasterly vectors (Ebels *et al* 1999). Elsewhere in Europe over 500 contributed to the largest invasion ever in Denmark, 100 were seen at Pape, Lithuania, and at least 29 were seen in Germany. Two even reached Hungary.

The first of the new Millennium was a male at Sedbergh, Cumbria, from 15th-17th May 2000. A small arrival of four in July 2001 was restricted to offshore islands (two on Shetland and singles on the Isle of May and Farne Islands). The only accessible individual was a juvenile to 1st-winter male on the Isle of May from 27th July-3rd August.

A more widespread arrival came late in 2002 and over the winter. No fewer than 18 were found in the latter part of the year and a further five were detected early in 2003. A male on Shetland on 13th July was the bellwether of 13 more to Shetland and Orkney between 2nd and 23rd August, though most arrived from 13th-23rd August (including four on Fair Isle from 16th-19th). The first to spill down to the mainland was a male at Bole Edge Plantation, South Yorkshire, from 1st to 5th September. It was quickly followed by the last on Shetland on 8th-

11th (Fair Isle). There were no further finds until a female at Bagley Wood, Oxfordshire, from 9th December-16th January 2003 and a male at Hedgerley, Buckinghamshire, from 27th January-14th March 2003. The last of the influx comprised an intriguing record of two males and two females at Morangie, Easter Ross, Highland, from 27th March-14th April. Although it might seem unlikely, breeding in Britain following an irruption is a possibility. Following the 1990 irruption breeding took place in Berlin, Germany, 1,700 km southwest of the usual breeding range.

Although the influxes in 1990 and 2002 were associated with large arrivals of Common Crossbills, that of 1987 was not. In Scotland, the mix of age classes in the influxes has varied. The majority has usually comprised juveniles or immatures with 87 per cent in 1987, 67 per cent in 1990 and 42 per cent in 2002.

Following the review period, an exceptional influx of Two-barred Crossbills took place in 2008. A wave of them passed west through Scandinavia, where the numbers were smaller than the 2002 influx. Between 20th July and 22nd August at least 57 birds were recorded on Shetland and Orkney, 40 of them between 5th and 11th August. These comprised 30 (possibly 32) on Shetland, a further 15 on Fair Isle and 12 on Orkney. On Shetland, the majority were at Sumburgh Head, where six on 5th August increased to 13 on 7th and peaked at 18 on 9th. Most were juveniles, though at least seven males and 11 females were recorded. The only records away from the islands during the initial influx were two in the Outer Hebrides (Harrop and Fray 2008).

Following the main arrival, there was just one additional record from the Northern Isles, a juvenile on Fair Isle from 8th-14th September. Three further records straggled south to form the tail of the influx, all of them amazingly found at garden feeders. A juvenile was photographed at Old Cassop, Co. Durham, on 17th September and a male was photographed at Oxenhope, West Yorkshire, on 29th October and then taken by a Eurasian Sparrowhawk *Accipiter nisus*. Fortuitously for the masses, an exceptionally obliging male frequented a garden at Bilsdale, North Yorkshire, from late October-23rd November.

Potential pitfalls for would-be finders include wing-barred Common Crossbills (Harrop and Millington, 1991). It is possible that some old sight records involve such eye-catching individuals. Hybrids between Common and White-winged Crossbills in America are reportedly unknown from the wild (Benkamn 1992), but an additional, albeit rare, hazard is presented by the specter of hybrids between Common and Two-barred Crossbills in Eurasia (Garner 1997; Proctor 1997).

Where: Shetland has acquired the lion's share of British sightings since 1950; its 45 per cent share of the archive is over five times that for Orkney at 8 per cent.

Elsewhere in Britain, news of birds arriving on the Northern Isles inspires observers to seek out and scrutinise their local Common Crossbill flocks with greater than normal intent. Away from the Northern Isles, records are noticeably scattered, due presumably to the birds flying far and wide in search of preferred food sources. There are few discernible peaks of newly detected sightings and no geographical association. In mainland Scotland five birds have been found in Highland, three in Dumfries and Galloway, two each in Forth and Fife and one in Moray and Nairn.

In England Norfolk is perhaps the most productive county with nine birds in all and five since 1990. In the northeast, there have been four birds each from Northumberland and Yorkshire. The Midlands has fared well over the years, with nine birds from seven counties.

In the southwest, four records are spread across Dorset, Somerset and Devon. There have been two from the northwest. One from Denbighshire in March 1991 represents the only recent record for Wales.

When: On the Scottish islands the earliest ever was a male shot on North Ronaldsay on 12th June 1894. The only other June sighting was a male on Fair Isle on 13th June 1908. The earliest of the modern recording era was an adult female on Fair Isle on 4th July 1953. Modest numbers have been recorded on the islands in July, but the vast majority produces a remarkably steep peak in mid-to-late August (Figure 92). Arrivals diminish during September and just one has been seen in October (Fair Isle on 20th in 1990.

Away from the Scottish islands, there is remarkably no evidence of a similarly or closely timed peak in sightings, a testament to the difficulty of finding a rare congener among its much commoner carrier species away from treeless landscapes.

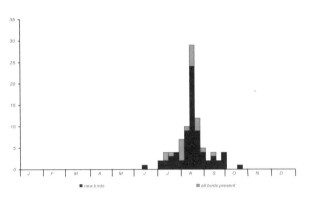

Figure 92: Timing of Two-barred Crossbills on the Scottish islands, 1900-2007.

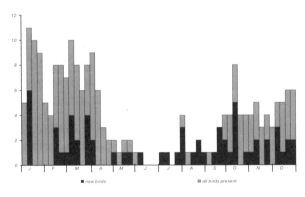

Figure 93: Timing of Two-barred Crossbills away from the Scottish islands, 1900-2007.

Discussion: Two-barred Crossbill feeds primarily on larch, especially so in the northern taiga zone. Larch cones remain on the tree with some of their seeds unshed for two to three years. Thus, compared with Common Crossbill, Two-barred is rendered less prone to extreme fluctuations in food supply. This may explain why Two-barred Crossbill is much less irruptive in the Western Palearctic, appearing infrequently in small numbers south of the boreal zone with

Common Crossbills. In the central and southern taiga zone where larch is less prevalent, the favoured food sources include the cones of spruce and pine trees. Irruptions into Scandinavia occur during larch cone failures farther east in Russia. As with many other irruptive species from Siberia there is a stronger westerly heading to their movements and so even during some influx years, records away from the Northern Isles can be rare or frustratingly absent. It has been proposed that there is a cyclic rhythm in the irruptions of seed-eating species, and a seven year cycle was observed for the autumn irruptions into Scandinavia (Larsen and Tombre, 1989).

There are no acceptable records of White-winged Crossbill *leucoptera* from Britain or elsewhere in Europe. In North America, *leucoptera* breeds in northern coniferous forests from Newfoundland to Alaska and undertakes irregular movements (at two to four year intervals) in response to food availability. Sometimes it undertakes continent-wide movements, so it has realistic potential as a vagrant to Britain. It has been reported over inshore waters and out over the western Atlantic on several occasions. There are four records from west Greenland, one from south Greenland, and at sea east of Newfoundland, though none have been recorded from Iceland (Robb and van den Berg, 2002).

Ship-assistance to Britain for *leucoptera* has been reported. In 1855 White-winged Crossbills were observed from one of the Cunard steamers in the Bay of Fundy, 1,000 km off Newfoundland crossing the Atlantic before a stiff westerly breeze. A number alighted in the rigging and eight were caught and held on board in a cage. Of these, one escaped when the ship was about 800 km west of Ireland and stayed on board until about 50 km from Ireland, when it left heading straight for land. Another escaped in Liverpool, with human intervention preventing the record from taking a place on the *British list* (Harrop *et al* 2007). A specimen from Worcestershire in 1838 was definitely of the race *leucoptera*, though the record is no longer accepted due to doubts about its provenance (Harrop *et al* 2007). There was an exceptionally tame and presumed escaped *leucoptera* in The Netherlands in 1963 (Robb and van den Berg 2002).

Compared with Two-barred, the bill of White-winged is much smaller and narrower, with a particularly long upper mandible, and is set upon a smaller and rounder head. Any crossbill with two wing bars in a location associated with Nearctic vagrants would merit close scrutiny.

Parrot Crossbill Loxia pytyopsittacus

Borkhausen 1793. Breeds from Scandinavia eastwards to the Kola Peninsula and Pechora River. Mainly resident, but also irruptive.

Monotypic. Britain's only endemic species, Scottish Crossbill *L. scotica* was long believed to be the only crossbill species that bred in the Scottish Highlands. However, Knox (1975; 1990) showed that Common Crossbill often bred there sympatrically with Scottish. More recently, Parrot Crossbill has also been proven to breed in the old pinewoods of the area (Marquiss and Rae 2002; Summers *et al* 2002; Summers *et al* 2007). Summers *et al* (2007) propose that strong assortative mating indicates that Parrot Crossbill, in addition to Scottish Crossbill and Common Crossbill, behaves as a good species when breeding sympatrically with the others. However, analysis of DNA has not revealed notable differences between the taxa.

It is thought that colonisation of the old Scottish conifer woods by Parrot Crossbills is a relatively recent event. The present population was put at 30 pairs in 2002, but is now thought to be in the region of 100 pairs (Summers *et al* 2002; Forrester *et al* 2007). Vagrants away from the breeding range were vetted by the *BBRC* until 1st January 2009, after which it was removed from the list of rarities.

Away from breeding areas a very rare irruptive vagrant, usually appearing later than other crossbill species. Parrot Crossbill has bred in England on occasion.

Status: A total of 504 birds have been recorded, 21 prior to the formation of the *BBRC* and 483 since.

Historical review: The first British record was a female taken at Blythburgh, Suffolk, in 1818. The first Scottish record was a male collected near Pitlochry, Perth and Kinross, in August 1928, an early date for a migrant of this species and suggestive of a breeding bird. Twenty one birds prior to 1950 are still considered acceptable by the *BBRC*.

Large invasions into the Netherlands in the 19th century (1867, 1868, 1877-78 and 1887-88) (van den Berg and Bosman 1999) appear to have passed unrecorded in Britain, with the exception of two females shot from a flock of seven crossbills at Earlham, Norfolk, on 22nd March 1888 (Catley and Hursthouse 1985).

The first birds of the modern recording period were a female trapped on the Isle of May on 18th September 1953 and an immature male killed by a car near Catcleugh, Northumberland, on 16th September 1954. These are the only accepted records prior to the 1962/1963 influx.

The majority of British records occurred in three influxes: 1962/1963, 1982/1983 and 1990/91 (the final documented influx). The accounts of the first two influxes are based on a summary by Catley and Hursthouse (1985).

The 1962/63 influx: This influx was largely restricted to the Northern Isles. A total of 33, including 16 trapped birds, were recorded on Fair Isle from 27th September to 10th October 1962. During this period, four also reached Iceland between 28th September and 14th October 1962. Further birds were recorded from Quendale, Shetland, on 7th October and two were seen on North Rona, Outer Hebrides, on 9th October, with one remaining to the 11th. Another arrival took place on Fair Isle on 11th October with 20 birds trapped and 25 present. Four were on Uig, Outer Hebrides, from 11th-13th October. Also on 11th, a 1st-winter male and a female were trapped at Spurn, with the male found dead there on 12th. A male was found dead at Spiggie, Shetland, on 12th October. A single tardy male was on Fair Isle on 29th. There were 66 birds accepted from this part of the influx, only two of them from England.

No further records occurred until a male was found dead at Hartsholme Gravel-pits, Lincolnshire, on 16th January 1963; up to nine were recorded at this site later in January, with three to four staying through February, others into March and a pair to at least 25th May. Significantly their identities were confirmed by a female found dead there on 17th March and another female with an injured wing taken into care in mid March. It escaped and was still in the area up to early 1964. Additional records from 1963 came from Fair Isle on 20th March and a female trapped (with two Common Crossbills) at Wisley, Surrey, on 15th May. There are 13 birds accepted for this period of the influx and only one was from Scotland.

All 1962-1963 records (79 birds) fell in the period 27th September-25th May (most during late September to the middle of October); most were on Fair Isle, including the only bird apparently making a spring return north in March.

Records during 1963-82: In the 19 years before the next influxes, only 11 Parrot Crossbills were accepted in Britain. In 1966, a male was found in Wells Wood, Norfolk, on 10th-12th November. In 1975 a well dispersed 10 birds were recorded in late October. The first three occurred within two days of each other: a male killed by a cat at Grutness, Shetland, on the 22nd, a male at Spurn on the same day; and a female found injured, at Tophill Low, East Yorkshire, on the 23rd, which subsequently died. A further seven were found at Gladhouse, Lothian, from 26th October feeding in the reservoirs surround of conifers until 3rd January 1976, some remaining to the 27th.

The 1982/83 influx: As is usually the case for crossbill influxes, the early birds were on the Northern Isles and along the east coast.

The first was a male on Fair Isle on 7th October 1982, followed by six there on 8th and another male on 10th. A 1st-winter trapped at Spurn on 11th died overnight. A male was found at Humberston Fitties, Lincolnshire, on the same day and there were two males there on 12th, the first staying to 15th and the second to 23rd. Others found on the 12th included males in Lincolnshire at Grainthorpe and Ingoldmells (both found exhausted and subsequently died), two females on Fair Isle, one of which was trapped and stayed until 29th; and another male trapped at Catfirth, Shetland. On the 14th, two corpses were found in Sea area Forties. On 16th, there was one on Burray, and a party of six at Voxter, Shetland, where three males and a juvenile were trapped. Two additional females were seen at Voxter on 18th, with a fourth male on 19th and at least one male to 25th. The only record from Norfolk was an immature male at Wells Wood from 16th-17th. Also on 17th, a male was trapped at Wick, Highland. On 18th there were females at Voe and Strand, both Shetland, the latter also seen on 19th. A rare record from the Outer Hebrides involved a female and four juveniles at Langass on 21st. A female was found dead at Lyrawa Plantation, Orkney, on 29th.

Subsequent searches of inland sites revealed flocks at Howden Reservoir and Hollingdale Plantation, Derbyshire/South Yorkshire, on 30th October, and at Wyming Brook, South Yorkshire, on 31st. At Howden, numbers increased from seven on 30th October to a maximum of 25 from 15th November to 10th January 1983; there were 20 until 3rd February and then a rapid dispersal, with the last (a male) on 13th February. At Hollingdale, there were 12 on 30th October, with up to 12 until 10th January and two pairs still present on 13th February. The Wyming Brook party numbered 14 (seven males and seven females) and stayed until 22nd December. After the main dispersal from these three closely associated sites, there was a series of records at Langsett, South Yorkshire, with a minimum of 11 individuals from 23rd February to 7th May.

Away from South Yorkshire, there were two in Hamsterley Forest, Co. Durham: a female from 28th December-2nd January and a male on the latter date. At North Winksley, North Yorkshire, another wintering party of up to 12 was found on 29th January, remaining until 24th

February. In Speyside, Highland, there was a male and two females on 11th April, though in light of more recent knowledge it seems more likely that these were resident birds rather than part of the influx.

Officially, a total of 109 birds was accepted for the 1982/1983 influx, although there could be an argument for revising the total to 106 in light of the trio in Speyside. One also reached Iceland on 14th November 1982.

Records during late 1983-89: Relatively few birds were seen during this period, but there were several breeding attempts in England.

A party of seven birds was discovered at Wells Wood on 26th October 1983 (four males, two females, one immature male). Most were present until 20th November and at least three (a male, a female and the immature) stayed into 1984. The adults paired and produced the first documented breeding for England at Wells Woods in 1984, when they successfully fledged one juvenile from two broods totaling four young.

The birds bred again at Wells in 1985 with broods of two and four fledged (Taylor *et al* 1999). Breeding probably occurred at Tunstall, Suffolk, in 1984 when two adults and three juveniles were present from 29th April to late summer. A pair with two juveniles was noted at the same site the following year (Piotrowski 2003).

There is also evidence that breeding took place near Langsett, South Yorkshire, in 1983 when at least two, possibly four, pairs were observed. One pair probably reared at least two young, but as the identity of the latter was unconfirmed, uncertainty remains over the breeding attempt (Lunn and Dale, 1993).

The last remnants of the 1982/83 influx were seen in Suffolk in March 1986 and in Norfolk in May 1986 where altogether 12 were present in early March. A pair was present at the breeding location at Wells on 4th May before deserting the area for good.

The 1990/91 influx: As usual, the harbinger of the 1990 influx arrived on Shetland, but surprisingly the islands went on to capture just one more during the whole influx. Unusually the main arrival vector deflected south along the northeast and East Anglia coasts. This coincided with a large winter invasion into the Netherlands where 302 individuals were recorded (van den Berg and Bosman 1999)

The first was an early female on Fair Isle from 23rd-24th September. After a three-week gap it was followed by two males and three females at Upper Sheringham and Weybourne, Norfolk, on 13th October (where birds lingered to 8th December and peaked at nine on 21st October). A female on Whalsay, Shetland, on 14th October was the second and last to be recorded in Scotland before the emphasis switched smartly to the northeast of England.

The main arrival took place from 18th October onwards. On that date five (including at least one male and two females) were at Donna Nook (with a male and a female to the 20th) and a female was at Stiffkey Fen, Norfolk. On the 19th, a male and a female were found at sea 32 km east of Spurn and a male plus two females appeared at Gibraltar Point (male to the 25th and females to the 26th). On the 20th a female was at Humberstone Fitties, Lincolnshire, (to 23rd) and a female dropped in to the Farne Islands. The run of coastal birds continued with two males and two females at Holme, Norfolk, on at least 21st October, a male trapped at Sandwich Bay on 27th October and a female at Holkham Meals, Norfolk, on 4th November.

Given a slowdown of coastal arrivals, observer attention naturally switched to inland sites. At Kirkby Moor, Lincolnshire, four on 4th November increased to a whopping 44 on 11th, with up to 35 to the end November and up to 20 through to January. At Hollingdale Plantation, South Yorkshire, two males were present on 4th November and at Lockwood Beck Reservoir, Cleveland, at least six were present from 11th November into 1991. In Co. Durham, up to 10 were in Hamsterley Forest from 7th November-10th March 1991 and 27 at Chopwell Woods from 17th November-26th December. At Brandon, Suffolk, there were up to 12 between 19th November and 2nd February, a pair from this group was seen displaying and nesting was suspected in the area (Piotrowski 2003). In Norfolk three were present at Sprowston between 29th November and 1st December. A female was at Lynford Arboretum on 5th December. A female was at Wakerley Great Woods, Northants, from 22nd November-16th March and a male was there from 25th November-9th January.

Good numbers continued to be found inland in Lincolnshire, perhaps not surprisingly as it had borne the brunt of coastal arrivals in the autumn. At Willingham Forest 14 on 2nd December peaked at 17 on 16th, with 16 still present on 21st January 1991. In the same county at Laughton Forest, 19 were on show from 2nd December to at least 13th March 1991. The Dukeries, Nottinghamshire, hosted seven from 14th December to 16th February. Two more parties were found at Castle Eden Dene in Co. Durham with seven from 16th December through to 10th March and nine from 19th December to 1st March. In North Yorkshire 12 were present at Winksley from 29th December to 4th January.

Surprisingly, a male at Kielder Forest, Northumberland, on 19th and 24th January was the only new bird found early in 1991. March produced a new flurry of finds as roving parties again became conspicuous, including up to 47 at Birk Brow, Cleveland, from 1st-18th March, two females on 17th March at Harwood Forest, Northumberland, three at Bourley Hill, Hampshire, on 29th March and seven at Osmotherley, North Yorkshire, between 31st March and 13th April.

Although birds were observed displaying during this influx (for example, in Co. Durham and Suffolk), none were proved to breed. In contrast with previous influxes, departure appears to have been rapid and none remained after early spring. Altogether, the influx of 1990/91 produced 264 acceptable individuals, but no doubt many more evaded detection.

1992 to present day: Since the superb influx of 1990/91 there have been just three migrant birds: a 1st-winter female was at Kergord, Shetland, from 19th-20th October 1994 and two females at Laughton Forest, Lincolnshire, on 8th March 1995. This trio was overshadowed by the discovery of a Scottish breeding population.

The first accepted occurrence of breeding in Scotland took place in 1991, in Abernethy Forest, Highland, where a pair nested and fledged four young (Ogilvie *et al* 1994). Whether this was associated with the irruption of the preceding winter or just coincidence is unknown, but recently it has become apparent that waiting for influxes is not the only way to see this impressive finch within our shores. Recent research into crossbills in Highland has changed our perception of this cryptic species totally. During a study in Abernethy Forest, Highland, from 1995 to 2001, Parrot Crossbill was found to be the most abundant crossbill nesting there. The species accounted for 74 per cent of the 64 crossbills trapped (11 were Common Crossbill, five Scottish Crossbill, 46 were Parrot and two were unidentified). It has since become apparent that Parrot Crossbills breed in several inland localities in Scotland (Summers *et al* 2002).

No one knows how long Parrot Crossbills have been nesting in Abernethy nor how widespread they are in the Highlands. Other sites in the region where Parrot Crossbills have been found recently in spring include Glenmore, Rothiemurchus and Curr Wood (Highland), Culbin Forest (Moray and Nairn), and Glen Tanar, Ballochbuie and the Mar Lodge woodlands (Northeast Scotland). Parrot Crossbills (identified by bill measurements) are uncommon among museum specimens, suggesting that their current presence in the Highlands may be a recent development. The Scottish population may now number around 100 pairs (Summers, 2002).

Several invasions into northwest Europe took place in the 19th century. If Parrot Crossbill had colonised Scotland after one of these invasions, its presence would presumably have been detected by the avid collectors of the time. It seems more likely that the current Scottish breeding population originated from one of the invasions in the late 20th century, possibly that of 1962/63 (which was centred more on Scotland than England).

Discussion: The periods of the three documented major influxes of Parrot Crossbills into Britain were very similar and rather precise, with most arriving during October and being concentrated along the east coast north to Shetland. The bulk (71 per cent) of the 1962/63 records were, however, on Fair Isle (only 13 per cent were found away from Shetland). By contrast, only 10 per cent of those in 1982/83 were on Fair Isle and 65 per cent came from inland localities. The 1990/91 influx occurred almost exclusively away from Shetland, where just two were recorded.

The two most recent influxes were also experienced in the Netherlands, where indications of breeding followed both invasions and seven were recorded in 1963 (van den Berg and Bosman 1999). It would appear that the influx of 1962/63 was not restricted to a trajectory across the north of Britain, so birds farther south may have been overlooked. A further smaller influx into the Netherlands took place over the winter of 2007/08, but none were detected in Britain. The differing geography of the British influxes most probably reflects birds originating from different sources.

Parrot Crossbill specialises on the large hard, thick-scaled cones of Scots Pine *Pinus sylvestris*. Invasions by Parrot Crossbills occur later in the year than those of Common Crossbill, which specialises on medium-coned conifers, and involve much smaller numbers. The three documented invasions of Parrots have all occurred during the same years as Common invasions, but Commons tend to arrive during the summer and early autumn; Parrots tend to appear during October.

The later invasions of Parrot are in keeping with the different phenology of their main food plant. Seeds in new Scots Pine cones do not normally develop before September, but provide a more consistent food source than that of the spruce loved by Common. Hence, the invasions of Parrots occur far less frequently. The next influx, when it comes, is to be savoured as it will be some time before another.

Trumpeter Finch Bucanetes githagineus

(M.H.C. Lichtenstein 1823). Largely resident from Canary Islands, southeast Spain and deserts of North Africa east through Middle East to southern Iran and Pakistan. Eastern populations in particular dispersive, some wintering east to deserts of northwest India.

Polytypic, four subspecies found across its extensive range. These include: nominate *githagineus* (Lichtenstein, 1823) of central and southern part of the Nile valley and south-eastern desert of Egypt, south to north-central and north-east Sudan; *amantum* (Hartert, 1903) in the Canary Islands; *crassirostris* (Blyth, 1847) from western Arabia, Sinai, Levant, and south-central Turkey east through Bahrain, Iran and Transcaucasia to Uzbekistan, Afghanistan, and western Pakistan; and *zedlitsi* (Neumann, 1907) in northwest Africa and, recently, southern Spain. All the races are generally regarded as residents, but birds are often nomadic and even dispersive, undertaking erratic movements. These lead to small-scale changes of range and sometimes in new areas local abundance one year followed by complete absence in the next (BWP). The race of British records is undetermined, but is most probably *zedlitsi*.

Very rare vagrant, typically in late spring, most probably from northwest Africa.

Status: 14 records; three from 2008 are yet to be assessed but all were well documented.

1971 **Suffolk:** Minsmere 1st-summer male 30th May-15th June
1971 **Highland:** Handa Island 8th-9th June
1981 **Orkney:** Sanday male 26th-29th May
1984 **West Sussex:** Church Norton 19th-23rd May, when taken by Eurasian Sparrowhawk
1985 **Essex:** Foulness 21st September
1987 **Northumberland:** Holy Island male 1st August
1992 **Highland:** Balnakeil male 4th June
2005 **Suffolk:** Landguard Point 1st-summer male 21st-26th May
2005 **Kent:** Tankerton male, 24th-25th May
2005 **Kent:** North Foreland male 9th June
2005 **Kent:** Dungeness male 11th-13th June
2008 **Outer Hebrides:** North Rona 25th May*
2008 **Norfolk:** Blakeney Point 1st-summer male 31st May-4th June*
2008 **East Sussex:** Telscombe Cliffs 1st-summer male 4th-6th June*

Discussion: After an expansion of its range in northwest Africa, *zedlitsi* is thought to have colonised Spain in the 1960s. Breeding was first presumed in 1968 and finally proven in 1971 (Wallace *et al* 1977). Despite the date coincidence, it is highly unlikely that the tiny Spanish population was responsible for the British sightings. But, it is conceivable that the origin of birds in both events was connected. Clusters of arrivals in 1971, 2005 and 2008 reflect the nomadic habits of the species.

Prior to the 2005 and 2008 influxes, records in northern European were very rare. These included sightings from Sweden (5th June 1966, 16th June 1971, 16th May-1st June 2001), the Channel Islands (29th October-1st November 1973), Denmark (22nd June-3rd July 1982, 2nd June 1997, 5th-9th August 2007), Germany (26th-27th July 1987; presently in category D due to the species' frequency in captivity, Peter Barthel *pers comm*), Austria (autumn 1907, 12th May 1989) and the Netherlands (31st May 2003). Records in southern Europe were more regular, with 21 occurrences comprising 29 birds from Cyprus, eight records from Greece, two for mainland Portugal and the Spanish birds.

In 2005, the four British records were part of a large influx across Europe. Six reached France (only four previous records) and others were found in Bulgaria (7th May), Switzerland (27th-29th April, 24th May, 4th July) and the Netherlands (12th June); there were additional records in Italy and Cyprus, plus an autumn one in Germany (26th-28th September). For the first time ever, birds reached Fenno-Scandia, with four in Sweden (two at different sites on 28th May, 4th June, 10th-12th June) and two Finland (4th June, 16th-21st July).

The influx of 2005 provided an opportunity for many to add the species to their British list. The first bird, in Suffolk, was a deservedly popular repeat of the first British record in the same county 34 years earlier. In contrast, the influx to Britain in 2008 was matched only by the third record for Finland (photographed on 7th June*). As with the British records, the majority of European sightings have been in spring, coinciding with post-breeding dispersal (Wallace *et al* 1977). Virtually all have been since the range expansion into southern Spain

Pine Grosbeak Pinicola enucleator

(Linnaeus 1758). Nominate enucleator resident or dispersive across northern Scandinavia east to Yenisey river. Other races largely resident in east Siberia and Kamchatka south to Hokkaido, Japan. Widespread across much of northern North America east to Newfoundland and south in mountains to central California and northern New Mexico, USA.

Polytypic, 10 or 11 subspecies. All records have been of nominate *enucleator* (Linnaeus, 1758) of northern Europe east to valley of lower and middle Yenisey and lowlands north of the Altai (*BWP*).

Very rare vagrant from Scandinavia and northeast Europe.

Status: 11 records, five of them since 1971.

pre-1831 Co. Durham: Bill Quay, Pelaw female shot, no date
pre-1843 Greater London: Harrow-on-the-Hill 1st-winter female shot, no date
c. 1861 North Yorkshire: Littlebeck, near Whitby immature male shot in winter
1890 Nottinghamshire: Watnall shot 30th October
1954 Fife: Isle of May adult female, trapped, 8th-9th November
1957 Kent: East Malling female or immature male 2nd November
1971 Kent: Maidstone adult male 15th May

1975 **Northumberland:** Holy Island male 11th-12th May
1992 **Shetland:** Lerwick, Mainland male, probably first-year, 25th March-25th April
2000 **Shetland:** Maywick, Mainland 1st-winter male 9th November
2004 **East Yorkshire:** Easington 1st-winter male 8th-10th November

Discussion: An escaped 'red' bird in July 2006 in Essex/Hertfordshire caused a modicum of commotion in some quarters before being established as a colour-fed female that had escaped nearby in late June. Likewise, a male in the Netherlands on 24th March 1996, which occurred following a large invasion into Denmark during the previous autumn, was shown by photographs to have deformed long claws, indicative of captive origin despite its natural plumage colours (van den Berg and Bosman 1999). Such records are further testament, if it were needed, that each potential rarity must be judged on an individual basis.

Pine Grosbeak is a largely sedentary species that is prone to occasional irruptions, which can bring large numbers south through Fenno-Scandia to southern Scandinavia. In Finland the heaviest documented irruption occurred in November 1976, when 1,200 migrating birds were counted from a single observation spot in Helsinki on 2nd November; several thousand birds were seen in total. Further irruptions were recorded in October-November 1981 (450 birds) and winter 1978/79 (more than 300 birds). The last invasion year was 2001, when over 1,000 birds were seen in Helsinki region, including a peak count of over 850 birds on 4th February; it was estimated that up to 2,000 were present during the winter (Aleksi Lehikoinen *pers comm*). Several thousand reached Norway in autumn 1995 (the largest-ever invasion there) and there have been several major irruptions since. An irruption into Denmark occurred in 1998, including at least 73 birds recorded at Skagen in November.

The large irruptions probably originate from farther east than the nearest breeding communities. Their scale often raises expectations in observers elsewhere in northwest Europe that are frequently dashed. Unlike Northern Bullfinch *Pyrrhula pyrrhula pyrrhula*, the species' momentum seems to run out of steam quickly. Very few have penetrated farther southwestwards into Europe. For example, in 2004 when 24 were found in Denmark only eight reached the Netherlands and just one was proved to cross the North Sea, to the Spurn area.

Records of this monster finch remain at an absolute premium. The only individual to have lingered was the well-watched Shetland bird of 1992. Sadly for many, the East Yorkshire bird in 2004 was identified late during its stay, at which point it departed. The next to grace the mainland will be an enticing draw for birders.

Evening Grosbeak *Hesperiphona vespertina*
(W. Cooper, 1825). Breeds from north-eastern USA and southeast Canada west across southern Canada to the Pacific coast and south through the Rockies and southwest USA to Mexico. Three subspecies are recognized. Northern/eastern nominate vespertina winters south to Texas and northern Florida. Extent of vagrancy not fully studied and varies according to irruptive factors.

Polytypic, three subspecies. Race undetermined, but most likely *vespertina* (W. Cooper, 1825).

Very rare vagrant from North America, both records in March.

Status: Two records.

1969 Outer Hebrides: St. Kilda adult male 26th March
1980 Highland: Nethybridge adult female 10th-25th March

Discussion: Northern/eastern *vespertina* is prone to regular autumn and winter irruptions in response to changes in food supply. Some east-west migration has been highlighted by ringing returns. In winter, males are more abundant in the more northeasterly parts of the range, with females commoner farther south (Gillihan and Byers 2001).

There are just two other Western Palearctic records, both from Norway (2nd-9th May 1973 and 17th-26th May 1975). That all Western Palearctic records are in March and May would suggest overshoots of wintering individuals, which in the USA depart for breeding grounds from March onwards but mainly in April.

Winter Evening Grosbeak counts have declined rapidly across much of the winter range since the 1970s (Bonter and Harvey 2008). Whether this is due to shifts in winter distribution or population declines is the subject of debate, but clearly the likelihood of this species reaching the British Isles is now much reduced.

Black-and-white Warbler Mniotilta varia

Julian R. Hough

(*Linnaeus, 1766*). *Breeds in central, southern and south-eastern Canada, south through most of central and eastern USA; isolated populations west of the usual range. Has one of the most extensive winter ranges of any North American wood-warbler: from southeastern USA through Central America and the West Indies to northern South America.*

Monotypic.

Very rare vagrant from North America, mostly in mid 1970s to mid 1980s, but in only one year since and none after 1996.

Status: 15 records, 13 in Britain and two in Ireland.

1936 Shetland: Tingwall, Mainland 1st-winter male found dead mid-October
1975 Scilly: St. Mary's female 27th-30th September
1977 Scilly: St. Mary's 29th September-1st October
1978 Devon: Tavistock 3rd March

1978 **Co. Cork:** Cape Clear 18th October
1980 **Pembrokeshire:** Skomer 10th September
1982 **Cornwall:** Mylor Bridge, near Penryn picked up dead 1st September
1983 **Cornwall:** The Lizard 1st-winter 24th September
1984 **Co. Londonderry:** Loughermore Forest female 30th September-2nd October
1985 **Norfolk:** How Hill, Ludham 3rd-15th December
1987 **Devon:** Prawle 8th-15th October
1996 **East Sussex:** Beachy Head 1st-winter female 2nd-3rd October
1996 **Scilly:** St. Mary's 1st-winter male 5th-14th October
1996 **Scilly:** Tresco 1st-winter female 20th-25th October
1996 **Norfolk:** Whitlingham Lane, Norwich 1st-winter male 9th-15th November

At about 10.45 on 29th September 1977, I was watching a rather sparse selection of birds along the Lower Moors nature trail on St. Mary's, Isles of Scilly. I had sat down to look at a Chiffchaff Phylloscopus collybita, which had shown itself briefly in the shadowy canopy of a large copse of mature sallow Salix, and was about to move farther down the trail, when I focused on a movement about 20 m away near the foot of a gnarled sallow trunk covered with ivy Hedera helix. There, slinking into view around the trunk, came a beautiful, unmistakable, near-apparition: a Black-and-white Warbler Mniotilta varia. It flew to a nearer trunk, where it continued to feed for a few seconds, pecking from the trunk and branches, then flew into the sallow canopy about 3 m above my head, and disappeared. I scribbled a sketch, noting its main features, waited a few minutes trying in vain to relocate it, then left the area to fetch others. By 11.30, about 20 observers had gathered at the spot. After a rather desperate ten-minute wait, someone glimpsed the warbler flying back into the copse, and soon it had shown itself to all of its much relieved and highly appreciative audience. It remained in or near the same copse for three days, until the afternoon of 1st October, and was seen by about 250 observers, some of whom had travelled from as far away as Fair Isle to see it!
Peter Grant. *British Birds 71:541-542*

Discussion: Elsewhere in Europe there have been two in Iceland (1st September 1970 and 19th-20th October 1991) and an unexpected summer record from the Faeroes (18th-20th July 1984).

This long-distance migrant with extensive breeding and wintering ranges performs an autumn migration that takes place from the Mississippi to the east coast. Some birds move directly to the Caribbean islands and others cross the Gulf of Mexico to Central and South America. Some individuals, presumably from the southern part of the breeding range, reach South America by late August, though migration of northern populations extends from late August to mid-September (Curson *et al* 1994; Kricher 1995).

Always a popular species, it seemed set to become a regular vagrant and failed to appear in just four years between 1975 and 1987. Inexplicably since then it has occurred only in 1996, when there was a deluge of four birds. Autumn records conform to expectation on date (1st September to 20th October) and location (most in the southwest extremities). Two winter birds in

Norfolk proved to be popular attractions, one in November and one in December, and the 1985 bird attracted more than 2,000 observers during its two week stay; both birds kept company with tit flocks (the species habitually attends multi-species flocks in winter in the Neotropics). One in Devon in March 1978 had presumably wintered.

Counts on North American BBS routes within the Atlantic flyway region have declined by an average of 4.2 per cent per annum since 1996 (Sauer *et al* 2008). Such a trend may go some way to explain the recent paucity of records in Europe. Perhaps more relevant is that the glut of records occurred during the "golden decades" for transatlantic vagrants, when ideal weather conditions were more prevalent. It is now 12 years since the last of these fantastic birds and this species is eagerly sought by a new generation of listers.

Golden-winged Warbler *Vermivora chrysoptera*

(Linnaeus, 1766). Breeds central North America east through the Great Lakes to New England and south through the Appalachians. Winters in southern Central America and northern South America.

Monotypic.

Status: One record.
1989 Kent: Larkfield, Maidstone male 24th January-10th April

Discussion: Golden-winged Warbler tends to arrive relatively early on the breeding grounds in spring and departs very early in the autumn, peaking from late August to early September. This timing means that migrants tend to avoid the Atlantic storms responsible for transporting other Nearctic landbirds.

Following a northward range extension through much of the early part of the 20th century, the breeding range of this attractive New World wood-warbler is presently contracting. Such fluctuations are frequent for the species, which is prone to local expansion and then extirpation (Confer 1992). Its preferred breeding habitat – shrubby vegetation typical of abandoned fields and regenerating forests – is becoming scarce in many parts of eastern North America due to successional change and maturation of forests. With a steady decline averaging 2.9 per cent per annum on its North American breeding range since the mid 1960s (Sauer *et al* 2008), the chances of further records of this species in Europe are diminishing with time.

Hybrids with Blue-winged Warbler *V. pinus* are frequent and produce the distinctively plumaged Brewster's and Lawrence's Warblers. Brewster's resembles a Blue-winged Warbler with a white chest. Lawrence's recalls an all-yellow Golden winged Warbler (Curson *et al* 1994).

The Golden-winged Warbler in Kent prompted one of the largest ever twitches after it was discovered in a housing estate and adjoining Tesco's supermarket car park. Its occurrence was reported widely on television and in the press, even making the front page news of one national newspaper. The date and location were unusual, but the bird could have made landfall in the southwest and headed eastwards during the autumn and early winter. It remains the sole record for the Western Palearctic.

Blue-winged Warbler *Vermivora pinus*

(Linnaeus, 1766). Breeds north-eastern USA and southernmost Canada. Winters from southeast San Luis Potosí, Mexico, south along Atlantic slope of Mexico and Central America to central Panama.

Monotypic.

Very rare vagrant from North America with just one Western Palearctic record.

Status: One record.
2000 Co. Cork: Cape Clear 1st-winter male 4th-10th October

This superb bird enticed many to travel to this famous island.

On Tuesday 3rd October 2000, the after-effects of Hurricane Isaac hit southwest Ireland, with torrential rain and winds reaching 80mph. Birders on Cape Clear Island, County Cork, spent the day sheltering; one or two efforts were made to venture out, but they were very decidedly short lived!

The following morning saw a complete change. The day dawned with a gentle breeze and blue skies and, it has to be said, a great deal of anticipation. Three of us left the bird observatory at about 8.00am, with plans to split up and cover as much of the island as possible. I went straight to the Waist, while Willie McDowell searched the bushes alongside Cotters Hill and Dennis Weir went into Cotters Garden.

It was Dennis that hit the jackpot almost immediately! As he entered the garden, he was greeted with a call that 'sounded not unlike a Myrtle Warbler'. After a few seconds, a bird appeared in the bushes, very briefly, showing a bright yellow front, a green back and bluish-grey wings. Dennis whistled to Willie, who was quickly at his side: "What have you got?" "I think it's an American!" "An American WHAT?" "Well, I reckon it may be a Blue-winged."

I received a very broken 'phone call on my mobile telling me to get down to Cotters to see a ……. The signal disappeared just as the species was to be mentioned, not just once but twice! I did not wait to hear the third effort, I just ran! By the time I reached them, Dennis and Willie were staring into a bank of bracken and not looking particularly excited. "What is it?" I demanded, "and where is it?!"

Having by now had good views and confirmed that it was indeed a Blue-winged Warbler, they told me what it was – and the sad news that it had dropped into the bracken and not seen since. We split again, to try to relocate the bird, which Dennis did within five minutes, back in the same place where he first found it. I have honestly never moved so fast!

The bird showed well and there was no mistaking it. It was a stunning, bright individual and, without doubt, a bird to be shared: we quickly telephoned other birders on the island and on the mainland.

The bird stayed until 10th October and gave excellent views to the many visiting bird-ers, although it did become more and more mobile as the days passed. Food appeared to be no problem for it, as it fed almost continuously before taking short breaks to digest the grubs it was catching. In all, about 600 birders visited the island to see this gem, the first Blue-winged Warbler to be recorded in the Western Palearctic. A collection taken raised just over £800 towards the restoration of the bird observatory library. Our thanks go to all contributors.

Steve Wing. *Birding World 13: 408-411.*

Discussion: Blue-winged Warbler extended its range north and northeast throughout the 20th century. In some areas it has replaced Golden-winged Warbler, with which it regularly hybri-dises to produce the distinctively plumaged Brewster's and Lawrence's Warblers (Curson *et al* 1994; Gill *et al* 2001). The reasons for the range extension are poorly understood, but are thought to be due to habitat changes induced by the early settlers and/or gradual climatic warming across the breeding range. In recent years populations have declined in the northeastern United States (Sauer *et al* 2008).

This species was among those predicted by Robbins (1980), but it was ranked with those least likely to occur. A medium-distance trans-Gulf migrant, it is regularly recorded northeast of the breeding range during late autumn in Newfoundland, the Maritime Provinces and islands off Maine. There is an increasing number of records from Bermuda (Dunn and Garrett 1997). Despite such hints at the possibility of more to come, this popular individual remains the only record for the Western Palearctic.

Tennessee Warbler *Vermivora peregrina*

(A Wilson, 1811). Breeds from Yukon across southern Canada to Newfoundland, Minnesota, New York and Maine. Winters in Central and northern South America to Colombia, Venezuela and northern Ecuador.

Monotypic.

Very rare autumn vagrant from North America.

Status: Four records.

1975 Shetland: Fair Isle 1st-winter 6th to 20th September, trapped 18th
1975 Shetland: Fair Isle 1st-winter trapped 24th September
1982 Orkney: Holm, Mainland 1st-year 5th to 7th September, trapped 7th
1995 Outer Hebrides: St. Kilda 20th September

Discussion: Elsewhere in the Western Palearctic there have been just three records: Iceland (14th October 1956), Faeroes (21st-28th of September 1984) and the Azores (21st November 2005).

Although the species was named from a migrant specimen on the banks of Tennessee's Cumberland River in 1811, its breeding range is restricted almost entirely to the boreal forest zone of Canada. Numbers fluctuate markedly from year to year in response to periodic outbreaks of spruce budworm *Choristoneura fumiferana* caterpillars, on which the species is a well-documented specialist predator. Breeding-population densities recorded during budworm epidemics may exceed 500 males per 100 ha, and Tennessee Warblers often rank as the most abundant breeding species in boreal forests of eastern Canada (Rimmer and Mcfarland 1998).

It is quite amazing that the first two British birds occurred in the same place just four days apart (Broad 1981). If the first bird had not been trapped, the observations would most certainly have been assumed to be a single record. Subtly plumaged, it is far less eye-catching than many of the other members of New World wood-warblers. Tennessee is the most *Phylloscopus*-like of all the Parulidae and a vagrant could be easily dismissed on poor views as a *Phylloscopus*.

Its rarity across this side of the Atlantic presumably results from its migration strategy. A long-distance migrant, it performs a protracted and variable autumn migration. Dispersal southwards of failed breeders and non-breeders begins in early to midsummer. Most migrants move through the eastern United States, primarily through inland states. The early movement of the species, and the fact that migration takes place on a broad front mostly inland, makes it less susceptible to autumnal displacement across the Atlantic. Some birds reach the wintering grounds via trans-Gulf flights and migration through the Greater Antilles, and it is perhaps these birds that are displaced in some years.

All the British records occurred relatively early in the autumn for a Nearctic vagrant. The locations in the Northern Isles and Outer Hebrides correlate well with the typical dates and occurrence patterns for other early departing migrants from North America. The northerly distribution of the other European records probably results from the more northerly tracks of Atlantic storms early in the season. The only southerly record, from the Azores, occurred much later than those farther north in Europe.

Northern Parula *Parula americana*

(Linnaeus, 1758). Breeds south-central to south-east Canada and New England; also south-eastern USA. Some winter in southern United States, mostly Florida, but main wintering range is through the Caribbean and Mexico and Central America, rarely to El Salvador.

Monotypic.

Very rare autumn vagrant from North America, now increasingly irregular.

Status: There have been 17 records, 14 in Britain and three in Ireland.

1966 **Scilly:** Tresco male 16th-17th October
1967 **Cornwall:** Porthgwarra 26th November
1968 **Dorset:** Portland 9th October
1982 **Greater Manchester:** Wigan moribund 2nd November

1983 **Scilly:** Tresco 1st-winter 1st October
1983 **Scilly:** St. Agnes 1st-winter 10th-13th October
1983 **Co. Cork:** Firkeel 1st-winter male 19th-24th October
1985 **Dorset:** Hengistbury Head 1st-winter female 30th September-12th October, trapped 9th
1985 **Scilly:** St. Mary's male 3rd-17th October; presumed same St. Agnes 18th-21st October
1985 **Cornwall:** Penlee Point, Cawsand 1st-winter female 17th-19th October
1987 **Cornwall:** Nanquidno male 13th-23rd October
1988 **Dorset:** Portland 1st-winter male 30th September-7th October, trapped 30th
1988 **Cornwall:** Cot Valley female 9th-19th October
1989 **Co. Cork:** Dursey Island female 25th September
1992 **Scilly:** St. Mary's 1st-winter 8th-10th October
1995 **Scilly:** St. Agnes 1st-winter 10th October
2003 **Co. Waterford:** Brownstown Head 1st-winter male trapped 5th October

The fifth British record was part of a glorious autumn for this gem of a bird.

I suppose every birder dreams of finding a major rarity in the Isles of Scilly. This must be even more so if, during that special October period when the cream of British birders are present, the bird discovered is one which the majority have not seen before in Britain. On 1st October 1983, a boatload of us crossed to Tresco to look for a reported Scarlet Rosefinch Carpodacus erythrinus. It was not relocated, so I strolled to the Borough Farm area southeast of Old Grimsby, where an Arctic Warbler Phylloscopus borealis had been observed previously. T. J. Wilson was just in front of me. He indicated that there were several warblers in the hedgerow. I raised my binoculars and latched on to a small warbler, which I immediately identified as a Northern Parula Parula americana. I amazed myself by only casually uttering the words 'I've got a Parula here'. Then the penny dropped, and I jumped in the air, clenched fist raised above my head, and screamed 'Yeah!'

I could hardly hold my binoculars steady as my blood pressure and excitement increased. After watching the bird for a few minutes, as it darted in and out of the hedge, to convince myself that my initial identification was correct, I then raced around to the farmer and told him to expect 'one or two' birders to arrive, since he had yet another rare bird on his land.

I then ran around the north end of the Great Pool spreading the good news. I burst into the local tavern and screamed at all the birders who were supping merrily and conversing expectantly. A moment's silence was followed by pandemonium as everyone dived out of the pub, grabbing their tripods and 'scopes on the way.

Back at the hedgerow, the scene was incredible: where there had before been a couple of birders, there were now masses, but all behaving impeccably as they caught glimpses of this avian jewel. The farmer was also on good terms and lost no time in collecting donations.

Robin Chittenden. British Birds 79:432-433.

Discussion: An additional category E record involved a bird at sea in 1962 which came aboard the RMS *Mauretania* in mid-Atlantic. It was caught on 19th September, fed in captivity and later died in Southampton, Hampshire. The specimen resides at Liverpool Museum.

Elsewhere in Europe there have been: seven from Iceland (found dead 25th October 1913, 21st October 1948, 28th October 1952, 24th October 1957, 8th October 1962, 27th September 1989, 29th September 1989); one from Germany (Baden-Württemberg on 25th October 1985); and two in France (17th-27th October 1987, 21st October 1995). The first for the Azores was photographed on Corvo on 11th October 2008.

Occurrences are almost exclusively confined to southwest England and southern Ireland. The sole record away from this area is an exceptional moribund bird in Greater Manchester in 1982; a ship-assisted vagrant into the Mersey perhaps? Eleven occurred between 1982 and 1989, with only 1984 and 1986 blank that decade. During this period the species rapidly became an expected autumn vagrant. Since those times of relative plenty, however, just three have occurred (1992, 1995 and 2003).

All have been in autumn, with 15 between 25th September-19th October and two in November. The peak occurrence period accords well with autumn departure from breeding areas, where most emigration occurs from September to mid-October.

Wintering primarily in eastern Mexico and the West Indies, this stunning species has a shorter migration route than many other New World wood-warblers. The main route passes through Florida rather than across the Gulf of Mexico, and this vector is reflected in higher densities of birds in the Caribbean than in Mexico and Central America (Moldenhauer and Regelski 1996). Species such as this, which undertake a sea crossing, are prone to sporadic displacements. The multiple arrivals during the glory years of the 1980s were well received. As with all New World wood-warblers, it is now difficult to believe that Scilly enjoyed six between 1983 and 1995. The next will surely receive a rapturous reception from a new generation of birders.

Yellow Warbler *Dendroica petechia*

(Linnaeus, 1766). The most widespread New World wood-warbler. The most northerly subspecies amnicola breeds in Canada from Newfoundland west to northeast British Columbia, Yukon territory and central Alaska north of aestiva, which occurs from southern-central Canada to central USA, including the Atlantic coast south to Georgia. Other races breed in western North America, Central America, the Caribbean and northern South America. Northern populations migrate to central Mexico to central Peru and northern Brazil.

Polytypic, over 40 subspecies. Classification of the broad range of morphological variation in the Yellow Warbler has ranged from the splitting of several species to the lumping of one highly polytypic species. Often divided into three racial groups: Yellow Warbler (*Aestiva* group; nine subspecies); Golden Warbler (*Petechia* group; 17 subspecies); and non-migratory Mangrove Warbler (*Erithachorides* group; 17 subspecies) (Browning 1994; AOU 1998). All European records have been of the race *aestiva* (Gmelin, 1789).

Very rare autumn vagrant from North America, typically in early autumn.

Status: Nine records, two from 2008 yet to be accepted*.

1964 Caernarfonshire: Bardsey 1st-year male trapped 29th August, died 30th August
1990 Shetland: Helendale, Lerwick, Mainland age/sex uncertain 3rd-4th November
1992 Orkney: North Ronaldsay 1st-winter male trapped 24th August
1995 Co. Waterford: Brownstown Head 1st-winter male 11th-12th October
1995 Co. Clare: Kilbaha trapped, 12th-31st October
2004 Outer Hebrides: Breibhig, Barra 1st-winter 2nd-7th October
2005 Shetland: Garths Ness, Mainland 1st-winter male 15th-17th September
2008 Co. Cork: Cape Clear 1st-winter female 24th-30th August*
2008 Co. Cork: Mizen Head 1st-winter male, 26th-28th August*

Discussion: Elsewhere in the Western Palearctic there are three from Iceland (5th October 1996, 25th August 2000, 10th-12th September 2003), two birds from the Selvagens Archipelago (Ilha Selvagem Grande, two on 10th and 12th September 1993) and two from the Azores (20th August 1995, 5th December 2001).

The autumn migration of eastern populations is among the earliest of any New World wood-warblers, getting underway by mid to late July in much of eastern part of the breeding range (Lowther *et al* 1999). The early dates of many of the Palearctic records, four of which are in August, correlate well with this departure pattern. The northerly bias to British and Western Palearctic records presumably stems from the more northerly track of Atlantic depressions in early autumn. As a general rule, migrants departing the northern Nearctic early in autumn are likely to be found farther north in Europe than those departing later.

Chestnut-sided Warbler *Dendroica pensylvanica*
(Linnaeus, 1766). Breeds primarily in northern hardwood and mixed forests of eastern North America from southern Canada east to Nova Scotia, southwards through the Appalachian Mountains to northern Georgia. Winters in Central America, from southern Nicaragua to Panama and casually south to northern Ecuador, western Venezuela and the Greater Antilles.

Monotypic.

Very rare autumn vagrant from North America.

Status: Two records.

1985 Shetland: Fetlar 1st-year 20th September
1995 Devon: Prawle Point 18th October

Unfortunately for the listing fraternity neither record was accessible to the masses, though some lucky souls managed to connect with the second.

On Wednesday 18th October 1995, during a family holiday in South Devon, my father and I decided to spend the afternoon at Prawle Point in the hope of discovering migrants as well as the resident Cirl Buntings.

At around 2.00pm, whilst having bite to eat in the car-park, I heard a Firecrest calling in the small trees that border the roadside. We soon found it, and as there were several other birds calling in the area I began to walk further down the road to check them out.

My attention was drawn to a bird in the tree canopy near to the entrance of the small nature reserve. It had pale unstreaked underparts and an obvious pale eye-ring. It was extremely difficult to see at first, as I was looking up at it through the trees and into the sun. It then dropped onto a lower branch and revealed two cream-yellow wing bars. I realised straight away that I was watching an American Dendroica warbler, but I had not seen any colour on the bird's upperparts so I was unsure of its identity. After following the bird for about 30 minutes, as it worked its way through the trees, it flew to the opposite side of the road where it could at last be observed clearly from the car-park as it caught flies and even perched on telegraph wires. The overall impression was of a bright lime green bird, with this colour being particularly noticeable on the crown. The face was grey, which highlighted the pale eye-ring and large dark eye, and the tail was grey with white on the outer feathers. The underparts were very pale, with the undertail-coverts pure white. The bill and legs were grey, with the lower mandible appearing paler. The bird called as it flew between the trees - a single repeated "chip", vaguely reminiscent of Crossbill.

Upon consulting consulting Rare Birds of Britain and Europe I was almost certain that I had found a 1st-winter Chestnut-sided Warbler. Realising the rarity of the species, I drove to the telephone in East Prawle village to release the news so that other birders could arrive and confirm my identification.

Eventually, at 4.30pm, the first birders appeared on the scene, and fortunately the bird continued to show itself well. By 6.15pm, when it was last seen, about 20 fortunate people had seen it and its identity was left in no doubt.

Despite the fact that it seemed settled and had been feeding on an obvious circuit, following a clear night, the bird had departed by dawn the next morning, causing disappointment to the several hundred people who had travelled to Prawle in the hope that it would still be present.

Andrew Brett. *Birding World 8: 391.*

Discussion: This species was predicted by Robbins (1980), but he ranked it with those least likely to occur. A long-distance migrant, leaving breeding grounds in late August and early September, it orientates mainly southwest in autumn. Its route largely avoids the Atlantic coast plain and lies primarily between the Blue Ridge Mountains and the Mississippi River, crossing the western Gulf of Mexico; a few move through Florida to the Greater Antilles (Richardson and Brauning 1995). Clearly the odds on trans-Atlantic displacement are low. There are autumn records for Greenland (1887 and September 1974), but no others for the Western Palearctic.

Blackburnian Warbler *Dendroica fusca*

James Gilroy

(*P.L.S. Müller, 1776*). *Breeds in eastern North America from south-central Canada east through the Great Lakes region to southern Newfoundland and New England south along eastern seaboard to Massachusetts and New York and along the Appalachian Mountains to Georgia. Winters from southern Central America to Peru and central Bolivia.*

Monotypic.

Very rare autumn vagrant from North America.

Status: Two records.

1961 Pembrokeshire: Skomer age and sex uncertain, 5th October
1988 Shetland: Fair Isle 1st-winter male 7th October

This stunning species is eagerly sought-after. The privileged few who happened to be on Fair Isle when the second British bird appeared are envied by many listers.

Friday 7th October 1988 dawned a wet, grey, and blustery morning on Fair Isle, Shetland. Having been kept awake during the night with very strong northwesterly winds and rain and having already "ticked" Lanceolated Warbler, Pallas's Grasshopper Warbler and Pechora Pipit in the south of the island earlier in the week, we decided we would walk north in the hope of catching up with a Rough–legged Buzzard reported in that area the previous day.

After breakfast our group of five set off at about 9.15 am, suitably attired in Wellingtons, waterproofs etc, as the weather was still pretty foul. About three quarters of a mile from the bird observatory Gordon Avery and I left the road to check out the cliff faces of the many "geos" that dominate the Fair Isle coastline.

I was thus looking into North Restensgeo when a flash of yellow, quickly going out of view behind a fold in the rock, caught my eye. Thinking it might be a Great or Blue Tit and realising even that would be a good bird on Fair Isle, I decided to check further. I had just trained my binoculars on the spot when a bird emerged and flew away from me giving excellent rear views; it was basically black and white. Expletives rained as I shouted to Gordon, almost certain that I had discovered a Black-and-white Warbler.

Almost immediately the bird then flew to the other side of the geo and displayed a golden yellow breast and face, broken only by dark marks through the eye and on the ear coverts. We called the other three in our party and Pete Massey, my son Matthew and

Gordon's wife Margaret quickly joined us on the cliff top. We were all virtually dumb-struck by the sheer beauty of the bird which was now hopping around on the cliff face. We realised it was an American Dendroica warbler, but which one? My mind raced through the pages of the National Geographic Guide to the Birds of North America. The bird appeared so bright we assumed it must be an adult male. No yellow rump was visible so that ruled out several species. We thought of Yellow-throated Warbler or even Blackburnian Warbler.

Realising that we needed advice quickly, we dispatched Matthew back to the "obs" to raise the alarm. I will never forget the sight of him taking off, weighted down with all his waterproofs billowing in the wind, with a speed of which an Olympic medallist would have been proud. Meanwhile Gordon raced ahead to alert a couple of birders who had been in front of us when we had set out earlier. Pete, Margaret and I stayed on the bird. Unfortunately, after a few minutes, it made its way to the seaward face of the crevasse and disappeared around the headland. I noticed Margaret draw her anorak hood tight over her ears as Pete and I gave vent to our pent up emotions in very ungentlemanly fashion!

Matthew now reappeared with several birders who had delayed setting out from the observatory until the weather improved (by now the rain had stopped and the sun was trying to break through). Gordon also reappeared with two other birders.

The bird had now disappeared completely. The usual doubting comments were now forthcoming: "Are you sure?" "Was that it?" (pointing to a Red-breasted Flycatcher!) If some people had realised how close they were to having an "accident" over the cliff edge they would have been far more believing!

Suddenly Chris Donald, to whom I will be eternally grateful, gave a shout. He had relocated he bird about 400 yards away from where we had found it. It was now in another crevasse at Furse. At last it could be seen well everyone present (only two birders on the island dipped) and it was quickly identified as an immature male Blackburnian Warbler Dendroica fusca. We all watched the bird on and off for about half an hour, in which time it was very active, moving along the cliff face to Kame O'Furse.

Then suddenly it flew up the cliff face and headed south. It veered east and then back north. For a while we hoped it was coming back to us, but then it gained height and veered east again towards Buness. Although searched for throughout the rest of the day it was never seen again.

Only then did we become aware of how close most of us had been standing to the cliff edges during this excitement, paying scant attention to Liz Riddiford's repeated warnings. Most birders then gradually trooped away towards Buness in the hope of relocating the bird. Me? I went back to the "obs" to change my - - ?

Jack Willmott. Birding World 1:355-356

Discussion: There has been just one other record for the Western Palearctic, a bird on a ship off the north coast of Iceland in October 1987. There is also an autumn record from Greenland.

This species was among those predicted by Robbins (1980), who ranked it with those least likely to occur. A long-distance migrant, its autumn departure takes place from August to

September. Most birds move through the Mississippi valley and along the Appalachian Mountains to the Gulf coast. Very few take routes farther east through the West Indies to the wintering areas (Morse 2004). The orientation of the autumn movements makes the species less susceptible to the Atlantic depressions that deliver other New World wood-warblers to this side of the Atlantic.

The Pembrokeshire bird was originally considered as a "*Dendroica* warbler, probably a Blackburnian Warbler but it was not definitely established as such" (Saunders 1963). However, the record was re-submitted in late September 1988, curiously just days before the second British record. The improved literature in the intervening 27 years had enabled the finder to be more confident of the identification (Saunders and Saunders 1992). The bird was duly accepted as the first record of the species for Western Palearctic (*BOURC* 1991). The coincidence with the arrival of the second is one of those quirks of birdwatching.

Cape May Warbler Dendroica tigrina

(J.F. Gmelin, 1789). Breeds across the Canadian boreal forest from Alberta to the northern Great Lakes and Nova Scotia, south to the northern United States including New England. Winters primarily on islands in the Caribbean, also casually from eastern Mexico to north Colombia and north Venezuela and southern Florida.

Monotypic.

Very rare autumn vagrant from North America.

Status: One record.
1977 Clyde: Paisley Glen male in song 17th June

Discussion: As with Tennessee Warbler *Vermivora peregrine*, the breeding population of this attractive wood-warbler fluctuates in response to periodic outbreaks of spruce budworm *Choristoneura fumiferana*. The average clutch size of six is greater than that of other New World wood-warblers and possibly allows Cape May Warbler populations to expand rapidly during such outbreaks.

This is a long-distance migrant. Most birds move south or southeast to the wintering areas (Baltz and Latta 1998), but extreme eastern breeders follow the Atlantic coast to Florida. In spring, many move up through Florida and most have departed the West Indies by early to mid-May. Perhaps these spring movements present an opportunity for a spring overshoot or ship-assisted vagrant to reach this side of the Atlantic. Maybe an autumn find awaits some lucky west coast birder; autumn migrants have been recorded in Bermuda from late August through to late November, peaking in October (Amos 1991).

In mid May 1977 many birds moving north along the eastern American seaboard were displaced by an Atlantic jet stream. Several Nearctic landbirds were noted in Scotland at the time (Elkins 1979). These included Britain's first White-crowned Sparrow *Zonotrichia leucophrys* on 15th-16th, Scotland's first Yellow-rumped Warbler *D. coronata* on 18th and a Dark-eyed Junco

Junco hyemalis on 19th. Although eventually found a month later at an unusual inland location, the Cape May Warbler could conceivably have arrived at the same time as the other three Nearctic passerines; the territorial behaviour of the bird and its healthy appearance might also suggest so (Byars and Galbraith 1980), though its proximity to the nearby port of Glasgow could equally indicate a ship-assisted journey.

An amazing Celtic circle in ornithological history was delivered by the Paisley gem. The species was first recognised in America by the Scottish ornithologist Alexander Wilson at Cape May, New Jersey (where it was not recorded again for more than 100 years). Remarkably, Wilson's Scottish home was visible from the trees in which the bird was singing!

Magnolia Warbler *Dendroica magnolia*

(A Wilson, 1811). Breeds across much of Canada and in the northeast USA from the Great Lakes east to New England and the northern Appalachians. Winters from eastern Mexico south to central Panama and in the West Indies.

Monotypic.

Very rare autumn vagrant from North America.

Status: One record.
1981 Scilly: St. Agnes adult male 27th to 28th September

Discussion: An additional category E record from Britain was of a dead desiccated 1st-winter male found on a tanker at Sullom Voe, Shetland, in mid-November 1993. The tanker had left Delaware, USA, and reached Shetland via Mexico and Venezuela. Durand (1963, 1972) logged three birds at sea on the RMS *Mauretania* about 640 km east of New York on 8th October 1962 and one on the *Queen Mary* about 1,500 km east of New York on 30th September 1963.

It has been recorded as a vagrant to Greenland, but the only other Western Palearctic records are two from Iceland (29th September-7th December 1995, 21st-23rd October 1995) and one from the Azores (21st-22nd September 1999).

A long-distance migrant, the Magnolia Warbler departs its northern breeding grounds from early to mid-August. Most birds follow the Appalachians and the Mississippi valley to cross the Gulf to Yucatan, and then through eastern Central America to the wintering grounds. A few move through Florida to the West Indies (Hall 1994). This species was predicted by Robbins (1980), but ranked with those least likely to occur. Its migration routes make the species unlikely to encounter the fast-moving Atlantic weather systems. As a consequence, it is an extremely rare vagrant to the Western Palearctic.

Occurring near the start of Scilly's golden period, this is another species that turns many modern listers green with envy.

Yellow-rumped Warbler *Dendroica coronata*

(Linnaeus, 1766). Nominate race breeds throughout much of North America from Alaska east through Canada to Newfoundland south to Michigan and Massachusetts. Migrates east of the Rocky Mountains to winter in most of the United States, Mexico, Central America and the Caribbean. Another race occurs in southwest Canada and western USA.

Ray Scally

Polytypic, four races. Race is undetermined, but *coronata* (Linnaeus, 1766) is most likely. Yellow-rumped Warbler traditionally comprises two distinct subspecies groups, formerly considered to be separate species: Myrtle Warbler (*coronata*; one race) of north and east North America and Audubon's Warbler (*auduboni* group; three races: *auduboni, nigrifrons and goldmani*) of the west and Mexican coniferous forests. Myrtle and Audubon's interbreed where they meet. This hybridisation led to the groups being combined into a single species named Yellow-rumped Warbler (*AOU* 1973). The western *auduboni* group would appear to be unlikely vagrants to Europe, with autumn migrations of most of this group made over shorter distances than *coronata*; some *auduboni* move to western Mexico and Guatemala and others to Costa Rica; *nigrifrons* and *goldmani* undertake only short-distance altitudinal movements (Curson *et al* 1994; Hunt and Flaspohler 1998).

Milá *et al* (2007) used mtDNA sequence data to reveal that the relatively sedentary Mesoamerican forms (*nigrifrons* of Mexico and *goldmani* of Guatemala) are reciprocally monophyletic to each other and to the migratory forms, from which they show a long history of isolation. In contrast, migratory Myrtle (*coronata*) and Audubon's (*auduboni*) form a single cluster, confirming their earlier lumping.

Rare vagrant from North America.

Status: 28 records; 17 from Britain and 11 from Ireland.

Historical review: The first British record was of a male frequenting the bird table of a garden at Newton St Cyres near Exeter, Devon, from 4th January–10th February 1955, when it was found dead. It was watched by at least 60 observers during its stay, many of whom travelled considerable distances (Smith 1955); this would appear to represent the first major twitch for a British first. This is the only American wood-warbler that winters regularly in northern USA, where it takes advantage of garden feeders. The next was in 1960, again in Devon, on Lundy from 5th-14th November. It was followed by a 1st-winter on St. Mary's from 22nd-27th October 1968. Four were seen between 1973 and 1977. One on Tresco from 16th-24th October 1973 proved, after it was re-found a mile away from its landfall spot, a popular quarry for large numbers of observers. Two in October 1976 included the first for Ireland on Cape Clear from 7th-8th October 1976; on 9th October, a pile of its feathers was found near its favourite feeding area and it was most probably killed by a Sparrowhawk *Accipiter nisus* (Burrows 1978). The first spring record was a male on Fair Isle on 18th May 1977. This bird was part of an exceptional spring displacement of Nearctic landbirds discovered in Scotland over a matter of days, including the first British White-crowned Sparrow *Zonotrichia leucophrys* on the same island three days earlier.

Tresco: *We landed at New Grimsby and immediately went north for Cromwell's Castle. Sadly within ten minutes it became evident that there was no Parulid feeding on the shore where David Hunt had found it. In the bracken and heather there were Redstarts and a Reed Warbler but no yellow-rumped beast. Depression set in and so a 'brew' at the café and then a search towards the hotel. This produced a flock of 250 Skylarks and a Siberian Chiffchaff but still nothing. Thence slowly down the north side of the Great Pool, once again many Goldcrests and the odd Chiffchaff, but otherwise a steady loss of morale. Our pace slackened but I kept going for the hide at the south end of the Great Pool. Once there the sun finally warmed up and Swallows began to appear. I decided to wait and watch. At 13:30 hrs, Bob Emmett went for a last try on Cromwell Castle. At 13:32 hours two northern Chiffchaffs flashed up into the willows opposite and suddenly with them a dancing sprite of a bird. A bright lemon-yellow rump clinched my first Myrtle Warbler in Britain (and first for 11½ years). It floated and danced after flies. I yelled, got no reply and shouted. At last Bob and Peter Grant appeared and for two hours, they and eventually Steve Madge and the others saw the "butterfly bird". It worked reeds (including their wind edge), willows, elms, with occasional short preens and once a listen to Peter's Audubon call gadget!*

Paraphrased from Ian Wallace's log for 17th October 1973.

Ten were seen during the 1980s: two in each of 1982 and 1983, with three of the four sightings coming from Cape Clear including two together on 10th October 1982 (one remaining to the 19th). The second spring record, trapped on the Isle of Man on 30th May 1985, was followed

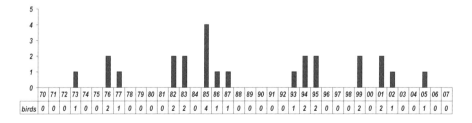

Figure 94: Annual numbers of Yellow-rumped Warblers, 1970-2007.

by three October records that year, including two on St. Mary's, Scilly. During the decade, five records came from Cape Clear, two from Scilly and singles from Co. Clare, Outer Hebrides and the Isle of Man.

The 1990s produced a further seven, the first of the decade coming from – where else? – Cape Clear in 1993. Two more in 1994 included the first for Wales, off the Pembrokeshire coast on Ramsey Island from 31st October-4th November, and the only appearance since the first to grace the mainland, in Eastville Park, Bristol, from 16th-17th November. Two more in early October 1995 were at opposite ends of the country (Orkney and Scilly). The last two of the decade in 1999 comprised another spring bird on the Northern Isles and an autumn bird on the Outer Hebrides.

Since 2001 just four more have been found, two from Cape Clear (2001 and 2005) and singles in Co. Kerry (2001) and an obliging bird on Orkney (31st October-6th November 2003). Just like the first, the Orkney individual frequented a garden. The local RSPB group leader had been putting apples on three sticks in his garden and noticed a Yellow-rumped Warbler on one of them. He telephoned Eric Meek to come and see the bird. When he arrived, there were just three birds on view, one on each of the apples: a Waxwing *Bombycilla garrulus*, a Barred Warbler *Sylvia nisoria* and the Yellow-rumped Warbler!

Where: As might be expected most have been found in Ireland and southwest England. No fewer than nine have been found on Cape Clear and singles in Co. Kerry and Co. Clare. Five appeared on the Isles of Scilly, two in Devon and one in Avon. In Scotland, two each have been found on Shetland, Orkney and the Outer Hebrides. One reached the Isle of Man. The only record for Wales was on Ramsey Island, Pembrokeshire, from 31st October-4th November 1994.

When: The seasonality of records is unique for a New World wood-warbler. The Devon record of 1955 is the only winter record, having presumably made landfall during the previous autumn. Three have been in spring: a male on Fair Isle on 18th May 1977; 1st-summer male on Calf of Man on 30th May 1985; and a 1st-summer male on Fair Isle from 3rd-5th June 1999.

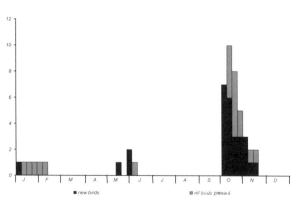

Figure 95: Timing of British and Irish Yellow-rumped Warbler records, 1955-2007.

These three birds were presumably overshoots, given that peak movements in northern USA and southern Canada occur from late April through mid-May.

As with nearly all New World wood-warblers, autumn is the time to connect with a Yellow-rumped. The earliest in autumn was from 2nd-6th October 2001 on Cape Clear. The period 2nd-

13th October has produced 13 (46 per cent) of all records, with three more between 16th and 19th, three from 22nd-26th and three on the 31st. Two have been found in November, one on Lundy from 5th-14th November 1960 and latest, at Eastville Park, Bristol, Avon, from 16th-17th November 1994. The sequence of dates and localities suggests a possible eastward progress following landfall.

Discussion: A bird considered to be this species was first noticed on the Empress of France on 10th September 1954, approximately 640 km east of the Straits of Belle Isle. It remained on board during most of the voyage, but was not seen after the ship came in sight of the Irish coast on 13th September. In January 1956, the observer of the bird was taken to the Natural History Museum and identified the bird as Yellow-rumped Warbler; she had noted all the field marks with the exception of the yellow on the crown (Tousey 1959).

One alighted on board the *Saxonia* shortly after 24th May 1955 in the Gulf of St. Lawrence. It remained on board until 30th May, when the boat passed between the visible coasts of Northern Ireland and Scotland en route for Liverpool. Several people scattered crumbs for it, but although it often hopped among them it was not seen to take any (Margeson 1959). Under present *BOURC* rulings, should this bird be considered a potential first, depending within whose waters it was seen? Two birds were with other migrants recorded at sea on the RMS *Mauretania* about 640 km east of New York on 8th October 1962 (Durand 1963). These records are not accepted as part of the national statistics, but illustrate clearly how some Nearctic vagrants reach Europe.

Elsewhere in the Western Palearctic this is one of the more regular New World wood-warblers. There have been 14 in Iceland (25th October 1964, 19th October 1976, 11th October 1976, 13th October 1976, two 26th September 1980, 1st October 1989, 13th October 1991, 25th September 1993, 16th October 1996, 19th October 1996, 10th October 1999, 21st October 2001, 17th October 2005) and in October 1996 singles were recorded in Norway (8th) and the Netherlands (13th-15th). Eight have reached the Azores, seven of them since 2005 (15th-16th October 2000, 21st November 2005, 21st October 2006, 25th October 2006, 26th October 2006, 17th October 2007, 13th October 2008, 24th-26th October 2008). In addition, there was an immature male at Maspalomas Gran Canaria, Canary Islands (25th February–2nd March 1984).

Yellow-rumped Warbler increased in numbers and expanded its range southwards in the northeast USA during the 1980s and 1990s (Sauer *et al* 2008), coinciding with the increase in records in Britain and Ireland.

Blackpoll Warbler Dendroica striata
(J.R. Forster, 1772). Breeds widely across North America from western Alaska east through Canada to Newfoundland and south to Maine. Migrates through eastern USA to Central and South America from Panama to Chile and eastern Argentina.

Monotypic.

Irregular but relatively frequent vagrant from North America; one ship-assisted spring record.

Status: There have been a total of 45 records, 38 in Britain and seven in Ireland, making this the commonest New World wood-warbler here by some margin; the next most frequent is Yellow-rumped Warbler *D. coronata* with 28.

Historical review: Surprisingly the first records occurred as recently as 1968, when the first English bird reached St. Agnes (12th-25th October) and the first Welsh bird was on Bardsey (22nd to 23rd October).

Further birds followed on St. Agnes from 20th-26th October 1970 and 19th-20th October 1975, with the latter bird again from 31st October-1st November. Autumn of 1976 will be remembered as an exceptional one for this species: 10 were delivered to our shores during an autumn deluge. The first was an early bird at Prawle Point from 18th-29th September, which was followed by no fewer than seven on Scilly between 4th and 20th October. Additional birds included the first for Ireland on Cape Clear from 6th-12th October and the second for Wales on Bardsey from 7th-9th October. The total for the decade was 13 birds

During the 1980s, 11 records were spread across six autumns. The peak count was four in October 1984 when two were found on Scilly and others were in Devon and Co. Cork. The first for Scotland was trapped on Whalsay, Shetland, on 30th September 1985 and remained to 3rd October.

The 1990s delivered 10 birds in seven autumns. These included the first for the east coast in East Yorkshire and the first winter record in East Sussex. The total was in keeping with the tallies achieved during the previous two decades, though records were more scattered with three each from Scotland and Scilly plus others from Co. Waterford and Cornwall.

Continuous easterly winds coming off a high pressure over the continent made for some exciting birdwatching on the Yorkshire coast during the autumn of 1993. Good numbers of common migrants were recorded along with the usual scattering of scarce and rare birds. A high degree of effort was, therefore, maintained at Flamborough and this was rewarded at 10.55hours on October 24th, not from the east as was expected, but from the west.

PJW and Sharon Chalders had spent the weekend looking for birds at Flamborough Head. Having seen D.I.M Wallace's Olive-backed Pipit, but not much else, we decided to have a look at the south end of Dane's Dyke, where a Firecrest had been reported on and off. Walking slowly down the track leading to the sea through some of the more extensive mature woodland on this part of the coast, PJW was immediately struck by the concentration of Chiffchaffs (many of eastern origin) feeding in the Sycamore, and was concentrating his efforts on the left hand side of the track, where the birds were feeding. SC was looking in the opposite direction, where there were less birds, but saw one which she did not recognise. "What's this?"

PJW broke off from tristis Chiffchaffs to look at the mystery bird. Initially it was hidden behind a Sycamore leaf and he expected it to be another Chiffchaff, but when it hopped into view he could not believe his eyes. After a few heart-stopping seconds the words "It's a Blackpoll Warbler .... A new bird for Flamborough ....A new bird for Yorkshire .... An American" emerged. He then gave a shriek of joy and hugged the finder before running off to alert other observers and the news lines. The bird remained in the immediate vicinity that afternoon and was seen by a large number of observers, but thereafter became more elusive, as its foraging took it over a large area of woodland. It was last seen on November 1st, but during its stay it was often seen only once or twice during the day and not at all on Saturday October 30th.

Paul Willoughby (www.birdholidays.co.uk). *Yorkshire Birding 2:117-119*

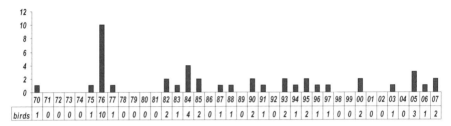

Figure 96: Annual numbers of Blackpoll Warblers, 1970-2007.

Since 2000, a total of nine birds have been found during five years, including the first spring record in Britain from Seaforth. Three have come from Scilly and three from Scotland.

Following the review period there were two records in autumn 2008 (awaiting acceptance, although both were photographed). One was at Marloes Mere, Pembrokeshire, on 7th October and another frequented St. Agnes, Scilly, from 8th-15th October.

Where: The prime area is the Isles of Scilly with 21 records (only four since 1990). Cornwall, Caernarfonshire and Devon each have two. In Ireland, four have been in Co. Cork and singles have been recorded in Co. Galway, Co. Waterford and Co. Wexford. Three each have reached the Outer Hebrides and Shetland and one Highland. After a first in 1985 the next six Scottish records occurred since 1990. With the Atlantic storm tracks tending now to run farther north than in previous decades, potential vagrants are pushed north towards Iceland in some years. It can not be an accident that most recent records of New World wood-warblers in Britain and Ireland have come from the Northern Isles and Outer Hebrides. Perhaps the epicenter of future records will switch permanently to the Scottish islands?

Away from the main areas, singles have been recorded in East Yorkshire (Flamborough Head, 1st-winter 24th October-1st November 1993), East Sussex (Bewl Water, 10th-20th December 1994) and Lancashire and North Merseyside (Seaforth, male, 2nd June 2000).

When: The sole spring record was the male at Seaforth, the first in Europe to be seen sporting the trademark black poll. Found next to Liverpool docks this bird had doubtless been ship-assisted, a matter of no consequence for those who were fortunate enough to see it. The only other New World wood-warblers to be seen in their spring finery here are one Cape May Warbler *D. tigrina* and three Yellow-rumped Warblers *D. coronata*.

All bar one have occurred in the autumn. The earliest date bird ushered in the influx of 1976, at Prawle Point from 18th-29th September. Four others have occurred in late September on Scilly (27th September 2005 on St. Agnes), Outer Hebrides (29th September 2005 on South

Uist) and Shetland (two on the 30th; Fair Isle in 1991 and Whalsay in 1995).

Most have been found between 2nd and 14th October (49 per cent) and 17th and 27th October (30 per cent). The two latest autumn records were both on 29th October (Bryher, Scilly in 1977 and Kenidjack, Cornwall, in 1995), though eight have stayed into early

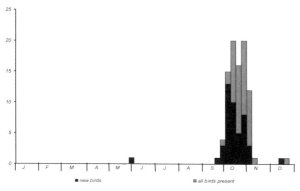

Figure 97: Timing of British and Irish Blackpoll Warbler records, 1968-2007.

November. The sole winter bird was at Bewl Water, East Sussex, from 10th-20th December 1994, so adding its species to the list of New World wood-warblers recorded at this season. Curiously this was one of just two species of American wood-warbler recorded during that particular year; the other was Yellow-rumped Warbler D. *coronata* (two birds).

Discussion: Records that do not form part of the national list begin with one remarkable bird that tried to stay on board RMS *Queen Elizabeth* for a full return voyage across the North Atlantic. On joining the boat in Southampton on 12th October 1961 for a westbound voyage, Durand (1972) was shown a bird that had completed the prior eastbound crossing and was tame yet quite lively. Durand tried to persuade it to fly ashore, but it merely ran up his arm to perch on the back of his collar. Sadly, about half-way back to America it died and its remains disposed of. Presumably it had been fed en route and so is not accepted as part of the *British List*. At least 10 joined migrants recorded at sea on the RMS *Mauretania* about 640 km east of New York on 8th October 1962 (Durand 1963). A desiccated corpse was found on a tanker at Sullom Voe, Shetland, on 28th January 1992. The tanker had left Porto Bolivia 24th December 1991, travelling to Shetland via Rotterdam, the Netherlands (Forrester *et al* 2007).

Elsewhere in the Western Palearctic there have been 13 from Iceland (28th September 1972, 18th October 1974, 22nd-24th October 1974, 3rd November 1974, 30th-31st October 1975, 23rd October 1979, 7th-8th October 1995, 28th-30th October 2001 and five between 16th-23rd October 2005). There are singles from Denmark (15th-16th October 2006), Norway (2nd-12th November 2006) and Finland (23rd October 2008).

There are four for France (9th-15th October 1990, 19th October 1995, 19th October-5th November 2000, 10th-21st October 2003) and one from the Channel Islands (26th October-4th November 1980). In spite of the frequency with which it is recorded from countries farther north, there have been just three on the Azores (24th October in 2006, 14th-15th October 2007, 13th October 2008).

Blackpoll Warbler undertakes the longest migration of any New World wood-warbler, with some individuals traveling over 8,000 km from Alaska to Brazil. Birds from the western breeding range travel southeast across Canada and the northern USA, collecting along the coastal plain south from Nova Scotia, where they meet migrants moving south or southwest from the eastern breeding range. There is considerable evidence that the Blackpoll Warbler undertakes a non-stop transoceanic autumn migration from New England and Atlantic Canada to South America. This unique migration strategy is not yet clearly understood (Nisbet *et al* 1995).

Most are believed to then head south or southeast over the Atlantic Ocean heading to Puerto Rico, the Lesser Antilles, or northern South America. These routes average 3,000 km over water, necessitating potentially a nonstop flight of up to 88 hours. To complete this gruelling flight Blackpoll Warblers nearly double their body mass and take advantage of shifts in prevailing wind direction to direct them to their destination (Hunt and Eliason. 1999). Butler (2000) reported a correlation between autumn storms over the western Atlantic and lower numbers of breeding Blackpoll Warblers in the following summer.

Of all New World wood-warblers occurring in Britain, this species shows the greatest tendency to linger in extended stays. On Scilly 75 per cent have remained for a week or more and 40 per cent for two weeks or more. Unlike some of their Nearctic counterparts they appear more than capable of making the Atlantic crossing in good shape, as they are well adapted to endure long flights.

This species has decreased in the southern part of its breeding range, as reflected in rapidly declining counts on BBS routes since the 1960s (Sauer *et al* 2008). There is little information on population trends in the core range within the Canadian boreal forests. It is difficult to ascertain whether range and population changes in North America have been sufficient to affect vagrancy patterns in Europe.

Bay-breasted Warbler Dendroica castanea

(A Wilson, 1810). Breeds in a narrow band from Nova Scotia and northern New England to the Northern Territories and northeast British Columbia. Population densities highest in the extreme east. Winters in Panama and northwestern South America, also in northwest and northern Colombia and northwest Venezuela.

Monotypic.

Very rare autumn vagrant from North America.

Status: One record.
1995 Cornwall: Land's End 1st-winter male 1st October

Discussion: On their long-distance migration, adult Bay-breasted Warblers appear to follow a more westerly migratory route south in the autumn than first-year birds. The main departure takes place during the first half of September. Most migrate through the Mississippi Valley and along the Appalachians to the Gulf coast before crossing to the Yucatan and passing through eastern Central America to their wintering areas. Some, mostly first-year birds, follow the Atlantic coast (Curson *et al* 1994; Williams 1996). This makes the species an unlikely candidate for frequent displacement across the Atlantic and there are no additional sightings from the Western Palearctic, though it is reported as a casual visitor to Greenland. This species was predicted by Robbins (1980), but ranked with the least likely to occur.

American Redstart *Setophaga ruticilla*

(Linnaeus, 1758). Breeds from southeastern Alaska to southern Labrador and Newfoundland, south to Utah, Louisiana and Georgia. Winters Mexico and Florida to northern South America; highest numbers in western Mexico and the West Indies.

Monotypic.

Very rare autumn vagrant from North America.

Ray Scally

Status: Eight records; five from Britain and three from Ireland, where one in 2008 is subject to acceptance.

1967 Cornwall: Porthgwarra 1st-winter male 21st October
1968 Co. Cork: Cape Clear male 13th-14th October
1982 Argyll: Portnahaven, Islay female or immature 1st November
1982 Lincolnshire: Gibraltar Point 1st-winter, probably male, 7th November-5th December, trapped 8th.
1983 Cornwall: St Just 1st-winter male 13th-24th October
1985 Hampshire: Winchester College Water Meadows 1st-winter male 4th-6th October
1985 Co. Cork: Galley Head 13th-15th October
2008 Co. Cork: Mizen Head 1st-winter female 18th September*

Discussion: A category E record involved a desiccated corpse found on a tanker at Sullom Voe, Shetland, in December 1992. The ship had travelled from Venezuela via Angola (Pennington *et al* 2004).

Elsewhere in the Western Palearctic there are single records from Iceland (10th-12th September 1975), France (Ouessant, 10th October 1961) and at sea off the Azores (female 5th October 1967, male 14th October 1967).

Despite being a medium- to long-distance migrant, it remains exceptionally rare here. The first departures from breeding areas begin in July, though many move much later, the bulk arriving on wintering grounds in October. Movement in autumn is on a broad front through eastern North America. Birds move through eastern Mexico across the Gulf and through the West Indies. Long-distance ringing recoveries suggest a tendency to winter directly south of breeding areas. Individuals that breed in eastern North America travel to the Greater Antilles and northern South America; those that breed in central and western North America migrate to Mexico and Central America. Calculations based on fat load suggest that, unlike Blackpoll Warbler, this species is not suited to long-distance flights over the western Atlantic Ocean (Morris *et al* 1996; Sherry and Holmes 1997). This physiological restriction may be the main reason for its rarity here.

This endearing *Paruline* put in five of its eight appearances between 1982 and 1985, during which period its true rarity status rather diminished for a generation of active birdwatchers. The Lincolnshire bird of 1982 coincided with the discovery of a Green Heron *Butorides virescens* and Great White Egret *Ardea alba* in nearby East Yorkshire, providing many observers with a superb haul of rarities for the day.

The pending record from Ireland in 2008 was the first for over two decades in the Western Palearctic. The species' highly desired status has been restored.

Ovenbird *Seiurus aurocapilla*

(Linnaeus, 1766). Breeds eastern North America from southeast British Columbia east to Newfoundland and south throughout central and eastern USA to northern Alabama and South Carolina. Winters from Florida and northern Mexico south to Panama and West Indies.

Polytypic, three subspecies. All records have been considered to be nominate *aurocapilla* (Linnaeus, 1766); *furvior* Batchelder, 1918, of Newfoundland, which winters in Cuba, Bahamas and eastern Central America, is a future possibility. Compared with *aurocapilla* it has, on average, darker green and less olive upperparts, a wider lateral crown-stripe, paler more brownish-orange or amber-brown central crown-stripe and more heavily streaked underparts (*BWP*; Curson *et al* 1994).

Very rare autumn vagrant from North America.

Status: Six records, including an exceptional inland bird in Herefordshire.

1973 Shetland: Out Skerries trapped 7th-8th October
1977 Co. Mayo: Lough Carra Forest found dead 8th December
1985 Devon: Spriddlestone, Wembury freshly dead 22nd October, probably 1st-winter
1990 Co. Cork: Dursey Island 1st-winter 24th-25th September
2001 Herefordshire: Dymock 20th December to 16th February 2002
2004 Scilly: St. Mary's 1st-winter 25th-28th October, when taken into care moribund

Discussion: A wing found on the tide line at Formby Point, Lancashire, on 4th January 1969 is no longer considered acceptable evidence for a first record, because it could not be ascertained that the bird died in British waters. Elsewhere in the Western Palearctic, one was found in Norway on 25th October 2003 and there have been three on the Azores (1st November-25th

December 2005, two on 21st October 2008 with one present to 18th November). It has also occurred as a vagrant to Greenland.

Ovenbird is a medium- to long-distance migrant, with ringing and tower kill data suggesting that most follow Atlantic coastal and Mississippi flyways during migration. Analysis of ringing recoveries indicates that Ovenbirds breeding east of the Appalachian Mountains follow the Atlantic flyway to Caribbean island wintering grounds and birds breeding west of the Appalachians follow the Mississippi flyway south to wintering grounds in Middle America (Van Horn and Donovan 1994). Ovenbird is the most numerous warbler in many northern hardwood forests in North America. It has shown relatively little overall population change since the 1960s (Sauer *et al* 2008).

The fact that several of the British birds have been found dead or in poor condition implies that this species is ill equipped for making the lengthy non-stop transatlantic crossing. This made the wintering bird that penetrated Herefordshire all the more amazing; its extended stay was exceptional and its occurrence stood out following an otherwise poor autumn for Nearctic passerines in Britain and Ireland. Understandably, the garden owners who found this bird wanted to retain their privacy in the face of what would have been a deluge of observers.

Lister frustration at missing out on the Hertfordshire bird was short-lived, being quickly removed by the bird on Scilly in 2004. This individual proved to be hugely popular, with large numbers of fans travelling to the islands until it was taken into care in a weakened state and sadly died.

Northern Waterthrush *Seiurus noveboracensis*

James Gilroy

(J.F. Gmelin, 1789). Breeds from west and north-central Alaska eastwards across Canada to the Atlantic coast and in north-western and north-eastern USA. Winters Mexico south through Central America to Colombia and Venezuela to northern Brazil, northern Ecuador and northeast Peru. Also winters in Bermuda and in the Caribbean in the Bahamas, Greater and Lesser Antilles, and the Netherland Antilles to Tobago and Trinidad.

Monotypic. Formerly considered to comprise three subspecies based on size and colour of upper and underparts (for example, *AOU* 1957; Godfrey 1986; *BWP*; Eaton 1995) but Molina *et al* (2000) found these differences to be clinal and recommended that the species be considered as monotypic.

Status: Eight records, one photographed in Ireland in 2008 still to be assessed*

1958 Scilly: St. Agnes 1st-winter 30th September-12th October
1968 Scilly: Tresco 3rd-7th October
1982 Scilly: Bryher 29th September-4th October
1983 Co. Cork: Cape Clear 10th-11th September
1988 Lincolnshire: Gibraltar Point 22nd-23rd October, trapped 22nd
1989 Scilly: St. Agnes 1st-winter 29th-30th August
1996 Dorset: Portland 14th-17th October, trapped 14th
2008 Co. Cork: Cape Clear 27th-30th August*

The Dorset bird of 1996 proved extremely popular, being a comparative long stayer on the mainland.

> *Monday 14th October dawned with classic 'fall' conditions at Portland Bird Observatory, but it came as something as a surprise that, amongst the wealth of typical mid October migrants like Redwings, Firecrests and Bramblings, one of the first birds that I came across in the half-light was a Northern Waterthrush! The bird was trapped shortly after my initial sighting and the identification quickly confirmed – both plumage features and measurements straight forwardly eliminated Louisiana Waterthrush, the only real confusion species. In the hand, the rufous tips to the tertials and the sharply pointed tail feathers enabled the bird to be aged as a 1st-winter.*
>
> *Initially, the bird weighed 14.2 gm, and this increased to 16.7 gm by 16th October when the bird was recaught by hand after entering the observatory garage. BWP indicates that these are mid range weights for the species and, taken with the 'fall' circumstances of its arrival, this perhaps indicates that it was not freshly arrived from an Atlantic crossing.*
>
> *News of the bird's discovery was quickly broadcast, and for four days this very confiding bird attracted a steady stream of admirers, with particularly stunning views being afforded to those who saw it feeding in the open on the observatory's front lawn. A collection raised nearly £500 for the observatory.*
>
> **Martin Cade.** *Birding World 9: 394.*

Discussion: This long-distance migrant moves on a broad front from the breeding grounds. Most southbound migration east of the Rockies commences as early as late July, is well underway by mid-August and peaks in September. Birds reach the wintering areas by heading across or around the Gulf of Mexico. It is also a common migrant on Bermuda (Eaton 1995).

The first Western Palearctic record came from France (17th September 1955). Other records are from Jersey, Channel Islands (17th April 1977) and the Azores (one found at sea on a ship on 4th November 1996). Any fortunate finder of a waterthrush in the Western Palearctic should be aware that Louisiana Waterthrush *S. motacilla* has also occurred (Canary Islands from 10th-26th November 1991).

The identification of the unusually pale bird on Scilly in 1968 was hotly disputed at the time as it was felt that Louisiana Waterthrush could not be eliminated. Criticism of the identification forced two re-circulations, but in March 1975 the Rarities Committee accepted it for the third and final time. Research prompted by this record led to the publication of a definitive paper on the separation of the two waterthrushes (Wallace 1976).

The two late-August birds are among the earliest New World wood-warbler sightings recorded on this side of the Atlantic and tie-in well with the early departure of birds from some parts of the breeding range. The most recent, on Cape Clear in August 2008, shared a small puddle on the island with a Solitary Sandpiper *Tringa solitaria*, while at the same time the island and nearby Mizen Head also hosted Yellow Warblers *Dendroica petechia*: no mean purple patch!

Common Yellowthroat *Geothlypis trichas*

Ray Scally

(Linnaeus, 1766). Widespread breeder across North America from southeast Alaska east to Newfoundland, south to central California and southern Texas, and to Oaxaca and Vera Cruz, Mexico. Winters southern USA south through Mexico to Panama; also commonly in Bermuda and the Bahamas, Greater Antilles and Cayman Islands.

Polytypic, 13 subspecies. Race undetermined, though nominate *trichas* (Linnaeus, 1766) from Ontario to Newfoundland and across much of the eastern USA would seem most likely. Nominate *trichas* is the smallest eastern race and has dull olive-green upperparts (*BWP*).

Very rare vagrant from North America, predominantly in autumn though one wintered and there is one spring record.

Status: Ten records.

1954 Devon: Lundy 1st-winter male trapped 4th November
1984 Shetland: Fetlar male 7th-11th June
1984 Scilly: Bryher 1st-winter male 2nd-17th October
1989 Kent: Near Sittingbourne 1st-winter male 6th January-23rd April
1996 Caernarfonshire: Bardsey female 27th September
1997 Shetland: Baltasound, Unst 1st-summer female 16th-23rd May, trapped 17th

1997 Scilly: St. Mary's 1st-winter male 9th October-2nd November
2003 Co. Clare: Loop Head 1st-year male 3rd-4th October
2004 Shetland: Foula 1st-winter male 9th-10th October
2006 Cornwall: Tremough Campus, Penryn 1st-winter male found dead 23rd October

Several birds have put in extended stays, including the only spring record.

On Friday 16th May 1997, on the island of Unst, Shetland, MP had just returned home from an evening walk searching from migrants around Baltasound, when we saw a small passerine fly in the willow outside our window. The bird was not immediately recognisable to the naked eye, and it was not immediately recognisable with binoculars either! In fact, we were faced with a bird which we could not place in an obvious genus! The bland face, with just a prominent eye-ring, did not seem to fit any bird from Europe, and the pale lemon-yellow throat made us wonder if it was something really special. We considered Common Yellowthroat, but decided to go calmly through the bird's features before coming to any rash conclusions.

The bird few into a small area of cover nearby and, while MP followed it, MM went to get a book; she was convinced that it must be rare if MP did not know what it was (a touching show of faith!) and selected Rare Birds of Britain and Europe (Lewington et al 1991). We continued watching the bird and MM pointed to the illustration in the book that fitted its appearance - female Common Yellowthroat!

It was settled in an area of stunted sycamores and rose bushes, and was apparently choosing a roost site, so we stayed for a few minutes to ensure it that had gone to roost, and then went inside to write notes and look through field guides. After writing notes, we went straight to the descriptions of Common Yellowthroat, and everything we read fitted the bird; it was indeed a Common Yellowthroat.

We were up at dawn the next morning to see if the bird was still present, and it appeared within a few metres of the roost site inside ten minutes. Other Unst birders saw the bird with in an hour or so, but it was to be another five hours before the ferries brought about 20 more birders from Mainland Shetland.

At about midday, the bird was caught in the mist-net that had been erected in the garden. An in-hand description was taken by Paul Harvey, Kevin Osborn and Dave Suddaby and it was ringed by MP. It was released into the rose bushes where it had roosted, and it remained around the garden and a small crop next to the adjacent Post Office for the rest of its stay. The first birders from mainland Britain saw it at about 7.30pm on the 17th and it remained in the area, being visited by a succession of birders, until 23rd May.

There will still be many hoping for a more accessible bird in the near future, but, for us, they do not come more accessible than seen from your own sitting room – a true armchair tick!

Michael G. Pennington and Margaret MacLeod. *Birding World 10: 185-186*

Discussion: Four birds were with migrants recorded at sea on the RMS *Mauretania* about 640 km east of New York on 8th October 1962 (Durand 1963).

Elsewhere in the Western Palearctic there have been six from the Azores (30th November 2005, 20th-21st October 2006 and 29th October 2006, 7th October 2008, 16th-30th October 2008, 17th October-11th November 2008) and one from Iceland (26th-27th September 1997). Vagrants have also been reported from Greenland.

A common bird in northeastern North America, this is a relatively short- to long-distance migrant. Some populations are partly migratory or sedentary. Movements tend to be directed mainly northeast to southwest in autumn (Guzy and Ritchison 1999), and it is unlikely to get caught up in the Atlantic storms that deliver other New World wood-warblers here. It is possible that some autumn records involve migrants on a reverse direction.

In spring Common Yellowthroats head northeast from their wintering grounds, so both British records (from Shetland) at this time are presumably overshoots that were heading along the eastern seaboard of the USA. The individual in Kent from January to April 1989 provided the sole wintering record to Britain, coinciding with the one and only Golden-winged Warbler *Vermivora chrysoptera* present over the same four months just a matter of miles away.

Canada Warbler *Wilsonia canadensis*

(Linnaeus, 1766). Breeds across Canada from eastern British Columbia to Nova Scotia, south to eastern Minnesota, northern Michigan and Connecticut, and the Appalachian Mountains to northern Georgia. Winters from Venezuela and Colombia south through eastern Ecuador to central Peru.

Monotypic.

Very rare vagrant from North America.

Status: One record.
2006 Co. Clare: Kilbaha 1st-winter 8th-13th October

The Co. Clare record was one of the birding highlights of 2006.

> *In County Clare, Ireland, Sunday 8th October 2006 was wet and windy. I contacted my good friend Seamus Enright and explained my plans to travel to look for the Red-eyed Vireo reported at Kilbaha, near Loop Head, and I suggested he might like to accompany me…. After all, it was October! Seamus agreed, and at 2.00pm we arrived at Kilbaha and immediately began searching for the vireo. With the assistance of another birder already present, we located our quarry in the pub garden. The vireo proved quite elusive for a while, but eventually it showed well. We watched it for some time and then I ventured off to check the other gardens towards Loop Head, while Seamus decided to stay at the pub garden and try to photograph the vireo.*
>
> *I decided to drive from the pub some 300m towards Loop Head (unfortunately for Seamus!) and first check the garden at Peter Gibson's farm. The weather began to clear, and the sun suddenly came out as if it had been trapped inside for days. Some birds began to appear with one Chiffchaff being the prize amongst the commoner garden species. I*

then heard a sound behind me. I have no idea what it was, but it made me look round, and then, at 4.45pm I saw a golden bird sitting in a small sycamore with the evening sun beaming upon it. Was I dreaming? No! There was a golden bird sitting there!

I moved closer and noted its obvious features. With one hand on my binoculars and the other on my mobile phone, I rang Seamus. I was met with a relaxed "hello", but I declared: "Seamus, I have a luminous American warbler! Whatever it is, we have a mega! Hurry!". I quickly described it to him: "it has a golden breast and throat, whitish eye-ring, white undertail-coverts and blue-grey upperparts". I ran to the car, hastily thumbed through the Sibley Guide, and returned to the Sycamore tree, while Seamus was on his way up the road at electric speed. "Where is it?" he gasped, to which I replied "I think it's a Canada Warbler"…. but the bird was missing!

The next few minutes seemed like hours. Would it ever be seen again? But then the welcome words "I have it" came from Seamus. What relief for both of us! It was time to ring out the news. I called Eric Dempsey and was greeted with the familiar healthy wel- come, but I had to interrupt him to explain "we are looking at a Canada Warbler on Loop Head". I was so excited, I could no longer talk and I handed the phone to Seamus to finish the conversation while I just leapt about! We made more calls, and from the reactions and congratulations we received, the enormity of the find began to sink in. It was not a dream: we were looking at Ireland's first (and the Western Palearctic's second) Canada Warbler.

We continued watching the bird, noting other features such as the pink legs, faint breast markings and unmarked tail etc, until about 6.00pm, by which time the first of the local birders had arrived and also seen the bird.

The bird remained in the sycamores and other trees of the garden over the next five days and hundreds of observers travelled to see it from all over Ireland, Britain and Europe. It could be quite elusive in the thickly wooded garden but, with patience, it always showed well and eventually a range of fine photographs was obtained.

Maurice Hanafin. *Birding World 19:429-434.*

Discussion: It has occurred in Greenland and is casual in Bahamas, Bermuda, Cuba, Jamaica, St. Croix, Guadeloupe and St. Lucia (Conway 1999). There is just one other record for the Western Palearctic, that of a moribund male from Iceland (29th September 1973).

Canada Warbler is a long-distance migrant, one of the latest New World wood-warblers to arrive in their breeding areas (in early June) and one of the first to depart in autumn in early August. Males spend just under 20 per cent of their annual cycle in breeding areas and females only 17 per cent (Tyler Flockhart 2007). Southbound migration commences in the first half of August. Northern parts of the range are vacated by early September. The main passage is rapid. The southerly migration route is mainly in and west of the Appalachian Mountains before fol- lowing the Gulf coast, generally avoiding the southeastern states (*BWP*; Conway 1999). Thus it is a less likely candidate for displacement by autumnal Atlantic weather systems than some other New World wood-warblers. Surprisingly, Robbins (1980) ranked this species with those most likely to occur, but it has not responded to his forecast.

Hooded Warbler Wilsonia citrina

(Boddaert, 1783). Breeds in southern North America west to Iowa and eastern Texas and north to the southern Great Lakes region and southern New England. Winters in Central America from south-east Mexico to Costa Rica, rarely to Panama; a few winter in northwest Mexico and the Caribbean.

Very rare vagrant from North America.

Monotypic.

Status: Two records.

1970 Scilly: St. Agnes female 20th-23rd September
1992 Outer Hebrides: St. Kilda male 10th September

Discussion: Elsewhere in the Western Palearctic there have been two 1st-year males from the Azores (26th-27th October 2005, 11th October 2008).

Medium to long-distance migrants, most Hooded Warblers move south to the Gulf coast and across the Gulf of Mexico to Yucatan and on to Central America. Vagrants have occurred northeast of the breeding range to northern New England and the Maritime Provinces. Departure from breeding areas takes place early, from late July, with arrival on the wintering grounds from early August (Ogden and Stutchbury 1994).

The timing of its migration from a relatively southern breeding range and its westerly migration route would appear to explain its rarity on this side of the Atlantic. The British records, and those to the northeast of the breeding range in North America, may involve reverse migrants.

Wilson's Warbler Wilsonia pusilla

(A Wilson, 1811). Breeds across Alaska and Canada east to the northern northeast states and south through the western United States to southern California and New Mexico; often the most abundant warbler in these regions especially in the west of the range. Winters in Middle America from northern Mexico south to western Panama. Also winters in USA in extreme southeastern Texas, along the upper Texas Gulf coast and throughout southern Louisiana.

Polytypic, three subspecies. Race of vagrants undetermined, but likely to be nominate *pusilla* (Wilson, 1811) of extreme northeast USA and eastern and central Canada west to Mackenzie and Alberta (*BWP*).

Very rare vagrant from North America.

Status: One record.
1985 Cornwall: Rame Head male 13th October

Discussion: There are no other records for the Western Palearctic. A single sighting is reported for Greenland.

The eastern populations of this medium- to long-distance migrant move south to the Gulf coast, which they then follow to wintering areas without normally crossing areas of sea. The breeding grounds are vacated in early August and arrival on the wintering grounds is from early September (Ammon and Gilbert 1999). Such a migration strategy offers little risk of diplacement and presumably accounts for its rarity on this side of the Atlantic. This species was amongst those predicted by Robbins (1980), ranked with the least likely to occur.

Summer Tanager *Piranga rubra*

(Linnaeus, 1758). Breeds in southern United States and northern Mexico north in east to southern Iowa and New Jersey. Winters from central Mexico south through Central America and northern South America to Bolivia and Brazil.

Polytypic, two subspecies. Race undetermined but likely to be *rubra* (Linnaeus, 1758) ofsoutheast USA west to Oklahoma and central Texas (*BWP*).

Very rare vagrant from North America.

Status: One record.
1957 Caernarfonshire: Bardsey 1st-winter male 11th-25th September, trapped 11th, 15th and 20th.

Discussion: During migration, nominate *rubra* of east and central populations crosses the Gulf of Mexico to Central America. Some from western populations of *rubra* may move overland through Mexico, as do all of the larger western race *cooperi*, which winters south to central Mexico; nominate *rubra* winters south to Peru and Ecuador (Robinson 1996).

There is a single record from Bermuda (Amos 1991) and casual but regular winter occurrences along Atlantic Coast (Robinson 1996). The species is apparently declining along the edges of its range in most areas of the eastern USA. It has been recorded in the eastern Caribbean. Autumn vagrants have reached Nova Scotia and Newfoundland, but are rare.

The only other record for the Western Palearctic was on the Azores (26th-28th October 2006). Its predominantly southern breeding range and short-range migration, compared with Scarlet Tanager *P. olivacea,* account for the paucity of Western Palearctic records.

Scarlet Tanager Piranga olivacea

(J.F. Gmelin, 1789). Breeds from southern Canada, Manitoba to Nova Scotia south to Arkansas and northern Georgia. Winters in northwest South America from western Columbia south to northwest Bolivia.

Monotypic.

Very rare vagrant from North America.

Status: Eight records, a well-watched bird from 2008 is yet to be assessed.

1963 Co. Down: Copeland female trapped 12th October
1970 Scilly: St. Mary's 1st-winter male 4th October
1975 Scilly: Tresco 1st-winter male 28th September-3rd October
1981 Cornwall: Nanquidno 1st-winter male 11th October
1982 Scilly: St. Mary's female 12th-18th October
1985 Co. Cork: Firkeel 1st-winter female 12th-14th October
1985 Co. Cork: Firkeel adult male 18th October
2008 Co. Cork: Garinish Point 1st-winter male 7th-11th October*

Discussion: Although the smallest of the four species in the genus *Piranga* breeding in North America, it travels a much greater distance between its more northerly breeding areas and its South American wintering grounds than the Summer Tanager *P. rubra*. As a trans-Gulf migrant, its eastern communities appear to pass through the West Indies in autumn and it is been suggested that the species takes a more easterly migration route to the wintering areas than Summer Tanager (Mowbray 1999). The differences in breeding ranges, wintering areas and migration strategies explain the difference in numbers between the two tanager species recorded in the Western Palearctic.

Elsewhere in the Western Palearctic, records are dated later on average than those in Britain and Ireland. There have been four from Iceland (November or December 1936, 7th-8th October 1967, 23rd October 1967 and 9th November 1992), one from France (12th-18th October 2000) and five from the Azores (29th October-5th November 2005, 20th October 2007 and 25th October 2007, 11th-16th October 2008, 30th October 2008).

Most birds do not depart breeding grounds until mid to late September, coinciding well with the eight British and Irish birds all of which arrived in a narrow 21-day period between the end of September and the middle of October.

Eastern Towhee Pipilo erythrophthalmus

(Linnaeus, 1758). Breeds in southeast Canada and east of the Great Plains in the USA. Mostly remains in breeding areas throughout the year; northeastern parts of the range are vacated in winter.

Polytypic, four subspecies are recognised. Variation in plumage and size within Eastern is clinal and subspecific limits may need redefining (*AOU* 1995; Greenlaw 1996a). Race undetermined, but *erythrophthalmus* (Linnaeus, 1758) most likely (other races are sedentary).

The taxonomy of towhees has been subject to debate for some time. Eastern Towhee and Spotted Towhee *P. maculatus* were until recently considered conspecific as the Rufous-sided Towhee. Eastern Towhee meets and hybridises with Spotted Towhee in a narrow north-south zone of contact in the central Great Plains. A recent re-evaluation of the Sibley and West (1959) hybrid evidence from the Great Plains contact zone led to a split by the *AOU* into two species: *P. erythrophthalmus* and *P. maculates*. Their divergence is supported by early molecular analysis (Ball and Avise 1992; Greenlaw 1996a).

Very rare vagrant from North America.

Status: One record.
1966 Devon: Lundy adult female, trapped, 7th-11th June

Discussion: An additional record from Spurn from 5th September 1975 to 10th January 1976 involved a Spotted Towhee of the western USA and adjacent parts of Canada, south through highlands of Mexico to southwest Guatemala. This is a short-distance migrant with highly variable migration strategies; some northern populations largely vacate their breeding grounds, other populations partly migrate, and some populations interchange with other populations (Greenlaw 1996b). Vagrants have been reported throughout the eastern USA and Canada, but this would appear to be an unlikely vagrant to Britain and this individual was considered to be of uncertain origin and placed on category E. The Spurn record coincided with known imports of Spotted Towhees in 1975. Given that distribution of *maculatus* does not reach the eastern seaboard, ship-assistance is unlikely and an escape from captivity seems the more probable origin of this record.

Most of the northern populations of Eastern Towhee migrate south in winter, although in recent years some have attempted to winter in the northern parts of the range; southern populations are resident. Birds from the northeast undertake a southwest and southerly movement between the Appalachian Mountains and the Atlantic Coast. There are extralimital records to the north, usually in autumn and early winter, to Newfoundland and Nova Scotia (Greenlaw 1996a).

The Lundy individual is the sole record outside of the Americas for a species that would appear to be an unlikely vagrant across the north Atlantic. Robbins (1980) attached a very low predictive value to 'Rufous-sided Towhee'. Its reluctance for offshore extralimital forays is illustrated by its absence from the avifauna of Bermuda (Raine 2002). It seems most probable that this individual was at best ship-assisted. It could perhaps have been an escape as the species has little reputation for extralimital occurrences elsewhere.

Lark Sparrow *Chondestes grammacus*

(Say, 1823). Breeds from southwest Canada across much of the USA east to the Mississippi and more locally east to Ohio, the western Carolinas and extreme northwestern Georgia; also south to northern Mexico. Northern birds move south and winter from southern part of the breeding range south to El Salvador, Cuba and the Bahamas.

Polytypic, two subspecies. Both records are attributable to nominate *grammacus* (Say, 1823).

Status: Two records.

1981 Suffolk: Landguard Point 30th June-8th July
1991 Norfolk: Waxham 15th-17th May

Discussion: These are the only records for the Western Palearctic. The first was originally admitted to category D of the *British List* due to the possibility of the bird being either an escape (perhaps from the Low Countries) or a ship-assisted bird and a seemingly unlikely vagrant. Subsequently, following new information about the species' distribution and migration in the USA combined with a modification of the *BOURC*'s position on ship assistance, Lark Sparrow was promoted to inclusion within category A of the *British List* (*BOURC* 1992).

It is rare away from its wintering areas in the southeastern USA, but records east of the breeding range extend to New Brunswick, Nova Scotia and Newfoundland as well as Central America, Cuba, Bahama Islands and Bermuda. However, the species is very rarely recorded along the Atlantic coast during spring, being noted with much greater frequency during autumn; in the West Indies records span autumn to early spring (Martin and Parrish 2000). Lark Sparrow populations declined by 61.2 per cent between 1966 and 1993, most obviously in the eastern and central regions of the US Breeding Bird Survey, but have subsequently stabilised (Sauer *et al* 2008). In the eastern part of its range, Lark Sparrow is threatened by agricultural intensification and suburbanisation (Sauer *et al* 1997, Martin and Parrish 2000).

The timing of both British birds correlates well with the established peak in records for Nearctic sparrow records on this side of the Atlantic in spring, but they do not accord with the rarity of birds along the east coast of the USA at that season. This, together with the late dates, suggests that both records relied on ship-assistance. The proximity of the first to a major container port is telling. The second, also in East Anglia, may have arrived via a similar route from a port somewhere in northwest Europe.

Savannah Sparrow *Passerculus sandwichensis*

(J.F. Gmelin, 1789). Breeds widely across North America from northern Alaska and Canada south to Baja California and western Virginia. Northern populations migrate to winter south of breeding range. Race P. s. princeps breeds only on tiny Sable Island off Nova Scotia, winters Atlantic coast from Nova Scotia to Georgia. Other races are resident in Mexico south to southwest Guatemala.

Polytypic, up to 28 subspecies, only about half of which are diagnosably distinct. The sub-species can be divided into five groups: *princeps* (Ipswich Sparrow); nominate *sandwichensis* and allies (typical Savannah Sparrow); *beldingi* (Belding's Sparrow and other dark saltmarsh taxa); *rostratus* (Large-billed Sparrow); and *sanctorum* (San Benito Sparrow) (Wheelwright and Rising 2008). Comparison of mtDNA sequences for Savannah Sparrow by Zink *et al* (2005) found that Savannah Sparrows of coastal southern California, Baja California and Sonora (*beldingi, sanctorum* and *rostratus* groups) form a clade distinct from other Savannah Sparrows. Birds of the *rostratus* group (Large-billed Sparrow) may be a distinct species, with differences in song as well as morphological and molecular distinctions. The larger and paler subspecies *princeps* Maynard, 1872 was originally described as a separate species and its present taxonomic position is uncertain. Zink *et al* (2005) concluded that the morphologically distinctive Ipswich Sparrow should remain classified as a subspecies.

Amazingly the first British record in 1982 was of the isolated and distinctive Ipswich Sparrow. Both Fair Isle birds were of one of the northeastern forms, probably either *labradorius* (Howe, 1901) or *oblitus* Peters and Griscom, 1938. The diagnosability of the latter is question-able and it was included within *labradorius* by Wheelwright and Rising (2008); both fall within the typical Savannah Sparrow group.

Very rare vagrant from North America.

Status: Three records.

1982 Dorset: Portland 11th-16th April, trapped 12th
1987 Shetland: Fair Isle 1st-winter 30th September-1st October, trapped 30th
2003 Shetland: Fair Isle 1st-winter 14th-19th October, trapped 14th.

The third record was part of an exciting period for the magical island.

> *Early October 2003 had seen northern Scotland blasted with strong north-westerlies as depression after depression swept across the Atlantic. Iceland was turning up several mouth-watering firsts for the Western Palearctic from North America, and a few American birds were also turning up in Ireland, but birding on Fair Isle was pretty slow – although a Pechora Pipit somehow managed to arrive on the island. Surely we were going to get something from the New World?*
> *A White-rumped Sandpiper duly arrived on 10th October – only the third record for the isle. "At last they are coming" we thought. Alas, hopes for a Nearctic warbler were dashed as the wind switched to the fabled southeast on 12th, and everyone's thoughts turned to vagrants from Siberia and Asia. A Radde's Warbler, three Pallas's Warblers, four Richard's Pipits and a Red-breasted Flycatcher arrived on 13th and so, with the wind still in the southeast, it was with enthusiasm that I ventured out to census the migrants in the southwest of the island on 14th. There were a fair few thrushes around, but I had seen nothing rarer than a single Yellow-browed Warbler and the first Woodcocks of the autumn by the time I reached Hesti Geo at 10.30am.*

I then received a mobile telephone call from Alan Bull at the Bird Observatory saying that John Walmsley, who was staying at Schoolton, had seen a "funny bunting" at Neder Taft. I headed across to check it out, and there I found Tony Quinn, one of our guests at the observatory, looking into the garden. John then appeared from nearby Upper Leogh and they both told me that "it looks like a Little Bunting but without the chestnut face". The bird in question then appeared and one look told me that it was no Little Bunting, but neither was it any other bunting that I could think of. "It's definitely not a Little Bunting, but I don't know what it is!" I informed my expectant companions.

It was obviously a small seed eating passerine and it was streaked and striped all over the place, especially on the head. "I think it's an American sparrow" I said. The excitement levels rose. Several other birders were quite close by and I waved them over, and began to make mobile telephone calls to others. A small crowd soon gathered, but no one had a definitive identification. We needed some literature. I telephoned the observatory and, after leaving several cursing messages because no one answered the 'phone, I asked Alan to get down the island with the Sibley guide (The North American Bird Guide, Sibley 2000). Simultaneously, we received an excited call from Harry Scott saying that he and Paul Baxter were watching a Paddyfield Warbler on the cliffs at Lerness (in the north of the isle)! He was quickly brought crashing down to earth by our news of an unidentified American sparrow and they were soon running towards the south of the isle!

A look through the Sibley guide only marginally helped matters as, although Savannah Sparrow seemed the closest fit (with binoculars in the dull light), no yellow could be seen in the supercilium and the breast streaking appeared black (whereas in the book it was brown). It was lunchtime and the bird seemed settled, so it was decided to take some notes and try and find some answers back at the observatory.

It was an exciting lunchtime as possible identifications were put forward, checked up on and rejected, using the available literature and even the internet. After an hour or so, I felt that Savannah Sparrow was still the most likely candidate – especially after seeing Sparrows of the United States and Canada: the Photographic Guide (Beadle and Rising 2002), and in particular the photograph on page 138 of a labradorius Savannah Sparrow which seemed to closely resemble our bird. However, more field views were needed and we would probably have to trap the bird to be sure.....

Back at the scene, we erected a net and tried to obtain better views and some photographs. The light was rather better and telescope views and (especially) digital photographs showed that the bird did indeed have some yellow in the fore-supercilium, and the streaking on the breast was actually brown – the feathers had blackish shaft streaks with a chestnut border and white fringes. I telephoned the news out that the identification had been clinched: it was a Savannah Sparrow: the second for Fair Isle and the third for Britain!

The bird then obligingly landed right next to the mist-net and was easily trapped. It was ringed, measured and photographed back at the observatory, and then released back in its favoured garden. I put some seed out in the garden in the hope that it may remain faithful to the site. Luckily it did, and it was successfully twitched by around 100 visitors,

some of whom came from as far away as the Isles of Scilly. On the bird's third day, some of the visitors landed to the news that there was also a Siberian Rubythroat on the isle. Oh, the looks on those faces! Some other people therefore made the journey from Scilly to Shetland twice in just a few days! Both birds were still present on 19th October, but both had departed by the morning to 20th.
Deryk Shaw. Birding World 16: 423-426

Discussion: East Coast birds are predominantly short-distance migrants. Those in northern inland breeding areas undertake long-distance migrations (Wheelwright and Rising 2008). The coastal populations are less likely to be susceptible to displacement resulting in an Atlantic crossing, and this could account for their rarity in the Western Palearctic. It is also possible that some are overlooked, as they are not so eye-catching as some Nearctic sparrows.

The species was recorded by Durand (1972) on four occasions involving at least 10 birds, including short-staying birds 1,300 km from New York on 8th September 1961 and 900 km from New York on 24th September 1964 (remained to 26th).

Elsewhere in the Western Palearctic there is a record from the Azores (31st October 2002). It is interesting that three of the four Western Palearctic records are in the autumn; the established pattern for Nearctic sparrows is predominantly of spring overshoots. Why this should be so is not immediately obvious.

Fox Sparrow *Passerella iliaca*
(Merrem, 1786). Breeds across much of Canada and Alaska except for far north and west USA. Winters from Pacific coast east across southern USA to the Atlantic coast.)

Polytypic, 15-18 subspecies recognised. The race involved was one of the *iliaca* group, either *iliaca* (Merrem, 1786) or *altivagans* Riley, 1911. On the basis of plumage coloration and mitochondrial DNA variation, three or four main groups are recognised: Red Fox Sparrow (*iliaca* group); Sooty Fox Sparrow (*unalaschcensis* group); Slate-Colored Fox Sparrow (*schistacea* group); and Large-billed or Thick-billed Fox Sparrow (*megarhyncha* group). Zink and Weckstein (2003) recommend that all four main groups be recognised as species based on molecular results: *P. iliaca*, *P. unalaschcensis*, *P. megarhyncha* and *P. schistacea*.

Very rare vagrant from North America.

Status: One record.
1961 Co. Down: Copeland Bird Observatory 3rd-4th June, trapped on both dates

Discussion: The heterogeneity of this species also extends to its migration patterns: those nesting in the Sierra Nevada, California, migrate short distances, mostly attitudinally; those in Alaska undertake long-distance migrations, with some undertaking open ocean crossings. The European records have been attributed to the *iliaca* group, Red Fox Sparrow, which breeds across northern Canada from Alaska and British Columbia to Newfoundland and winters principally east of the Great Plains. Breeders from Canada's Maritime Provinces move southwest to winter on the mid-Atlantic coastline, with movements of over 2,000 km recorded. Birds from these populations are late autumn migrants and early spring migrants; the earliest arrive in Newfoundland by early April (Weckstein *et al* 2002).

Elsewhere in the Western Palearctic there is one record from Iceland (5th November 1944). Two from Germany (13th May 1949 and 24th April 1977) are considered to have been ship-assisted, as is one from Italy in 1936. It was not among the species recorded by Durand (1972) on transatlantic crossings. The record from Northern Ireland appears late; despite the favourable geographical location, it too may have been ship-assisted rather than a late spring overshoot.

Song Sparrow *Melospiza melodia*

(A Wilson, 1810). Breeds across North America; one of the most abundant birds in eastern North America. Nominate melodia breeds eastern North America from eastern Ontario east to Newfoundland and south to Virginia; atlantica breeds in the eastern USA from Long Island south to North Carolina. Populations from central and eastern Canada and the northern USA are migratory, reaching the southern part of the breeding range and beyond to wintering in the US Gulf Coast and Florida, where they do not breed. Elsewhere largely sedentary.)

Polytypic, up to 52 subspecies have been proposed, but fewer than half are thought diagnosable enough to merit subspecific recognition and 24 are currently recognised (Patten 2001). Molecular analysis of the forms shows little differentiation, suggesting that plumage and morphology have evolved rapidly since the last post-glacial expansion in range (Fry and Zink 1998). The subspecies are arranged in five groups based on size and plumage that correlate with geography: Great Basin to Eastern North America group; Alaska and Pacific Northwest group; Cismontane California group; Desert Southwest and Northwestern Mexico group; and the Mexican Plateau group (Arcese *et al* 2002).

The racial attribution of the British records is undetermined, but most probably involves those from the Eastern group (three subspecies that are small, relatively brown, long-winged and black-streaked). This includes nominate *melodia* (Wilson, 1810), *atlantica* Todd, 1924 and the less likely *montana* Henshaw, 1884 from west of the Rocky Mountains.

Very rare vagrant from North America.

Status: Seven records.

1959 Shetland: Fair Isle male 27th April-10th May, trapped 27th April and 6th May
1964 East Yorkshire: Spurn trapped 18th May
1970 Caernarfonshire: Bardsey 5th-8th May, trapped 5th
1971 Isle of Man: Calf of Man 13th May-3rd June, trapped 13th May
1979 Shetland: Fair Isle male 17th April-7th May, trapped 17th; same, Sumburgh, Mainland 10th June
1989 Shetland: Fair Isle male 11th-26th April, trapped 11th
1994 Lancashire and North Merseyside: Seaforth 15th-17th October

Discussion: That six of the seven British records are in spring fits well with expectations for the arrival patterns of Nearctic sparrows, as do the small number of records from northern Europe. The only additional Western Palearctic records are from Norway (11th May 1975), Belgium (trapped and ringed on 30th September 2004) and the Netherlands (30th April 2006).

Occurrences appear to involve overshoots returning to breeding areas; the April and May dates are indicative of a species that re-occupies its breeding areas early. Departure from wintering areas peaks between mid-March to mid-April; arrivals in Newfoundland occur in May. From ringing studies most autumn individuals migrating along the east coast of the USA travel in a southwesterly direction to their wintering grounds in the southeast USA, where the largest winter concentrations occur. Movements involve mean distances of nearly 900 km. (Arcese *et al* 2002).

This species was one of the many that Durand (1963) encountered on board the RMS *Mauretania*. At least nine were seen at sea between 8th and 12th October 1962 as part a fall encountered 640 km east of New York. At dusk on 12th October, a few hours prior to passing the Fastnet Light at the approaches to the south Irish coast, two Song Sparrows were still on board. One of these was still present when the boat berthed at Southampton on 14th October and was seen again the following day. After docking, the food and water supply was discontinued. This individual was accepted in category E because it was provisioned during the journey, but it clearly illustrated the route that some birds use to get here.

The Lancashire bird of 1995 is the odd one out with respect to date, being one of just two autumn records in Europe and the only British bird to be found away from a bird observatory. It was undoubtedly ship-assisted, being found adjacent to a major shipping port. A similar method of transport would seem likely for the records from Belgium and the Netherlands.

Those who saw the Seaforth bird will be thankful for the opportunity, as sightings have been few and far between in recent years. Once here, the species exhibits a propensity for extended stays. Many will hope that the next will linger at least as long.

White-crowned Sparrow *Zonotrichia leucophrys*

(J.R. Forster, 1772). Breeds across Canada from northern Newfoundland to Alaska; also along the west of North America through British Columbia and Alberta to California and New Mexico. Most northern populations are entirely migratory; those from the western USA less so. Winters southern USA to Mexico.

Polytypic, five subspecies. Race undetermined but most likely to be nominate *leucophrys* (J R Forster, 1772), which breeds from north-central Ontario to Newfoundland. Eastern *leucophrys* winters from central Texas to lower Ohio River Valley. The subspecies differ in colour of bill, lores, flanks and back (Chilton *et al* 1995).

Very rare vagrant from North America.

Status: Six records; two well documented birds from 2008 await formal publication.

1977 **Shetland:** Fair Isle 15th-16th May, trapped 15th
1977 **East Yorkshire:** Hornsea Mere 22nd May
1995 **Lancashire and North Merseyside:** Seaforth 1st-winter 2nd October
2003 **Co. Cork:** Dursey Sound 20th-27th May
2008 **Norfolk:** Cley male 3rd January-14th March*
2008 **Fife:** St Michaels, near Leuchars 17th May*

Discussion: There are two additional ship-assisted category E records. On 30th May 1948, a day out from Newfoundland on board the SS *Nova Scotia* bound for Liverpool, one landed on the ship. It remained for several days until the morning that Ireland appeared on the horizon. During its crossing, it was supplied with food and water (Parish 1961). A similar sighting involved a bird first noted on board the QE2 on the first morning out of New York on 21st September 1988. It remained on board until arrival at Southampton on the 26th (Frankland 1989). The species was recorded on three occasions by Durand (1972), with at least seven birds noted including one 900 km from New York on 24th September 1964 on an eastbound crossing.

Other Western Palearctic records come from France on 25th August 1965, Iceland 4th-6th October 1978 and the Netherlands from mid-December 1981 to mid-February 1982 at a garden feeder; the last was re-identified following publication of a photograph of the bird captioned as a Rock Bunting *Emberiza cia* (van den Berg and Bosman 1999). One has reached the Azores (25th-27th October 2005). An inland bird from Germany (17th August 1969) was considered to have been an escape. One from Iceland (28th May 2002) had been ship-assisted and cared-for on ship. Migration takes place on a broad front. Eastern *leucophrys* migrates southwest to the wintering grounds. Midwestern *leucophrys* and *gambelii* migrate south to their wintering grounds. Eastern *leucophrys* arrives back on the breeding grounds in late May; *gambelii* slightly earlier in mid-May (Chilton *et al* 1995). These dates accord well with the British and Irish records, suggesting that some of these birds are overshoots heading for their breeding areas, in keeping with other Nearctic sparrows. As with many of the other Nearctic sparrows, most are likely to have completed some or all of their trip on ships during spring movements.

The Seaforth bird was almost certainly ship-assisted. The wintering individual from Norfolk in 2008 could have made landfall elsewhere in northwest Europe by whatever means in the preceding autumn. It became one of the most watched birds of all time during its stay of just over two months. On-site donations contributed £6,500 for local charities in the process. Whether the Fife bird that followed was the same individual is impossible to say, but it is possible.

Ian Wallace

White-throated Sparrow
Zonotrichia albicollis
(J.F. Gmelin, 1789). Breeds North America from south-east Yukon east to Newfoundland, south to Great Lakes and northern USA to New Jersey. Winters southeast USA from Massachusetts south to Florida, Texas and into north Mexico and California.

Monotypic. Adults during the breeding season are polymorphic, with white or tan head-stripes arising from a difference in their chromosomes. The differences in plumage and genetic material are maintained by negative assortative mating, whereby each morph nearly always mates with its opposite. White-striped males are more aggressive and territorial, and tend to seek matings outside the pair bond more; tan-striped females provide more parental care than their white-striped counterparts (Falls and Kopachena 1994).

Rare vagrant from North America, most often in spring.

Status: White-throated is one of the commonest of the Nearctic sparrows in Britain and Ireland with 34 records; 32 in Britain and one each from the Republic and Northern Ireland to the end of 2007. Both morphs have occurred. The most recent tan morph was in May 2008; the popular wintering individual in Lincolnshire in 1992/93 was also of this morph.

Historical review: The first was a male shot on the Flannan Isles, Outer Hebrides, on 18th May 1909. The next was not until 1961, when one was at Needs Ore Point, Hampshire, on 19th May. Six more followed between 1965 and 1968, three in spring and three in autumn. Two in 1967 included national firsts for Ireland (immature male at Cape Clear on 3rd April) and Wales (1st-year on Bardsey from 15th October-7th November, trapped on 21st). One of the 1968 autumn birds was the first to attempt to winter. It survived at Herringfleet, Norfolk, from 16th November until 1st January 1969, when it was found dead from pneumonia (Taylor *et al* 1999).

On 18th October 1968 I had a week to finalise a career and family move to Nigeria and my bird boots had been hung up. Late in the evening, the phone rang. It was Bob Emmett.

"Ian, the Holy Grail-bird is at Beachy".

"What?!"

"Yes, Pallas's Leaf, trapped today, released in Belle Tout Wood"

"Oh God, alright, a last try. See you at 0400 hours."

Bob and I reached the car park before dawn. The wood loomed above us. As the light came up and I prayed for my most desired ever lifer, it became apparent that we weren't the only travellers of the 19th. Birds started to move in flocks, mostly to the east. We weren't going to be bored but how to find one small bird in one large wood?

Bob took the lee edge, I the east path up within the wood. The result: nothing but flickering leaves, but at the top looking at last onto canopy, I spotted a bright flash of movement. Glasses hit eyes, eyes locked onto the most contrastingly-striped sprite ever. It was 0730 hours. "Bob, up here, quick" I yelled.

An hour later, we were broken men, equipped with good views of a striped sprite but given only six pale lines, it was definitely only a Yellow-browed. Yet another failure with proregulus jarred and Nigeriana began to disturb my concentration.

"Have you got it?" a voice called out from below. "No, we (expletive deleted) haven't," we answered, making out parts of Pete Colston in the top bushes. Oddly he stopped dead in his tracks ...."Cheer up chaps, there's a Yank sparrow in the bottom of the elder"

Instantly brought back into at least British ticking mode, we got to our knees, shuffled over and peered in. There, grubbing about in the leaf litter, was indeed a young White-throated Sparrow, only the sixth for Britain, three times as rare as the 19th Pallas's but not the one that I really wanted. Still it was the best of 3000 migrants of 28 species in, around, and over, the magic wood in just two hours.

Cruelly the seven-striped sprite showed for all-comers on the 20th and I flew to Nigeria still without the then (and still) best gem of all rarities.
Paraphrased from Ian Wallace's log for 19th October 1968.

Fewer were seen in the 1970s, with just four between 1970 and 1978 all in Scotland. These included three from Shetland and an individual that remained in Caithness for about four months from early May 1970.

A slight improvement in fortunes produced five in the 1980s, including another wintering bird, this time in Belfast, Co. Antrim, from 1st December 1984 until May 1985. This individual proved to be hugely popular as it was the first long-stayer for a new generation of increasingly mobile birdwatchers who considered Irish Sea crossings fair game for the acquisition of new birds. Two in 1987 was the first multiple arrival for 20 years; it involved two late spring sightings from Shetland a month apart.

The 1990s returned the same number of birds as the previous decade, but included two crowd-pullers in 1992. The first, a spring bird, was present at Trimley St Mary, Suffolk, for a week from 31st May to 8th June. The second frequented a small clump of hawthorns in a clearing in the middle of a huge inland Forestry Commission plantation at Willingham, Lincolnshire, from 5th December through to 28th March 1993. Two in 1996 and one in 1998 were on the Northern Isles.

There has been a noticeable upsurge since 2000, with 12 to the end of 2007. Records in the North Sea 2001 and 2002 (on the oil installation Maersk Curlew in sea area Forties and Uisge Gorm in sea area Dogger) were followed by an obliging October bird at Flamborough Head from 22nd to 29th October 2002, the first mainland sighting for 10 years. It proved to be an attractive draw over its week-long stay. Three spring birds in 2003 set a new annual record and once again included a mainland bird, albeit just for just three days, in Cheshire and Wirral.

Four birds occured following the review period: a tan-striped bird was caught at Heysham, Lancashire, on 6th May 2008; a 1st-winter was on Cape Clear from 12th-18th October 2008; one was

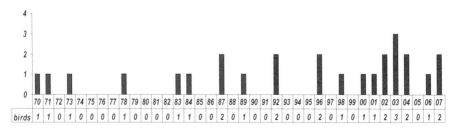

Figure 98: Annual numbers of White-throated Sparrows, 1970-2007.

at Helsby, Cheshire, from December 2008 to March 2009, and in April 2009 a male was at Old Winchester Hill, Hampshire (but was first seen in November 2008).

Where: Records are well spread, but the absence of sightings in the southwest and just three in Ireland to the end of 2008 is notable. Shetland accounts for 13 birds (38 per cent) and East Yorkshire three. Six have been found on the east coast between Yorkshire and Suffolk, plus two in the North Sea and three on the south coast (East Sussex and Hampshire, not including the 2009 bird). In the northwest, singles have reached Bardsey, Cheshire, Isle of Man and Cumbria. The only Scottish records outside of Shetland were on the Outer Hebrides and Highland.

When: The New World sparrows have a characteristic arrival pattern, with most in spring and a large proportion of them on the Northern Isles. This is the case for the current species: 76 per cent of records coming in spring and 24 per cent between late September and early December. Spring overshoots are thought to account for this distinctive pattern; ship-assistance

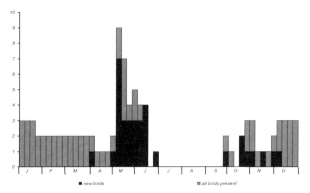

Figure 99: Timing of British and Irish White-throated Sparrow records, 1909-2007.

presumably plays a role in a proportion of records, perhaps accounting for many of the disproportionate number of east coast records.

The earliest spring record was on Cape Clear on 3rd April 1967. Around two-thirds fall between 10th May (Fetlar, Shetland in 2003) and 17th June (Walney Island, Cumbria, in 1965 and Fair Isle in 1978). One on Foula on 1st July 2004 is outside of the expected dates, but is presumably a late spring find.

Autumn records straddle the period late September through to early December. The earliest, and only September record, was at Voe, Shetland, from 26th September to 7th October 1996. Three October records fall between 15th and 22nd (Bardsey in 1967, Beachy Head in 1968 and Flamborough Head in 2002); there are November records from Shetland and Norfolk (latter remained to January of the following year). The two long-staying wintering birds were first located in December: Belfast, Co. Antrim, from 1st December 1984 to May 1985 and inland at Willingham, Lincolnshire, from 5th December 1992 to 28th March 1993.

Discussion: A number of records are within a relatively short distance of busy ports and at least some presumably involve ship-assisted individuals. Even those elsewhere may have made the crossing on a boat and resumed their migration on this side of the Atlantic. A number of records from the 1950s and 1960s of birds that arrived via such routes are not accepted as part of the national archive.

This is one of the commonest species to be reported at sea during Atlantic crossings. Among the avian passengers aboard the *Statendam*, which left New York on 27th September 1957 and arrived in Southampton on 4th October, were two White-throated Sparrows. Both were fed with crumbs by passengers and stewards and were seen to be alive and apparently healthy on the day when the ship came within sight of the British Isles (MacArthur and Klopfer 1958). In 1958, four were on board a Cunard ship that docked in Southampton, Hampshire, in October or November; they were placed in East Park aviary in the city, where the last died in January 1964 (Sharrock, 1965). One was found on board the RMS *Queen* Elizabeth on 27th April 1961, some 1120 km out, and slightly north of east, from New York. The bird was approachable and roosted in the garden lounge. It was fed by various stewards, being present at least until the boat was off the Isles of Scilly on 30th April. The observer considered that it may have remained until docking at Southampton, where one turned up in a nearby park three days later (Durand, 1961).

On 8th October 1962 Durand (1963) estimated that among a large fall of migrants onto RMS *Mauretania* some 640 km out from New York at least 20 White-throated Sparrows came aboard. Two remained until the ship docked in Southampton from 14th-15th October; one was found in a nearby park on the 16th. In 1998 a 1st-summer or female and two males landed on RV *Akademik Ioffe* on 1st May, c250 km SSW of Avalon Peninsula, Newfoundland, Canada. The female was last seen 1,280 km southwest of the Westman Islands, Iceland, but the two males progressed far enough to enter British waters. One died in the North Sea, but the other survived to reach Kiel, Germany. These birds were known to have been fed on board ship and so could only be admitted to Category E (Cook 1998). Another present in Southampton dock on 12th and 13th May 2007 is accepted in the national totals. How many more must arrive via this route undetected? The almost exclusively north European distribution of the records suggests that a proportion may involve long-distance overshoots, particularly in the case of the spring birds reaching the offshore islands of Scotland.

Elsewhere in the Western Palearctic this is also the commonest of the Nearctic sparrows, with six from Iceland (10th-17th June 1964, 18th June 1974, 5th November 1979, 21st-29th November 1981 and presumably the same 13th-14th April 1982, 18th-20th January 1990 and 12th November 2007-21st April 2008), five from the Netherlands (28th September 1967, 8th October 1967, 24th April 1977, 10th June 1989 and 30th April 2001), two from Norway (10th-19th July 2002 and 17th-28th June and 19th-24th September 2003) and three from Finland (23rd June-20th July 1967, 2nd June 1972 and 29th May-8th June 1999). There are also singles from Sweden (1st-winter male caught on 5th December 1963 after flying into a window), Denmark (23rd May 1976) and France (20th April-29th June 2008). Not surprisingly, the only record for southern Europe was near another popular harbour, Gibraltar (18th-25th May 1986). There was a breeding attempt by a presumed escaped pair in Hamburg, Germany, in 1972.

Whatever their route here, these attractive birds are always capable of drawing a crowd. Quite what is causing the present upsurge in records is unknown, but their presence here is most welcome.

Dark-eyed Junco *Junco hyemalis*

(Linnaeus, 1758). Breeds in North America from tree line of northern Alaska and Canada, south to southern California, northern Texas and northern Georgia, USA. British occurrences relate to form previously separated as Slate-coloured Junco, breeding in north and east of range south to Georgia. Northern populations migrate south of breeding range in winter.

Polytypic, 15 subspecies. Prior to 1973, split into five separate species, three of which comprised two subspecies each. Currently juncos with dark eyes are classified as a single species (Nolan *et al* 2002). Slate-colored Junco (*hyemalis* group) comprises three weakly to moderately differentiated subspecies. Nominate *hyemalis* (Linnaeus, 1758) breeds from northern Alaska east to Newfoundland and south to south-central Alaska, northeast British Columbia and from central Alberta east to northern Pennsylvania, southeast New York and southern New England. Commences spring migration from the beginning of March; most movement taking place from mid-March to early April. (Nolan *et al* 2002). A recent molecular analysis found little differentiation among North American subspecies of the entire Dark-eyed Junco complex (Milá *et al* 2007). All British and Irish records have involved nominate *hyemalis*.

Status: There have been 34 records, 31 of them in Britain and three in Ireland.

Historical review: The first was a male shot at Loop Head, Co. Clare, on 30th May 1905. The next, the first British record, was 55 years later: a male trapped at Dungeness on 26th May 1960. Three more were recorded between 1966 and 1969, all of them one-day birds on Shetland with dates ranging from 1st-10th May.

Another one-day bird in 1972 in East Sussex continued the short-stay theme for the species, but two in 1975 included the first collectable long-stayer, a male at Haresfield, Gloucestershire, from 1st-12th April and the first and so far only Welsh bird on Bardsey from 25th April-3rd May. Two birds in 1977 were one-day returns to form and were the last of the five for the decade.

The first of seven in the 1980s was caught exhausted on 24th May 1980 on an oil supply vessel in the Leman Bank gas field 55 km off the Norfolk coast. It was transported by helicopter to the mainland on 28th May and watched by over 50 observers when it was released at Holme reserve on 31st May. It stayed until at least 6th June. Three birds in 1983 were all found between 20th and 27th May in Hampshire, Somerset and Cornwall.

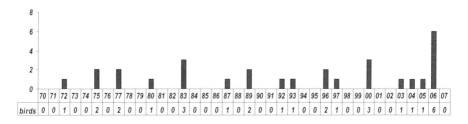

Figure 100: Annual numbers of Dark-eyed Juncos, 1970-2007.

The most notable occurrence of all was a first-summer male at Church Crookham, Hampshire, from 30th May to 7th June 1987. It was there, singing, on 20th May 1988 and was seen again on 7th February 1989. Yet again it returned, this time to winter from 26th December 1989 to 9th March 1990. This was the first (and only) instance of a Nearctic passerine returning to a British (or Palearctic) location over three annual cycles. 1989 also produced a brief male on the Isle of Wight and later in the year another long-staying bird, at Portland from 3rd December 1989-8th April 1990.

Five in the 1990s were all in 1992-1997. These included three one-day birds and a male in Dorset from 7th-19th November 1993 and a very popular bird in Chester from 15th December 1997 to 19th April 1998.

An above average 12 were logged between 2000 and 2007. Three in 2000 was the best year's showing at the time and included two spring records in Scotland, one of which was on the North Sea oil installation Maersk Curlew, and the second for Ireland in Co. Wicklow for a day in early August. Singles in 2003, 2004 and 2005 were all May records, two of them from Scotland plus the third Irish record on 30th May 2004, 99 years to the day after the first. The 2005 bird was another from the oil installation Maersk Curlew. It was photographed, but is yet to be assessed by the *BBRC*. An exceptional six were recorded in 2007, easily the largest annual number ever, including one on 12th May in Cornwall, three Scottish birds between 30th May and 23rd June and, unexpectedly, two in Norfolk on 14th July.

Following the review period there were four records in 2008. These included a well-watched 1st-winter female at Dungeness from 7th-9th April and a male at East Coker, Somerset, from 16th-18th November. There were less obliging birds photographed inland on 13th April at Ingleton, North Yorkshire and a male at Hayle, Cornwall, on 26th November.

Where: The distribution of records recalls in some respects that of the other new world sparrows, including the high proportion from Scotland (over 40 per cent). Four have reached Shetland, two Highland and, astonishingly, two on the oil installation Maersk Curlew in Sea area Forties. Elsewhere there were singles from Orkney and Caithness and inland birds in Highland and Clyde. In the west, there have been further singles from Highland and the Outer Hebrides.

In distinct contrast to other Nearctic sparrows, there are a high number of records from along the south coast. Eleven of these fall between Cornwall and Kent and include three in Dorset and two each in Hampshire and Cornwall, plus singles in Somerset, Isle of Wight, East Sussex and Kent. The proximity of many of these sightings to ports and busy shipping lanes offers a likely pointer to their origin.

Elsewhere, there are inland birds from Gloucestershire and South Yorkshire plus two inland birds in Norfolk. In the northwest, there are singletons from Lancashire and Cheshire and the sole Welsh record, on Bardsey in 1975. In Ireland, records have come from Co. Clare, Co. Wicklow and Co. Antrim.

When: Around 75 per cent of records have occurred in the spring, the earliest a male at Haresfield, Gloucestershire, from 1st-12th April 1975. Just one more has occurred in early April (male at Wooton, Isle of Wight, 8th-9th April 1989), and the overwhelming majority have been between

25th April and 30th May. Summer records comprise two found in late June 2007 and two further birds on 14th July of the same year in Norfolk. Apart from these, a male at Ballygannon, Co. Wicklow, on 10th August 2000 looks distinctly lonely sitting between the late spring records and the relatively few winter sightings. There are five winter records between November and February including two long-staying wintering birds: a male at Portland, Dorset, from 3rd December 1989-8th April 1990 and one at Vicar's Cross, Cheshire from 15th December 1997-19th April 1998. A male loitered at Dorchester, Dorset, from 7th-19th November 1993. The two remaining records are one-day birds in South Yorkshire (3rd January 1977) and East Sussex (12th February 1972).

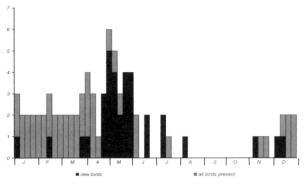

Figure 101: Timing of British and Irish Dark-eyed Junco records, 1909-2007.

Discussion: As with other seedeaters there is a high number of records recorded at sea. The most notable occurred in 1962 as part of the large arrival of migrants at sea aboard the RMS *Mauretania* and remained on the ship until docking in Southampton on 14th October (Durand 1963).

The spring bias to British and Irish sightings corresponds with expectations for Nearctic sparrows and presumably relates to overshoots, although the relative proximity of many records to major south coast ports suggests that most did not make the entire crossing under their own power. That relatively few records involve young birds is consistent with the shorter and later movements of birds of this age relative to adults, which winter farther south (Ketterson and Nolan 1982).

The small cluster of winter records might relate to new arrivals, rather than birds shifting after crossings made in the previous autumn. It has been suggested that unfavorable midwinter weather in North America causes a renewed southward migration, since the species is more numerous in the southern part of wintering range during harsh winters (Nolan *et al* 2002).

Elsewhere in the Western Palearctic further records involve birds from Iceland (1st-year female on 6th November 1955), the Netherlands (trapped February 1962 and kept in captivity until it died on 7th November 1968), Poland (4th May 1963), Denmark (captured on 13th December 1980 and kept in captivity until it died on 18th March 1993), and three birds from Norway (two records comprising three birds: two on 4th December 1987 and one on 18th May 1989). From southern Europe there is a record from Gibraltar (18th-25th May 1986, which coincided with the arrival of a White-throated Sparrow) and a bird of unknown origin from Italy (28th November 1914). This is one rarity that birders can anticipate finding at their garden feeders. Many of the records here are of birds frequenting such food sources, their use learnt in North America where it is a common feeder bird.

Black-faced Bunting *Emberiza spodocephala*

Julian R. Hough

Pallas, 1776. Breeds from Russian Altai and Ob river east across Siberia to Sea of Okhotsk, south to Baikal region, northern Mongolia, Amurland, northern Korean peninsula and extreme northeast China. Winters South Korea and through much of eastern China to Hong Kong. Other races breed in Japan and central China.

Polytypic, three subspecies. All records have been of nominate *spodocephala* Pallas, 1776.

Very rare vagrant from Siberia, all in autumn except for a presumed overwintering bird.

Status: Five records.

1994 Greater Manchester: Pennington Flash 1st-winter male 8th March-24th April, trapped 8th March
1999 Northumberland: Newbiggin-by-the-Sea female or 1st-winter 24th October
2001 Devon: Lundy 12th October
2001 Shetland: Fair Isle probably 1st-winter male, 20th-24th October
2004 East Yorkshire: Flamborough Head 15th October

Discussion: There is also a category E record of a male of the nominate race with an elongated upper mandible at Spurn on 13th-14th May 2000. Elsewhere in Europe there have been two from Germany (1st-winter female 5th November 1910, male 23rd-26th May 1980; one on 18th February 2005 is pending and another on 29th July 2006 was a probable escape and not accepted), three from the Netherlands (1st-winter male caught on 16th November 1986 and kept in an aviary before being released at the end of the winter, 1st-winter male caught on 28th October 1993, 1st-winter female caught on 18th November 2007) and singles from Finland (male 2nd November 1981), Norway (2nd October 1999 and another from 18th July 2005 placed on category D) and France (8th-14th April 2006).

The western edge of the range reaches the southeast of western Siberia, where it is locally very common. Vagrants have been recorded in the wooded tundra of western Siberia (Ryabitsev 2008). In autumn, nominate *spodocephala* leaves central Siberia chiefly in September, with the main passage through northeast China and Mongolia during September (*BWP*).

Long ago predicted as a potential vagrant to Britain by Wallace (1980), this species was largely ignored in the British literature until an identification paper by Bradshaw (1992). Shortly after its publication the first British record was discovered. Reportedly rare in captivity, the presumed escapes recorded in Europe suggest nevertheless that it may be more frequently kept as a

417

cage bird than is acknowledged. One escaped from the National Exhibition of Cage and Aviary Birds in Birmingham in December 1989 (Bradshaw 1992).

Many will be relieved that they saw the Pennington Flash individual, albeit in a location that few would have expected to be watching a British passerine first. Subsequent records have been either less obliging or less accessible, but surely another is not too far away in light of the increased number of European records? The upsurge in sightings accords with that of several other very rare vagrants that were formerly considered once-in-a-lifetime birds. This phenomenon may be due partly to greater observer awareness, but that alone is unlikely to account fully for the increasing frequency of sightings in Europe.

Pine Bunting Emberiza leucocephalos

S.G. Gmelin, 1771. Breeds across temperate Russia from west Urals to upper Kolyma River, south to southern Siberia, Mongolia, lower Amur River and Sakhalin. Isolated population breeds Qinghai and Gansu provinces, central China. Small isolated wintering populations occur regularly in western Italy and central Israel. Otherwise winters south of breeding range from Turkestan east through Himalayan foothills to central and eastern China north of Yangtze.

Polytypic, two subspecies. All records have been of nominate *leucocephalos* S.G. Gmelin, 1771.

Rare vagrant from Siberia, mostly in October and November, although wintering records regular.

Status: There have been 50 records of this attractive bunting to the end of 2007, 48 in Britain and two in Ireland.

Historical review: The first was a male obtained from Fair Isle on 30th October 1911. The next was 32 years later when a male was seen on Papa Westray, Orkney, on 15th October 1943.

The first of the modern recording era was not until 1967, when an unseasonal male was present on North Ronaldsay from 7th-11th August. Just two more followed in the 1970s (male at Portland on 15th April 1975 and Golspie, Highland, from 6th-8th January 1976).

In 1980 two were seen on Fair Isle in the autumn, these birds acting as the harbingers of increasing numbers during the decade. The decade went on to amass 13 birds, a rapid ascent in numbers considering that just five had been recorded up to the start of the decade. This included

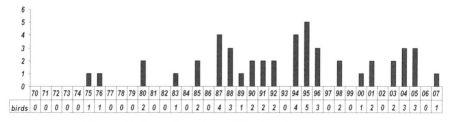

Figure 102: Annual numbers of Pine Buntings, 1970-2007.

no less than four on the Northern Isles in 1987, with two each on Fair Isle and North Ronaldsay between 11th October and 3rd November. Three followed in late October 1988, one from East Yorkshire and two from Orkney.

The 1990s kept up the pace, accumulating 20 records between 1990 and 1998, including four from the Northern Isles between 22nd October and 7th November 1994 and an as yet unsurpassed five in 1995. This record number was spread throughout the year and included the first for Ireland at North Slob, Co. Wexford, from 20th January-19th February, a spring bird in Lincolnshire, a June bird on Orkney and October birds in Suffolk/Norfolk and on Fair Isle.

Between 2000 and 2007, there were a further 12, including three each in 2004 and 2005. These included the first and second for Wales, males on Skokholm on 28th April 2000 and Bardsey on 30th April 2001. Given the increased observer emphasis on the Northern Isles, this relatively small crop might represent a slowdown in the numbers occurring here.

Where: Not surprisingly there is a bias towards the Northern Isles and east coast, especially during autumn. Shetland has accounted for 13 (26 per cent) birds (10 in October, three in November), nine of them on Fair Isle. Orkney has produced 10 birds (one in each of June and August, six in October and two in November), eight of them on North Ronaldsay. The only other Scottish records have come from Highland (January) and the Outer Hebrides (November).

Along the east coast, counties between Northumberland and Norfolk have attarcted 13 birds. Six of these have been from Yorkshire (two were well inland), three from Northumberland and two in Norfolk. In these areas, the detection pattern differs markedly from that in the Northern Isles, with three found in January-February, six in March and April and just three in October and one in November.

In the southwest there have been six records, two from Scilly and singles in Devon, Dorset, Pembrokeshire and Bardsey, with one in each of January and November and four in April. In Ireland, records have been in January (Co. Wexford) and March (Co. Dublin). Elsewhere, singles have reached Surrey (January) and London (February) and two have been found in Worcestershire (January and February).

When: The earliest wintering bird to be detected was in the first week of January; nine have been discovered between 6th January and 18th February. Three between 28th February and 3rd March might well be termed forerunners of a protracted spring exodus that is accounted for by 12 birds between late February and the latest spring

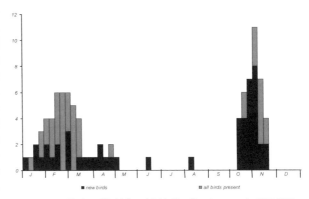

Figure 103: Timing of British and Irish Pine Bunting records, 1911-2007.

419

bird, a male on Bardsey on 30th April 2001. Summer males on North Ronaldsay on 17th June 1995 and from 7th to 11th August 1967 fall outside the typical patterns.

The peak period for occurrence is early October to mid November. The earliest was a female on Fair Isle from 8th-11th October 1998. Seven more have occurred between 11th and 21st October. A third of all records have been found between 22nd October and 7th November. Just three autumn birds fall after this date, all females, the two latest first found on 16th November on St. Mary's (to 18th) in 1983 and on North Uist in 2005.

As would be expected given the difficulty of identification, most have been males, with only a fifth of all records relating to females or immatures. Presumably these very subtle birds are easily overlooked in inland bunting flocks. The most well-watched was the 'Big Waters bunting', present in Northumberland from 18th February-16th March 1990. It provided a reappraisal of what to look for with potential females (Lewington, 1990; Bradshaw and Gray, 1993; Shirihai *et al* 1995).

Comparison of the sex-ratio of British records with that elsewhere in Europe suggests that a good number of less obvious females are being overlooked here. In the Netherlands, for example, roughly a third of the 34 birds accepted records involve females, many trapped at ringing stations. In Italy, 70 per cent (of 110 birds) were 1st-winters and there was a ratio of two females to every male (Occhiato 2003). By contrast Swedish records compare with those from Britain in that over 80 per cent have been males. A refresher course of what to look for in females, especially 1st-winter birds, and close scrutiny of Yellowhammer *E. citrinella* flocks at inland localities would make useful pursuits for self-finders intent on clarifying the status of the species in Britain (and elsewhere).

An area of hybridisation with Yellowhammer, with which Pine Bunting has sometimes been considered conspecific, occurs in western Siberia. Hybrids exhibit a variety of characters of both parent species and are not infrequent in Europe. Some most closely resemble one of the parent species and are less obvious to separate (Byers *et al* 1995). Mixed pairings have also occurred in Europe, for example in Poland in summer 1994. Questionable males exhibiting the tricky plumage spectrum have been documented in Britain (for example, Lansdown and Charlton 1990). Anyone who spends time watching Yellowhammner flocks encounters oddly marked birds from time-to-time, with males often sporting small chestnut moustachial marks. Formerly any Pine Bunting showing traces of yellow in the pale fringes to the primaries was deemed to be a hybrid and unacceptable. However, the *BBRC* sensibly redefined the parameters for what are now considered to be acceptable individuals (Rogers *et al* 2004). The 1st-winter male accepted from Fair Isle in 2007 showed a faint trace of yellow in the primary fringes and small underwing-coverts (Hudson *et al* 2008).

Carefully weighing up the available options, BBRC has decided to accept those Pine Buntings which show yellow on the primary fringes only, provided that the head pattern and other characters show no clear evidence of hybridisation. Equally, we will accept those birds where the primary colour was not seen, as long as the head pattern was described in sufficient detail to exclude any other hybrid features.

The key head-pattern features required to separate pure Pine Buntings from the majority of hybrid males includes:

- The presence of chestnut (not grey or black) on the lores;
- The presence of extensive chestnut on the throat region – the whole throat and malar area must be virtually concolorous, and there should be no prominent dark malar line or pale submoustachial stripe;
- The absence of a white supercilium - the supercilium should be chestnut or, at most, grey;

The presence of any yellow on the head (or any other part of the bird, other than along the primary fringes) confirms hybrid origin.

Discussion: Pine Bunting was first recorded in Crimea on 13th November 1987, but it is now regular in winter. It is also a regular migrant in recent years in the northwest Black Sea area. These developments are thought to be linked to a westward expansion of the breeding range (Tsvelykh *et al* 1997). The Urals steppe zone has been suggested as the peripheral part of the Pine Bunting's passage area and there have been a few cases of breeding west of the Urals (Morozov & Kornev 2000; Ryabitsev 2008).

The records, in terms of distribution and timing, point towards birds arriving via the Northern Isles and perhaps the east coast during the autumn and filtering through to inland areas, often farther south and southwest. Midwinter appearances within bunting flocks appear to be the norm and are often well inland. Birds exiting the country via the east coast present themselves during the early spring period, as do birds moving through the southwest.

Elsewhere in most of Europe Pine Bunting is a scarce vagrant, predominantly in autumn. More than 40 have been recorded in Scandinavia; 22 in Sweden, 12 from Norway, 10 in Finland and just two from Denmark; one has reached Iceland. There have been about 15 in Germany, 34 in the Netherlands and c20 in Belgium.

Small numbers have been discovered wintering in the eastern Mediterranean. Flocks have been recorded wintering in the Camargue, France, and Italy, where it is annual in varying numbers (a flock of 50+ was recorded in 1995/96). It is also a scarce, but regular winter visitor to northern and central Israel and the second record for Syria comprised a flock of seven in March 2008 at the same site as the first a year earlier, possibly a regular wintering site. There are also 10 records from Turkey and four records from Greece, plus three recent records from Cyprus, including a flock of 12 birds in late December. There is a small number of records from Croatia and the first two records for Romania occurred in April 2008 and two recent records from Hungary constitute the first national sightings. Five autumn and winter records from northeast Spain, all associated with Yellowhammer flocks, indicate that the species may be overlooked there (Ricard Gutiérrez *pers comm*).

It was suggested by Gilroy and Lees (2003) that Pine Bunting is best treated as a pseudo-vagrant, or very rare passage migrant. Perhaps the birds passing through Britain are not only on their way to wintering grounds here but also to locations much farther south.

Rock Bunting *Emberiza cia*

Linnaeus, 1766. Largely sedentary through Mediterranean basin from North Africa and Iberia north to central France, southern Germany and east through Turkey, Caucasus region and Iran to western Himalayas. Some populations dispersive or short-distance migrants.

Polytypic, six subspecies. Race undetermined but likely to be *cia* Linnaeus, 1766 of central and southern Europe, North Africa east to Tunisia, north-west and western Asia Minor and Levant (*BWP*).

Very rare vagrant from central or southern Europe.

Status: Four records of five birds.

1902 **West Sussex:** Near Shoreham-by-Sea two caught at the end of October
1905 **Kent:** Perry Woods, near Faversham caught alive on 14th February
1965 **East Yorkshire:** Spurn 19th February-10th March
1967 **Caernarfonshire:** Bardsey male 1st June

Of the four, only the date of Bardsey record accords with extralimital records elsewhere in Europe. The elusive Spurn bird remains the most tantalising example of a species that has not put in an appearance for over 40 years.

At 1240hrs on 19th February, R.G.Pearson and J.R.Preston started to walk south along the sea wall (it is now collapsed and 100yds out on the sea beach) south of Sandy Beaches Caravan Park, when they flushed a passerine from an area of rubble. It had the appearance of a largish bunting. It only flew c. 20yds before alighting and started to forage amongst some rubble. It allowed the observers to approach to within c. 15yds. before repeating its actions, doing this all along the area behind the sea wall until it reached Big Hedge (there were still anti tank blocks beside Big Hedge at that time). At this point it flew up and disappeared towards the Crown and Anchor. During this period, both observers had reasonably good views, enough to identify the bird as a Rock Bunting (European, not African!!!).

Later R.G.P., J.R.P. and B.R.Spence, went back to where it had first been seen, but there was no sign of it. The three observers then walked along the road to the Crown and Anchor and along the Humber Bank as far as Sammy's, then back and south along the tide wrack on the Humber shore opposite the Canal Zone. After c. 50 yds walking along the tide wrack opposite the Canal Zone, a bird was flushed giving a Yellowhammer type call. It landed further along giving B.R.S. some brief views as it fed along the tide wrack. However, it soon flew off again, apparently into the triangle and could not be found again that day. It was subsequently seen by quite a number of observers, mainly south of Cliff Farm plus in the original area where it was first found, both on 20th and 22nd February. Then on 27th February, it was seen briefly at Sammy's before it flew off and appeared to drop into the stack yard of the first farm north of the Crown and Anchor. The next sighting was on the 4th March, when it was seen, albeit briefly, in the area where it was first

found and next day, again briefly, along the tide wrack opposite the Canal Zone. It was last seen on 10th March in its original area, though yet again only briefly. Although traps were put out for it on at least two occasions, it was never caught, contrary to some published accounts.
Barry Spence. *Spurn Bird Observatory website (www.spurnbirdobservatory.co.uk)*

Discussion: Rock Bunting has decreased recently at the northwest limit of its range. Breeding densities vary greatly, being very low to the north of the Alps due to a lack of suitable habitat. Most populations are sedentary or dispersive, but it is a partial migrant or migrant over short or medium distances from those parts of the range with the coldest winters. It is a rare but regular passage migrant at Swiss alpine passes from October to November (Hagemeijer and Blair 1997).

A number of putative records have been found to be *E. tahapisi*, known variously as African Rock Bunting or Cinnamon-breasted Bunting, a favourite cage bird.

Elsewhere in Europe, away from breeding areas this remains a rare vagrant. Extralimital records in the last decade or so to northern Europe include singles in Belgium (2nd-21st April 1997), Denmark (31st May-1st June 2003), the Netherlands (30th May 2004) and on Helgoland, Germany (11th May 2006). In addition there have been four males from Sweden (21st June 1972, 5th June 1973, 6th April 1978, 7th-11th May 1997).

The north European records give hope for observers of another here. No doubt if one does grace a mainland site, it will prove exceptionally popular.

Cretzschmar's Bunting Emberiza caesia
Cretzschmar, 1828. Breeding range restricted to southeast Europe and the Middle East. Breeds Greece, Albania, western and southern Turkey, Syria, Lebanon, Israel and Jordan. Winters Sudan and Eritrea.

Monotypic.

Status: Four records, one of which from 2008 is yet to be assessed.

1967 Shetland: Fair Isle 1st-summer male 10th and 14th-20th June, trapped 14th
1979 Shetland: Fair Isle male 9th-10th June
1998 Orkney: Stronsay male 14th-18th May
2008 Orkney: North Ronaldsay male 19th-21st September*

The species' penchant for Scottish offshore islands is clearly demonstrated by the quartet of records.
Autumn 2008 on North Ronaldsay had already got off to a cracking start with multiple Two-barred Crossbills, Pacific Golden Plovers, a Citrine Wagtail and Greenish Warbler all in August. September meanwhile had, up to mid month, been relatively quiet for rarities but had been remarkable for the largest fall of trans-Saharan drift migrants in over

15 years; predominantly Common Redstarts, Pied Flycatchers, Whinchats, Northern Wheatears, Garden and Willow Warblers. With just six (semi-) resident birders covering the island (and no visitors) we were never going to find as many rarities as were found on Fair Isle, not to mention the latter's other advantages of greater isolation, sparser cover and smaller size, but by mid month we were a little down-hearted that we hadn't had anything to set the grapevine alight. Sitting in the pub on the night of the 17/18th September Pete Donnelly allayed my fears with the comment that the 'biggie' always comes way after the fall, but noted that he was too busy with work commitments to have a chance at finding it himself.

Two days later Pete arrived home mid afternoon from checking the lobster creels and flushed a bird off the track in front of his house. He jumped out of the car to see a bunting with a blue head, but lost it in turning around to get his camera. The rest of the obs crew were summoned with a report of 'probable Cretzschmar's Bunting, Grey-necked not eliminated' – enough to get anyone moving in a hurry. After a brief search the bird was relocated and duly posed on the wall - a stunning adult male type Cretzschmar's Bunting, with Fair Isle in the backdrop! The relief was palpable, if a little galling that the rest of us had been thrashing around day after day with only scarcities to our name and Pete meanwhile had just rolled up at his house and found a mega that was about to cause as much a stir as a 1st for Britain.

There had been three previous British records of Cretzschmar's Bunting, all clustered within a 35 km radius – the first two on Fair Isle in June 1967 and June 1979, followed by one on Stronsay, Orkney, in May 1998. This was therefore the first autumn record for the UK, although there have been other extralimital records in Western Europe at this season, including one in the Netherlands on 11th October 1859 and one in Finland on the 30th September 1990. With the bird returning to seed outside Pete's house at Sangar we set about releasing the news and making arrangements for the ensuing twitch.

We refound the bird at first light the following morning, and by 9.15 the first charter flights were arriving. The bird showed on and off for two days, and everyone that arrived over the weekend managed to see it - a total of about 80 birders. The bird was last seen going to roost on the Sunday night after which there was a significant clear-out of migrants, including the Cretz, which could not be found in any of its favourite haunts.

Alexander C. Lees (www.nrbo.co.uk)

Discussion: Northward migration begins in February and peaks in the eastern Mediterranean in March and April, with birds back on eastern Mediterranean breeding grounds from March (*BWP*). The spring records are relatively late and presumably represent dispersing failed breeders overshooting to northwest Europe. The absence of females and immatures may be because they are overlooked as Ortolan Buntings *E. hortulana*.

Elsewhere in northwest Europe there were six spring records from France in the 19th century (and one in 1917). A similar pattern of records exists for Helgoland, Germany, over a similar period (between 1848 and 1879 about 10, mainly in May). There have been two records from the Netherlands (11th October 1859 and 7th-11th May 1994), one from Sweden (29th-30th May

1967) and four from Finland (19th May 1981, 30th September 1990, 8th-10th May 2007, 14th May 2007), plus one from Austria (1st May 1995).

The recent bird on North Ronaldsay was only the second autumn vagrant for northwest Europe during recent times. The opportunity to see such a rarity was gratefully received by a good number of birders. With a breeding range restricted to southeast Europe and the Middle East, records are always likely to be at a premium, as with other species from the same region.

Yellow-browed Bunting *Emberiza chrysophrys*

Pallas, 1776. Breeding range uncertain, but from central Siberia south to Lake Baikal and east to the Vilyuy River and Yakutsk. Winters in central and southeastern China.

Monotypic.

Very rare vagrant from Siberia.

Status: Five records.

1975 Norfolk: Holkham Meals female or immature 19th October
1980 Shetland: Fair Isle male, age uncertain, 12th-23rd October, trapped 12th
1992 Orkney: North Ronaldsay 22nd-23rd September
1994 Scilly: St. Agnes male 19th-22nd October
1998 Orkney: Hoy 4th-5th May

Discussion: This species has a small range in central Siberia. The limits are poorly known, but extend west to the Middle Yenisey, where it is fairly common (Ryabitsev 2008). Southward migration commences in August or September, with the main passage from September into November, according well with the four British autumn records; the spring bird had presumably overwintered somewhere in western Europe and was heading north. The main spring passage through northeast China is during late April and May (*BWP*). The Fair Isle bird of 1980 was originally accepted as the first British record, but the belatedly accepted Norfolk individual pre-dated it.

The first record for the Western Palearctic was a 1st-winter male captured in northern France in autumn 1827. There are just three others away from Britain: a male captured in Belgium (20th October 1966), a 1st-winter male trapped in the Netherlands (19th October 1982) and a male in Sweden (3rd January-10th February and 23rd-27th February 2009). In Germany a very confiding male (18th April 2004) in a garden in Mecklenburg was considered to have been a probable escape and placed in category E. The quintet of British records remains surprising in light of its extreme rarity elsewhere in Europe.

Although the full extent of the breeding range is uncertain, this attractive bunting is also most certainly one of the most easterly distributed passerines to be accepted on category A of the *British List*. A stunning bird and a cracking find, this must rank highly on any birder's self-find wish list. Not surprisingly the Scilly bird of 1994 was hugely popular. Three in the space of seven years in the 1990s hinted at an impending descent in status, perhaps indicating some distant change in range. The subsequent dearth of records ensures that it remains a rarity of the greatest calibre.

Chestnut-eared Bunting *Emberiza fucata*

Pallas, 1776. Nominate fucata breeds from Lake Baikal to the Amur River, north-east China and Japan; winters in northern Thailand, southern China, southern Korea and southern Japan. Form kuatunensis occurs in south-east China and winters in the southern part of the range and from Bangladesh to northern Indo-China; arcuata is found from the Himalayas to Yunnan and is an altitudinal migrant.

Polytypic, three subspecies. The Fair Isle bird was nominate *fucata* Pallas, 1776 of the northern part of the range.

Very rare vagrant from Siberia.

Status: One record.
2004 Shetland: Fair Isle 1st-winter male 15th-20th October

Discussion: This is the only European record of a seemingly unlikely vagrant that few would have predicted, even on a sweepstake, given its restricted far eastern distribution. It is among the most distant passerines to reach Britain. However, it was difficult to argue with the supporting cast for this individual: record numbers of White's Thrush *Zoothera dauma*, an Eastern Crowned Warbler *Phylloscopus coronatus* in Finland and, just one week later, the first Western Palearctic Rufous-tailed Robin *Luscinia sibilans* during an amazing purple patch for the most famous island of rarities.

Biometrics and plumage characters established it as nominate *fucata* (the migratory subspecies). This, its 1st-winter age and the species' rarity in captivity (and a ban on imported birds from the Far East at the time) makes it reasonable to assume that the bird was a genuine vagrant. Yet the question remains: had it not had such strong supporting credentials, would it have made category A?

Yellow-breasted Bunting *Emberiza aureola*

Pallas, 1773. European range restricted to small and declining population in central Finland, centred on Gulf of Bothnia. Farther east breeds widely across Russia and Siberia to Kamchatka, south to northeast China and northeast Hokkaido, Japan. Winters locally from eastern Nepal through Himalayan foothills to northeast India and widely in southeast Asia.

Polytypic, two subspecies. All records have related to nominate *aureola Pallas, 1773.*

Once scarce, now increasingly rare migrant during early autumn; exceptionally rare in spring.

Status: There has been a total of 236 records to the end of 2007, four of which were in Ireland; eight were recorded prior to 1950.

Historical review: The first was a 1st-winter female obtained at Cley on 21st September 1905. Two September records followed in 1907, from Norfolk (immature female shot at Wells on 5th September) and Fair Isle (shot on 28th September). Another reached Fair Isle in 1909, followed by one on the Outer Hebrides in 1910 and then another from Norfolk in 1913. Just two more were added prior to the present recording period (Isle of May in 1936 and Fair Isle in 1946), bringing to eight the total number of accepted records prior to 1950.

Four were added during the 1950s. The first of these was an adult male shot on Fair Isle on 13th July 1951, which was the first to be recorded outside of September; the carcass was presented to *FIBO* staff after no one came to check the sighting (Pennington *et al* 2004). Consecutive September records between 1957 and 1959 included singles on the Isle of May and Fair Isle, plus the first for Ireland from Tory Island, Co. Donegal, on 18th September 1959.

The 1960s added a further 18 records, all of them on Scottish islands. There were 13 from Shetland (10 on Fair Isle), four from the Isle of May and one from Orkney. Annual totals of three in 1963 and 1967 included two together on Fair Isle on 10th September 1967, but these totals were bettered by four Shetland birds in 1969.

Increased observer numbers and coverage, accumulated 55 more during the 1970s, including the only Welsh record (Bardsey 4th-5th September 1973). Four graced Fair Isle in September 1971 and numbers soared to a since unequalled 11 in 1974. This bumper crop comprised four on Out Skerries between 12th and 14th September, four on Fair Isle, two from the Isle of May and the first English record for over 60 years on St. Mary's from 25th-27th September. Just three were recorded the following year, but they included the first spring record with a male at Spurn from 14th-15th June 1975. Sightings in the rest of the decade were dominated by Shetland. Ten in 1977 remains the second-highest annual total and included another east coast spring male, this time at Gibraltar Point on 15th May 1977.

There was a slight increase in the number of birds recorded during the 1980s, with 62 logged. Again most were on Shetland, especially Fair Isle. Nine were recorded in 1980 and 1983. Singletons in Somerset in 1980, Argyll in 1981, Devon in 1982 and 1989 and on Cape Clear in 1983 and 1985 confirmed that detection farther west and southwest was still possible.

During the 1990s, the number of birds recorded held its own at 59. However, in terms of the increasing attention being paid to Shetland and along the east coast, the total surely represented

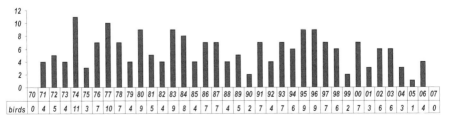

Figure 104: Records of Yellow-breasted Buntings, 1970-2007.

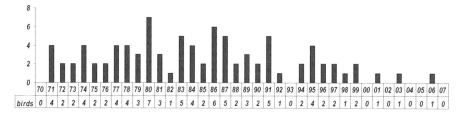

Figure 105: Records of Yellow-breasted Buntings on Fair Isle, 1970-2007.

a decline in real terms. Some high scores were still noted, especially 1995 and 1996 with nine records apiece, and three were present together on Fair Isle on 12th September 1991.

A further 30 birds have been added since 2000, including seven in 2000 and six in both 2002 and 2003, but an overall trend to falling numbers since 1995 and 1996 became evident. In 2007 no birds were recorded for the first time since 1970 and only the second time since 1960.

The Northern Isles hold on Yellow-breasted Bunting sightings has also weakened significantly. The trend for finds to be made away from there is illustrated by the increasing proportion of records found at locations to the south. During the 1980s, just 18 per cent of sightings were found outside the islands; during the 1990s and since 2000 this quotient rose to 34 per cent and then 40 per cent.

Since 1970 there has been a national average of 5.4 birds per annum, dropping slightly to 4.9 since 1990. On the face of it, this does not appear to equate to a large decline, but the numbers are surely skewed by increasing observer effort away from the usual locations for the species. An analysis of records for Fair Isle (Figure 105), which has received constant coverage over the period, reveals the dramatic extent of the decline for this popular bunting. This portrays a sad state of affairs for one of the specialty birds of the island; no longer is its autumn presence there guaranteed.

On a decade by decade basis, numbers on Fair Isle peaked during the 1970s and 1980s, and have since plummeted. The decline presumably represents the break up of the westerly range and the reduced populations of the nearest communities. An additional factor might be a signif-icant reduction in standing oat crops, which are the favoured habitat of the species on Shetland, affording good recovery niches and allowing easy searching. On Stronsay, 64 km southwest of Fair Isle, oats were planted in the hope of attracting this, its logo species, to the newly created bird reserve. The tactic worked within five years (Holloway 1997).

Table 25: Average number of birds per annum by decade on Fair Isle

Period	1960-1969	1970-1979	1980-1989	1990-1999	2000-2007
Average per annum	1.0	2.7	3.8	2.1	0.4

Where: Shetland has produced 157 birds (66 per cent), concentrated mostly on Fair Isle with 104 birds (44 per cent of all records). Just over a quarter of sightings have now been found away from the Northern Isles, but a strong island link is evident still: 13 birds on the Isle of May, nine from Orkney, 10 on the Farne Islands, six on Scilly, two on St. Kilda, two on Lundy, one on Tiree (Argyll), one (the only Welsh record) on Bardsey and one on Hilbre (Cheshire), as well as three on the Isle of Portland (Dorset). The four Irish records have all been on islands (two on Cape Clear and two on Tory Island) and one has been found at sea (Sea area Humber on Loggs Gas Platform).

Excluding Portland, just 26 have been found at mainland sites, all strictly coastal except in Wester Ross (Highland) where a male sang in June 1982. Other mainland records have come from Yorkshire (eight, including five at Spurn), Norfolk (six) and Fife (three), with singles in six other east-coast English counties (Northumberland, Co. Durham, Lincolnshire, Suffolk, Essex and Kent). In the southwest, birds have occurred in Somerset and Devon.

When: There have been four males in spring and two in midsummer. Two were in both May (Gibraltar Point 15th May 1977 and Sumburgh 20th-21st May 1996) and June (Wester Ross, Highland, in song, 14th June 1982 and Spurn 14th-15th June 1975). Two midsummer birds were both on Fair Isle in early July, a female on 4th July 1980 and a male shot on 13th July 1951.

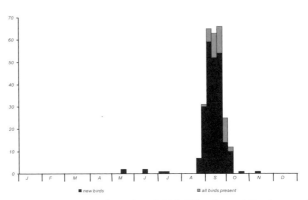

Figure 106: Timing of British and Irish Yellow-breasted Bunting records, 1905-2007.

The earliest autumn record was one on the Farne Islands from 20th-23rd August 1996 and the next earliest on Fetlar, Shetland, on 22nd August 1997. A male trapped at Landguard on 12th August 1999 was initially accepted before being considered to be an escape. It had a missing toe and claws, plus worn outer primary tips, with one snapped off part way down its length. (Odin et al 1999).

September is the main month for sightings, with 70 per cent of all records in the first three weeks of the month. Sightings diminish rapidly during the last week of September, with just 14 seen. Early October records are at a premium: just 10 seen during the first week, the latest on the Isle of May on 7th October 1961. Two late records involve one at Fife Ness on 18th October 1997 and an exceptionally late bird at Bolberry Down, Devon, on 7th November 1993.

Discussion: Yellow-breasted Bunting leaves its breeding grounds early, with departure commencing at the beginning of July and involving firstly adults and then juveniles. Two of the

429

Fair Isle records involve early July birds and these presumably relate to early dispersing adults. The peak of records from late August and into mid-September corresponds with the main departure from west Siberian breeding grounds. Western birds head east from breeding grounds then south, and eastern birds south or south-west, to reach winter quarters via Mongolia, south-east Russia and China (*BWP*).

Yellow-breasted Bunting has extended its range west since the 19th century, reaching as far west as Lake Onega by 1875 and eastern Karelia and Lake Ladoga by 1900. Southeast Finland was invaded by 1925. Around 100 pairs were believed to be nesting by 1958 and up to 200 pairs in the late 1980s. Numbers have declined rapidly since then to the point where there was an average of eight observations there in 2000-2002, three in 2003 and only two birds recorded in Finland since 2004 (Aleksi Lehikoinen *pers comm*). Climatic amelioration from 1850 onwards is believed to be responsible for the range extension (Burton, 1995) and the westward growth was clearly associated with the upsurge in records here. However, the stronghold population in Russia declined markedly, by 20-29 per cent, during 1990-2000 and the species underwent a moderate decline of more than 10 per cent overall (BirdLife International 2004). The decline correlates with the smaller numbers now arriving here.

Formerly one of the most abundant breeding passerines across vast swathes of Siberia this species has been uplisted to vulnerable by BirdLife International, as evidence suggests that it has undergone a very rapid population decline. This is thought to result from excessive trapping at migration and, in particular, wintering sites. Roosting flocks in reed beds are disturbed and then caught in mist nets to be cooked and sold as "sparrows" or "rice-birds". This practice was formerly restricted to a small area of southern China, but has become more widespread. Over a million individuals are killed annually to be sold as snacks. Thousands of males are stuffed and sold as mascots since their presence in Chinese homes is thought to confer happiness. Agricultural intensification, the shift to irrigated rice production and consequent loss of winter stubble has reduced the quality and quantity of wintering habitat, and the loss of reed beds has reduced the number of available roost sites (BirdLife International 2008).

Explaining the northerly bias in the British distribution of this species is not easy. The dominance of Shetland has been used to support the reverse migration theory for this species (Vinicombe and Cottridge 1996) and the preponderance of sightings there is associated with easterly airflows off anticyclonic conditions from Scandinavia eastwards. There would appear to be little significant evidence of the trickle down of records in Britain and Ireland evident in the distribution patterns of many other eastern vagrants, which could be used to lend credence to the application of the reverse migration theory in this instance. However, it may be that much of the bias in sightings is due to habitat factors. The trend of increased southern detections in recent years may simply reflect increased observer coverage.

Elsewhere in Scandinavia, this species appears to be oddly quite rare. There are only 39 records from Norway, 33 records from Sweden (only one since 1996) and just six from Denmark (all between 1984-1996), yet more than 25 records have come from Germany. One has reached Iceland (22nd September 1958). These areas would presumably lie within the vagrancy shadow proposed by Vinicombe and Cottridge (1996).

It is apparent that dispersal into Europe takes place on a broader front than has been suggested in some of the literature. For example, in The Netherlands there are 13 records (two since 1996: 11th September 2001 and 12th September 2003) and there have been four from Belgium (three between 1991 and 2003). Records are spread across Europe, with a number from southern countries. There have been 23 in Italy, four in Greece (including one in January 1956 and another in May 1961), two from Spain (7th November 1987, 4th-11th October 1995) and singles from Portugal (14th September 1996) and Cyprus (May 1974).

Farther east there are 11 records from Israel in August-November (most recently August 2007) and four records from the United Arab Emirates (singles in October and November plus two in December). This spread of records tends to refute the idea that reverse migration is the sole vector for the species to reach Western Europe.

The number of spring records in Britain is very low, with just four in May-June and two in July on Fair Isle. A number of eastern vagrants are increasingly regular in spring, implying the use of new wintering areas in Europe or Africa; these species are popularly being referred to as pseudo-vagrants (Gilroy and Lees 2003). Despite the geographical spread of European sightings, there seems to be no evidence that Yellow-breasted Bunting is such a species. All spring and summer records in Britain came at a time when population levels in northeastern Europe were greater. Such records presumably referred to late spring overshoots or, in the case of the summer records, early departing adults. Population declines and a reduction in habitat quality on the breeding grounds in parts of its range (BirdLife International 2008) look likely to ensure that this is one species that we can expect to become much rarer. The halcyon days of the 1970s and 1980s when an early autumn visit to Fair Isle was all that was needed to bag the species may soon be a distant memory.

Pallas's Reed Bunting *Emberiza pallasi*

Ian Wallace

(Cabanis, 1851). Birds of the race polaris breed from northeast European Russia across Siberia to the Sea of Okhotsk and Chutotskiy Peninsula. Two other races breed from the Tien Shan mountains to Lake Baikal and Mongolia. Winters in China and Mongolia.

Polytypic, three subspecies. Race undetermined, but both Fair Isle birds were considered to show characters of the highly migratory *polaris* Middendorff, 1851, the race to be expected as a vagrant in western Europe (Pennington *et al* 2004).

Status: Three records.

1976 **Shetland:** Fair Isle adult female 29th September-11th October, trapped 10th
1981 **Shetland:** Fair Isle 1st-winter 17th-18th September, trapped 17th
1990 **East Sussex:** Icklesham 1st-winter male trapped 17th October

Discussion: This species is expanding its range west. The westernmost extension of the range is the Bol'shezemel'skaya tundra in extreme northeast European Russia. The range limits have not been determined and fluctuate strongly (Ryabitsev 2008).

It seems likely that this enigmatic species is overlooked here. That two of the three records are from Fair Isle is not surprising. What is truly telling is that the Sussex bird was trapped and yet not identified until a revelation two years later when the ringers read an identification paper on Black-faced Bunting *E. spodocephala* and Pallas's Reed Bunting (Bradshaw 1992).

Elsewhere in the Western Palearctic there are singles from Portugal (1st-winter female on three dates from 12th January-4th February 1997) and Italy (female, 13th January 1997). A record from Spain (10th April 1996) is no longer acceptable, as an eastern race Common Reed Bunting *E. schoeniclus* could not be eliminated. Two claims from Norway in 1996 were not accepted. The winter records from Iberia and Italy provide another hint that birds are missed elsewhere. It is even possible that small numbers winter in southern Europe.

The dearth of records from Europe may well reflect the difficulty of identification from Common Reed Bunting. All three British records were trapped at some point during their stay, yet even they were subject to protracted identification processes.

Black-headed Bunting *Emberiza melanocephala*
Scopoli: 1769. Breeds from central Italy east to Greece, Turkey, northern Iraq and western Iran, and north through Caucasus to Ukraine and southern Russia. Winters western and central India.

Monotypic.

Scarce spring migrant from southeast Europe or southwest Asia, much rarer in autumn.

Status: There have been 194 records to the end of 2007, 185 in Britain and nine in Ireland. Of these, six were prior to 1950 and 188 since.

Historical review: The first was an adult female shot at Brighton Racecourse, East Sussex, around 3rd November 1868. The five further records prior to 1950 were from Nottinghamshire (male caught June or July 1884), Fair Isle (21st September 1907, 25th August 1910 and 27th May 1929, all shot) and the Isle of May (22nd September 1949).

Seven were recorded in the 1950s, all bar one on offshore islands (including the first for Ireland, a male on Great Saltee Island, Co. Wexford, on 31st May 1950). Two appeared in early October 1951 (Devon and Shetland), singles were recorded in 1954 (Shetland) and 1957 (Lundy), and two more were logged in 1958 (Scilly and Co. Wexford).

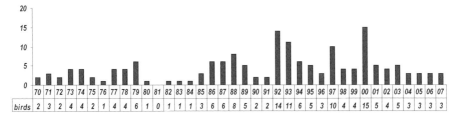

	70	71	72	73	74	75	76	77	78	79	80	81	82	83	84	85	86	87	88	89	90	91	92	93	94	95	96	97	98	99	00	01	02	03	04	05	06	07
birds	2	3	2	4	4	2	1	4	4	6	1	0	1	1	1	3	6	6	8	5	2	2	14	11	6	5	3	10	4	4	15	5	4	5	3	3	3	3

Figure 107: Annual numbers of Black-headed Buntings, 1970 to 2007.

A slight increase was evident through the 1960s, with 14 birds recorded, again mostly on islands. Three were on Bardsey (a male from 27th-29th May 1963 was the first for Wales), two on Skokholm, four in Shetland, two on Orkney and one on Islay, Argyll. Mainland records came from Nottinghamshire and Leicestershire.

The 1970s yielded 32 birds, just over a third of them at mainland sites including five on Portland. Offshore islands continued to take the majority share, including 10 from Shetland. There was just one inland bird, a tame male in Nottinghamshire that was suspected of being an escape. Four were recorded in each of four years; six, including three during 7th-13th May, delivered a then record tally in 1979.

The 1980s also saw 32 recorded. None were seen in 1981 (only the second blank return since 1961). Most of the seven seen between 1980 and 1985 were inaccessible to the masses, and an air of rarity returned to the species for a generation of observers. However, an influx of six in 1986 included an accessible male in Essex. Another six appeared in 1987 and then a record eight in 1988, six of them in Scotland and two in Wales.

Records increased dramatically during the 1990s. The decade produced 61 birds (almost 40 per cent of all records). The first double-figure tally was achieved through a bumper influx of 14 in 1992, six of which arrived between 4th and 14th June, five of them on Fair Isle (including a male and female on the 10th). A further influx in 1993 comprised eight males between 24th May and 2nd June – seven of them along the south coast between Cornwall and Kent and another inland in Cambridgeshire – and three stragglers to 1st July.

A total of 42 during 2000-2007 contained a record annual total of 15 in 2000. Records were dispersed widely between late May and August and lacked a discernable focus of arrival.

There is no doubt that increased numbers are being detected, even allowing for the odd escapee that finds its way into the totals. The last blank year was as long ago as 1981 and the average since 1970 is just over four per annum, increasing since 1990 to nearly six per annum. Signs of a recent downturn are hopefully just a blip, but could be indicative of problems on the breeding areas due to the increasing application of intensive agricultural practices (Hagemeijer and Blair 1997).

Where: The geographical spread of records is interesting. Shetland accounts for just under a quarter of all records with occurrences spread evenly between spring and autumn. Nationally, over half of all autumn records have come from Scotland, particularly Shetland.

The propensity of birds to overshoot west and northwest in spring is illustrated in Table 26. Nearly two-fifths of all spring records come from westerly locations, 27 per cent from the south coast and 21 per cent from the Northern Isles. Eighty five per cent of west coast records have been in spring. Birds overshooting their breeding grounds on a northwesterly orientation presumably realise the error of their judgment and make landfall at west coast locations.

In the autumn, the Northern Isles dominate records and account for half of all sightings. Its rarity on the east coast at all seasons is illustrated by the fact that just one has been found in the well-watched county of Norfolk and three in Yorkshire.

Table 26: Distribution of Black-headed Buntings by region, by season, 1950-2007.

Region	Spring	Autumn*
Northern Isles	21%	50%
English east coast (London to NE Scotland)	8%	12%
Inland counties	4%	0%
South coast (Scilly to Kent)	27%	21%
West coast (northwest, Wales, Ireland and west Scotland)	39%	17%
Total number of birds	**135**	**52**

* Autumn is defined as 1st August onwards.

When: Four April males constitute the earliest arrivals. The two earliest were forerunners of an earlier than usual arrival in 1979 (St. Agnes 15th to at least 26th and Cley 20th-3rd May). The two other April records were also noted during years of above average numbers, in 1992 and 1997. Small numbers have been recorded in early May, but the period from late May to

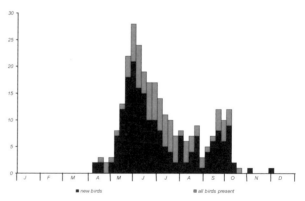

Figure 108: Timing of British and Irish Black-headed Bunting records, 1868 to 2007.

mid-June is the best time to find this species in Britain and Ireland; over a third of all records fall between 23rd May and 17th June.

Arrivals continue through the summer months. Slight peaks in records are evident in early August and again at the end of August. Autumn waifs are spread throughout, with the peak period between mid-September and 7th October. The only accepted October records after this date in recent times were a female or immature on St. Mary's on 14th October 1977 and an adult male at Loch of Strathbeg, Northeast Scotland, from 13th-18th October 2005.

Two accepted November birds include a female shot at Brighton, East Sussex, about 3rd November 1868 and one at Ballymore, Co. Wexford, on 28th November 1958. Both records are well outside the expected arrival times for this species and the possibility is that they were escapes or frauds.

Of the 32 records after 1st September since 1970 the age composition is: 34 per cent males; 19 per cent females; 6 per cent females or 1st-winters; and 41 per cent 1st-winters. The males have been in a broad geographical spread of largely mainland locations; all bar two of the 1st-winters and females have been on offshore islands, with the only recent record for the mainland from Gunton, Suffolk, from 24th-25th September 2002. Of those aged as 1st-winters, seven have been from the Northern Isles and two from the Isles of Scilly; others have come from Skomer, Bardsey, Tory Island and Suffolk.

Discussion: Not surprisingly most records have involved the eye-catching males. The problem of separating 1st-winter Red-headed *E. bruniceps* and Black-headed Buntings in autumn has always been problematic (for example, Vinicombe 2007). It is likely that many birds in such plumage elude detection, especially away from offshore hotspots.

Return to southeast European breeding areas takes place in April and early May; it is one of latest migrant breeders within the area. Birds in northwest Europe are often assumed to be spring overshoots, and for a proportion that would appear to be so. Summer peaks are less easy to explain, but presumably involve the arrival of dispersing failed breeders and non-breeding birds.

Departure from breeding grounds takes place from late July to August, again correlating with a slight peak of records here. Birds later in the autumn, mostly 1st-winters, are presumably dispersing migrants that, instead of moving south-east or east-southeast to their wintering grounds, take a northwesterly trajectory. Such a heading would account for the high proportion of birds on the Northern Isles (over 50 per cent of all autumn sightings).

Rose-breasted Grosbeak Pheucticus ludovicianus
(Linnaeus, 1766). Breeds central Canada to Nova Scotia and through mid-west and northeast USA to Maryland. Migrates through eastern states to winter from central Mexico through Central America to northern South America.

Monotypic.

Rare vagrant from North America in autumn; has also occurred in winter and spring.

Status: There have been 31 records, 23 in Britain and eight in Ireland.

At about mid-day on 6th October 1966, J. R. H. Clements found a strikingly unfamiliar bird in one of the small fields below the parsonage on St. Agnes, Isles of Scilly. In brief views before it flew off he noted that it was a large, bunting-like bird with a dark brown head relieved by prominent whitish supercilia and a creamy central crown-stripe.

Later, at about 2.30 p.m., E. Griffiths, D. J. Holman, P. A. Dukes and I, searching near the spot where it was first seen, flushed the bird, which then flew over to my side of a hedge and began feeding amongst the ground cover of weeds about twenty yards away. I was able to get the first good views which, after later reference to Roger Tory Peterson's A Field Guide to the Birds (1947), were sufficient to identify it as a female Rose-breasted Grosbeak Pheucticus ludovicianus. After another lengthy disappearance it was eventually seen satisfactorily by all observers on the island during the afternoon, including Mr. and Mrs. K. Allsopp, Miss J. M. Glibbery and Mr. and Mrs. N. J. Westwood. At about 6.15 p.m. N.J.W. drove the bird into a mist-net set in a field near where it was first found. As it was nearly dark, we kept it overnight at the observatory after weighing it. A full description was taken in the hand the next morning and the bird was released at 6.45 a.m. It remained in the area below the parsonage, and could be seen there regularly until 10th October. It was last recorded on the 11th, in tamarisks in Covean on the other side of the island.

Peter Grant. British Birds 61:176-180

Historical review: The first was a 1st-winter male on Cape Clear from 7th-8th October 1962. Others followed in 1966 and 1967. These included the first British record (St. Agnes 6th-11th October 1966) and the first Welsh record (female trapped on Skokholm 5th October 1967).

The next was a real surprise: a wintering 1st-winter male at Leigh-on-Sea, Essex, from 20th December 1975 to 4th January 1976. Two on different Scilly islands in October 1976 provided the first multiple record. Two more in 1979 made landfall a day apart in mid-October on Cape Clear and Scilly.

A bumper showing of six 1st-winter males between 7th and 19th October 1983 kicked off the 1980s in style. The first of these was on the Outer Hebrides on 7th, followed by one on Cape Clear on 9th. Two on the 10th included singles from Scilly and another in Co. Cork. One was trapped on Bardsey on 14th. The last of the autumn was a third from Co. Cork on the 19th.

Three more in 1985 included early October birds in Co. Wexford and Scilly plus a late October bird on Lundy. Singles were found on Scilly in 1986 and 1987, the latter year also

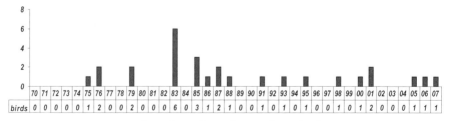

Figure 109: Annual numbers of Rose-breasted Grosbeaks, 1970 to 2007.

delivering another to Co. Cork. One on Skomer in 1988 was the last of a whopping 13 birds for the decade.

As for many Nearctic landbirds, the pace slackened in the 1990s with just four recorded. The first of these was a rare mainland sighting, again on the east coast, in East Yorkshire in early November 1991. Scilly accommodated singletons in 1993 and 1998. One from the Isle of Wight at Ventnor from 30th October to 1st November 1995 constitutes the only south coast record for the species.

Fortunes improved slightly between 2000 and 2007 with six records: one from Co. Kerry and three in the southwest, including two October birds in 2001. The second-ever Scottish record was found on Barra, Outer Hebrides, in 2005. The first spring record, a 1st-summer female, hit a window at Holme, Norfolk on 4th May 2006; it was taken into care and released next day.

Following the review period one was present at Kilbaha, Co. Clare, from 22nd-23rd October 2008.

Where: The distribution of records has a south-west bias, with 12 on the Isles of Scilly and six in Co. Cork. Two have reached Devon (both on Lundy). Three Welsh birds have been found on the islands of Bardsey, Skomer and Skokholm. East coast records include a wintering immature male at Leigh-on-Sea, Essex, from 20th December 1975-4th January 1976 and a similarly aged bird at Bridlington, East Yorkshire, from 5th-6th November 1991. Both Scottish records originate from the Outer Hebrides (1983 and 2005). There have been singles on the Isle of Wight and Co. Wexford.

The occurrence pattern accords well with unaided passage over the Atlantic. There is evidence from ringing studies that individuals wintering in Panama and northern South America are more likely to originate from eastern part of range (Wyatt and Francis 2002). It may be these longer-distance migrants are prone to displacement and supply the transatlantic waifs.

When: The only winter record was at Leigh-on-Sea, Essex, from December 1975 to January 1976. One spring record involved a 1st-summer female at Holme, Norfolk, from 4th-5th May 2006.

Most records are from late September to late October. The earliest both made landfall on 29th September (1st-winter female on Skomer in 1988 and female on Great Blasket Island, Co. Kerry, in 2000). Early October represents the peak period for

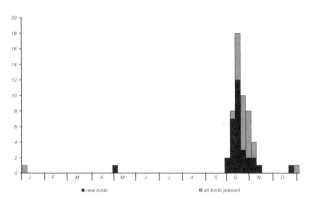

Figure 110: Timing of British and Irish Rose-breasted Grosbeak records, 1962 to 2007.

arrivals, with 60 per cent of all records arriving between 3rd and 14th October. There have been seven more through the rest of October, including two late birds present from 30th October-1st November (Ventnor in 1995 and Bryher in 1998). The sole November record was the individual in Bridlington, East Yorkshire, from 5th-6th.

Discussion: This medium- to long-distance migrant travels from its breeding grounds south or southeast through much of eastern North America, primarily from the eastern edge of the Great Plains to the Atlantic coast. Most individuals have left western Canada by early to mid September and departed eastern Canada and northeast USA by late September. Birds breeding in the northwestern parts of the range presumably migrate southeast initially before shifting southwards, thus avoiding the Great Plains. Most are thought to cross the Gulf of Mexico. Ringing data suggests that wintering birds in Panama and northern South America most likely originate from the eastern part of range; those wintering in Mexico and western Central America are more likely to come from western parts of breeding range (Wyatt and Francis 2002).

Elsewhere in the Western Palearctic, there is one from Iceland (20th-29th October 2001), two in Norway (13th-19th May 1977 and 1st October 1977) and one each from Sweden (10th October 1988) and Denmark (1st-winter male November-20th December 2008). Another from Sweden (10th-11th May 1992) and one from Germany (28th September 1993) were dismissed as escapes.

There are two from the Channel Islands (26th September 1975 and 10th-13th October 1987) and one each France (15th-22nd October 1985), Spain (17th October 1982) and Slovenia (29th October 1976), plus three from Malta (April 1977, October 1978 and October 1979). There have also been eight from the Azores since 1999 (20th October 1999, 23rd-31st October 2005, 29th October-9th November 2005, 19th November 2005, 23rd-24th October 2006, 24th October 2006, 17th October 2008, 19th-20th October 2008).

The southern bias to British records could arise from the late migration of this species. Earlier departing migrants from North America tend to have more northerly occurrence patterns in Britain. This presumably reflects the more northern trajectory of Atlantic depressions earlier in the autumn. Interestingly, for the few years in which multiple arrivals have occurred, records have mostly involved birds of the same sex.

Indigo Bunting *Passerina cyanea*
(Linnaeus, 1766). One of the most abundant songbirds in eastern North America. Breeds in the eastern USA and southeastern Canada. Winters from southern Florida to northern South America.

Monotypic.

Status: Two records.

1985 Co. Cork: Cape Clear 1st-winter 9th-19th October
1996 Pembrokeshire: Ramsey Island 1st-winter male 18th-26th October

Discussion: Medium-distance migrant. Departure from northern breeding areas takes place in September, with nearly all birds vacating the breeding grounds by mid-October. The first arrivals in the Neotropics occur in mid September (Payne 2006).

Status elsewhere in the Western Palearctic is clouded by the possibility of escapes, though 15 from the Azores (10 in 2005, three in 2006 and two in 2007; earliest 20th October 2007) and two from Iceland (27th October 1951 and 20th October 1985) would appear to be vagrants.

There is one record from Gibraltar (26th April 2004). Elsewhere two are accepted for the Netherlands (8th June-15th July 1983 and 10th-23rd March 1989). Others of uncertain origin or probable escapes include records from Germany (four category E records), Poland (18th-26th June 1982), Finland (five, all May), Denmark (5th August 1987) and Norway (male on 22nd November 1987).

Further records, currently in Category E, were reviewed in light of the Pembrokeshire record, but none were felt suitable for a Category A upgrade. A male on Fair Isle, from 3rd-7th August 1964 was felt to be a likely addition to the *British List* at the time, but was exceptionally early for a transatlantic vagrant as departure from the northern breeding areas usually takes place in September. A second on the island in 1974, a female on 20th May, was considered an escape when examined in the hand. Two other records generated much discussion at the time: a male at Holkham Meals, Norfolk, from 21st-30th October 1988 and a 1st-summer male at Flamborough Head from 23rd-25th May 1989. The Rarities Committee of the time felt that they were probable escapes, on the grounds that both exhibited clear signs of abnormal moult.

The Norfolk bird was initially identified as a 1st-winter male, but was found to be an adult male in subordinate plumage; a phenomenon that occurs when a dominance hierarchy builds up in an aviary. This species was frequent in aviculture at this time, and even the impressive coincident influx of Nearctic birds around the North Sea (Cliff Swallow in Cleveland, Northern Waterthrush in Lincolnshire and a Northern Mockingbird in the Netherlands) could not influence the committee after the bunting was found to be an adult sporting abnormal moult consistent with an escape from captivity (Tim Melling *pers comm*). Adult Nearctic passerines are almost unheard of in autumn.

This is a species whose occurrences are always hotly debated regarding their provenance. Even records from migration hotspots such as Fair Isle have failed to impress. Age, location and timing are key factors in securing acceptance in Category A.

Bobolink *Dolichonyx oryzivorus*

(Linnaeus, 1758). Breeds widely across southern Canada and northern USA, south to northeast California and New Jersey. Winters Peru to southern Brazil and northern Argentina.

Monotypic.

Status: There have been 29 records, 26 in Britain and three in Ireland

Historical review: The first two were on Scilly: St. Agnes from 19th-20th September 1962 and St. Mary's on 10th October 1968.

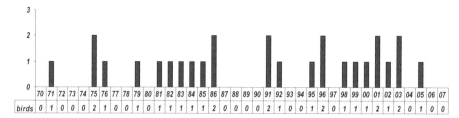

Figure 111: Annual numbers of Bobolinks, 1970 to 2007.

birds	70	71	72	73	74	75	76	77	78	79	80	81	82	83	84	85	86	87	88	89	90	91	92	93	94	95	96	97	98	99	00	01	02	03	04	05	06	07
	0	1	0	0	0	2	1	0	0	1	0	1	1	1	1	1	2	0	0	0	2	1	0	0	1	2	0	1	1	1	2	1	2	0	1	0	0

A third, and the first for Ireland, was at Hook Head, Co. Wexford, on 12th and 14th October 1971. The 1970s produced four more between 1975 and 1979, three of them on Scilly and the first Scottish record from Out Skerries on 18th September 1975 (the first year in which two were recorded in an autumn).

Seven more followed from 1981 to 1986. These were mostly in the southwest (three on Scilly; one each on Cape Clear and Lundy), but two Scottish birds on consecutive days arrived in late September 1986 (St. Kilda and Fair Isle).

There was a five-year gap until the first of the 1990s, singles in Devon and Scilly in 1991. The Devon bird at East Soar from 17th-21st September was the first to grace the mainland; the following autumn one was at Portland from 14th-18th September. Eight during the 1990s included four on Scilly, including birds simultaneously on different islands in early October 1996. Additional records came from Shetland. The first for Wales was on Skokholm from 13th-14th October 1999.

The good showing has continued since 2000, with seven more recorded by 2005 including two each in 2001 and 2003. Another mainland bird at Prawle Point from 9th to 15th October 2001 proved popular, as did one the following year at Hengistbury Head from 1st-23rd November. In contrast to earlier records, just one has reached Scilly. Two records have come from Shetland and one from Co. Cork, plus two from south coast. The first east coast record made landfall in East Yorkshire at Easington on 27th October 2001.

Where: Occurrences have been noticeably concentrated in Ireland and the southwest of Britain, with no fewer than 13 birds recorded on Scilly, two in Co. Cork, singles in Co. Wexford and Pembrokeshire, plus three in Devon and two in Dorset. In northern Britain, Shetland has five records and there is one from the Outer Hebrides. The sole record from eastern Britain is the one from East Yorkshire.

When: All records have been in the autumn period between mid-September and early November. The earliest record was one on Cape Clear from 13th-24th September 1982, followed by one at Portland from 14th-18th September 1992. Peak periods for arrivals occur in two periods with peaks of 11 birds between 17th and 30th September and nine more between 9th and 13th October. Just four have arrived after mid-October. The latest was a long-staying bird at Hengistbury Head from 1st-23rd November 2002.

Comparing the Shetland and Scilly records there is a difference in median arrival dates between the two (28th September on Shetland (n=5) versus 9th October on Scilly (n=13)), suggesting that those reaching the north cross the Atlantic earlier in the autumn than those farther south.

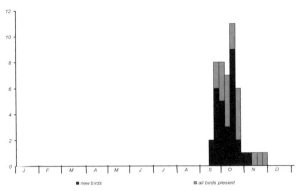

new birds ■ all birds present

Figure 112: Timing of British and Irish Bobolink records, 1962-2007.

Discussion: Vacation of the breeding grounds commences from late July and early August. Birds then congregate for several weeks in freshwater or coastal marshes in North America to complete their post-nuptial moult. Bobolinks breeding in the western USA and Canada take a route eastwards to the Atlantic coast before turning south. Migration proceeds through Florida and across the Caribbean. Some birds stop over in Cuba and Jamaica before continuing to South America. Bobolinks sighted over Bermuda are thought to be on a non-stop flight from the Atlantic coast of Nova Scotia to Virginia and South America.

Formerly abundant, Bobolinks used to be called rice birds when rice production was widespread across the southeastern United States. They were killed in large numbers as a pest to protect crops and for food. Early in the 20th century, the "bobolink route" over Jamaica was so named because of the numbers of birds passing through that island. In Jamaica they are called butter birds, a testament to their high fat levels when they pass through the island on migration (Martin and Gavin 1995).

The migratory route involves crossing of large expanses of water, making them prone to displacement by Atlantic depressions. The outfall from hurricane Emily grounded 10,000 Bobolinks in Bermuda on 25th September 1987 (yet only one bird reached Europe that autumn). In common with other grassland obligates, this species has declined throughout its North American range since at least the 1960s (Sauer *et al* 2008)

Elsewhere in the Western Palearctic just 14 birds have been recorded. Eight of these are from the Azores, all since 1999 with two in September and six in October (3rd September 1998, 21st-22nd September 1999, 5th October 2002, 24th October 2005, 25th October 2005, two on 4th October 2007 with one to 20th October, 14th-20th October 2008). There are three from France (15th-16th October 1987, 17th August 1995 and 25th September 2005) and singles from Norway (6th-8th November 1977), Gibraltar (11th-16th May 1984) and Italy (18th September 1989). The record from Gibraltar is the only spring record for the Western Palearctic.

The distribution of British and Irish occurrences is more in keeping with that of the New World wood-warblers than the Nearctic sparrows. This suggests that this accomplished long-distance migrant can comfortably make the Atlantic crossing without assistance. This is not surprising when account is taken of their spring and autumn trans-equatorial migration of 20,000 km between breeding and wintering grounds, which is one the longest annual migrations of any New World passerine.

Brown-headed Cowbird

Molothrus ater

Ray Scally

(Boddaert, 1783). Breeds from central British Columbia, southeastern Yukon, east to Prince Edward Islands, Nova Scotia, and southern Newfoundland south to central Mexico and northern Florida. Winters Pacific Coast of United States and southern and eastern United States south to southern Florida and southern Mexico.

Polytypic, three subspecies. Race undetermined; more likely nominate *ater* (Boddaert, 1783) or *artemisiae* Grinnell, 1909, than *obscurus* (Gmelin, 1789) based on size.

Status: Three records, two photographed in 2009 pending acceptance.

1988 Argyll: Ardnave, Islay male, probably 1st-year, 24th April
2009 Northumberland: Belford male, 1st-2nd May
2009 Shetland: Fair Isle male, 8th-10th May

Discussion: Formerly a grassland specialist that followed bison herds, this nest parasite expanded its range with the advent of cattle farming in North America and the cutting of forests. It has been shown to cause declines in populations of some host species. Benefitting greatly from forest fragmentation, numbers have recently declined in some areas due to forest regeneration (Smith and Cook 2000.)

A short-distance migrant sometimes present year-round in parts of their range. Dolbeer (1982) documented movements from northeast USA of up to 850 km between breeding and wintering sites. Migrant males arrive back on the breeding grounds first, from late March to late April (Lowther 1993).

Although a seemingly unlikely vagrant, two in 2009 mean that four have now made an Atlantic crossing by some means or other. The other Western Palearctic record was one from Norway on 1st June 1987. It was trapped and died shortly after.

Baltimore Oriole *Icterus galbula*

(Linnaeus, 1758). Breeds southern Canada from central Alberta east to central Nova Scotia, south through eastern USA from northern Texas to western South Carolina. Winters from southern Mexico to Colombia and Venezuela.

Monotypic.Baltimore Oriole was formerly treated as conspecific with Bullock's Oriole *I. bullockii,* of western America. Baltimore Oriole hybridises extensively with Bullock's Oriole where their ranges overlap in the Great Plains (Sibley and Short 1964) and they were formerly merged under the name Northern Oriole (*AOU* 1983). However, Rising (1970) considered that there was a stable hybrid zone and restricted gene flow. This recommendation led to the *AOU* treating the taxa as separate species again (Monroe *et al* 1995). They also have different moult timing: Baltimore moults on the breeding grounds before migration and Bullock's during or after autumn migration (Rising and Flood. 1998). Molecular analysis indicates that the two are not each other's closest relatives (Omland *et al* 1999). Baltimore Oriole is a sister species to Black-backed Oriole *I. abeillei* and Bullock's Oriole *I. bullockii* is a sister species to Streak-backed Oriole *I. pustulatus* (Omland and Kondo 2006).

Rare vagrant from North America, mostly during autumn; has overwintered and occurred in spring.

Status: There have been 25 records of this gaudy icterid to the end of 2007, 23 in Britain and two in Ireland.

Historical review: The first was an immature male caught on Unst, Shetland, on 26th September 1890. It died on the 28th. This record has only recently been reassessed and accepted, and represents the first record of a North American passerine in Europe (*BOURC* 2003). It was originally dismissed as an escape, because observers at the time did not believe that passerines could cross the Atlantic.

The route that at least some birds take across the Atlantic is illustrated by one that boarded RMS *Mauretania* in 1962. Departing her berth in the Hudson River, New York City, on 7th October the *Mauretania*'s eastbound crossing met the southern edge of hurricane Daisy and provided a temporary life raft for over 130 land birds of 34 species, including a Northern Oriole. Although it does not form part of the national statistics, the oriole left in very good shape within an hour or two of the Irish coast to make a very probable, though unrecorded, landfall (Durand, 1963; Durand, 1972).

> *The last new bird, surprisingly, turned up as late as 17.00 hours on the 10th, when we were nearly 1,600 miles out at about 45° 50'N, 40° 38'W. This was a female or imma-ture Baltimore Oriole (Icterus galbula), which joined the 'stewards' feast regularly on the First-Class Sundeck (where the birds became noticeably plumper daily!) and rev-elled in the soft fruit provided, though on occasions it pecked away happily at toast like any sparrow.*
> **Alan Durand.** British Birds 56: 157-164

The next was not until 1958, when an immature female graced Lundy from 2nd-9th October. No less than 10 were found between 1962-1968, of which an adult male at Beachy Head from 5th-6th October 1962 was unusual regarding both its age and location. An influx of four occurred in October 1967. The first, on Skokholm from 5th-10th October, was the first for Wales. Three more

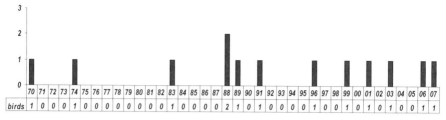

	70	71	72	73	74	75	76	77	78	79	80	81	82	83	84	85	86	87	88	89	90	91	92	93	94	95	96	97	98	99	00	01	02	03	04	05	06	07
birds	1	0	0	0	1	0	0	0	0	0	0	0	0	1	0	0	0	0	2	1	0	1	0	0	0	0	1	0	0	1	0	1	0	1	0	1	0	1	1

Figure 113: Annual numbers of Baltimore Orioles, 1970 to 2007.

arrived on 17th and 18th October, two of which made landfall on Lundy on 17th and one on St. Agnes, Scilly, from 18th October to 26th. In 1968 three more followed, including the first spring record for Europe, a 1st-winter male at Bodmin Moor, Cornwall, from 11th-13th May. Later that year a 1st-winter male was found dead in Coventry, Warwickshire, on 16th December. It was the first ever winter record.

In contrast, the 1970s mustered only singletons in 1970 (the second spring record, this time a male from Hook, Pembrokeshire, from 6th-7th May) and 1974 (on Fair Isle). The 1980s fared little better with four; one in 1983 was followed by three more in 1988-1989. Two singletons from the Isles of Scilly and one from the Outer Hebrides were eclipsed by a well-watched 1st-winter female in Pembrokeshire from 2nd January-23rd April 1989.

Three in the 1990s comprised singletons in 1991, 1996 and 1999. The first of the decade was another wintering individual, a 1st-year male at Westcliff-on-Sea, Essex, from 2nd December 1991-24th March 1992. The two additional records came from Bryher, Isles of Scilly.

Four more were found between 2001 and 2007. These include an overdue first for Ireland, a 1st-winter male, which quite appropriately took up temporary residence at Baltimore, Co. Cork, from 7th-8th October 2001. It was followed quickly by another, at nearby Cape Clear in 2006. A third spring bird in 2007 reached Caithness for three days in late May, belatedly identified from photographs after it had departed. One of the most popular birds was found inland in Oxfordshire from 10th December 2003 to 16th January 2004.

Sunday 14th December 2003 could only have improved. I had co-led a group of Oxford Ornithological Society members on a fruitless morning's trudge around Foxholes Nature Reserve, Oxfordshire. Finishing at lunchtime, I thought I would save something of the day by calling in at my favourite site in the county, Dix Pit. A couple of Caspian Gulls had been seen recently and, sure enough, shortly after I arrived, I picked out a small but classic adult. Warwickshire birders John Judge and Mike Doughty were also there and we talked gulls whilst concentrating on the cachinnans.

Just as they were about to leave, I received a mobile telephone call from Pete Allen with some quite extraordinary news. A resident of Headington, Oxford, had contacted local birder Tony Morgan to help identify a bird that had been frequenting the feeders in his garden for much of the day. Tony believed the bird to be a Baltimore Oriole, but wanted confirmation and so summoned the advice of Dave John, Bob Hurst and Pete. Upon their arrival at the site, they were invited into the resident's conservatory and were confronted with point-blank views of a stunning Baltimore Oriole gorging itself on sunflower hearts.

Overhearing "Baltimore Oriole?!" in my telephone conversation, John and Mike stopped in their tracks. Pete explained that the owner of the house was apprehensive of the news being released and wanted to speak to me (in my capacity as county bird recorder) first. After a long chat with the householder, during which he made it clear that the bird could not be seen from anywhere other than his living room and that the news was not to be broadcast nationally, he suggested that I popped round to evaluate the site myself. John and Mike tagged along. Well, you would, wouldn't you?

The twenty-minute drive through the delights of north Oxford dragged rather. The last two reported 'Baltimore Orioles' that I had gone to check out in Oxfordshire had turned out to be escaped Village Weavers, but this was different. The quiver in Pete's voice had had the ring of controlled panic.

Upon arrival, I made a quick recce of the site. It seemed that much of the back garden in question could be seen from an adjacent minor road, but I wanted to hear what the owner had to say. I declined his offer to wait for the bird in his conservatory for as long as I wished (very laudable I know, but I never take advantage of County Recorders privilege!) and explained that bird's status and the likely interest that it would generate. I felt the location could handle a good crowd, although space for parking was going to be a slight problem. The fact that the following day was Monday decreased the likelihood of a deluge of visitors, however. I telephoned a couple of friends for a bit of counselling and advice, but was completely distracted when the Baltimore Oriole then popped up in a bush in the next door garden.

I described the possible successful scenarios if the news was broadcast, but the resident was adamant that it should not be released nationally. He added, however, that it would be alright to tell a few local birders. I explained to him the impossibilities of this halfway house and that the news would inevitably leak, but he stood firm. It was a difficult situation, but we had to respect his wishes.

As it happened, things turned out for the best. Next morning, news of the oriole's presence spread like wildfire. The bird attracted a stream of admirers over the following week and there were no problems with the estate residents. Feeders and fruit were put out in the neighbouring gardens and around the adjacent recreation ground and a collection was taken for local charities. The bird remained until at least the end of December and looked set to enjoy the winter in Headington rather than Honduras.

Ian Lewington. Birding World 16: 503-505.

Where: The records have predominantly a southwesterly bias, six of them on Scilly (between 23rd September and 18th October), three each in Devon (all on Lundy in October) and Pembrokeshire (October, May and January), plus two each in Cornwall (May and October) and Co. Cork (October).

In Scotland there are two from Shetland (September), one from the Outer Hebrides (September) and one in Highland (May). One has reached the Isle of Man (October). Singles in December have come from Warwickshire, Oxfordshire and Essex; one from East Sussex was in October.

When: Overwintering birds in backgardens have been detected on three occasions. The most recent was a hugely popular 1st-winter male at Headington, Oxfordshire, from 10th December 2003-16th January 2004, which echoed extended stays of a 1st-year male at Westcliff-on-Sea, Essex, from 2nd December 1991-24th March 1992 and a 1st-winter female at Roch, Pembrokeshire, from 2nd January-23rd April 1989. The first of the quartet of the winter records was a 1st-winter male found dead at Coventry, Warwickshire, on 16th December 1968.

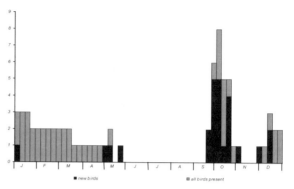

Figure 114: Timing of British and Irish Baltimore Oriole records, 1890-2007.

Three May records may be either birds on the move from Britain after having made landfall the previous autumn or fresh spring overshoots. Three records involve a 1st-winter male near Bodmin, Cornwall, from 11th-13th May 1968, a male at Hook, Pembrokeshire, from 6th-7th May 1970 and a male at John O' Groats, Highland, from 24th-27th May 2007.

The expected arrival period is in the autumn from mid-September. The earliest in autumn was a 1st-winter on Fair Isle from 19th-20th September 1974, several days earlier than one on St. Agnes, Scilly, from 23rd September-4th October 1983. Ten have occurred between 26th September and 7th October, and six more between 10th and 18th October. The latest autumn find was on St. Agnes from 18th-26th October 1967.

Discussion: A medium- to long-distance migrant, wintering primarily in Central America and northern South America. It moves early in autumn. Some vacate breeding areas in July and early August; most depart by mid-September. Most migrate overland through Mexico, principally along the Atlantic Slope. Smaller numbers take the over-sea route to the Caribbean islands (Rising and Flood 1998). It is birds taking the latter route that presumably account for British and Irish records, many in the wake of Atlantic depressions. In the USA, it frequently overwinters north to New England, most records coming from feeders as has been the case with the wintering British records.

A further record at sea was reported by Frankland (1989) on board the QE2 in late September 1988. A very tired female joined the ship half way across the Atlantic heading for Southampton. It flew into a glass partition, but was picked up and later recovered sufficiently to fly several times around the decks; she was still to be seen on 24th and 25th prior to docking on 26th. A further record that does not comprise part of the archive was a 1st-winter female that reportedly killed itself against a window of a ship in the English Channel in early October 1994 (a photo of the mounted specimen is in *Birding World 8:6*).

Elsewhere in Western Palearctic there are four from Iceland (1st-year male found dead on 8th November 1955, 8th October 1956, adult female collected on 15th October 1971 and 7th-13th October 2003) and singles from Norway (male on 13th May 1986), the Netherlands (trapped on 14th October 1987 and kept in captivity until released on 18th and present until 20th). Given the number of records from Britain and Ireland it is remarkably rare in the Western Palearctic. Even the Azores can muster just one bird (6th November 2006).

Several obliging birds have put on lengthy performances over the past couple of decades, each gratefully received by the masses. Populations in North America have been stable during recent times (Sauer *et al* 2008), suggesting that we can expect more occurrences in the near future. The southwest records conform to expectations, but the scatter of birds elsewhere is surprising for such a rare species. The several inland records have been most unexpected.

CATEGORY D & SELECTED CATEGORY E SPECIES

BOU definitions for Category D and E are:

Category D *is for species that would otherwise appear in Category A except that there is reasonable doubt that they have ever occurred in a natural state. Species placed in Category D only form no part of the British List and are not included in the species totals.*

Category E *is for species that have been recorded as introductions, transportees or escapees from captivity, and whose British breeding populations (if any) are thought not to be self-sustaining. Breeding escapees are identified with an asterisk (E*) in the Rare Breeding Bird Reports in British Birds.*

The *Irish Rare Birds Committee* uses the following classifications:

Category D1 *is for species that would otherwise appear in Categories A or B except that there is a reasonable doubt that they have ever occurred in a natural state.*
Category D2 *is for species that have arrived through ship or other human assistance.*
Category D3 *is for species that have only ever been found dead on the tideline.*
Category D4 *is for species that would otherwise appear in Category C1 except that their feral populations may or may not be self-supporting.*

The following species are present on both Category D and E or have recently been removed from Category D (*BOURC* 2009). All are possible vagrants or are accepted elsewhere in Europe as genuine vagrants.

Yellow-headed Blackbird by Julian R. Hough

Northern Flicker Colaptes auratus

Breeds from southern Canada eastwards through the Great Lakes and northeast USA. Winters in central and southern USA, Central America south to Panama and the Greater Antilles.

Status: One record.
1962 Co. Cork: Cobh Harbour 13th October

> *At dusk on 12th October, a few hours before we passed the Fastnet Light at the approaches to the south Irish coast, there were at least nine birds still alive in a free-flying state on board, including the last Yellow-shafted Flicker, the Baltimore Oriole, one Slate-coloured Junco, three White-throated Sparrows, two Song Sparrows and a Field Sparrow. At dawn the following morning we dropped anchor just inside Cobh Harbour, Co. Cork, and at 8.00 hours the Flicker, after circling the ship two or three times, flew strongly away on to the eastern headland at the entrance to the harbour, behind Roche Point.*
> **Alan Durand.** *British Birds 56:163*

Discussion: Northern Flicker comprises 11 subspecies in four morphologically distinct subspecies groups, of which *auratus*, part of the yellow-shafted group, along with *luteus*, occurs from Alaska across Canada and in the eastern United States (Wiebe and 2008). Yellow-shafted Flicker would seem to be the only likely candidate for reaching Europe.

Durand (1972) documented six records involving at least 20 birds on his transatlantic crossings, making it one of the most numerous species encountered. In May 1964 a Yellow-shafted Flicker survived a crossing to Liverpool aboard RMS *Sylvania*. In partial captivity some of the time, with food and water provided, it was eventually presented alive by the ship's Captain to Chester Zoo.

A category E record involved one found dead on board a ship that arrived at Caithness in July 1981. It was presumed to have died outside British waters. The only category A record in Europe was one from Denmark (18th May 1972) which was presumed to have been ship assisted. A female, the bird frequented a garden near the harbour of Ålborg.

Daurian Redstart Phoenicurus auroreus

Nominate auroreus breeds from South-Central Siberia and Mongolia east to Amurland, south to Korea and Northeast China and winters in Japan, Taiwan and Southeast China. The race leucopterus breeds in Northeast India and Central and Eastern China, wintering from the Eastern Himalayas east to Northern Indochina.

Polytypic, two subspecies.

Status: One record from Category E.
1988 Fife: Isle of May male 29th-30th April, later died; possibly seen on 23rd March

Discussion: The Fife bird, which died after being caught, was identified as belonging to the more southern race *leucopterus*. This form is less likely to occur naturally than the nominate race. The record was not accepted as a genuinely wild individual by the *BOURC*, with reasons outlined by Knox (1993).

Wallace (1980) included this species with those he predicted to occur in Britain. On first examination the inclusion of this attractive redstart seems unlikely. There is no doubt that the vagrancy potential of the species has been sullied by the Isle of May bird. There is a category D record from Sweden (male 22nd-26th September 1997). However, northern *auroreus* is a long-distance migrant that shares its breeding areas with species that have reached the Western Palearctic, so it is a possible vagrant. The possibility is further supported by a record from a site west of the Urals at c61°46'N, 56°43'E, within the Western Palearctic: a male photographed in the Pechoro-Ilychskiy Reserve on the Upper Pechora and present from 18th-30th September 2006 was the first for European Russia.

Additional records of note from the east side of the Urals (outside the Western Palearctic) include one in the Novosibirsk Region (near Barabinsk) in 1937 and a (possibly nesting) pair in the in the Salair Mountains, Kemerovo Region (Gyngazov & Milovidov 1977; Belyankin 2002; Neyfel'd & Teplov 2007; Ryabitsev 2008).

Mugimaki Flycatcher *Ficedula mugimaki*

Breeds in Russia from the northeast Altai east through Transbaikalia to the Sea of Okhotsk, lower Amur, Ussuriland, Sakhalin and Japan. Winters in southeast Asia to western Indonesia, Sulawesi and the Philippines.

Status: The sole British record was, and still is, the source of great debate over its origin.
1991 East Yorkshire: Stone Creek 1st-winter male 16th-17th November

Discussion: This bird occurred during a good autumn for eastern vagrants and was felt by many to involve a genuine vagrant. Unexpectedly for such a long-distance vagrant, there is little evidence of vagrancy to the west of its range. A widely quoted claim from Treviso, Italy, on 29th October 1957 was not admitted to the Italian List. To the east of its range there is a record from Shemya Island, Alaska, in May 1985 (West 2002).

Despite being an autumn find and a 1st-winter, the *BOURC* felt that the bird was later in the autumn than would be expected for a species with a similar breeding range to Pallas's Leaf Warbler *Phylloscopus proregulus* (Parkin and Shaw 1994). Date alone is not sufficient grounds upon which to have doubts regarding the veracity of the record; far more damning was the fact that the species had, just prior to this record, been advertised for sale in trade and small numbers had been imported (for example, 16 in 1986, seven in 1989 and 10 in 1990). Mugimaki Flycatcher was first advertised in *Cage and Aviary Birds* in autumn 1989. Several further advertisements appeared in 1990 (Parkin and Shaw 1994). So, its only appearance in the Western Palearctic coincided with the brief period that the species was known to be imported.

However, in Russia vagrant records are known as far west as the Urals (Ryabitsev 2008). Strel'Nikov and Strel'Nikova (1998) stated that Mugimaki Flycatcher may be extending its range westwards. They trapped a juvenile male in the autumn of 1995 and an adult female in 1996 in the Yugan Nature Reserve in the basin of the Bol'shoy Yugan at c74°41'E. In 2002, near the village of Saygatina (24 km west of Surgut on the River Ob' at 61°13'N, 73°22'E), there were two records in the first half of July. A male was observed in song and some days later a presumed female was observed (Yemtsev and Drenin 2006). More importantly from a Western Palearctic perspective, an adult male was seen near Neftekamsk on 2nd August 2007. The site is just east of the Kama (an east-bank tributary of the Volga) in Bashkortostan, at 56°06'N, 54°14'E, and is within the Western Palearctic (Fominykh 2007). This individual conceivably represents the first for the region.

Despite the increased vagrancy of far-eastern species to Europe during recent times, this species has not yet made another appearance. In many quarters there are calls for the record to be re-assessed in light of increased knowledge over vagrancy patterns, but in the absence of further records it is difficult to believe that a different conclusion over this bird could be reached.

Daurian Starling *Sturnus sturninus*
Breeds in central and eastern Asia from Transbaikalia, Amurland, and Ussuriland, south to northern China and Korea. Migrates south through southeast Asia to winter mostly in Peninsula Malaysia, Sumatra and Java.

Status: Three British claims, both Scottish birds generated much discussion. The Dutch CSNA have reassigned Daurian Starling to *Agropsar* as opposed to *Sturnus*.

1985 Shetland: Fair Isle male 7th-28th May
1997 Northumberland: Ponteland male 26th August-5th September
1998 Sutherland: Balnakeil age uncertain, 24th-27th September

Discussion: The Fair Isle bird was originally accepted as a new species for the Western Palearctic (Riddiford *et al* 1989; *BOU* 1991). It was followed by a juvenile shot in Norway later the same year (29th September 1985), which was also placed on category A, as was a 1st-winter male in the Netherlands (11th-12th October 2005). Two previous records from the Netherlands (15th May 1999 and 5th November 1999) are under review to assess whether they also involved birds of a wild origin.

The *BOURC* subsequently reviewed the Fair Isle bird and concluded that natural vagrancy for the species in spring was unlikely. The species was kept in captivity in Britain and Europe at the time. The sighting was placed on category D. Knox (1993) alludes to the opening of a market in a considerable number of Chinese species from the mid 1980s, thus there is clear evidence that trade birds were available at the time. The Ponteland bird was in suburban gardens and was generally accepted as an escape, with both the early date and its age not supportive of wild origin. The Sutherland bird was seen at a good time of year for a genuine vagrant. It was more problematic to age, but based upon photographs it seems most likely that the bird was an adult.

It is possible that a bird with seemingly good vagrancy credentials will put in an appearance at some point; for example, a late-autumn bird with the tell-tale retained secondaries of a 1st-winter. Even when faced with a potentially credible 1st-winter, what is the likelihood of the species arriving in Western Europe as a genuine vagrant?

Daurian Starling is not a Siberian passerine, but an Oriental one with its movements confined to the far east of Asia. As such, without a superb supporting cast to enhance its credentials, even a bird of the correct age and in a good place for vagrants would have to be considered unlikely and a potential escape.

A number of other Oriental starlings have been recorded in Britain (for example, White-cheeked *S. cineraceus* and White-shouldered *S. sinensis* Starlings) which, when found, were acting as though wild. Both were clearly escapes and most unlikely vagrants. Spring Daurian Starlings would appear to be non-starters. There are instances of Siberian vagrants being recorded in Western Europe on spring passage that have presumably overwintered, but records of such species from the Oriental region are almost non-existent, the exception being Amur Wagtail *Motacilla (alba) leucopsis*.

White-winged Snowfinch Montifringilla nivalis

Breeds across mountain chains of southern Europe and central Asia to western China. Mainly resident, making short winter movements.

Status: One record; recently deleted from Category D and moved to Category E (*BOURC* 2009).
1969 Suffolk: Lakenheath adult June 1969 to June 1972

Discussion: The species was formerly on the *British list*, but was removed as part of the Hastings Rarities scandal (Nicholson and Ferguson-Lees 1962). One of these was said to be shot from a flock of four or five similar-looking birds at Paddock Wood, Kent, on December 28th, 1906 and the bird was examined on 2nd January (Ticehurst 1908). The prolonged stay of the Lakenheath bird in a lowland site at a time when it was available in trade ensured that little credibility could be attached to the sighting. Initially identified as a 1st-year bird when it was first seen by bird-watchers, a photograph taken when it first arrived showed that it was an adult. The Cumbrian Spanish Sparrow *Passer hispaniolensis* is the only passerine with a comparable length of stay.

Movements outside of the breeding areas are rare, but two at Cape St Vincent were present from 4th-29th December 1998, with one bird seen again on 2nd January 1999, were the first for Portugal. Others have been recorded from the Balearic Islands, southern Spain and Morocco. There is a recent record of an adult male from Helgoland (21st March 2005). The subspecies could not be determined from the photograph, but the bill colour was wrong for a wild bird at that time of year and the record was placed in category D. There are two further records from the north of Germany, both from the Harz mountains (13th April 2008 and 7th May 2008). The identification is accepted (both were photographed), but at the moment they are not assigned to a category (Peter H. Barthel *pers comm*). A 1st-year ringed in Austria (3rd June 2005) was recaptured in Girona province, Catalonia region, northeast Spain (14th January 2006), a distance of 1,065 km (241° WSW) (Juan Carlos Fernández-Ordóñez *pers comm*).

Palm Warbler *Dendroica palmarum*

Breeds from central Mackenzie east to Newfoundland, south to central Alberta, north-east Minnesota, northern Wisconsin and Maine. Winters from South Carolina south along Atlantic and Gulf of Mexico coasts to Louisiana, from Yucatán (eastern Mexico) south along coast to northeast Nicaragua, and in the Greater Antilles and Bahamas.

Status: One record
1976 Cumbria: Walney Island tideline remains of an adult male found on 18th May

Discussion: This bird was originally accepted onto the now defunct category D3. All D3 species were reviewed with the view to including them on Category A if it was considered they died within territorial waters. However, when this record was reviewed it was found to be a highly desiccated corpse, suggesting it had died outside territorial waters. Moreover, there were tiny spots of red paint on the plumage, suggesting it may have been dead on a ship for a long period, long enough for maintenance painting to take place nearby (Tim Melling *pers comm*).

Despite its name, the Palm Warbler is among the northernmost of any *Dendroica* species. There is one accepted record for the Western Palearctic from Iceland (5th-10th October 1997), which involved a bird of the eastern race *hypochrysea*. This is another Nearctic species that could be expected in Britain or Ireland in future.

Yellow-headed Blackbird *Xanthocephalus xanthocephalus*

Breeds west of the Great Lakes north to central Canada. Winters in southwestern USA and Mexico.

Status: Five records, all placed on category E, except for one in Shetland in May 1987 which is on Category D.

1964 Lancashire and North Merseyside: Leighton Moss adult male 4th-10th August
1965 Northumberland: Seaton Burn adult male 17th-29th July
1970 Cheshire and Wirral: Watchlane Flash, Sandbach adult male 20th September
1987 Shetland: Norwick, Unst adult male 10th May; Burrafirth, Unst 11th-12th May; Cullivoe, Yell 12th May; Mid Yell 13th May
1990 Shetland: Fair Isle adult male 26th-30th April

Discussion: The Unst bird occurred following persistent northwesterly winds and three days before a White-throated Sparrow *Zonotrichia albicollis* was found at Norwick. Following review the *BOURC* considered that the bird had some of the credentials of a truly wild bird and, on the basis of this record, the species was promoted from Category E to Category D (*BOU* 2006). All other record are considered to involve escapes; the Fair Isle bird had a damaged eye.

The only European record accepted onto Category A was from Iceland (male 23rd-24th July 1983, when collected). Another male from Iceland mentioned in Proctor and Donald (2003) on 15th July 2002 was never submitted. A record from Denmark (2nd October 1918) was formerly on Category A, but is now placed on category E. Other European records of unknown origin or known escapes include three from the Netherlands (18th-20th May 1982, 14th June 1982, 2nd-3rd July 1982) with singles from Sweden (23rd May 1956), Channel Islands (Guernsey on 27th July 1978), Norway (30th May 1979) and France (23rd August-15th September 1979).

Yellow-headed Blackbird is a medium- to long-distance migrant. Ringing studies have shown Yellow-headed Blackbirds to have moved up to 3,500 km (Royall *et al* 1971).

Extralimital autumn records include three on Bermuda in September-October (Amos 1991). Vagrants have reached Greenland in autumn (2nd September 1840 and 7th August 1900) and it is increasingly detected in spring in eastern USA and Canada (Twedt and Crawford 1995). It seems possible that the species could occur as a genuine vagrant; the 1987 bird arguably has better credentials than some Nearctic species that already reside on Category A of the *British List*.

Chestnut Bunting Emberiza rutila

Breeds eastern Siberia, from northwest Irkutsk region east to Sea of Okhotsk, south to Baikal region and probably northern Mongolia and northern Manchuria. Winters in southern China, Indochina and Myanmar.

Status: Eight records; all were recently deleted from Category D and moved to Category E (*BOURC* 2009).

1974 Shetland: Foula male 9th-13th July
1985 Fife: Isle of May 1st-year trapped 11th June
1986 Caernarfonshire: Bardsey immature male 18th-19th June, trapped 19th
1986 Shetland: Fair Isle 1st-summer male 15th-16th June, trapped 15th
1994 Shetland: Out Skerries adult female 2nd-5th September, trapped 3rd
1998 Norfolk: Salthouse male 30th May-1st June
2000 Co. Durham: Whitburn 1st-summer male 17th-20th May
2002 Shetland: Fair Isle adult female 4th-7th September, trapped 5th

Discussion: Despite the fact that all but two have been on famous islands with first-rate rarity pedigrees, none have impressed with their credentials and all exhibit an occurrence pattern at odds with those of other accepted far-eastern vagrants. Both autumn records were very early in September and both involved adults, though the Out Skerries bird occurred after a prolonged period of easterly winds (Osborn and Harvey 1994).

There are several Category A records from elsewhere in western Europe. These included records from the Netherlands (1st-winter female on 5th November 1937), Norway (1st-winter on 13th October 1974, not to 15th as frequently cited, Vegard Bunes *pers comm*) and Finland (1st-winter male 30th September-1st October 2002). Additional records include birds from Malta (1st-winter male November 1983) and Slovenia (1st-winter male 10th October 1987).

Adults in autumn and those of any age in spring and summer are likely to be considered as escapes. However, should a 1st-winter occur in the right place and during the classic late September or October period, it could certainly help enable this species to make the transition to Category A.

Red-headed Bunting *Emberiza bruniceps*

Breeds from Caspian Sea east through Kazakhstan to the Altai Mountains and into the western part of the Xinjiang Uygur Autonomous Region, China. To the south, the breeding range extends into eastern Iran, Afghanistan and northern Pakistan. Winters west-central India.

Status: This species was formerly imported into Britain in great numbers. Since 1950 there have been 374 birds recorded in Britain (Keith Naylor *pers comm*) and 19 in Ireland between 1951-1997. Doubtless many more birds will have passed unreported as they will have been assumed to be escapes.

Historical review: It seems probable that Red-headed Bunting has arrived under its own steam, but the difficulty faced by records committees is deciding which, if any, have impeccable credentials. It is a medium-distance migrant with a north or northwest orientation to its spring migration, suggesting that spring overshoots are a possibility.

The first British record was an adult male found on North Ronaldsay on 19th June 1931. At the time it was considered to involve a genuine vagrant though subsequent records were treated as escapes. Growing skepticism over the records led to a re-examination of the skin of the 1931 specimen. It was found to show a scarred-over injury to the base of the culmen and front of the forehead. This individual was removed from the *British List* (*British Birds* 61: 43).

The number of birds recorded per decade peaked in the 1960s and 1970s. A total of 36 were reported in 1950s, 132 in the 1960s, 127 in the 1970s, 34 in the 1980s and 42 in the 1990s. The upsurge in records during the 1960s and 1970s presumably reflected a greater level of reporting activity, rather than any realistic change in status. The species has long been popular in the cage bird trade. One on Fair Isle in May 1966 bore an aviculture ring (Forrester *et al* 2007). In keeping with expectations of genuine vagrants, many have come from Shetland (75 birds) and there have been a good number from Dorset (30 birds) and Scilly (20 birds).

A noteworthy record from Walberswick, Suffolk, occurred in 1966. A male was observed singing and displaying to a female bunting, most probably a Yellowhammer *E. citrinella*, on 4th July. It was subsequently flushed from a nest with three eggs on the 8th, and one egg hatched on the 10th. The 'pair' was last seen on 29th July (Benson 1967; Piotrowski 2003).

The decline during the 1980s coincides with an export ban imposed by the Indian Government in 1982. Despite the ban, significant numbers continued to be recorded in the 1980s and 1990s. These included some individuals that were clearly escapes, suggesting that the trade had continued despite the ban.

There have been just three since 1998, comprising single males from 16-17th June 2001 at Baldhoun, Isle of Man, on 21st May 2002 near Cattawade, Essex, and a singing male at Monreith, Dumfries and Galloway, on 9th June 2004 (Vinnicombe, 2007). The BOURC (2009) recently reviewed records for this species and concluded that none merited upgrading to Category A.

Discussion: The range of Red-headed Bunting in the European part of Russia is considered inaccurate for a number of recent handbooks (Morozov and Kornev 2000). The distribution is mainly in the eastern part of Central Asia, but extends north-west into the steppes of the Urals/Western Siberia region. In Russia the range was considered to extend north to 50°N in the Volga-Ural interfluve, roughly to the southern edge of the Trans-Volga part of the Saratov Region (Stepan-yan 1990). A range expansion since the mid-1990s has extended the northern limit of the breeding range in the Volga-Ural interfluve to 51°N (Zav'yalov and Tabachishin 1999). Thus, there is evidence of a slow northward spread (Ryabitsev 2008).

Vinicombe (2007) reviewed the British records and made a number of recommendations. He concluded that records up to 1982 should all be treated as escapes as there was no way to assess which, if any, may have involved vagrants. The present arrival rate of Red-headed Buntings was felt to be compatible with natural vagrancy and Vinicombe (2007) proposed that these should be treated in the same way as other rarities from the same region of central Asia and the species be reinstated to the *BBRC* list. He also felt that contemporary late-autumn records of 1st-winters were much more likely to relate to wild birds than captive ones on the basis that the last known case of Red-headed Bunting breeding in captivity in Britain was in 1989 and the last known case in Germany was in 1994. However, young birds in autumn are the most frequently caught by trappers and illegal importers may release dull birds that they consider unsaleable and so not worth the risk of smuggling past customs.

The records in Italy and France were reviewed by Yésou *et al* (2003), who concluded that records in both countries probably related to wild birds based on the occurrence pattern, with the result that Red-heading is now in Category A (and D) of the respective national lists. Dier-schke (2007) analysed records from Helgoland, Germany, and found an occurrence pattern similar to that of the British records. Records on the island peaked in the 1960s, since when there have been far fewer and none during 1997-2006.

Blue Grosbeak *Passerina caerulea*

Breeds in southern North America from California to New Jersey and south to Central America and Costa Rica. Winters from Middle America and Mexico south to central Panama; northern breeders winter within the range and slightly south of southern residents' breeding range.

Status: Four records, three from Scotland; all recently deleted from Category D and moved to Category E (*BOURC* 2009).

1970 Shetland: Out Skerries adult male 17th-26th August
1972 Highland: Kiltarlity adult male 10th-11th March
1977 Borders: Inner Huntly, Ettrick male, dead, 22nd May
1986 Gloucestershire: Newent caught by a cat 9th May

Discussion: Elsewhere in Europe there are two Category D records from Sweden (male 5th July 1980 and 2nd-calendar year male 8th May 1983). A previous record from Norway (male 22nd November 1987) that was originally published as this species actually referred to an Indigo Bunting *Passerina cyanea* (Vegard Bunes *pers comm*). Although probably capable of occurring as a vagrant to Europe, the likelihood of one being an escape ensures that all British records so far have been placed on either Category D or E. Interestingly, the 1977 bird occurred just after the first and second White-crowned Sparrows *Zonotrichia leucophrys* for Britain, a Y ellow-rumped Warbler *Dendroica coronata* and a Dark-eyed Junco *Junco hyemalis*. This individual may have had a good claim as a potential candidate as a wild bird. However, it was assumed to be an escape so no description or photographs were obtained and the corpse was discarded (*BOURC* 2009).

House Crow Corvus splendens

Widespread in southern Asia. It has been introduced to East Africa around Zanzibar and Port Sudan, and arrived in Australia on a ship, but has up to now been exterminated. Has bred Hoek van Holland since 1998.

Status: One record placed on category D2 by the *Irish Rare Birds Committee*.
1974 Co. Waterford: Dunmore East 3rd November to autumn 1980

Discussion: This is something of an anomaly when it comes to migration, as its preferred method is to travel by ship. The sole Irish record arrived by ship to Dunmore East and is placed on Category D2.

In the Netherlands two 1st-summer birds travelled aboard ship to Hoek van Holland, Rotterdam, in April 1994. On 17th August 1997 they appeared to have raised at least one young and this constituted the first breeding record for Europe. Young were once again observed in 1998. The population now numbers a few tens of birds, all at Hoek van Holland and all presumably descended from the original pair. Unlike in the 1990s, there are no longer reports of individuals from other Dutch sites (Arnoud van den Berg *pers comm*).

American Goldfinch Carduelis tristis

Breeds from southwest Newfoundland west through southern Manitoba, north into central Saskatchewan and Alberta, and south along the Rocky Mountains through southern British Columbia to the coast. Winter and breeding ranges overlap, with populations generally shifting southwards in winter.

Status: One record, placed on category D1 by the *Irish Rare Birds Committee*.
1894 Co. Mayo: Achill Island shot 6th September

Discussion: This bird was widely considered to have been a probable escape from captivity (Humphreys 1937; Kennedy 1961). The Irish Bird Report for 1974 dismissed the record as unlikely vagrant as it is a short-distance migrant.

Some have been documented as covering large distances. For example, a ringed 1st-winter male covered 1,626 km in eight months. Migratory activity peaks mid-April to early June and late October to mid-December. Birds from maritime Canada and New England states follow a coastline route to wintering areas (Middleton 1993). The Irish record would appear exceptionally early for a species that normally moves late in the year. Indeed, the date would place it amongst the earliest ever Nearctic landbirds. Vagrants have been recorded from Bermuda and Cuba (Middleton 1993).

REFERENCES

Addinall, S. 2005. The Amur Wagtail in County Durham – a new Western Palearctic bird. *Birding World* 18: 155-158.

Adriaens, P., & Vandegehuchte, M. 2007. Zeldzame vogels in België in 2005 Drieëndertigste rapport van het Belgisch Avifaunisch Homologatiecomité. *Oriolus* 73: 52-61.

Albrecht, J.S.M. 1984. Some notes on the identification, song and habitat of the Green Warbler in the western Black Sea coastlands of Turkey. *Sandgrouse* 6: 69-75.

Alder, J. 1957. Brown Flycatcher in Northumberland. *Brit. Birds* 50: 125-126

Aldrich, J.W. 1993. Classification and distribution. Pages 47-54 in *Ecology and management of the Mourning Dove.* (Baskett, T. S. and Sayre, M. W. and Tomlinson, R. E. and Mirarchi, R. E., Ed.).Stackpole Books, Harrisburg, PA.

Alerstam, T. 1990. Ecological causes and consequences of bird orientation. *Experientia* 46: 405-415.

Alexander, H.G. 1955. Field-notes on some Asian leaf-warblers. *Brit. Birds* 48: 293-299.

Alexander, W.B & Fitter, R. S. R. 1955. American Land Birds in Western Europe. *Brit. Birds* 48: 10.

Aliabadian, M., Kabolic, M., Prodonc, R., Nijmana, V., & Vences, M. 2007. Phylogeny of Palaearctic wheatears (genus *Oenanthe*) - Congruence between morphometric and molecular data. *Molecular Phylogenetics and Evolution* 42: 665-675.

Alström, P. 1988. Identification of Blyth's Pipit. *Birding World* 1:268-272.

Alström, P. 2006. Species concepts and their application: insights from the genera *Seicercus* and *Phylloscopus*. *Acta Zoologica Sinica* 52: 429-434.

Alström, P., & Mild, K. 1987. Mystery photographs 122: Blyth's Pipit. *Brit. Birds* 80: 50-52.

Alström, P., & Mild, K. 1987. Some notes on the taxonomy of the Water Pipit complex. Proceedings of the 4th International Identification Meeting, Eilat, 47-48.

Alström, P., & Mild, K. 1988. Calls of Blyth's Pipit. *Brit. Birds* 81: 655.

Alström, P., & Olssön, U. 1988. Taxonomy of Yellow-browed Warblers. *Brit. Birds* 81: 656-657.

Alström, P., & Mild, K. 1996. The identification of Rock, Water and Buff-bellied Pipits.*Alula* 2: 161-175.

Alström, P., & Ödeen, A. 2002. Incongruence between mitochondrial DNA, nuclear DNA and non-molecular data in the avian genus Motacilla: implications for estimates of species phylogenies. In: Species Limits and Systematics in Some Passerine Birds (ed. Alström P). Acta Universitatis Upsaliensis, Uppsala.

Alström, P., Mild, K., & Zetterström, B. 1991. Identification of Lesser Short-toed Lark.*Birding World* 4: 422-427.

Alström, P., Mild, K., & Zetterström, B. 2003. *Pipits and Wagtails of Europe, Asia and North America.* Christopher Helm, London.

American Ornithologists' Union*(AOU)*. 1989. Thirty-seventh supplement to the *AOU*'s check-list of North American Birds. *Auk* 106: 532-538.

American Ornithologists' Union*(AOU)*. 2003. Forty-fourth supplement to the American Ornithologists' Union Check-list of North American Birds. *Auk* 120: 923–932.

American Ornithologists' Union Committee on Classification and Nomenclature. 1973. Thirty-second supplement to the American Ornithologists' Union check-list of North American birds. *Auk* 90: 411–419.

American Ornithologists' Union*(AOU)*. 1957. Check-list of North American birds. 5th ed. Am. Ornithol. Union, Washington, D.C.

American Ornithologists' Union*(AOU)*. 1983. Check-list of North American birds. 6th Edition. American Ornithologists' Union.

American Ornithologists' Union*(AOU)*. 1995. Fortieth supplement to the American Ornithologists' Union check-list of North American birds. *Auk* 112: 819–830.

American Ornithologists' Union*(AOU)*. 1998. *The Check-list of North American Birds: the species of birds of North America from the Arctic through Panama, including the West Indies and the Hawaiian Islands.* 7th Edition. American Ornithologists' Union, Ithaca.

Ammon, E. M., & Gilbert, W. M. 1999. Wilson's Warbler (*Wilsonia pusilla*), *The Birds of North America Online* (A. Poole, Ed.). Ithaca: Cornell Lab of Ornithology; Retrieved from *The Birds of North America Online*: http://bna.birds.cornell.edu/bna/species/478.

Amos, E.J.R. 1991. *A Guide to the Birds of Bermuda.* Warwick, Bermuda.

Andersson, R. 1988. Revirhävdande rödstjärthybrid. *Vår Fågelvärld* 47: 149-150.

Anon. 1960. Lesser Short-toed Larks in Cos. Kerry, Wexford and Mayo: a bird new to Britain and Ireland *Brit. Birds* 53: 241-243.

Anon. 1996. From the rarities Committee's files: Whitewinged Lark in Norfolk. *Brit. Birds* 89: 232-234.

Anon. 2008. The list of all the records accepted by the Estonian RC by 26.06.2008. http://www.eoy.ee/yhing/hk/hk_koik20080628.pdf

Arcese, P., Sogge., M.K., Marr., A. B., & Patten, M. A. 2002. Song Sparrow (Melospiza melodia), *The Birds of North America Online* (A. Poole, Ed.). Ithaca: Cornell Lab of Ornithology; Retrieved from *The Birds of North America Online*: http://bna.birds.cornell.edu/bna/species/704

Argeloo, M., & Meijer, A., 1997. Balearische Roodkoplauwier bij Voorhout in juni 1993. *Dutch Birding* 19: 65-67.

Arkhipov, V.Yu., Wilson, M.G., & Svensson. L. 2003. Song of Dark-throated Thrush. *Brit. Birds* 96: 79-83.

Arnaiz-Villena, A., Moscoso, J., Ruiz-del-Valle, V., Gonzalez, J., Reguera, R., Wink, M., & I. Serrano-Vela, J. 2007. Bayesian phylogeny of *Fringillinae* birds: status of the singular African oriole finch *Linurgus olivaceus* and Evolution and heterogeneity of the genus Carpodacus. *Acta Zoologica Sinica*, 53:826 - 834

Arnaiz-Villena, A., Guillén, J., Ruiz-del-Valle, V., Lowy, E., Zamora, J., Varela, P., Stefani, D., & Allende, L. M. 2001. Phylogeography of crossbills, bullfinches, grosbeaks, and rosefinches. *Cellular and Molecular Life Sciences* 58: 1159–1166.

Ash, J.S. 1956. Female Pied Wheatear: the problem of identification. *Brit. Birds* 49: 317-322.

Avilés, J.M., & Parejo, D. 2004. Farming practices and Roller *Coracias garrulus* conservation in south-west Spain. Bird Conservation International. 14: 173-181.

Axel, H.E., & Jobson, G.J. 1972. Savi's Warblers breeding In Suffolk. *Brit. Birds* 65: 229-232.

Azzopardi, J. 2006. A review of the status of Black-eared Wheatear in the Maltese Islands. *Brit. Birds* 99: 484-489.

Baccetti, N., Massa, B., Violani, C., 2007. Proposed synonymy of *Sylvia cantillans moltonii* Orlando, 1937, with *Sylvia cantillans subalpina* Temminck, 1820. *Bull.Br. Ornithol. Club* 127: 107–110.

Baines, R. 2007. The Brown Flycatcher in East Yorkshire. *Birding World* 20: 425-428.

Ball, R.M., & Avise, J.C. 1992. Mitochondrial DNA phylogeographic differentiation among avian populations and the Evolutionary significance of subspecies. *Auk*. 109: 626-636.

Baltz, M.E., & Latta, S.C. 1998. Cape May Warbler (*Dendroica tigrina*), *The Birds of North America Online* (A. Poole, Ed.). Ithaca: Cornell Lab of Ornithology; Retrieved from *The Birds of North America Online*: http://bna.birds.cornell.edu/bna/species/332.

Barbier, P.G.R. 1967. Red-headed Buntings in Britain and Ireland. *Brit. Birds* 60: 344–347.

Bardin, A.V. 1998. [Two records of Booted Warbler in the Leningrad Region]. *Russ. J. Orn. Express.* 47: 16–17. (In Russian.)

Bauer, H-G., & Kaier, A. 1991. Herbstfangdaten, Verweildauer, Mauser und Biometrie teilziehender Gartenbaumläufer (*Certhia brachydactyla*) in einem südwestdeutschen Rastgebiet. Vogelwarte 36:85-98.

Belyankin, A.F. 2002. [New data on rare and little-studied bird species of the Kemerovo Region]. Pp.25–31 in Ryabitsev, V.K. (ed.) Materialy k rasprostaneniyu ptits na Urale, v Priural'ye i Zapadnoy Sibiri. [Distribution of birds in the Ural Mountains, their environs and Western Siberia.] Ekaterinburg. (In Russian.)

Beaman, M., & Madge, S. 1998. *The Handbook of Bird Identification for Europe and the Western Palearctic*. Helm.

Benkman, C.W. 1992. White-winged Crossbill (*Loxia leucoptera*), *The Birds of North America Online* (A. Poole, Ed.). Ithaca: Cornell Lab of Ornithology; Retrieved from *The Birds of North America Online*: http://bna.birds.cornell.edu/bna/species/027

Benoit, F., & Märki, H. 2004. Nouvelles données sur les quartiers d'hiver du Venturon montagnard Serinus citrinella en Espagne. Nos *Oiseaux* 51:1–10.

Bensch, S., & Pearson, D.J. 2002. The Large-billed Reed Warbler *Acrocephalus orinus* Revisited. *Ibis.* 144: 259–267.

Benson, G.B.G. 1967. Red-headed Bunting breeding in Suffolk. *Brit. Birds* 60:343-344.

Bergier, P., Franchimont, J., Thévenot, M., & CHM. 2002. Les *Oiseaux* rares au Maroc. Rapport de la Commission d'Homologation Marocaine numéro 7. www.go-south.org.

Bergier, P., Franchimont, J., Thévenot, M., & CHM. 2006. Les *Oiseaux* rares au Maroc. Rapport de la Commission d'Homologation Marocaine numéro 11. *Go-South Bull.* 3: 31-42

Bevier, L.R., Poole, A.F., & Moskoff, W. 2005. Veery (*Catharus fuscescens*), *The Birds of North America Online* (A. Poole, Ed.). Ithaca: Cornell Lab of Ornithology; Retrieved from *The Birds of North America Online*: http://bna.birds.cornell.edu/bna/species/142

BirdLife International. 2008. Species factsheet: *Emberiza aureola*. Downloaded from http://www.birdlife.org on 23/12/2008.

Bishop, J., & Gray, M. 2002. The Rufous Turtle Dove in Orkney. *Birding World* 15: 501-505.

Blondel, J., Catzeflis, F., & Perret, P. 1996: Molecular phylogeny and the historical biogeography of the warblers of the genus *Sylvia* (Aves). *J. Evol. Biol.*

Boertmann, D. 1994. An annotated checklist to the birds of Greenland. Meddelelser om Grønland, *BioScience* 38.

Bonter, D.N., & Harvey, M.G. 2008. Winter Survey Data Reveal Rangewide Decline in Evening Grosbeak Populations. *Condor* 110: 376-381.

Borodin, O.V. 2004. [Penetration of the Paddyfield Warbler (*Acrocephalus agricola*) and the Eurasian Reed Warbler (*Acrocephalus scirpaceus*) into the Middle Volga region, Russia]. *Ornithologia* 31: 212–213. (In Russian.)

Bradshaw, C. 1992. Field identification of Black-faced Bunting. *Brit. Birds*. 85:653-665.

Bradshaw, C. 1994. Blyth's Pipit identification. *Brit. Birds* 87: 136-142.

Bradshaw, C. 1996. The Scilly eastern Nightingale. *Birding World* 9: 197

Bradshaw, C. 2000a. From the Rarities Committee's files: The occurrence of Moustached Warbler in Britain. *Brit. Birds* 93:29–38.

Bradshaw, C. 2000b. Separating *Acrocephalus* and *Hippolais* warblers. *Brit. Birds* 93:277.

Bradshaw, C. 2001. Blyth's Reed Warbler: problems and pitfalls. *Brit. Birds* 94: 236-245.

Bradshaw, C. 2001. Two-barred Greenish Warbler' on Scilly: new to Britain and Ireland. *Brit. Birds*. 94: 284-288.

Bradshaw, C., & Gray, M. 1993. Identification of female Pine Buntings. *Brit. Birds* 86: 378-386.

Brady, A. 1992. Major offshore nocturnal migration of Blackpoll Warblers, herons, and shorebirds. *Cassinia* 64: 28–29.

Brady, A. 1994. Offshore nocturnal migration of warblers, herons, and shorebirds no. 2 (1992). *Cassinia* 65: 15.

Brambilla, M., Janni, O., Guidali, F., & Sorace, A., 2008. Song perception among incipient species as a mechanism for reproductive isolation. *J. Evol. Biol.* 21: 651–657.

Brambilla, M., Vitulano, S., Spina, F., Baccetti, N., Gargallo, G., Fabbri, E., Guidali, F., & Randi, E. 2008. A molecular phylogeny of the *Sylvia cantillans* complex: Cryptic species within the Mediterranean basin. *Mol Phylogenet Evol.* 48: 461-72.

Bräunlich, A., & Steiof, K. 2001. Recent record of Eastern Olivaceous Warbler *Hippolais pallida elaeica* in Germany. *Limicola* 15: 147-155.

British Ornithologists' Union *(BOU)*. 1950. Twenty-first Report of the Committee on Nomenclature and Records of the occurrence of rare birds in the British Isles and on certain necessary changes in the nomenclature of the B.O.U. List of British Birds. *Ibis* 92: 132-141.

British Ornithologists' Union *(BOU)*. 1971. The Status of Birds in Britain and Ireland. Blackwell Scientific Publications, Oxford.

British Ornithologists' Union *(BOU)*. 1980. British Ornithologists' Union Records Committee: 10th Report. *Ibis* 122: 564-568.

British Ornithologists' Union *(BOU)*. 1984. British Ornithologists' Union Records Committee: 11th Report. *Ibis* 126: 440-444.

British Ornithologists' Union *(BOU)*. 1986. British Ornithologists' Union Records Committee: 12th Report. *Ibis* 128: 601-603.

British Ornithologists' Union *(BOU)*. 1991. British Ornithologists' Union Records Committee: 14th Report. *Ibis* 133: 218-222.

British Ornithologists' Union *(BOU)*. 1992a. Checklist of Birds of Britain and Ireland. 6th Edition. Helm Information Ltd, London.

British Ornithologists' Union *(BOU)*. 1992b. British Ornithologists' Union Records Committee: 16th Report. *Ibis* 134: 211-214.

British Ornithologists' Union *(BOU)*. 1992c. British Ornithologists' Union Records Committee: 17th Report (May 1992). *Ibis* 134: 380-381.

British Ornithologists' Union *(BOU)*. 1993a. British Ornithologists' Union Records Committee: 18th Report (December 1992). *Ibis* 135: 220-222.

British Ornithologists' Union *(BOU)*. 1993b. British Ornithologists' Union Records Committee: 19th Report (May 1993). *Ibis* 135: 493-499.

British Ornithologists' Union *(BOU)*. 1997. British Ornithologists' Union Records Committee: 23rd Report (July 1996). *Ibis* 139: 197-201.

British Ornithologists' Union *(BOU)*. 1998. British Ornithologists' Union Records Committee: 24th Report (October 1997). *Ibis* 140: 182-184.

British Ornithologists' Union *(BOU)*. 1999. British Ornithologists' Union Records Committee: 25th Report (October 1998). *Ibis* 141: 175-180.

British Ornithologists' Union *(BOU)*. 2001. British Ornithologists' Union Records Committee: 27th Report (October 2000). *Ibis* 143: 171-175.

British Ornithologists' Union *(BOU)*. 2002. British Ornithologists' Union Records Committee: 28th Report (October 2001). *Ibis* 144: 181-184.

British Ornithologists' Union *(BOU)*. 2003. British Ornithologists' Union Records Committee: 29th Report (October 2002). *Ibis* 145: 178-183.

British Ornithologists' Union *(BOU)*. 2004. British Ornithologists' Union Records Committee: 30th Report (October 2003). *Ibis* 146: 192-195.

British Ornithologists' Union *(BOU)*. 2005. British Ornithologists' Union Records Committee: 31st Report (October 2004). *Ibis* 147: 246-250.

British Ornithologists' Union *(BOU)*. 2006a. British Ornithologists' Union Records Committee: 32nd Report (October 2005). *Ibis* 148: 198-201.

British Ornithologists' Union *(BOU)*. 2006b. The British List: a checklist of birds of Britain (7th edition). *Ibis* 148: 526-563. Compiled by S.P. Dudley, M. Gee, C. Kehoe, T.M. Melling and the BOURC.

British Ornithologists' Union *(BOU)*. 2006c. British Ornithologists' Union Records Committee: 33rd Report (April 2006). *Ibis* 148: 594.

British Ornithologists' Union *(BOU)*. 2007. British Ornithologists' Union Records Committee: 34th Report (October 2006). *Ibis* 149: 194-197.

British Ornithologists' Union *(BOU)*. 2008. *British Ornithologists' Union Records Committee:* 35th Report (April 2007). *Ibis* 149: 652-654.

British Ornithologists' Union *(BOU)*. 2008. *British Ornithologists' Union Records Committee*: 36th Report (November 2007). *Ibis* 150: 218–220.

British Ornithologists' Union *(BOU)*. 2009. British Ornithologists' Union Records Committee: 37th Report (October 2008). *Ibis* 151: 224–230.

Broad, R.A. 1981. Tennessee Warblers: new to Britain and Ireland. *Brit. Birds* 74: 90-94.

Broad, R.A., & Hawley, R.G. 1980. White-crowned Sparrows: new to Britain and Ireland. *Brit. Birds* 73: 466-470.

Broad, R.A., & Oddie, W.E. 1980. Pallas's Reed Bunting: new to Britain and Ireland. *Brit. Birds* 73: 402-408.

Broad., R.A. 1981. Tennessee Warblers: new to Britain and Ireland. *Brit. Birds* 74:90-94.

Brodie Good, J. 1991. Philadelphia Vireo in Scilly: new to Britain. *Brit. Birds* 84: 572-574.

Brown, B.J. 1986. White-crowned Black Wheatear: new to Britain and Ireland. *Brit. Birds* 79: 221-227.

Brown, C.R. 1997. Purple Martin (*Progne subis*), *The Birds of North America Online* (A. Poole, Ed.). Ithaca: Cornell Lab of Ornithology; Retrieved from *The Birds of North America Online:* http://bna.birds.cornell.edu/bna/species/287.

Brown, C.R., & Brown, M.B. 1995. Cliff Swallow (*Petrochelidon pyrrhonota*), *The Birds of North America Online* (A. Poole, Ed.). Ithaca: Cornell Lab of Ornithology; Retrieved from *The Birds of North America Online:* http://bna.birds.cornell.edu/bna/species/149.

Browning, M.R. 1993. Comments on the taxonomy of *Empidonax traillii* (Willow Flycatcher). *Western Birds* 24: 241–257.

Browning, M.R. 1994. A taxonomic review of *Dendroica petechia* (Yellow Warbler; *Aves: Parulinae*). *Proceedings of the Biological Society of Washington.* 107: 27–51.

Bruce, M.D. 1999. Family *Tytonidae* (Barn-owls). Pages 34-75 in *Handbook of the birds of the world. Vol. 5. Barn-owls to hummingbirds.* (Hoyo, J. del, A. Elliott, and J. Sargatal, Eds.) *Lynx Edicions*, Barcelona, Spain.

Brugger, K.E., Arkin, L.N., & Gramlich, J.M. 1994. Migration Patterns of Cedar Waxwings in the Eastern United States. *J. Field. Ornithol.* 65: 381-387.

Burrows, I. 1978. Yellow-rumped Warbler in Co. Cork. *Brit. Birds* 71: 224.

Burton, J.F. 1995. *Birds and Climate Change.* A&C Black.

Butler, R.W. 2000. Stormy seas for some North American songbirds: are declines related to severe storms during migration? *Auk* 119: 518-522.

Byars, T., & Galbraith, H. 1980. Cape May Warbler: new to Britain and Ireland. *Brit. Birds* 73: 2-5.

Cade, M., & Walker, D. 2004. Eastern Subalpine Warblers in spring 2004. *Birding World* 17: 202–203.

Castell, P & Kirwan, G.M. 2005. Will the real Sykes's Warbler please stand up? Breeding data support specific status for *Hippolais rama* and *H. caligata*, with comments on the Arabian population of 'booted warbler'. *Sandgrouse* 27: 30–36.

Catley, G. 1994. The Alpine Accentor in Lincolnshire *Birding World* 7: 436-437.

Catley, G.P., & Hursthouse, D. 1985. Parrot Crossbills in Britain. *Brit. Birds* 78: 482-505.

Cavitt, J.F., & Haas, C.A. 2000. Brown Thrasher (*Toxostoma rufum*), *The Birds of North America Online* (A. Poole, Ed.). Ithaca: Cornell Lab of Ornithology; Retrieved from *The Birds of North America Online*: http://bna.birds.cornell.edu/bna/species/557.

Cederroth, C., Johansson, C., & Svensson L. 1999.Taiga Flycatcher *Ficedula albicilla* in Sweden: the first record in western Europe. *Birding World* 12: 460–468.

Chapman, M. 1987. Yellow-browed Warblers of the race *humei* in Europe. *Brit. Birds* 80: 578-580.

Cherry, J.D., Doherty, D.H., & Powers, K.D. 1985. An offshore nocturnal observation of migrating Blackpoll Warblers. *Condor* 87: 548–549.

Chilton, G., Baker, M.C., Barrentine, C.D., & Cunningham, M.A. 1995. White-crowned Sparrow (*Zonotrichia leucophrys*), *The Birds of North America Online* (A. Poole, Ed.). Ithaca: Cornell Lab of Ornithology; Retrieved from *The Birds of North America Online*: http://bna.birds.cornell.edu/bna/species/183

Christensen, R. 1996. A Red-necked Nightjar in Denmark. *Birding World* 9: 152-152.

Cimprich, D.A., Moore, F.R., & Guilfoyle, M.P. 2000. Red-eyed Vireo (*Vireo olivaceus*), *The Birds of North America Online* (A. Poole, Ed.). Ithaca: Cornell Lab of Ornithology; Retrieved from *The Birds of North America Online*: http://bna.birds.cornell.edu/bna/species/527.

Cink, C.L., & Collins, C.T. 2002. Chimney Swift (*Chaetura pelagica*), *The Birds of North America Online* (A. Poole, Ed.). Ithaca: Cornell Lab of Ornithology; Retrieved from *The Birds of North America Online*: http://bna.birds.cornell.edu/bna/species/646

Clement, P. 1987. Field identification of West Palearctic wheatears. *Brit. Birds* 80: 137-157, 187-238.

Clement, P., Harris, A., & Davis, J. 1993. *Finches and Sparrows: an identification guide.* Christopher Helm, London

Clement, P., Helbig, A. J., & Small, B. 1998. Taxonomy and identification of chiffchaffs in the Western Palearctic. *Brit. Birds* 91: 361-376.

Cobb, P.R, Rawnsley, P., Grenfell, H.E., Griffiths, E., & Cox, S. 1996. Northern Mockingbirds in Britain. *Brit. Birds* 89: 347-356.

Collar, N., 2005. Family *Turdidae* (Thrushes). In: del Hoyo, J., Elliott, A., Christie, D.A. (Eds.), *Handbook of Birds of the World, Cuckoo-shrikes to Thrushes*, Vol. 10. Lynx Edicions, Barcelona, pp. 514–807.

Collinson, M. 2001. Greenish Warbler, 'Two-barred Greenish Warbler', and the speciation process. *Brit. Birds* 94: 278-283.

Collinson, M., Knox, A G., Parkin, D.T., & Sangster, G. 2003. Specific status of taxa within the Greenish Warbler complex. *Brit. Birds* 96: 327-331.

Collinson, J.M., & Melling, T. 2008. Identification of vagrant Iberian Chiffchaffs - pointers, pitfalls and problem birds. *Brit. Birds* 101: 174–188.

Conder P.J., & Keighley J. 1949. First record of Bonelli's Warbler in the British Isles. *Brit. Birds* 42: 215-216.

Conder, P. 1979. Britain's first Olive-backed Pipit. *Brit. Birds* 72: 2-4.

Confer, J.L. 1992. Golden-winged Warbler (*Vermivora chrysoptera*), *The Birds of North America Online* (A. Poole, Ed.). Ithaca: Cornell Lab of Ornithology; Retrieved from *The Birds of North America Online*: http://bna.birds.cornell.edu/bna/species/020.

Conway, C.J. 1999. Canada Warbler (*Wilsonia canadensis*), *The Birds of North America Online* (A. Poole, Ed.). Ithaca: Cornell Lab of Ornithology; Retrieved from *The Birds of North America Online*: http://bna.birds.cornell.edu/bna/species/421

Cook, S. 1998. White-throated Sparrows at sea. *Birding World* 11: 269-270.

Cooke, M.T. 1946. Wanderings of the mockingbird. *Bird-Banding* 17: 784.

Corso, A. 1997. Balearic Woodchat Shrikes in Britain. *Birding World* 10:152-153.

Corso, A. 2001. Plumages of Common Stonechats in Sicily, and comparison with vagrant 'Siberian Stonechats'. *Brit. Birds* 94:315-318.

Cottridge, D., & Vinicombe. K. 1996. *Rare Birds in Britain and Ireland: A. Photographic Record.* HarperCollins Publishers, London

Coues, E.1895 Gatke's "Heligoland." *Auk*, 12: 322-346

Coyle, S.P., Grant, T.C.R., & Witherall, M.J. 2007. Purple Martin on Lewis: new to Britain. *Brit. Birds* 100: 143–148.

Cramp, S. 1988. *The birds of the Western Palearctic, vol 4.* Oxford University Press, Oxford.

Crochet, P-A, Dubois, P.J., Jiguet, F, Le Maréchal, P., Pons, J-M., & Yésou, P. 2006. From the CAF files: recent decisions, 2004 and 2005. *Ornithos* 13: 244 – 257.

Curson, J. 1994. Identification forum: separation of Bicknell's and Grey-cheeked thrushes. *Birding World* 7: 359–365.

Curson, J., Quinn, D., & Beadle, D. 1994. *New World Warblers.* Helm, London.

Davis, P. 1963. The Parrot Crossbill irruption at Fair Isle. *Bird Migration* 2: 260-264.

Davis, P. 1964. Crossbills in Britain and Ireland in 1963. *Brit. Birds* 57: 477-495.

Davis, P. 1966. The great immigration of early September 1965. *Brit. Birds* 59:353-376.

De Smet, G. 2001. A Moltoni's Subalpine Warbler in Belgium. *Birding World* 14:250.

Degnan, L., & Croft, K. 2005. Black Lark: new to Britain. *Brit. Birds* 98:306-313

Dement'ev, G.P. & Gladkov, N.A. (eds) 1954. Ptitsy Sovetskogo Soyuza 6. Moscow: Sovetskaya Nauka. [English translation: Dement'ev, G.P. & Gladkov, N.A. (eds) 1968. *Birds of the Soviet Union, vol.6.* Jerusalem: Israel Program for Scientific Translations.]

Dennis, R.H., & Wallace, D.I.M. 1975. Field identification of Short-toed and Lesser Short-toed Larks *Brit. Birds* 68: 238-241.

Dernjatin, P & Vattulainen, M. 2005. Black-eared and Pied Wheatear – a continuing identification problem. *Alula* 11:98-107.

Derrickson, K.C., & Breitwisch, R. 1992. Northern Mockingbird (*Mimus polyglottos*), *The Birds of North America Online* (A. Poole, Ed.). Ithaca: Cornell Lab of Ornithology; Retrieved from *The Birds of North America Online*: http://bna.birds.cornell.edu/bna/species/007.

Devillers, P. 1980. Project de nomenclature française des *Oiseaux* de monde. Gerfaut 70:121-146.

Dickerman. R.W. 1990. Geographic Variation in the Juvenal Plumage of the Common Nighthawk (Chordeiles Minor) in North America. *Auk* 107:610-613.

Dickie, I. R., & Vinicombe, K. E. 1995. Lesser Short-toed Lark in Dorset: new to Britain. *Brit. Birds* 88: 593-599.

Dickinson, E. C., (Ed.). 2003. *The Howard and Moore complete checklist of the birds of the world.* revised and enlarged 3rd ed. Princeton Univ. Press, Princeton, NJ.

Dierschke, J. 2001. Erstnachweis des Fahlseglers Apus pallidus für Helgoland. *Ornithologischer Jahresbericht Helgoland* 11: 71-75.

Dierschke, J. 2007. The status of Black-headed and Red-headed Buntings on Helgoland, Germany. *Brit. Birds* 100: 554–557.

Dinsmore, S.J. 2006. The changing Seasons: Weatherbirds. *North American Birds* 60:14-26.

Dolbeer, R.A. 1982. Migration patterns for age and sex classes of blackbirds and starlings. *J. Field Ornithol.* 53: 28–46.

Donald, P.F. 2007. Adult sex-ratios in wild bird populations. *Ibis* 149: 671-692.

Drovetski, S.V., Zink, R.M., Fadeev, I.V., Nesterov, E.V., Koblik, Ye.A, Red'kin, Ya.A., & Rohwer, S. 2004. Mitochondrial phylogeny of *Locustella* and related genera. *J. Av. Biol* 35: 105-110.

Dubois, P. 2007. Yellow, Blue-headed, 'Channel' and extralimital Wagtails: from myth to reality. *Birding World* 20: 104-112.

Dubois, P.J. 2001. Les formes nicheuses de la Bergeronnette printaniére *Motacilla flava* en France. *Ornithos* 8:44-71.

Duncan, J.R., & Duncan, P.A. 1998. Northern Hawk Owl (*Surnia ulula*), *The Birds of North America Online* (A. Poole, Ed.). Ithaca: Cornell Lab of Ornithology; Retrieved from *The Birds of North America Online*: http://bna.birds.cornell.edu/bna/species/356

Dunn, J.L., & Garrett, K.L. 1997. *A field guide to warblers of North America.* Houghton Mifflin Co., Boston, MA.

Duquet, M. 2006. Eastern Nightingale – identifying a potential split. *Birding World* 19:171-173.

Durand, A.L. 1961. White-throated Sparrow and American Robin crossing Atlantic on board ship. *Brit. Birds* 54:439-440.

Durand, A.L. 1963. A remarkable fall of American landbirds on the 'Mauretania', New York to Southampton, October 1962. *Brit. Birds* 56: 157-164

Durand, A.L. 1972. Landbirds over the North Atlantic: unpublished records 1961-65 and thoughts a decade later. *Brit. Birds* 65: 428-442.

Eaton, S.W. 1995. Northern Waterthrush (*Seiurus noveboracensis*), *The Birds of North America Online* (A. Poole, Ed.). Ithaca: Cornell Lab of Ornithology; Retrieved from *The Birds of North America Online*: http://bna.birds.cornell.edu/bna/species/182

Ebels, E.B., & Halff, R. 2003. Groenlandse Witstuitbarmsijs bij Huisduinen. *Dutch Birding* 25:439-440. (In Dutch, with English summary.)

Ebels, E., 1997. Balearische Roodkoplauwier bij Knardijk in juni 1983. *Dutch Birding* 19: 64-65. (In Dutch, with English summary.)

Edelaar, P. 2008. Assortative mating also indicates that common crossbill Loxia curvirostra vocal types are species. *J. Av. Biol.* 39:9-12.

Eigenhuus K.J. 1992. The Irish Lesser Short-toed Larks. *Birding World* 5: 66.

Elias, G. 2004. Aspects of Iberian Chiffchaff *Phylloscopus ibericus* distribution in Spain and Portugal. *Ibis*, 146: 685–686

Elkins, N. 1979. Nearctic landbirds in Britain and Ireland: a meteorological analysis. *Brit. Birds* 72: 417–433.

Elkins, N. 1999. Recent records of Nearctic landbirds in Britain and Ireland. *Brit. Birds* 92: 83–95.

Elmberg, J. 1993. Song differences between North American and European White-winged Crossbills. *Auk* 110: 385.

Ennis, T., & Dick, H. 1959. Breeding of the Ashy-headed Wagtail and Yellow Wagtail in Northern Ireland. *Brit. Birds*. 52:10-12.

Ertan, K.T. 2002. *Evolutionary Biology of the Genus Phoenicurus*: Phylogeography, Natural Hybridisation and Population Dynamics. Marburg: Tectum Verlag.

Ertan, K.T. 2006. The Evolutionary history of Eurasian redstarts, *Phoenicurus*. *Acta Zoologica Sinica* 52: 310–313.

Falls, J.B., & Kopachena, J.G. 1994. White-throated Sparrow (*Zonotrichia albicollis*), *The Birds of North America Online* (A. Poole, Ed.). Ithaca: Cornell Lab of Ornithology; Retrieved from *The Birds of North America Online*: http://bna.birds.cornell.edu/bna/species/128.

Ferguson-Lees, I.J & Sharrock J.T.R. 1977. When will the Fan-tailed Warbler colonise Britain? *Brit. Birds*. 70:152-158.

Finlayson, C., & Tomlinson, D. 2003. *Birds of Iberia*. Santana, Malaga.

Fisher, D. 1988. Beware Mongolian Lark. *Brit. Birds* 81:652-653.

Flood, R.L., Hudson, N., Thomas, B. 2007. *Essential Guide to Birds of the Isles of Scilly*. Cornwall.

Fominykh, M.A. 2007. [On rare birds in the Republic of Bashkortostan]. Pp.74–75 in Ryabitsev, V.K. (ed.) Materialy k rasprostraneniyu ptits na Urale, v Priural'ye I Zapadnoy Sibiri. Ekaterinburg. (In Russian.)

Forrester, R.W. on behalf of the Scottish Birds Records Committee. 2003. Amendments to Scottish List – species and subspecies. *Scottish Bird Report* 2000: 5-9.

Forrester, R.W. on behalf of the Scottish Birds Records Committee. 2004. Ammendments to the Scottish List – species and subspecies. *Scottish Bird Report* 2001: 7-12.

Forrester, R. W., Andrews, I. J., McInerny, C. J., Murray, R. D., McGowan, R. Y., Zonfrillo, B., Betts, M. W., Jardine, D. C. & Grundy, D. S (eds) 2007. *The Birds of Scotland*. The Scottish ornithologists Club, Aberlady.

Förschler, M. I., & Kalko, E.K.V. 2006. Breeding ecology and nest site selection in allopatric mainland Citril Finches *Carduelis citrinella citrinella* and insular Corsican Finches *Carduelis citrinella corsicanus*. *J. Ornithol* 147: 553-564

Förschler, M. I., & Kalko, E.K.V. 2007. Geographical differentiation, acoustic adaptation and species boundaries in mainland citril finches and insular Corsican finches, superspecies *Carduelis citrinella*. *J. Biog* 34: 1591-1600

Förschler, M.I., Förschler, I., & Dorka, U. 2006. Flowering intensity of spruces *Picea abies* and the population dynamics of Siskins *Carduelis spinus*, Common Crossbills *Loxia curvirostra*, and Citril Finches *Carduelis citrinella*. *Ornis Fennica* 83:91-96.

Foster, K. 2006. Siberian Blue Robin at Minsmere: new to Britain. *Brit. Birds* 99: 517–520.

Frankland, J.B. 1989. North American landbirds on the 'QE2'. *Brit. Birds* 82:568-569.

Fraser, P.A., & Rogers, M.J. 2004. Report on scarce migrant birds in Britain in 2002. Part 1: European Bee-eater to Little Bunting. *Brit. Birds* 97:647-664.

Fraser, P.A., & Rogers, M.J. 2006. Report on scarce migrant birds in Britain in 2003. Part 2: Short-toed Lark to Little Bunting. *Brit. Birds* 99:129-147.

French, P.R. 2006. Dark-breasted Barn Owl in Devon. *Brit.Birds* 99: 210–211.

Fry, A.J., & Zink, R.M. 1998. Geographic Analysis of Nucleotide Diversity and Song Sparrow Population History. *Molecular Ecology*. 7:1303-1313.

Fry, C.H., Fry, K., & Harris, A. 1992. *Kingfishers, Bee-Eaters, & Rollers: A Handbook.* Christopher Helm, London.

Gantlett, S. 2001. Subalpine Warbler forms in Britain. *Birding World* 14: 482–483.

Gantlett., S.J.M., & Millington, R.G. 1983. Rock Sparrow: new to Britain and Ireland. *Brit. Birds* 76: 245-247

Gardali, T., & Ballard, G. 2000. Warbling Vireo (*Vireo gilvus*), *The Birds of North America Online* (A. Poole, Ed.). Ithaca: Cornell Lab of Ornithology; Retrieved from *The Birds of North America Online:* http://bna.birds.cornell.edu/bna/species/551

Gargallo, G., 1994. On the taxonomy of the western Mediterranean islands populations of Subalpine Warbler *Sylvia cantillans.* Bull. Br. Ornithol. Club 114, 31–36.

Garner, M. 1997. An apparent hybrid wing-barred Crossbill. *Birding World* 10:71-72.

Garner, M. 2006. Swift Revelations: Identifying Swifts including Pallid Swift and Asian Common Swift.http://www.birdguides.com/webzine/article.asp?a=790.

Gauthier, J., Dionne, M., Potvin, J., Cadman, M., & Busby, D. 2007. *Unsolicited COSEWIC Status Report on Chimney Swift Chaetura pelagica.* Prepared for the Committee on the Status of Endangered Wildlife in Canada.

Gehring, W . 1963. Radar- und Feldbeobachtungen über den Verlauf des Vogelzuges im schweizerischen Mittelland: der Tagzug im Herbst (1957-1961). *Ornithol. Beob.* 60:35-68.

Gellert, M., & Laird, B. 1990. Eastern Yellow Wagtails. *Birding World* 3:277-280.

George, T.L. 2000. Varied Thrush (*Ixoreus naevius*), *The Birds of North America Online* (A. Poole, Ed.). Ithaca: Cornell Lab of Ornithology; Retrieved from *The Birds of North America Online:* http://bna.birds.cornell.edu/bna/species/541

Geroudet, P. 1973. Notes sur le Pouillot de Bonelli oriental sa distribution et sa voix. *Oiseau* 43: 75–79.

Ghalambor, C.K., & Martin, T.E. 1999. Red-breasted Nuthatch (*Sitta canadensis*), *The Birds of North America Online* (A. Poole, Ed.). Ithaca: Cornell Lab of Ornithology; Retrieved from *The Birds of North America Online:* http://bna.birds.cornell.edu/bna/species/459.

Gibbs, D., Barnes, E., & Cox, J. 2001. *Pigeons and Doves: a guide to the pigeons and doves of the world.* Pica Press, Robertsbridge.

Gill, F.B., Canterbury, R.A., & Confer, J.L. 2001. Blue-winged Warbler (*Vermivora pinus*), *The Birds of North America Online* (A. Poole, Ed.). Ithaca: Cornell Lab of Ornithology; Retrieved from *The Birds of North America Online:* http://bna.birds.cornell.edu/bna/species/584.

Gill, F.B., Slikas, B., & Sheldon, F.H. 2005. Phylogeny of titmice (*Paridae*): II. Species relationships based on sequences of the mitochondrial cytochrome b gene. *Auk* 122:121-143.

Gillihan, S.W., & Byers, B. 2001. Evening Grosbeak (*Coccothraustes vespertinus*), *The Birds of North America Online* (A. Poole, Ed.). Ithaca: Cornell Lab of Ornithology; Retrieved from *The Birds of North America Online:* http://bna.birds.cornell.edu/bna/species/599.

Gilroy, J. J., & Lees, A. C. 2003. Vagrancy theories: Are autumn vagrants really reverse migrants? *Brit. Birds* 96: 427-438.

Giralt, D., & Valera, F. 2007. Population trends and spatial synchrony in peripheral populations of the endangered Lesser Grey Shrike in response to environmental change. *Biodiversity and Conservation* 16:841-856.

Golley, M. 2007. The Moltoni's Subalpine Warbler in Norfolk. *Birding World* 20:459-463.

Golley, M.A., & Millington, R. 1996. Identification of Blyth's Reed Warbler in field. *Birding World* 9:351-353

Gordeev, Yu.I. 1977. [Recent materials on distribution of birds in the Khanty-Mansiysk district]. *Ornitologiya* 13: 33-39 (In Russian.)

Grant, K., & Small, B. 2004. A Caucasian Stonechat at Paphos, Cyprus. *Birding World* 17:154-156.

Grant, P.J. 1989. The Portland Pipit, a personal assessment. *Birding World* 2: 178-179.

Grantham, M. 2008. Dark-breasted Barn Owl tragically found dead. http://www.birdguides.com/webzine/article.asp?a=1386.

Gray, M. 1996. Richard's Pipit or Blyth's? *Brit. Birds* 89: 144-146.

Greenlaw, J.S. 1996a. Eastern Towhee (*Pipilo erythrophthalmus*), *The Birds of North America Online* (A. Poole, Ed.). Ithaca: Cornell Lab of Ornithology; Retrieved from *The Birds of North America Online:* http://bna.birds.cornell.edu/bna/species/262

Greenlaw, J.S. 1996b. Spotted Towhee (*Pipilo maculatus*), *The Birds of North America Online* (A. Poole, Ed.). Ithaca: Cornell Lab of Ornithology; Retrieved from *The Birds of North America Online:* http://bna.birds.cornell.edu/bna/species/263.

465

Grieve, A. 1992. First record of Thick-billed Warbler *Acrocephalus aedon* in Egypt. *Sandgrouse* 14:123-124.

Guzy, M.J., & Ritchison, G. 1999. Common Yellowthroat (*Geothlypis trichas*), *The Birds of North America Online* (A. Poole, Ed.). Ithaca: Cornell Lab of Ornithology; Retrieved from *The Birds of North America Online*: http://bna.birds.cornell.edu/bna/species/448.

Gyngazov, A.M. & Milovidov, S.P. 1977. Ptitsy Zapadno-Sibirskoy ravniny. [The birds of the West Siberian Plain]. Tomsk: *Tomsk University Press*. (In Russian.)

Hagemeijer, W.J.M., & Blair, M.J. (eds.). 1997. *The EBCC Atlas of European Breeding Birds: their distribution and abundance*. Poyser, London.

Hall, G. A. 1994. Magnolia Warbler (*Dendroica magnolia*), *The Birds of North America Online* (A. Poole, Ed.). Ithaca: Cornell Lab of Ornithology; Retrieved from *The Birds of North America Online*: http://bna.birds.cornell.edu/bna/species/136.

Hamas, M.J. 1994. Belted Kingfisher (*Ceryle alcyon*), *The Birds of North America Online* (A. Poole, Ed.). Ithaca: Cornell Lab of Ornithology; Retrieved from *The Birds of North America Online*: http://bna.birds.cornell.edu/bna/species/084.

Hampe A., Heinicke T., & Helbig A. J. 1996. Erste brut der Zitronenstelze Motacilla citreola in Deutschland. *Limicola* 10:311–316

Hanski, I., Hansson, L., & Henttonen, H. 1991. Specialist predators, generalist predators and the microtine cycle. *J. Anim. Ecol* 60:353-367.

Harrap, S., & Quinn, D. 1996. *Tits, Nuthatches and Treecreepers*. Christopher Helm/A&C Black, London.

Harris, T., & Franklin, K. 2000. *Shrikes and Bush-Shrikes*. Christopher Helm, London.

Harrop, H.R., Mavor, R., & Ellis, P.M. 2008. Olive-tree Warbler on Shetland: new to Britain. *Brit. Birds* 101: 82-88.

Harrop, H., & Fray, R. 2008. Two-barred Crossbills in the Northern Isles in July and August 2008. *Birding World* 21:329-339.

Harrop, S., & Millington, R. 1991. Identification forum: Two-barred Crossbill. *Birding World* 4:55-59.

Harrop., A.H.J., Knox, A.G., & McGowan, R.Y. 2007. Britain's first Two-barred Crossbill. *Brit. Birds* 100: 650-657.

Harvey, P. 1992. The Brown Flycatcher on Fair Isle: a new British bird. *Birding World* 5:252-255.

Heard, C D.R. 1990. Blyth's Pipit in Cornwall. *Birding World* 3: 375-378.

Helb, H.-W.; Bergmann, H.-H., & Martens, J. 1982. Acoustic differences between populations of western and eastern Bonelli's Warblers (*Phylloscopus bonelli, Sylviidae*). *Cellular and Molecular Life Sciences* 38(3): 356–357.

Helbig, A.J., Salomon, M., Wink, M., & Martens, J. 1993. Absence of mitochondrial gene flow between European and Iberian "chiffchaffs" (*Aves: Phylloscopus collybita collybita, P. (c.) brehmii*). The taxonomic consequences. Results drawn from PCR and DNA sequencing. Comptes Rend. *Acad. Sci. Paris*, t. 316, Série III, (1993): 205-210.

Helbig, A.J., Seibold, I., Martens, J., & Wink, M. 1995. Genetic differentiation and phylogenetic relationships of Bonelli's Warbler *Phylloscopus bonelli* and Green Warbler *P. nitidus*. *J. Av. Biol* 26: 139-153.

Helbig, A.J., Martens, J., Seibold, I., Henning, F., Schottler B., & Wink, M. 1996. Phylogeny and species limits in the Palearctic Chiffchaff *Phylloscopus collybita* complex: mitochondrial genetic differentiation and bioacoustic evidence. *Ibis* 138: 650-666.

Helbig, A.J., & Seibold, I. 1999. Molecular phylogeny of Palearctic–African *Acrocephalus* and *Hippolais* warblers (Aves: Sylviidae). *Mol. Phyl., & Evol.* 11: 246–260.

Helbig, A.J. 2001. The molecular identification of the Eastern Olivaceous Warbler *Hippolais pallida elaeica* from Berlin. *Limicola* 15: 155-156.

Helbig, A.J., Knox, A. G., Parkin, D.T., Sangster, G., & Collinson, M. 2002. Guidelines for assigning species rank. *Ibis* 144:518-525.

Hernández, M.A, Campos F, Gutiérrez-Corchero, F., & Amezcua A. 2004. Identification of Lanius species and subspecies using tandem repeats in the mitochondrial DNA control region. *Ibis* 146: 227–230

Herremans, M. 1990. Taxonomy and Evolution in Redpolls *Carduelis flammea – hornemanni*; a multivariate study of their biometry. *Ardea* 78: 441–458.

Herremans, M. 1998. Monitoring the world population of the Lesser Grey Shrike (*Lanius minor*) on the non-breeding grounds in southern Africa. *J. Ornithol.* 139:485-493.

Hill, J.R., III. 2002. A guide to sexing and aging Purple Martins, with some notes on martin rarities in the ABA Area. *Birding* 34: 247–257.

Hinchon, G. 2008. The *'vittata'* form of Pied Wheatear. *Birding World* 21:121-122.

Hipkiss, T., Hörnfeldt, B., Lundmark, A., Norbäck, M., Ellegren, H. 2002. Sex ratio and age structure of nomadic

Tengmalm's Owls: a molecular approach. *J. Av. Biol* 33: 107–110.

Hirschfeld, E. 1992. Identification of Rufous Turtle Dove. *Birding World* 5: 52-57.

Holloway, J. 1997. Stronsay Bird Reserve. *Alula* 3:122-125.

Hollyer, J.N. 1970. The invasion of Spotted Nutcrackers in autumn 1968. *Brit. Birds* 63: 353-373.

Holt, C., & Turner, S. 1999. The Desert Lesser Whitethroat on Fair Isle. *Birding World* 12:281-283.

Hourlay, F., Libois, R., D'Amico, F., Sarà, M., O'Halloran, J., & Michaux, J.R. 2008. Evidence of a highly complex phylogeographic structure on a specialist river bird species, the dipper (*Cinclus cinclus*). *Molecular Phylogenetics and Evolution*, In Press, Corrected Proof, Available online 14 August 2008.

Howell, S.N.G., & Webb, S. 1995. *The birds of Mexico and northern central America*. Oxford Univ. Press, New York.

Howell, T R. 1953. Racial and sexual differences in migration in *Sphyrapicus varius*. *Auk* 70: 118–126.

Hughes, J.M. 1999. Yellow-billed Cuckoo (*Coccyzus americanus*), *The Birds of North America Online* (A. Poole, Ed.). Ithaca: Cornell Lab of Ornithology; Retrieved from *The Birds of North America Online*: http://bna.birds.cornell. edu/bna/species/418.

Hughes, J.M. 2001. Black-billed Cuckoo (*Coccyzus erythropthalmus*), *The Birds of North America Online* (A. Poole, Ed.). Ithaca: Cornell Lab of Ornithology; Retrieved from *The Birds of North America Online*: http://bna.birds.cornell.edu/bna/species/587.

Hume, R.A. 1995. Blue Rock Thrush in Strathclyde: new to Britain and Ireland. *Brit. Birds* 88: 130-132.

Hume., R., & Parkin, D. 1995. Red-breasted Nuthatch in Norfolk: new to Britain and Ireland. *Brit. Birds* 88:150-153.

Humphreys, G.R. 1937. *A List of Irish Birds*, The Stationery Office, Dublin.

Hunt, J.S., Bermingham, E., & Ricklefs, R.E. 2001. Molecular systematics and biogeography of Antillean thrashers, tremblers, and mockingbirds (Aves: *Mimidae*). *Auk* 118:35-55.

Hunt, P.D., & Flaspohler, D.J. 1998. Yellow-rumped Warbler (*Dendroica coronata*), *The Birds of North America Online* (A. Poole, Ed.). Ithaca: Cornell Lab of Ornithology; Retrieved from *The Birds of North America Online*: http://bna.birds.cornell.edu/bna/species/376.

Hunt, P.D., & Eliason, B.C. 1999. Blackpoll Warbler (*Dendroica striata*), *The Birds of North America Online* (A. Poole, Ed.). Ithaca: Cornell Lab of Ornithology; Retrieved from *The Birds of North America Online*: http://bna.birds. cornell.edu/bna/species/431.

Incledon, C.S.L. 1968. Brown Thrasher in Dorset: a species new to Britain and Ireland. *Brit. Birds* 61: 550-553.

Irwin, D.E. 2000. Song variation in an avian ring species. *Evolution* 54: 998–1010.

Irwin, D.E., Alström, P., Olsson, U., & Benowitz-Fredericks, Z.M. 2001. Cryptic species in the genus *Phylloscopus* (Old World leaf warblers). *Ibis* 143: 233–247.

Irwin, D.E., Bensch, S., & Price, T.D., 2001. Speciation in a ring. *Nature* 409: 333–337.

Irwin, D.E., Bensch, S., & Price, T.D., 2005. Speciation by Distance in a Ring Species. *Science* 307: 414 – 416.

Irwin, D.E., & Hellström, M. 2007. Green Warbler *Phylloscopus (trochiloides) nitidus* recorded at Ottenby, Öland: a first record for Scandinavia *Ornis Svecica* 17: 75–80.

Isenmann, P., & Bouchet, M.A. 1993. L'aire de distribution française et le statut taxonomique de la Pie-grièche Grise Méridionale *Lanius elegans meridionalis*. *Alauda* 61: 223–227.

Isenmann, P., & Lefranc, N. 1994. Le statut taxonomique de la Pie-grièche méridionale. *Alauda* 62: 138.

Isenmann, P., Gaultier,T., El Hili, A., Azafzaf, H., Dlensi, H., & Smart, M. 2005. *Oiseaux de Tunisie*. *Société d'Études Ornitholoiques de France*, Paris.

James, R.D. 1998. Blue-headed Vireo (*Vireo solitarius*), *The Birds of North America Online* (A. Poole, Ed.). Ithaca: Cornell Lab of Ornithology; Retrieved from *The Birds of North America Online*: http://bna.birds.cornell.edu/bna/species/379.

Jännes, H. 1995. Rufous Turtle Dove. *Alula* 1: 56-65.

Johnson, N.K., Zink, R.M., & Marten, J.A. 1988. Genetic evidence for relationships in the avian family *Vireonidae*. *Condor* 90:428-445.

Jones, P.W., & Donovan, T.M. 1996. Hermit Thrush (*Catharus guttatus*), *The Birds of North America Online* (A. Poole, Ed.). Ithaca: Cornell Lab of Ornithology; Retrieved from *The Birds of North America Online*: http://bna.birds.cornell.edu/bna/species/261

Kehoe, C. 2006. Racial identification and assessment in Britain: a report from the RIACT subcommittee. *Brit. Birds* 99: 619-645.

Keith, A.R. 1968. A summary of the extralimital records of the Varied Thrush, 1848 to 1966. *Bird-Banding* 29: 245–276.

Kelly, J.F., Bridge, E.S., & Hamas, M.J. 2009. Belted Kingfisher (*Megaceryle alcyon*), *The Birds of North America Online* (A. Poole, Ed.). Ithaca: Cornell Lab of Ornithology; Retrieved from *The Birds of North America Online*: http://bna.birds.cornell.edu/bna/species/084

Kennedy, P.G., 1961. *A List of the Birds of Ireland*. The Stationery Office, Dublin.

Ketterson, E.D. & Nolan, V. 1982. The role of migration and winter mortality in the life history of a temperate-zone migrant, the Dark-eyed Junco, as determined from demographic analyses of winter populations. *Auk* 99: 243–259.

King, J. 1996. Identification of nightingales. *Birding World* 9:179-189.

Kirwan, G.M., Boyla, K.A., Castell, P., Demirci, B., Ozen, M. et al. 2008. *The Birds of Turkey* (Helm Field Guides).

Kitson, A.R. 1979a. Identification of Olive-backed Pipit, Blyth's Pipit and Pallas's Reed Bunting. *Brit. Birds* 72: 94-100

Kitson, A.R. 1979b. Identification of Isabelline Wheatear, Desert Warbler and three *Phylloscopus* warblers. *Brit. Birds* 72: 5-9.

Klicka, J., Voelker, G., & Spellman, G.M. 2005. A molecular phylogenetic analysis of the "true thrushes" (Aves: Turdinae). *Molecular Phylogenetics and Evolution* 34: 486-500.

Knox, A.G. 1988a. Taxonomy of the Rock/Water Pipit superspecies *Anthus petrosus, spinoletta* and *rubescens*. *Brit. Birds* 81:206-211.

Knox, A.G. 1988b. The taxonomy of redpolls. *Ardea* 76: 1–26.

Knox, A.G. 1990. The sympatric breeding of Common and Scottish Crossbills *Loxia curvirostra* and *L. scotica,* and the Evolution of crossbills. *Ibis* 132: 454–466.

Knox, A.G. 1993. Daurian Redstart in Scotland: captive origin and the British List. *Brit. Birds* 86:359-366.

Knox, A.G. 1994. Removal of Citril Finch from the *British & Irish List*. *Brit. Birds* 87:471-473.

Knox, A.G. 1996. Gray-cheeked and Bicknell's thrushes: Taxonomy, identification and the British and Irish records. *Brit. Birds* 89: 1–9.

Knox, A.G., Collinson, M., Helbig, A.J., Parkin, D.T., & Sangster, G. 2002. Taxonomic recommendations for Brit. Birds. *Ibis* 144: 707–710.

Koistinen, J. 2002. Vagrancy and weather. *Alula* 8: 28.

Kolbeinsson, Y. 2003. Iceland in October 2003. *Birding World* 16:435-440.

König, K., Weick., F., & Becking, J.H. 1999. *Owls: A Guide to the Owls of the World*. Pica / Christopher Helm.

Korpimäki, E. 1988. Effects of Age on Breeding Performance of Tengmalm's Owl *Aegolius funereus* in Western Finland. *Ornis Scandinavica* 19:21-26.

Kricher, J.C. 1995. Black-and-white Warbler (*Mniotilta varia*), *The Birds of North America Online* (A. Poole, Ed.). Ithaca: Cornell Lab of Ornithology; Retrieved from *The Birds of North America Online*: http://bna.birds.cornell.edu/bna/species/158/

Kvist, L., Martens, J., Ahola, A., & Orell, M. 2001. Phylogeography of a Palearctic sedentary passerine, the Willow Tit (*Parus montanus*). *J. Evol. Biol* 14:930–41.

Lansdown, P., & Charlton, T. D. 1990. 'The Sizewell bunting': a hybrid Pine Bunting X Yellowhammer in Suffolk. *Brit. Birds* 83: 241-242.

Lansdown, P.G. and the Rarities Committee. 1991. Status of Spectacled Warbler in Britain. *Brit. Birds* 84: 431-432.

Lansdown, P. 1995. Ages of Great Spotted Cuckoos in Britain and Ireland. *Brit. Birds* 88:141-149.

Lansdown, P. 1999. Comparison of Short-toed and Lesser Short-toed Larks. *Brit. Birds* 92: 308–312.

Larkin, R.P., & Thompson, D. 1980. Flight speeds of birds observed with radar – evidence for 2 phases of migratory flight. *Behav. Ecol. Sociobiol*. 7: 301–317.

Lassey, P.A. 2005. Taiga Flycatcher in East Yorkshire: new to Britain. *Brit. Birds* 98:542-546.

Lauga, B., Cagnon, C., D'Amico, F., Karama, S., & Mouchés, C. 2005. Phylogeography of the white-throated dipper *Cinclus cinclus* in Europe. *J. Ornithol* 146: 257–262

Lea, R.B. 1942. A study of the nesting habits of the Cedar Waxwing. *Wilson Bull*. 54:225-237.

Leck, C.E, &. Cantor, F.L. 1979. Seasonality, clutch size, and hatching success in the Cedar Waxwing. *Auk* 96:196-197

Lee, C., & Birch, A. 2002. Notes on the distribution, vagrancy and field identification of American pipit and 'Siberian Pipit'. *North American Birds* 56: 389-398.

468

Lefranc, N. 2007. Isabelline Shrike : taxonomy, identification and status in France. *Ornithos* 14: 201-229

Lefranc, N., & Worfolk, T. 1997. *Shrikes: A Guide to the Shrikes of the World.* Robertsbridge: Pica Press.

Lehikoinen, A. (ed.), Ekroos, J., Jaatinen, K., Lehikoinen, P., Piha, M., Vattulainen, A., & Vähätalo, A. 2008: Bird population trends based on the data of Hanko Bird Observatory (Finland) during 1979–2007. - *Tringa* 35: (in press).

Leisler, B., Heidrich, P., Schulze-Hagen, K., & Wink, M. 1997. Taxonomy and phylogeny of reed warblers (genus *Acrocephalus*) based on mtDNA sequences and morphology. *J. Ornithol* 138: 469-496.

Lewington, I. 1990. Identification of female Pine Bunting. *Birding World* 3: 89-90.

Lewington, I., Alström, P., & Colston, P. 1991. *A Field Guide to the Rare Birds of Britain and Europe*, London.

Lewington, I. 1999. Separation of Pallid Swift and *pekinensis* Common Swift. *Birding World* 12:450-452.

Leyshon, O., & Kerr, A. 1996. Another difficult 'Collared Flycatcher'. *Birding World* 9:115.

Limbert, M. 1984. Vagrant races of Willow Tit in Britain. *Brit. Birds* 77:123.

Lindblom, K. 2008. Booted Warbler and Lanceolated Warbler in Finland. *Alula* 2:84-90.

Lindholm, A. 2001. Apparent hybrid redstarts in Finland resembling Black Redstart of eastern subspecies *phoenicuroides*. *Brit. Birds* 94: 542-545.

Lindroos, T., & Tenovuo, O. 2000. White-winged lark - Field identification and European distribution. *Alula* 6:170-177.

Lindroos, T., & Tenovuo, O. 2002. Black Lark – its identification in the field and distribution in Europe. *Alula* 8:22-28.

Lowery, G.H. Jr. 1943. The dispersal of 21,414 Chimney Swifts banded at Baton Rouge, Louisiana, with notes on probable migration routes. Louisiana Academy of *Science* 7:56-74.

Lowther, P.E. 1993. Brown-headed Cowbird (*Molothrus ater*), *The Birds of North America Online* (A. Poole, Ed.). Ithaca: Cornell Lab of Ornithology; Retrieved from *The Birds of North America Online*: http://bna.birds.cornell. edu/bna/species/047.

Lowther, P.E. 1999. Alder Flycatcher (*Empidonax alnorum*), *The Birds of North America Online* (A. Poole, Ed.). Ithaca: Cornell Lab of Ornithology; Retrieved from *The*

Birds of North America Online: http://bna.birds.cornell. edu/bna/species/446.

Lowther, P.E., Celada, C., Klein, N.K., Rimmer, C.C., & Spector, D.A. 1999. Yellow Warbler (*Dendroica petechia*), *The Birds of North America Online* (A. Poole, Ed.). Ithaca: Cornell Lab of Ornithology; Retrieved from *The Birds of North America Online*: http://bna.birds.cornell.edu/bna/ species/454.

Lowther, P.E., Rimmer, C.C., Kesel, B., Johnson, S.L., & Ellison, W.G. 2001. Gray-cheeked thrush: *Catharus minimus*. In A. Poole and F. Gill (eds.), The birds of North America, 591. The Birds of North America, Inc., Philadelphia, PA.

Lundberg, A., & Alatalo, R.V. 1992. *The Pied Flycatcher*. T and A Poyser, London, England.

Lunn, J., & Dale, J.E. 1993 Breeding activities of Parrot Crossbills *Loxia pytyopsittacus* in South Yorkshire in 1983. *The Naturalist,* 118: 9-12.

Luukkonen, A. 1995. Ruostepyrstö - sai ''arktiset massat''liikkeelle *Alula* 1: 68-69. (In Finnish with English summary)

Mack, D.E., & Yong, W. 2000. Swainson's Thrush (*Catharus ustulatus), The Birds of North America Online* (A. Poole, Ed.). Ithaca: Cornell Lab of Ornithology; Retrieved from *The Birds of North America Online*: http://bna.birds.cornell.edu/bna/species/540

Madge, S.C. 1985. Vocalisations and *Phylloscopus* taxonomy. *Brit. Birds* 78: 199-200.

Madge, S.C., Hearl, G.C., Hutchings S.C., & Williams, L.P. 1990. Varied Thrush: new to the Western Palearctic. *Brit. Birds* 83: 187-195.

Madge, S. 1991. Splitting and lumping in the Soviet Union. *Birding World* 4:401-404.

Margeson, J.M.R. 1959. Myrtle Warblers crossing the Atlantic on board ship. *Brit. Birds* 52: 237.

Marquiss, M., & Rae, R. 2002. Ecological differentiation in relation to bill size amongst sympatric, genetically undifferentiated crossbills Loxia spp. *Ibis* 144:494–508.

Marr, A., & Porter, R. 1995. The White-winged Lark in Britain. *Brit. Birds* 88: 365-371.

Marshall, J.T. 2001. The Gray-cheeked Thrush, *Catharus minimus*, and its New England subspecies, Bicknell's Thrush, *Catharus minimus bicknelli*. Publ. *Nuttall Ornithol. Club*, no. 28.

Martens, J., & S. Eck, 1995. Towards an ornithology of the Himalayas: Systematics, Ecology and Vocalizations of Nepal Birds. Bonner Zool. Monogr., 38: 1-445.

Martin, J.W., & Parrish, J.R. 2000. Lark Sparrow (Chondestes grammacus), *The Birds of North America Online* (A. Poole, Ed.). Ithaca: Cornell Lab of Ornithology; Retrieved from *The Birds of North America Online*: http://bna.birds.cornell.edu/bna/species/488

Martin, S.G., & Gavin, T.A. 1995. Bobolink (*Dolichonyx oryzivorus*), *The Birds of North America Online* (A. Poole, Ed.). Ithaca: Cornell Lab of Ornithology; Retrieved from *The Birds of North America Online*: http://bna.birds.cornell.edu/bna/species/176.

Maumary, L., Vallotton, L., & Knaus, P. 2007. Les *Oiseaux* de Suisse. Station ornithologique suisse, Sempach, et Nos *Oiseaux*, Montmollin. 848 p.

McGowan, R.Y. 2002. Racial identification of Pallid Swift. *Brit. Birds* 95: 454-455.

McKay, C.R. 2000. Cedar Waxwing in Shetland: new to the Western Palearctic. *Brit. Birds* 93: 580-587.

McLaren, I.A., Lees, A.C., Field, C., & Collins, K.J. 2006. Origins and Characteristics of Nearctic Landbirds in Britain and Ireland in Autumn: a Statistical Analysis. *Ibis*, 148: 707-726

McShane, C. 1996. Eastern Phoebe in Devon: new to the Western Palearctic. *Brit. Birds* 89:103-107.

McSorley, C. A., Noble, D. G., & Rehfisch, M. M. 2006. Population estimates of birds in Great Britain and the United Kingdom. *Brit. Birds* 99: 25-44.

Mead, C.J., & Wallace, D.I.M. 1976. Identification of European treecreepers. *Brit. Birds* 69: 117-131

Meissner W., & Skakuj M. 1997. First broods of the Citrine Wagtail *Motacilla citreola* in Poland and changes in the species breeding range in Europe. *Not Orn* 38:51-60

Melling, T. 2004. The White's Thrush in East Yorkshire. *Birding World* 17: 432-434.

Melling, T. 2006. Time to get rid of the Moustache: a review of British records of Moustached Warbler. *Brit. Birds* 99:465-478.

Melling, T. 2009. Should Red-necked Nightjar be on the British List?. *Brit. Birds* 102:110-115.

Middleton, A.L. 1993. American Goldfinch (*Carduelis tristis*), *The Birds of North America Online* (A. Poole, Ed.). Ithaca: Cornell Lab of Ornithology; Retrieved from *The Birds of North America Online*: http://bna.birds.cornell.edu/bna/species/080

Mikkola, H. 1973. The Red-flanked Bluetail and its spread to the west. *Brit. Birds* 66:3-12.

Mikkola, H. 1983. *Owls of Europe*. T & A.D. Poyser, Calton

Milá, B., McCormack, J.E., Castañeda, G., Wayne, R.K., & Smith, T.B. 2007. Recent postglacial range expansion drives the rapid diversification of a songbird lineage in the genus *Junco*. Proc. R. Soc. Lond. B 274:2653-2660.

Milá, B., Smith, T.B., & Wayne, R.K. 2007. Speciation and rapid phenotypic differentiation in the Yellow-rumped Warbler *Dendroica coronata* complex. *Molecular Ecology* 16:159-173.

Millington, R. 1989. The Portland Pipit - Britain's third Blyth's? *Birding World* 2: 90.

Milne, P., McAdams, D.G., & Dempsey, E. 2002. Forty-eighth Irish Bird Report, 2000. *Irish Birds* 7: 79-110.

Milright, R.D.P. 1994. Fieldfare *Turdus pilaris* ringing recoveries during autumn winter and spring, analysed in relation to river basins and watersheds in Europe and the Near East. *Ringing & Migration* 15: 129-189.

Milright, R.D.P. 2002. Redwing *Turdus iliacus* migration and wintering areas as shown by recoveries of birds ringed in the breeding season in Fennoscandia, Poland, the Baltic Republics, Russia, Siberia and Iceland. *Ringing & Migration* 21: 5-15.

Mineyev [Mineev], Yu.N. 2003. Guseobraznye ptitsy vostochnoevropeyskikh tundr [Wildfowl (*Anseriformes*) of the East European tundras]. Ekaterinburg: Urals Branch of the Russian Academy of *Sciences*. (In Russian.)

Moldenhauer, R.R., & Regelski, D.J. 1996. Northern Parula (*Parula americana*), *The Birds of North America Online* (A. Poole, Ed.). Ithaca: Cornell Lab of Ornithology; Retrieved from *The Birds of North America Online*: http://bna.birds.cornell.edu/bna/species/215.

Molina, P., Ouellet, H., & McNeil, R. 2000. Geographic variation and taxonomy of the Northern Waterthrush. The *Wilson Bulletin* 112:337-346.

Money, D. 2000. The Desert Lesser Whitethroat on Teesside. *Birding World* 13:451-453.

Monroe, B.L., Banks, R.C., Fitzpatrick, J.W., Howell, T.R., Johnson, N.K. et al. 1995. Fortieth supplement to the American Ornithologists' Union Check-list of *North American Birds. Auk* 112: 819-830.

Moon, S.J., & Herbert, R.A. 1989. The Penllergaer *Phylloscopus* puzzle. *Brit. Birds* 82: 275-277.

Moore, F.R. 1990. Ecophysiological and behavioral response to energy demand during migration. Acta XX Congressus Internationalis Ornithologici. Vol. II, Christchurch , New Zealand.)

Moores, N. 2004. Brown Shrike *Lanius cristatus*: appearance and variability of individuals seen on migration in South Korea. Wesbite: www.birdskorea.org/Birds/Identification/ID_Notes/BK-ID-Brown-Shrike.shtml. Update November 2004.

Morozov, V.V. & Kornev, S.V. 2000. [Supplementary data on the avifauna of the steppe zone in the Urals region]. *Russ. J. Orn.* Express. 88: 15–22. (In Russian.)

Morozov, V.V. & Syroechkovskiy, E.E. Jr. 2004. [On the bird fauna of Kolguev Island]. *Ornithologia* 31: 9–50. (In Russian.)

Morris, S.R., Holmes, D.W., & Richmond, M.E. 1996. A ten-year study of the stopover patterns of migratory passerines during fall migration on Appledore Island, Maine. *Condor* 98: 395–409.

Morse, D.H. 2004. Blackburnian Warbler (*Dendroica fusca*), *The Birds of North America Online* (A. Poole, Ed.). Ithaca: Cornell Lab of Ornithology; Retrieved from *The Birds of North America Online*: http://bna.birds.cornell.edu/bna/species/102.

Mowbray, T.B. 1999. Scarlet Tanager (*Piranga olivacea*), *The Birds of North America Online* (A. Poole, Ed.). Ithaca: Cornell Lab of Ornithology; Retrieved from *The Birds of North America Online*: http://bna.birds.cornell.edu/bna/species/479.

Nankinov, D.N. 2000. Expansion of the Paddyfield Warbler in Europe in the second half of the XX century. Berkut 9 (1–2): 102–106.

Navarrete, J., Martin, J., Jiménez, J., & Berrai, D. 1991. Noticario ornitologico. *Ardeola*, 38: 349.

Newton, I. 2002. Population limitation in Holarctic owls. In Newton, I., Kavanagh, R., Olsen, J., & Taylor, I. Ecology and Conservation of Owls: 3–29. Collingwood, Australia: CSIRO Publishing.

Newton, I. 2007 *The Migration Ecology of Birds*. London, Academic Press/Elsevier.

Neyfel'd, N.D. & Teplov, V.V. 2007. [A vagrant Daurian Redstart on the Upper Pechora]. P.198 in Ryabitsev, V.K. (ed.) Materialy k rasprostraneniyu ptits na Urale, v Priural'ye i Zapadnoy Sibiri. Ekaterinburg. (In Russian.)

Nicholson, E.M., & Ferguson-Lees, I.J. 1962.The Hastings Rarities. *Brit. Birds* 55: 299-384.

Nicolai, B., Schmidt, C., & Schmidt, F.U. 1996. Gefiedermerkmale, Maße und Alterskennzeichen des Hausrotschwanzes (Phoenicurus ochruros). *Limicola* 10: 1–41. (in German with English summary).

Nieuwstraten, E. 2004.Vermoedelijke Italiaanse Kwikstaart bij Makkum Presumed Ashy-headed Wagtail near Makkum. *Dutch Birding* 26: 220–222. (in Dutch with English summary).

Nisbet, I.C.T. 1963. Quantitative study of migration with 23-centimetre radar. *Ibis* 105: 435–460.

Nisbet, I.C.T., McNair, D.B., Post, W., & Williams, T.C. 1995. Transoceanic migration in the Blackpoll Warbler: a summary of scientific evidence and response to criticism by Murray. *J. Field. Ornithol.* 66:612-622.

Nolan., V, Ketterson., E.D., Cristol, D.A., Rogers, C.M., Clotfelter., E.D., Titus, R.C., Schoech, S.J., & Snajdr, E. 2002. Dark-eyed Junco (*Junco hyemalis*), *The Birds of North America Online* (A. Poole, Ed.). Ithaca: Cornell Lab of Ornithology; Retrieved from *The Birds of North America Online*: http://bna.birds.cornell.edu/bna/species/716 doi:10.2173/bna.716)

Occhiato, D. 2003. Pine Bunting in Italy: status and distribution. *Dutch Birding* 25:32-39.

Ödeen, A., & Alström, P. 2001. Evolution of secondary sexual traits in wagtails (genus *Motacilla*). In: Effects of Post-Glacial Range Expansions and Population Bottlenecks on Species Richness (ed. Ödeen A). Acta Universitatis Upsaliensis, Uppsala.

Ödeen, A., & Björklund, M. 2003. Dynamics in the *Evolution* of sexual traits: losses and gains, radiation and convergence in yellow wagtails (*Motacilla flava*). *Molecular Ecology* 12:2113–2130.

Odin, N., James, M., & Holmes, P. 1999. The Yellow-breasted Bunting in Suffolk. *Birding World* 12: 317.

Ogden, L.J., & Stutchbury, B.J. 1994. Hooded Warbler (*Wilsonia citrina*), *The Birds of North America Online* (A. Poole, Ed.). Ithaca: Cornell Lab of Ornithology; Retrieved from *The Birds of North America Online*: http://bna.birds.cornell.edu/bna/species/110.

Ogilvie., M and the Rare Breeding Birds Panel. 1994. Rare breeding birds in the United Kingdom in 1991. *Brit. Birds* 87:366-393.

Olsson, U., Alström, P., Ericson, G. P., Sundberg, P. 2005. Non-monophyletic taxa and cryptic species - Evidence from a molecular phylogeny of leaf-warblers (*Phylloscopus*, Aves). *Molecular Phylogenetics and Evolution* 36: 261–276.

Omland, K.E., & Kondo, B.K. 2006. Phylogenetic studies of plumage Evolution and speciation in New World orioles (*Icterus*). *Acta Zoologica Sinica*, Proceedings of the 23 International Ornithological Congress. 52 (supplement), 320-326.

Omland, K.E., Lanyon, S.M., & Fritz, S.J. 1999. A Molecular Phylogeny of the New World Orioles (*Icterus*): the Importance of Dense Taxon Sampling. *Molecular Phylogenetics and Evolution* 12:224-239.

Osborn, K., & Harvey, P. 1994. The Chesnut Bunting in Shetland. *Birding World* 7:371-373.

Osborn, K., & Suddaby, D. 1990. Pallas's *Sandgrouse* in Shetland. *Birding World* 3: 161-163.

Ostrowski, S., &. Guinard, E. 2002. First record of Green Warbler *Phylloscopus nitidus* in western Saudi Arabia *Sandgrouse* 24: 58-59

Otis, D.L., Schulz, J.H., Miller, D., Mirarchi, R.E., & Baskett, T.S. 2008. Mourning Dove (*Zenaida macroura*), *The Birds of North America Online* (A. Poole, Ed.). Ithaca: Cornell Lab of Ornithology; Retrieved from *The Birds of North America Online*: http://bna.birds.cornell.edu/bna/species/117.)

Ouellet, H. 1993. Bicknell's Thrush: taxonomic status and distribution. *Wilson Bull.* 105: 545–572.

Outlaw, D.C., Voelker, G,. Mila, B., Girman, D. J. 2003. Evolution of long-distance migration in and historical biogeography of Catharus thrushes: A molecular phylogenetic approach. *Auk* 120:299-310.

Ovaa, A., van der Laan, J., Berlijn, M., & CDNA. 2008. Rare birds in the Netherlands in 2007. *Dutch Birding* 30: 369–389.

Page, D. 1997. From the rarities Committee's files: Problems presented by a pale Blyth's Pipit. *Brit. Birds* 90:404-409

Page, D. 1999. Identification of Bonelli's Warblers. *Brit. Birds* 92:524-531.

Pan, Q.-W., Lei, F.-M., Yin, Z.-H., Krištín, A., & Kaňuch, P. 2007. Phylogenetic relationships between *Turdus* species: Mitochondrial cytochrome b gene analysis. *Ornis Fennica* 84: 1-11.

Panov E.N. 1995. Superspecies of shrikes in the former USSR. *Proc. West. Found. Vert. Zool.* 6: 26-33.

Panov, E.N. 2005. Wheatears of Palearctic: Ecology, Behaviour and Evolution of the Genus *Oenanthe*. Sofia. Panow E.N. 1996. *Die Wtirger der PaHiarktis. Gattung Lanius*. Second edition. Neue Brehm Bticherei, Magdeburg.

Paradis, E., Baillie, S.R., Sutherland, W.J., & Gregory, R.D. 1998. Patterns of natal and breeding dispersal in birds. *J. Anim. Ecol* 67: 518–536.

Parish, A.L. 1961. White-crowned Sparrow crossing Atlantic on board ship. *Brit. Birds* 54:253-254.

Parkin, D.T., & Shaw, K.D. 1994 Asian Brown Flycatcher, Mugimaki Flycatcher and Pallas's Rosefinch. *Brit. Birds* 87:247-252.

Parkin, D.T., Collinson, M., Helbig, A.J., Knox, A.G., Sangster, G., & Svensson, L. 2004. Species limits in *Acrocephalus* and *Hippolais* warblers from the Western Palearctic. *Brit.Birds* 97: 276–299.

Patten, M.A. 2001. The roles of habitat and signalling in speciation: evidence from a contact zone of two Song Sparrow (*Melospiza melodia*) subspecies. Ph.D. diss., Univ. of California, Riverside.

Pavlova, A., Zink, R.M., Drovetski, S.V., Red'kin, Y., & Rohwer, S. 2003. Phylogeographic patterns in *Motacilla flava and M. citreola*: species limits and population history. *Auk*, 120, 744–758.

Payne, R.B. 2006. Indigo Bunting (*Passerina cyanea*), *The Birds of North America Online* (A. Poole, Ed.). Ithaca: Cornell Lab of Ornithology; Retrieved from *The Birds of North America Online*: http://bna.birds.cornell.edu/bna/species/004.)

Pearson, D.J. 2000. The races of the Isabelline Shrike *Lanius isabellinus* and their nomenclature. *Bull. Br. Orn. Club* 120: 22-27.

Pennington, M., & Maher, M. 2005. Greenland, Iceland and Hornemann's Redpolls in Britain. *Birding World* 18: 66–78.

Pitt, R.G. 1967. Savi's Warblers breeding in Kent. *Brit. Birds* 60:349-355.

Portenko, L.A. 1972: Die Schnee-eule. Die Neue Brehm-Bucherei 454.

Poulin, R.G., Grindal, S.D., & Brigham, R.M. 1996. Common Nighthawk (*Chordeiles minor*), *The Birds of North America Online* (A. Poole, Ed.). Ithaca: Cornell Lab of Ornithology; Retrieved from *The Birds of North America Online*: http://bna.birds.cornell.edu/bna/species/213.)

Prescott, D.R., & Middleton, A.L.A. 1990. Age and sex differences in winter distribution of American Goldfinches in eastern *North America*. *Ornis. Scand.* 21: 99–104.

Preston, M. 1999.The first Pallid Swift to be ringed in Britain. *Birding World* 12: 448-449.

Proctor, B. 1997. Discussion, based on biometric data, of an apparent hybrid Common Crossbill x Two-barred Crossbill. *Birding World* 10: 152.

Proctor, B., & Donald, C. 2003. Yellow-headed Blackbirds in Britain and Europe. *Birding World* 16: 69-81.

Pyle, P. 1997. *Identification guide to North American birds* - part 1. Slate Creek Press.

MacArthur,R., & Klopfer, P. 1958. North American birds staying on board ship during Atlantic crossing. *Brit. Birds* 51: 358

Raffaele, H., Wiley, J., Garrido, O. H., Keith, A., & Raffaele, J I. 1998. *A Guide to the Birds of the West Indies*. Princeton Univ. Press, Princeton, NJ.

Raine, A. 2002. *A Field Guide to the Birds of Bermuda*. Macmillan, Oxford.

Raines, R J., & Bell, A. A. 1967. Penduline Tit in Yorkshire: a species new to Britain and Ireland. *Brit. Birds*. 60:517-520.

Red'kin, Ya.A. 1998. [Notes on grasshopper warblers *Locustella (Sylviidae)* of the Arkhangel'sk Region]. *Russ. J. Orn.* Express 32: 3-7. (In Russian.)

Reid, J.M., & Riddington, R. 1998. Identification of Greenland and Iceland Redpolls. *Dutch Birding* 20: 261-269.

Remsen, J.V.Jr. 2001. True winter range of the Veery (*Catharus fuscescens*): lessons for determining winter ranges of species that winter in the tropics. *Auk* 118: 838-848.

Rich, T.D et al. 2004. *Partners in Flight North American Landbird Conservation Plan*. Cornell Lab of Ornithology. Ithaca, NY.

Richardson, M., & Brauning, D. W. 1995. Chestnut-sided Warbler (*Dendroica pensylvanica*), *The Birds of North America Online* (A. Poole, Ed.). Ithaca: Cornell Lab of Ornithology; Retrieved from *The Birds of North America Online*: http://bna.birds.cornell.edu/bna/species/190.

Riddiford, N. & Harvey, P. V. 1992. Identification of Lanceolated Warbler. *Brit. Birds* 85: 62-78.

Riddiford, N., Harvey, P. V., & Shepherd, K. B. 1989. Daurian Starling: new to the Western Palearctic. *Brit. Birds*. 82:603-612.

Riddington, R., & Votier, S. 1997. Redpolls from Greenland and Iceland. *Birding World* 10: 147-149.

Riddington, R., Votier, S., & Steele, J. 2000.The influx of redpolls into Western Europe, 1995/96. *Brit. Birds* 93: 59-67.

Ridgely, R.S., & Gwynne, J.A. 1989. *A guide to the birds of Panama, with Costa Rica, Nicaragua, and Honduras*. 2nd ed. Princeton Univ. Press, Princeton, NJ.

Rimmer, C.C., & Mcfarland, K.P. 1998. Tennessee Warbler (*Vermivora peregrina*), *The Birds of North America Online* (A. Poole, Ed.). Ithaca: Cornell Lab of Ornithology; Retrieved from *The Birds of North America Online*: http://bna.birds.cornell.edu/bna/species/350.

Rimmer, C.C., Mcfarland, K.P., Ellison, W.G., & Goetz, J.E. 2001. Bicknell's Thrush (*Catharus bicknelli*), *The Birds of North America Online* (A. Poole, Ed.). Ithaca: Cornell Lab of Ornithology; Retrieved from *The Birds of North America Online*: http://bna.birds.cornell.edu/bna/species/592

Rising, J.D. 1970. Morphological variation and Evolution in some North American orioles. Syst. Zool. 19: 315-351.

Rising, J.D., & Flood, N.J. 1998. Baltimore Oriole (*Icterus galbula*), *The Birds of North America Online* (A. Poole, Ed.). Ithaca: Cornell Lab of Ornithology; Retrieved from *The Birds of North America Online*: http://bna.birds.cornell.edu/bna/species/384

Rissanen, E., Keskitalo, M., LAukkanen, S., & Ohtonen, M. 2002: Minor rarities in Finland in 2000. - Linnutvuosikirja 2001: 97-106.

Robb, M.S., & van den Berg, A.B. 2002. Presumed escaped White-winged Crossbill at De Zilk in 1963. *Dutch Birding* 24: 215-218.

Robbins, C.S. 1980. Predictions of future Nearctic landbird vagrants to Europe. *Brit. Birds* 73: 448-457.

Robertson, I.S. 1986. Identification of White-winged Lark. *Brit. Birds* 79: 332-335

Robertson, I.S. 1977. Identification and European Status of eastern Stonechats. *Brit. Birds* 70:237-245

Robertson, R.J., Stutchbury, B.J., & Cohen, R.R. 1992. Tree Swallow (*Tachycineta bicolor*), *The Birds of North America Online* (A. Poole, Ed.). Ithaca: Cornell Lab of Ornithology; Retrieved from *The Birds of North America Online*: http://bna.birds.cornell.edu/bna/species/011.

Robinson, M., & Becker, C.D. 1986. Snowy Owls on Fetlar. *Brit. Birds* 79:228-242.

Robinson, W.D. 1996. Summer Tanager (*Piranga rubra*), *The Birds of North America Online* (A. Poole, Ed.). Ithaca: Cornell Lab of Ornithology; Retrieved from *The Birds of North America Online*: http://bna.birds.cornell.edu/bna/species/248.

Robson, J. 1911. *Canaries, hybrids and British Birds in cage and aviary*. Edited by Lewer, S. H. Cassell, London.

Rodewald, P.G., & James, R.D. 1996. Yellow-throated Vireo (*Vireo flavifrons*), *The Birds of North America Online* (A. Poole, Ed.). Ithaca: Cornell Lab of Ornithology; Retrieved from *The Birds of North America Online*: http://bna.birds. cornell.edu/bna/species/247

Rogers, M.J. 1998. Records of Western Bonelli's Warbler in Britain, 1948-96. *Brit. Birds* 91: 122-123.

Rogers, M.J. & The Rarities Committee. 1999. Report on rare birds in Great Britain in 1999. *Brit. Birds* 92: 554-609.

Rogers, M.J. & The Rarities Committee. 2003. Report on rare birds in Great Britain in 2002. *Brit. Birds* 96: 542-609.

Rogers, M.J. & The Rarities Committee. 2004. Report on rare birds in Great Britain in 2003. *Brit. Birds* 97: 558-625.

Roselaar, C.S. (Kees), van den Berg, A.B., van Loon, A.J. & Maassen, E. 2006. [Pallas's Grasshopper Warblers in Noord-Holland in September 2005]. *Dutch Birding* 28: 273-283. (In Dutch, with English summary.)

Roth T., & Jalilova G. 2004. First confirmed breeding record of Pallas's Grasshopper Warbler *Locustella certhiola* in Kyrgyzstan. *Sandgrouse* 26: 141-143.

Roth, R.R., Johnson, M.S., & Underwood, T.J. 1996. Wood Thrush (*Hylocichla mustelina*), *The Birds of North America Online* (A. Poole, Ed.). Ithaca: Cornell Lab of Ornithology; Retrieved from *The Birds of North America Online*: http:// bna.birds.cornell.edu/bna/species/246

Round, P.D., Hansson, B., Pearson, D.J., Kennerley, P.R., & Bensch, S. 2007. Lost and found: the enigmatic large-billed reed warbler *Acrocephalus orinus* rediscovered after 139 years. *J. Av. Biol* 38:2 133.

Rowlands, A. 2003. From the Rarities Committee's files: 'Black-headed Wagtail' in Essex in 1999 – a suspected *feldegg* intergrade *Brit. Birds* 96: 291-296.

Royall, W.C., Guarino, J.L., de Grazio, J.W., & Gammell, A. 1971. Migration of banded Yellow-headed Blackbirds. *The Condor* 73:100-106.

Rufray, X., & Rousseau, E. 2004. *Oiseau* de France - La Pie-grièche à poitrine rose (*Lanius minor*) - une fin annoncée. *Ornithos* 11:36-38.

Ryabitsev, V.K. 2008. Ptitsy Urala, Priural'ya i Zapadnoy Sibiri: Spravochnik-opredelitel'. [*Birds of the Ural Mountains, their Environs and Western Siberia: an Identification Handbook*]. 3rd rev. & enl. edn. Ekaterinburg: Urals University Press. (In Russian.)

Salewski, V., Herremans, M., & Stalling, T. 2005. Postjuvenile and postbreeding wing moult of Eastern Olivaceous Warbler *Hippolais pallida reiseri* at stopover sites at the southern fringe of the Sahara. *Ringing & Migration* 22: 185-189.

Salewski, V., & Herremans, M. 2006. Phenology of Western Olivaceous Warbler *Hippolais opaca* and Eastern Olivaceous Warbler *Hippolais pallida reiseri* on stopover sites in Mauritania *Ringing & Migration* 23:15- 20.

Salomon, M. 1987. Analyse d'une zone de contact entre deux formes parapatriques: le cas des Pouillots véloces *Phylloscopus c. collybita* et P. c. brehmii. *Rev. d'Ecologie (Terre Vie)* 42: 377-420.

Salomon, M. 1989: Song as a possible reproductive isolating mechanism between two parapatric forms. The case of the chiffchaffs *Phylloscopus c. collybita* and P. c. brehmii in the western Pyrenees. *Behaviour* 111: 270-290.

Salomon, M., & Hemim, Y. 1992. Song variation in the chiffchaffs (*Phylloscopus collybit*a) of the western Pyrenees – the contact zone between the *collybita* and the *brehmii* forms. *Ethology* 92: 265-272.

Salomon, M., Voisin, J.-F., & Bried, J. 2003. On the taxonomic status and denomination of the Iberian Chiffchaffs *Ibis* 145: 87-97.

Salzburger, W., Martens, J., Nazarenko, A.A., Sun, Y. H., Dallinger, R., & Sturmbauer, C. 2002. Phylogeography of the Eurasian Willow Tit (*Parus montanus*) based on DNA sequences of the mitochondrial cytochrome b gene. *Molecular Phylogenetics and Evolution*. 24:26-34.

Sangster, G. 1997. *Acrocephalus* and *Hippolais* relationships: shaking the tree. *Dutch Birding* 19: 294-300.

Sangster, G. 2000. Genetic distance as a test of species boundaries in the Citril Finch *Serinus citrinella*: a critique and taxonomic reinterpretation. *Ibis* 142:487–490.

Sangster, G., Collinson, J.M., Helbig, A.J., Knox, A.G., & Parkin, D.T. 2004. Taxonomic recommendations for British Birds: second report. *Ibis* 146: 153–157.

Sangster, G., Collinson, J.M., Knox, A.G, Parkin, D.T., & Svensson, L. 2007. Taxonomic recommendations for British Birds: Fourth report. *Ibis* 149: 853–857.

Sangster, G., Hazevoet, C.J., van den Berg, A.B., & Roselaar, C. S. K. 1998. CSNA-mededelingen Dutch avifaunal list: species concepts, taxonomic instability, and taxonomic changes in 1998. *Dutch Birding* 20: 22-32.

Sangster, G., Hazevoet, C.J., van den Berg, A.B, Roselaar, C.S., & Sluys, R. 1999. Dutch Avifaunal List: Species Concepts, Taxonomic Instability, and Taxonomic Changes in 1977-1998. *Ardea* 87: 139-165.

Sangster, G., Knox, A.G., Helbig, A.J., & Parkin, D.T. 2002. Taxonomic recommendations for European birds. *Ibis* 144:153–159.

Sauer, J.R., Hines., J.E. Gough., G., Thomas, I., & Peterjohn. B.G. 1997. *The North American Breeding Bird Survey results and analysis*, version 96.4. Patuxent Wildlife Research Center, Laurel, Maryland.

Sauer, J.R., Hines, J.E., & Fallon, J. 2008. *The North American Breeding Bird Survey, Results and Analysis 1966 - 2007*. Version 6.2.2008. USGS Patuxent Wildlife Research Center, Laurel, Maryland.

Saunders, D.R. 1963. Notes on a probable Blackburnian Warbler on Skomer and other American birds in Wales, Autumn 1961. *Nat. in Wales* 8: 155-157.

Saunders, D., & Saunders, S. 1992. Blackburnian Warbler: new to the Western Palearctic. *Brit. Birds*. 85:337-343.

Schipper, W.J.A., & Vegten van, J.A. 1968. Two American Redstarts (*Setophaga ruticilla*) near the Azores. *Ardea* 56: 195-196.

Schmitz, E. 1903. Tagebuch-Notizen aus Madeira. *Ornithol. Jahrb. XIV*: 206-211.

Scholander, S. I. 1955. Land birds over the western North Atlantic. *Auk* 72: 225–239.

Ściborska, M. 2004. Breeding biology of the citrine wagtail (*Motacilla citreola*) in the Gdańsk region (N Poland) *J Ornithol* (2004) 145: 41–47.

Scott, R.E. 1976. Short-toed Treecreeper in Kent: a species new to Britain and Ireland. *Brit. Birds* 69: 508-509.

Sehhatisabet, M.E. 2006. Possible first Asian Brown Flycatcher *Muscicapa dauurica* in Iran and the Middle East. *Sandgrouse* 28: 179-180.

Sharrock, J.T.R. 1965, White-throated Sparrows in Hampshire *Brit. Birds* 58: 230

Sharrock, J.T.R. 1972. Fan-tailed Warbler in Co. Cork: a species new to Britain and Ireland. *Brit. Birds* 65: 501-510.

Sharrock, J.T.R. 1976. *The Atlas of Breeding Birds in Britain and Ireland*. T&AD Poyser.

Sharrock, J.T.R. and the Rare Breeding Birds Panel. 1982. Rare breeding birds in the United Kingdom in 1980. *Brit. Birds* 75:154-178.

Shaw, D. 2006. Rufous-tailed Robin on Fair Isle: new to Britain. *Brit. Birds* 99: 236-241.

Shchegolev, I., Gerzhik, I., Korzyukov, A., Pirogov, N. & Potapov, O. 1996. [Paddyfield Warbler in the north-west-ern Black Sea coastal area]. *Russ. J. Orn.* Express-issue 4: 8–11. (In Russian.)

Sherry, T.W., & Holmes, R.T. 1997. American Redstart (*Setophaga ruticilla*), *The Birds of North America Online* (A. Poole, Ed.). Ithaca: Cornell Lab of Ornithology; Retrieved from *The Birds of North America Online*: http://bna.birds.cornell.edu/bna/species/277

Shirihai, H. 1990. Possible hybrid Yellow and Citrine Wagtail in Israel. *Dutch Birding* 12:18-19.

Shirihai, H. 1996. *The Birds of Israel*. Academic Press. London. UK.

Shirihai, H, Christie, D.A., & Harris, A. 1995. Field identification of Pine Bunting. *Brit. Birds* 88: 621-626.

Shirihai, H., & Colston, P.R. 1987. Siberian Water Pipits in Israel. *Dutch Birding* 9: 8-12.

Shirihai, H., & Golan, Y. 1994. First records of Long-tailed Shrike *Lanius schach* in Israel and Turkey. *Sandgrouse* 16:36-40.

Shirihai, H., & Madge, S. 1993. Identification of Hume's Yellow-browed Warbler. *Birding World* 6: 439-443.

Shirokov, Yu.V. & Malashichev, E.B. 2001. [Nesting of Booted Warbler *Hippolais caligata* near Zaostrov'ye (Lodeynoe Pole District, Leningrad Region)]. *Russ. J. Orn.* Express-issue 135: 201–202. (In Russian.)

Sibley, C.G. and West, D.A. 1959. Hybridization in the Rufous-sided Towhees of the Great Plains. *Auk* 76: 326–338.

Sibley, C.G., & Monroe, B.L. 1993. *A supplement to Distribution and taxonomy of birds of the world*. Yale University Press, New Haven & London.

Sibley, C.G., & Short, L.L. 1964. Hybridization in the orioles of the Great Plains. *Condor* 66: 130-150.

Slaterus, R. 2007. Iberian Chiffchaffs in the Netherlands. *Dutch Birding* 29: 83-91.

Small, B.J. 2009. From the Rarities Committee's files: The identification of 'Ehrenberg's Redstart' with comments on British claims. *Brit. Birds* 102:84-97.

Small, B.J., & Walbridge, G. 2005. A review of the identification of 'Balearic' Woodchat Shrike, and details of three British records. *Brit. Birds* 98: 32-42

Smiddy P. 1992. Irish Lesser Short-toed Larks. *Birding World* 5: 395-396.

Smith, F.R. 1955. Myrtle Warbler in Devon: a new British bird. *Brit. Birds* 48:204-207

Smith, J.N., & T. L. Cook (eds). 2000. Ecology and Management of Cowbirds and Their Hosts: Studies in the Conservation of North American Passerine Birds. University of Texas Press

Smith, P. 1996. The Cedar Waxwing in Nottingham – a new British bird. Birding World 9:70-73.

Smith, P.W. 1997. The history and taxonomic status of the Hispaniolan Crossbill Loxia megaplaga. Bulletin B.O.C. 117, 4:264-271.

Sokolov, V.A. & Sokolov, A.A. 2005. [Interesting bird records in south-west Yamal in 2005]. Pp.243–246 in Ryabitsev, V.K. (ed.) Materialy k rasprostraneniyu ptits na Urale, v Priural'ye i Zapadnoy Sibiri. Ekaterinburg. (In Russian.)

Sokolov, V.A. 2003. [On the bird fauna of the south-western Yamal Peninsula]. Pp.167–169 in Ryabitsev, V.K. (ed.) Materialy k rasprostraneniyu ptits na Urale, v Priural'ye i Zapadnoy Sibiri. Ekaterinburg. (In Russian.)

Sokolov, V.A. 2006. [Distribution of Pechora Pipit in western Yamal]. Pp.192–193 in Ryabitsev, V.K. (ed.) Materialy k rasprostraneniyu ptits na Urale, v Priural'ye i Zapadnoy Sibiri. Ekaterinburg. (In Russian.)

Soler, M., Palomino, J., Martinez, J.G., & Soler, J.J. 1994. Activity, Survival, Independence and Migration of Fledgling Great Spotted Cuckoos. Condor 96: 305-809.

Sotnikov, V.N. 1996. [Paddyfield Warbler in the Kirov Region]. Russ. J. Orn. Express-issue 3: 15–18. (In Russian.)

Spahn, R., & Tetlow, D. 2006. Observations on the Cave Swallow Incursion of November 2005. The Kingbird. 56:216-225.

Spencer, R and the Rare Breeding Birds Panel. 1993. Rare breeding birds in the United Kingdom in 1990. Brit. Birds 86:62-90.

Steijn, L. 2003. Oosterse Zwarte Roodstaart te Ijmuiden. Dutch Birding, 25: 441-442.

Steijn, L.B. 2005. Eastern Black Redstarts at IJmuiden, the Netherlands, and on Guernsey, Channel Islands, in October 2003, and their identification, distribution and taxonomy. Dutch Birding 27: 171–194.

Stepanyan, L.S. 1978. Structure and distribution of bird fauna in the USSR. Vol. 2. Nauka, Moscow. In Russian.
Stepanyan, L.S. 1983. Superspecies and sibling species in the avifauna of the USSR. Nauka., Moscow. In Russian.

Stepanyan, L.S. 1990. Konspekt ornitologicheskoy fauny SSSR [Conspectus of the ornithological fauna of the USSR]. Moscow: Nauka. (In Russian.)

Stepanyan, L.S. 2003. Konspekt ornitologicheskoy fauny Rossii i sopredel'nykh territoriy (v granitsakh SSSR kak istoricheskoy oblasti) [Conspectus of the ornithological fauna of Russia and adjacent territories (within the borders of the USSR as a historic region]. Moscow: Academ-kniga. (In Russian.)

Stevenson, A. 2005. Redpolls in the Outer Hebrides. Birding World 18: 124.

Stoddart, A. 2008. The 'vittata' Pied Wheatear in Britain. Birding World 21:156-157.

Stoddart, A., & Joyner, S. 2005. The Birds of Blakeney Point. Wren Publishing.

Strel'nikov, E.G. & Strel'nikova, O.G. 1998. [Brief comments on the distribution of some bird species in the basin of the Bol'shoy Yugan]. Pp.173–180 in Ryabitsev, V.K. (ed.) Materialy k rasprostraneniyu ptits na Urale, v Priural'ye i Zapadnoy Sibiri. Ekaterinburg. (In Russian.)

Summers, R.W. 2002. Parrot Crossbills breeding in Abernethy Forest, Highland. Brit. Birds. 95: 4-11.

Summers, R.W., Dawson, R.J., & Phillips, R.E. 2007. Assortative mating and patterns of inheritance indicate that the three crossbill taxa in Scotland are species. J. Av. Biol. 38:153-162.

Summers, R.W., Mavor, R.A., Buckland, S.T., & MacLennan, A.M. 1999. Winter population size and habitat selction by Crested Tits Parus cristatus in Scotland. Bird Study 46:230-234.

Summers-Smith, J.D. 1988. The Sparrows. Poyser, London.

Svensson, L. 1987. More about Phylloscopus taxonomy. Brit. Birds 80: 580-581.

Svensson, L. 1988a. Field identification of black-headed Yellow Wagtails Brit. Birds 81: 77-78.

Svensson, L. 1988b. Identification of black-headed Yellow Wagtails. Brit. Birds 81:655-656

Svensson, L. 1992. Identification Guide to European passerines. Fourth edition, Stockholm.

Svensson, L. 2001. "The correct name of the Iberian Chiffchaff Phylloscopus ibericus Ticehurst 1937, its identification and new evidence of its winter grounds". Bulletin of the British Ornithologists' Club 121: 281–296.

Svensson, L. 2001. Identification of Western and Eastern Olivaceous, Booted, and Sykes's Warbler. Birding World 14: 192-219.

Svensson, L. 2003. Hippolais update: identification of Booted Warbler and Sykes's Warbler. *Birding World* 16: 470-474.

Svensson, L. 2004. One steppe beyond. *Birdwatch* 146: 34-38.

Svensson, L., Collinson, M., Knox, A. G., Parkin, D.T., & Sangster, G. 2005. Species limits in the Red-breasted Flycatcher. *Brit. Birds* 98:538-541.

Svensson, L & Millington, R. 2002. Field Identification of Sykes's Warbler. *Birding World* 15: 381-382.

Svingen, P. 1995. The Varied Thrush in Minnesota. *Loon* 67: 129-136.

Szentirmai, I., Székely, T., & Komdeur, J. 2007. Sexual conflict over care: antagonistic effects of clutch desertion on reproductive success of male and female penduline tits. *J. Evol. Biol.* 20: 1739-1744.

Taylor I. 1994. *Barn Owls.* Cambridge University Press. Cambridge.

Tenovuo, J., & Varrela, J. 1998. Identification of the Great Grey Shrike complex in Europe. *Alula* 4: 2-11.

Thelwell, D. 2006. White-throated Robin on Skokholm: new to Britain. *Brit. Birds* 99:361-364.

Thévenot, M., Vernon, R., & Bergier, P. 2003. The Birds of Morocco. British Ornithologists' Union Check List, Tring. 594 pp.

Thibault, J.C., & Bonaccorsi, G. 1999. The Birds of Corsica. BOU Checklist No. 17. British Ornithologists' Union, Tring.

Thomas, C. 2006. Blyth's Reed Warbler at Filey. Yorkshire *Birding* 3:98-100.

Thorup, K. 2001. First record of Pallid Swift *Apus pallidus* in Denmark and of ssp. *illyricus* in northern Europe. *Dansk Orn. Foren.Tidsskr.* 95: 169-172.

Thorup, K. 2004. Reverse migration as a cause of vagrancy. *Bird Study* 51, 228-238

Ticehurst, N.F. 1908. Snow-Finch in Kent. *Brit. Birds* 1:189.

Töpfer, T. 2006. The taxonomic status of the Italian Sparrow – *Passer italiae* (Vieillot, 1817): Speciation by stabilised hybridisation? A critical analysis Zootaxa 1325: 117-145.

Tousey, K. 1959. Myrtle Warblers crossing the Atlantic on board ship. *Brit. Birds* 52: 236-237.

Tsvelykh, A.N., Astakhov, A.I. & Panyushkin, V.E. 1997. [Records of rare bunting species in Crimea]. *Russ. J. Orn. Express* 16: 20-22. (In Russian.)

Tulloch, R.J. 1968. Snowy Owls breeding in Shetland in 1967. *Brit. Birds* 61:119-132.

Turchin, P., & Hanski, I. 1997. An empirically-based model for the latitudinal gradient in vole population dynamics. *Am. Naturalist*, 149, 842–874.

Turner, A., & Rose, C. 1989. *A Handbook to the Swallows and Martins of the World.* Christopher Helm, London.

Twedt, D.J., & Crawford, R.D. 1995. Yellow-headed Blackbird (*Xanthocephalus xanthocephalus*), *The Birds of North America Online* (A. Poole, Ed.). Ithaca: Cornell Lab of Ornithology; Retrieved from *The Birds of North America Online*: http://bna.birds.cornell.edu/bna/species/192

Tye, A., & Tye, H. 1983. Field identification of Wheatear and Isabelline Wheatear. *Brit. Birds* 76: 427-437.

Tyler Flockhart, D.T. 2007. Migration Timing of Canada Warblers near the Northern Edge of their Breeding Range. *The Wilson J. Ornithol* 119:712–716.

Tyler, S., & Ormerod, S. 1994. *The Dippers.* Poyser, London.

U. S. Fish and Wildlife Service. 2007. Migratory bird hunting activity and harvest during the 2005 and 2006 hunting seasons: Preliminary estimates. U.S. Dept. of the Interior, Washington, D.C.

Ukkonen, M. 1992. Pohjois-Savon linnuston faunistinen katsaus 1991. Siivekäs 13: 56.

Ullman, M. 1994. Identification of Pied Wheatear and Eastern Black-eared Wheatear. *Dutch Birding* 16: 186-194.

Ullman, M. 2003a. Separation of Western and Eastern Black-eared Wheatear. *Dutch Birding* 2: 77–97.

Ullman, M. 2003b. Separation of Western and Eastern Black-eared Wheatear. *Dutch Birding* 25: 77-97.

Urquhart, E. 2002. Stonechats: *A guide to the genus Saxicola.* Christopher Helm, London.

Valera, F., Rey, P., Sanchez-Lafuente, A.M., & Muñoz-Cobo J. 1990. The situation of penduline tit (*Remiz pendulinus*) in southern Europe: A new stage of its expansion. *J. Ornithol* 131:413-420.

Valera, F., Rey, P., Sanchez-Lafuente, A.M., & Muñoz-Cobo J. 1993. Expansion of penduline tit (*Remiz pendulinus*) through migration and wintering. *J. Ornithol* 134: 273-282.

Valyuev, V.A. & Valyuev, K.V. 2003. [On some rare bird species of Bashkiriya]. Pp.73-74 in Ryabitsev, V.K. (ed.) Materialy k rasprostraneniyu ptits na Urale, v Priural'ye i Zapadnoy Sibiri. Ekaterinburg. (In Russian.)

Van den Berg, A., & Bosman, C. 1999. *Rare Birds of the Netherlands*. Pica Press, Robertsbridge.

Van den Berg, A., Ebels, E. E., & Robb, M. S. 2007. Groenlandse Witstuitbarmsijs te Huisduinen in oktober 2003 en determinatie, taxonomie en voorkomen / Hornemann's Redpoll near Huisduinen in October 2003 and its identification, taxonomy and occurrence. *Dutch Birding* 29:25-31.

Van den Berg, M., & Oreel, G.J. 1985. Field identification and status of black-headed Yellow Wagtails in Western Europe *Brit. Birds* 78: 176-183

Van der Laan, J. and CDNA. 2008. Occurrence and identification of 'isabelline shrikes' in the Netherlands in 1985-2006 and records in Europe. *Dutch Birding* 30:78-92.

Van der Vliet, R.E., Kennerley, P.R., & Small, B.J. 2001. Identification of Two-barred, Greenish, Bright-green and Arctic Warblers. *Dutch Birding* 23: 175-191.

Van Horn, M.A., & Donovan, T.M. 1994. Ovenbird (*Seiurus aurocapilla*), *The Birds of North America Online* (A. Poole, Ed.). Ithaca: Cornell Lab of Ornithology; Retrieved from *The Birds of North America Online*: http://bna.birds.cornell.edu/bna/species/088.

Van IJzendoorn, E.J., & Nuiver, R. 1987. Alpenheggemussen in Nederland in voojaar van 1986. *Dutch Birding* 9:162-167.

Vinicombe, K. 2003. Red-headed Bunting revisited. *Birdwatch* 137: 32.

Vinicombe, K. 2004. Trans Europe Express. *Birdwatch* 144: 24-27.

Vinicombe, K.E. 2007. The status of Red-headed Bunting in Britain. *Brit. Birds* 100: 540-551.

Voelker, G. 1999. Molecular *Evolutionary Relationships in the Avian Genus Anthus (Pipits: Motacillidae). Molecular Phylogenetics and Evolution*. 11:84-94.

Voelker, G. 2002. Systematics and historical biogeography of wagtails: dispersal versus vicariance revisited. *Condor*, 104: 725-739.

Voelker, G., & Klicka, J. 2008. Systematics of *Zoothera* thrushes, and a synthesis of true thrush molecular systematic relationships. *Molecular Phylogenetics and Evolution* 49: 377-381. doi:10.1016/j.ympev.2008.06.014

Voelker, G., Rohwer, S., Bowie, R.C.K., & Outlaw, D.C. 2007. Molecular systematics of a species, cosmopolitan songbird genus: Defining the limits of, and relationships

among, the *Turdus* thrushes. *Molecular Phylogenetics and Evolution*. Volume 42(2), P.422-434. doi:10.1016/j.ympev.2006.07.016

Volet, B., & Burkhardt, M. 2006. Seltene und bemerkenswerte Brut- und Gastvögel und andere ornithologische Ereignisse 2005 in der Schweiz. *Der Ornithologische Beobachter* 103: 257-270.

Voous, K.H. 1977. List of recent Holarctic bird species. Passerines. *Ibis* 117: 381.

Voous, K.H. 1979. Capricious taxonomic history of Isabelline Shrike. *Brit. Birds* 72: 573-578.

Votier, S.C., Steele, J., Shaw, K D., & Stoddart, A.M. 2000. Arctic Redpoll *Carduelis hornemanni exilipes*: an identification review based on the 1995/96 influx. *Brit. Birds* 93: 68-84.

Walbridge, G. 1999. Egyptian Nightjar in Dorset: The Second British Record. *Brit. Birds* 92: 155-161.

Walker, D. 2001. Apparent Continental Stonechats in England. *Birding World* 14: 156-158.

Walker, D. 2006. A Pallid Swift in the hand in Kent. *Birding World* 19:473-474.

Wallace, D.I.M. 1976. A review of waterthrush identification with particular reference to the 1968 British record. *Brit. Birds* 69:27-33.

Wallace, D.I.M. 1964. Field identification of *Hippolais* warblers. *Brit. Birds* 57: 282-301.

Wallace, D.I.M. 1973. Identification of some scarce or difficult west Palearctic species in Iran. *Brit. Birds* 66: 376-390.

Wallace, D.I.M. 1979. *Discover Birds*, Deutsch.

Wallace, D.I.M. 1980. Possible future Palearctic passerine vagrants to reach Britain. *Brit. Birds* 73:388-397.

Wallace, D.I.M., Cobb, F.R., & Tubbs, C.R. 1977. Trumpeter Finches: new to Britain and Ireland. *Brit. Birds* 70:45-49.

Wallin, K, & Andersson, M. 1981. Adult Nomadism in Tengmalm's Owl *Aegolius funereus*. *Ornis Scandinavica* 12:122-126

Walters, E.L., Miller, E.H., & Lowther, P.E. 2002. Yellow-bellied Sapsucker (Sphyrapicus varius), *The Birds of North America Online* (A. Poole, Ed.). Ithaca: Cornell Lab of Ornithology; Retrieved from *The Birds of North America Online*: http://bna.birds.cornell.edu/bna/species/662.

Wardlaw Ramsay, R.G. 1923. *Guide to the Birds of Europe and North Africa*. Gurney and Jackson, London.

Wassink, A & Oreel, G.J. 2007. *The birds of Kazakhstan*. De Cocksdorp.

Wassink, A. 2004. Pale-throated ('*vittata*') morph of female Pied Wheatear. *Dutch Birding* 26: 43-44.

Wassink, A., & Ebels, E. B. 2005. Waarschijnlijke Oosterse Gekraagde Roodstart op Texel. *Dutch Birding* 27: 366-367.

Watson, A. 1957. The behavior, breeding, and food ecology of the Snowy Owl (*Nyctea scandiaca*). *Ibis* 99:419-462

Weckstein, J.D., Kroodsma, D.E., & Faucett, R.C. 2002. Fox Sparrow (*Passerella iliaca*), *The Birds of North America Online* (A. Poole, Ed.). Ithaca: Cornell Lab of Ornithology; Retrieved from *The Birds of North America Online*: http://bna.birds.cornell.edu/bna/species/715

Weeks H.P. 1994. Eastern Phoebe (*Sayornis phoebe*), *The Birds of North America Online* (A. Poole, Ed.). Ithaca: Cornell Lab of Ornithology; Retrieved from *The Birds of North America Online*: http://bna.birds.cornell.edu/bna/species/094.

Wernham, C.V.,Toms, M.P., Marchant, J.H., Clark, J.A., Siriwardena, G.M., & Baillie, S.R. (eds.) 2002. The Migration Atlas: movements of the birds of Britain

West, G.C. 2002. *A Birder's Guide to Alaska*. American Birding Association.

West, S. 1995. Cave Swallow (*Petrochelidon fulva*), *The Birds of North America Online* (A. Poole, Ed.). Ithaca: Cornell Lab of Ornithology; Retrieved from *The Birds of North America Online*: http://bna.birds.cornell.edu/bna/species/141.

Wheelwright, N.T., & Rising, J.D. 2008. Savannah Sparrow (*Passerculus sandwichensis*), *The Birds of North America Online* (A. Poole, Ed.). Ithaca: Cornell Lab of Ornithology; Retrieved from *The Birds of North America Online*: http://bna.birds.cornell.edu/bna/species/045

White, C.A. 1967. Red-headed Buntings in Britain and Ireland. *Brit. Birds* 60: 529–530.

Wiebe, K.L., & Moore, W.S. 2008. Northern Flicker (*Colaptes auratus*), *The Birds of North America Online* (A. Poole, Ed.). Ithaca: Cornell Lab of Ornithology; Retrieved from *The Birds of North America Online*: http://bna.birds.cornell.edu/bna/species/166a

Williams, J.M. 1996. Bay-breasted Warbler (*Dendroica castanea*), *The Birds of North America Online* (A. Poole, Ed.). Ithaca: Cornell Lab of Ornithology; Retrieved from

The Birds of North America Online: http://bna.birds.cornell.edu/bna/species/206.

Williamson, K. 1953. Rare Larks and Pipits at Fair Isle in 1952. *Brit. Birds*. 46:210-212.

Williamson, K. 1955. Two Yellow-headed Wagtails at Fair Isle: A new British bird. *Brit. Birds* 48:26-29.

Williamson, K. 1959. Meadow Pipit migration. Bird Migration 1:88-91.

Williamson, K. 1977. Blyth's Pipit in the Western Palearctic. *Bull. BOC* 97; 60-61.

Willmott, J. 1988. Blackburnian Warbler on Fair Isle – A New Western Palearctic Bird. *Birding World* 1:355-356.

Willoughby, P., & Chalders, S. 1994. Blackpoll Warbler at Flamborough Head. Yorkshire. *Birding World* 2:117-119

Wilson, A.M., & Slack, R. 1996. *Rare and Scarce Birds in Yorkshire*. Thetford.

Wilson, K. 2008. The Alder Flycatcher in Cornwall – a new British bird. *Birding World* 21:425-431.

Wilson, M. 1979. Further range expansion by Citrine Wagtail. *Brit. Birds* 72:42-43.

Wilson, M.G., & Korovin, V.A. 2003. Oriental Turtle Dove breeding in the Western Palearctic. *Brit. Birds* 96: 234–241.

Wilson, T. J & Fentman, C. 1999. Eastern Bonelli's Warbler in Scilly: New to Britain and Ireland. *Brit. Birds* 92:519-523.

Wink, M., Sauer-Gürth, H., & Gwinner E. 2002. Evolutionary relationships of stonechats and related species inferred from mitochondrial-DNA sequences and genomic fingerprinting. *Brit. Birds* 95:349-355.

Winker, K., & Pruett, C. L. 2006. Seasonal migration, speciation, and morphological convergence in the Catharus thrushes (Aves: Turdidae). *Auk* 123:1052-1068.

Winters, R. 2006. Head pattern of some 'Yellow Wagtails' in the Netherlands. *Dutch Birding* 28: 232–234.

Witherby, H.F. 1912. The Northern and the Central-European Crested Tits as British Birds. *Brit. Birds* 5:109-110.

Witherby, H.F., Jourdain, F.C.R., Ticehurst, N.F., & Tucker, B.W. 1938. *The Handbook of British Birds*. H.F., & G. Witherby, London.

Witmer, M.C., Mountjoy, D.J., & Elliot, L. 1997. Cedar Waxwing (*Bombycilla cedrorum*), *The Birds of North America Online* (A. Poole, Ed.). Ithaca: Cornell Lab of

Ornithology; Retrieved from *The Birds of North America Online:* http://bna.birds.cornell.edu/bna/species/309.

Wittmann, U., Heidrich, P.,Wink, M., & Gwinner, E. 1995. Speciation in the Stonechat (*Saxicola torquata*) inferred from nucleotide sequences of the cytochrome b gene. *J. Zool. Syst. Evol. Research* 33: 116-122.

Worfolk, T. 2000. Identification of Red-backed, Isabelline and Brown Shrikes. *Dutch Birding* 22: 323-362.

Wyatt, V.E., & Francis, C.M. 2002. Rose-breasted Grosbeak (*Pheucticus ludovicianus*), *The Birds of North America Online* (A. Poole, Ed.). Ithaca: Cornell Lab of Ornithology; Retrieved from *The Birds of North America Online:* http://bna.birds.cornell.edu/bna/species/692.

Yakushev, N.N., Zav'yalov, E.V. & Tabachishin, V.G. 1998. [Range dynamics of Paddyfield Warbler *Acrocephalus agricola* in the north of the Lower Volga region during the 20th century]. *Russ. J. Orn.* Express-issue 47: 18–22. (In Russian.)

Yeatman-Berthelot, D. 1991. *Atlas des Oiseaux de France en hiver.* Société Ornithologique de France, Paris.

Yemtsev [Emtsev], A.A. & Drenin, A.A. 2006. [A record of Mugimaki Flycatcher near Surgut]. Pp.74–75 in Ryabitsev, V.K. Materialy k rasprostraneniyu ptits na Urale, v Priural'ye i Zapadnoy Sibiri. Ekaterinburg. (In Russian.)

Yésou, P., Duquet, M., & Corso, A. 2003. Le Bruant à tête rousse Emberiza bruniceps en France et en Italie: statut et origine. *Ornithos* 10: 249–251.

Yong, W., & Moore, F.R. 1994. Flight morphology, energetic condition, and the stopover biology of migrating thrushes. *Auk* 111: 683-692.

Zakharov, V.D. 2003. [On the avifauna of the southern Chelyabinsk Region.] P.102 in Ryabitsev, V.K. (ed.) Materialy k rasprostraneniyu ptits na Urale, v Priural'ye i Zapadnoy Sibiri. Ekaterinburg. (In Russian.)

Zav'yalov, E.V. & Tabachishin, V.G. 1999. [New data on the distribution of the Red-headed Bunting Emberiza bruniceps in the north of the Lower Volga region]. *Russ. J. Orn. Express* 71: 22–23. (In Russian.)

Zink, R.M., Dittmann., D.L. Cardiff., S.W., & Rising, J.P. 1991. Mitochondrial DNA variation and the taxonomic status of the large-billed Savannah Sparrow. *Condor* 93(4):1016-1019.

Zink, R.M., Rising., J.D. Mockford, S., Horn, A.G., Wright, J.M., Leonard, M., & Westberg, M.C. 2005. Mitochondrial DNA variation, species limits, and rapid Evolution of plumage coloration and size in the savannah sparrow. *Condor.* 107:21–28.

Zink, R.M., Rohwer, S., Andreev, A.V., & Dittmann, D.L. 1995. Trans-Beringia comparisons of mitochondrial DNA differentiation in birds. *Condor* 97: 639–649.

Zink., R.M., & Weckstein, J.D. 2003. Recent Evolutionary history of the Fox Sparrows (Genus: *Passerella*). *Auk* 120: 522–527.

INDEX OF SPECIES' ENGLISH NAMES

INDEX OF SPECIES' SCIENTIFIC NAMES